HANDBOOK OF FIBER OPTICS

HANDBOOK OF FIBER OPTICS: THEORY AND APPLICATIONS

Edited by

HELMUT F. WOLF

Garland STPM Press

New York & London

15 14 13 12 11 10 9 8 7 6 5 4 3 2 1

Library of Congress Cataloging in Publication Data

Main entry under title:

Handbook of fiber optics: theory and applications.

Includes bibliographical references and indexes.
1. Optical communications. 2. Fiber optics.
I. Wolf, Helmut F.
TK5103.59.H36 621.38′0414 78-31977
ISBN 0-8240-7054-2

Published by Garland STPM Press
136 Madison Avenue, New York, New York 10016

Printed in the United States of America

CONTRIBUTORS

John Joseph Esposito	EMR Photoelectric, Princeton, New Jersey
Roger A. Greenwell	Air Surveillance Systems Project Office, Naval Ocean Systems Center, San Diego, California
Oskar Krumpholz	AEG-Telefunken Research Institute, Ulm, Germany
Stefan Maslowski	AEG-Telefunken Research Institute, Ulm, Germany
Jeff D. Montgomery	Gnostic Concepts, Inc., Menlo Park, California
Tadashi Morokuma	Olympus Optical Company, Ltd., Tokyo, Japan
Tsuneo Nakahara	Sumitomo Electric Industries, Ltd., Osaka, Japan
Klaus Petermann	AEG-Telefunken Research Institute, Ulm, Germany
Arjun N. Saxena	International Science Company, Palo Alto, California
Helmut F. Wolf	Gnostic Concepts, Inc., Menlo Park, California

CONTENTS

PREFACE — ix
Helmut F. Wolf

1. INTRODUCTION — 1
Helmut F. Wolf

2. OPTICAL WAVEGUIDES — 43
Helmut F. Wolf

3. OPTICAL SOURCES — 153
Helmut F. Wolf

4. OPTICAL DETECTORS — 203
Arjun N. Saxena and Helmut F. Wolf

5. OPTICAL CONNECTORS, COUPLERS, AND SWITCHES — 241
John Joseph Esposito

6. ECONOMIC ASPECTS — 305
Roger A. Greenwell

7. COMPONENT APPLICATIONS — 357
Jeff D. Montgomery

8. SYSTEM APPLICATIONS — 367
Jeff D. Montgomery

9. SYSTEM ASPECTS — 377
Helmut F. Wolf

10. ENDOSCOPY — 429
Tadashi Morokuma

11. OPTICAL COMMUNICATIONS ACTIVITIES — 465

OVERVIEW — 465
Helmut F. Wolf

UNITED STATES ACTIVITIES — 468
Helmut F. Wolf

EUROPEAN ACTIVITIES — 473
Oskar Krumpholz, Stefan Maslowski, and Klaus Petermann

JAPANESE ACTIVITIES 488
Tsuneo Nakahara

12. REVIEW AND OUTLOOK 513
Helmut F. Wolf

APPENDIX: ABBREVIATIONS 525

SUBJECT INDEX 527

AUTHOR INDEX 541

PREFACE

Optical communication is a relatively young data transmission technique. Aside from smoke fires in times past and light signals on the seven seas, little technologic innovation was reported during earlier centuries. After the principle of total reflection was understood, the concept of light confinement was demonstrated in a stream of flowing water about one hundred years ago; practical optical transmission through a stationary medium, however, became possible only after glass fibers with adequate performance had been developed in the 1960s. After that significant achievement, various existing components were quickly put together to form rudimentary optical communications systems that promised a communications revolution; however, these systems were not yet optimized in most respects. Today, this optimization has been nearly achieved owing to the diligent and often stupendous efforts of a large number of researchers throughout the world.

Although attempts to optimize performance and to reduce component and systems costs continue, the time to review the progress and to assess the possibilities of optical communications has come. In this book, several well-qualified authors give an overview of the foundation, the achievements, and the potential of fiber optics. The emphasis in these treatments is on guided transmission of intelligent information, although the first use of fibers was mainly in "nonintelligent" display applications. The term "intelligent" refers to the modulation of light by an information-carrying signal, but fibers are also important in the medical and industrial fields, where the optical beam is normally not modulated. The progression from optical systems consisting of discrete components to those with monolithically integrated counterparts is similar to the advance from discrete semiconductor components to complex integrated circuits. All the possible and often still unknown implications of these systems' new features have not yet arrived at a stage where inclusion in a book such as this one seems justified.

Research and development efforts in optical communications are worldwide. Considerable activity is found in all major laboratories. The United States research centers are looking apprehensively toward new advances coming from Japan, the Japanese industry is attentively watching progress reported from the United States and Europe, and Western Europe is advancing quite rapidly toward a technology in its own right, with encouraging field trials everywhere pointing the way toward early actual use. In all of these areas, considerable progress has been achieved in recent years, and little can be said of who is ahead of whom. In actuality, the question of national superiority is of minor relevance, although much of this effort is supported and encouraged by the governments.

As a result of these advances, the use of fiber optics systems and components will be staggering in the coming decade. Worldwide, the total use of fiber optics systems will increase from slightly over $100 million in 1980 to more than $1.5 billion in 1990. By that time, about 10% of the total communications wire and cable market in the United States will be penetrated by fiber optics cables. Perhaps this is still a modest number in relative terms, but it is not modest if compared in an absolute sense. On the one hand there are the difficulties of implementation such a revolutionary change will engender, and, on the other, there is the vastly superior data-transfer capability that a dollar's worth of fiber cable will be able to provide compared to a dollar's worth of a more conventional metal cable of any type.

This handbook does not purport to be a thorough, theoretical analysis of all aspects of signal transmission through fibers. There are enough excellent publications on that subject. Rather, it attempts to be a practical and useful guide to the understanding and use of fibers, sources, detectors, and the other components that form an optical communications system. It also stresses the decisive impact of economic factors on the fate of most technical innovations.

The handbook is intended to be a reference source for scientists and engineers engaged in all phases of optical and other communications and for senior students of communications, materials, and electrical engineering. We hope that it fills a gap by providing a collection of the latest information on optical communications in general and fiber optics in particular. As such, it stresses both theory and practice and attempts to give a balanced perspective on calculated data and practical applications.

To familiarize the reader with all aspects of fiber optics technology, the chapters following a general overview by H. F. Wolf (chapter 1) will give a detailed analysis of the various facets of this increasingly important field. The discussion is divided into a treatment of the major component families, to be followed by an analysis of systems considerations, including installation, maintenance, applications, and cost estimation, and by a review of nonmodulated transmission. Finally, an overview of current fiber optics activities in the major geographic areas will be presented.

Chapter 2, by H. F. Wolf, gives a comprehensive review of the theory and practice of optical waveguides. The various types of optical fibers, their characteristics, their operating principles and limitations, and their anticipated advances are covered. An in-depth analysis of theoretical and practically achieved performance data is given for step-index, graded-index, monomode, and multimode fibers. The salient features of all fiber types are reviewed, and the meaning of these with regard to the overall system performance is discussed. Also, all aspects of fiber fabrication and cabling, environmental testing, and compatibility of fibers with other systems components are reviewed. Environmental aspects include a discussion of the effects of temperature, pressure, radiation, humidity, lifetime, etc. on the performance of the system components and their interaction.

Chapter 3, by H. F. Wolf, presents an up-to-date survey of all applicable optical sources, their characteristics, their applications, and their relative advantages and disadvantages, including light-emitting diodes and semiconductor and nonsemiconductor lasers. Emphasis is placed on the compatibility of optical waveguides and sources as well as on future trends. The chapter also discusses the modulation characteristics of light-emitting diodes and semiconductor lasers. Electrical–optical and optical–electrical conversions are analyzed, including a treatment of multiplexing and demultiplexing techniques applicable to optical data transmission. The discussion includes a detailed analysis of the prevalent failure modes and the responsible mechanisms as well as proposed ways to increase source lifetime under high current density.

Chapter 4, by A. N. Saxena and H. F. Wolf, is an analysis of the various optical detectors that interface with the waveguide at the receiving end. Again, a detailed overview is given of all pertinent detectors, with theory and practice given appropriate weight. The discussion includes the various performance criteria, such as the signal-to-noise ratio, frequency response, and efficiency.

In chapter 5, by J. J. Esposito, the multitude of optical connections that are found throughout an optical system are discussed. This includes engageable connections, such as connectors; fixed connections, such as splices; and couplers, such as T and star couplers. The theoretical and practical requirements and the characteristics of these connections are treated. The types of optical switches available and under development are compared with current requirements and theoretical limits.

Chapter 6, by R. A. Greenwell, deals with systems economy, with a particular focus on competing systems. The analysis encompasses discussions of some applicable models that can be used to describe the system economy and that allow optimization of system cost. The discussion evaluates system cost versus system length and bandwidth or data rate for different approaches, and gives expected trends over the next few years. The treatment covers the trade-offs, recognizing that few systems can simultaneously optimize both cost and performance.

The applications of optical communications components are reviewed in chapter 7 by J. D. Montgomery, where the components are considered for applications requiring low-, medium-, and high-loss communications links. It is recognized that applications and component economy are elements of a feedback loop and are in mutual interaction.

System applications are described by J. D. Montgomery in chapter 8. Applications of entire optical communications systems are discussed, the analysis concentrating on the seven major electronic equipment categories (business, communications, computer, consumer, government and military, industrial, and instrumentation). The use of optical data links in each of these categories is reviewed.

In chapter 9 H. F. Wolf discusses the performance aspects of optical communications systems. This analysis establishes the relationship between cost and performance in a quantitative manner. It reviews appropriate multiplexing schemes as well as noise and dispersion limitations of the system, giving the dependence of the data transfer rate on the parameters.

The chapter on nonmodulated transmission or endoscopy (chapter 10), by T. Morokuma, gives a comprehensive review of the field of endoscopic fiber uses, including the use of fibers and fiber bundles in medical research and in industrial applications as well as in all other areas where inaccessibility prevents other means of inspection. Since nonmodulated light is employed in these uses, the requirements for these systems deviate considerably from those applicable to modulated optical uses.

Optical communications activities throughout the United States, Japan, and Western Europe are rather diversified and generally at a high level. To give an overview of these efforts, H. F. Wolf, S. Maslowski, and T. Nakahara discuss in chapter 11 present and anticipated activities of the major industrial, academic, and government organizations involved in the research, design, and implementation of optical communications components and systems in the major developmental centers. Although emphasis is on current achievements, anticipated activities are included wherever applicable.

Chapter 12, by H. F. Wolf, deals with a projection for the remainder of this century and attempts to anticipate the optical systems and components of the future. The expected component and system performance as well as the structure of future components and systems are described on the basis of reasonable extrapolations of present analogues.

The appendix gives a list of the abbreviations used most frequently in the book since they are not generally accepted or applied in the field.

Much of the writing of this book would not have been possible without the creative atmosphere at Gnostic Concepts, which provided the needed stimulus to the Editor to bring such an undertaking to fruition. Although there are a number of people responsible for the outcome, my particular thanks go to my secretary, Linda Savage, who untiringly typed and retyped most of the manuscript and who often gave encouragement when there was occasion for despair. Thanks are also due to my son, Michael Wolf, who provided most of the art work. Finally, although not least, I appreciate the wide-ranging contributions by the authors, whose chapters may well become authoritative and widely used reference sources in the years to come.

Helmut F. Wolf

Menlo Park, California
June, 1979

HANDBOOK OF FIBER OPTICS

INTRODUCTION

Helmut F. Wolf

Optical communication is the transmission of signals over a specified distance by modulation of an optical wave, either in air, vacuum, or a transparent dielectric medium, in contrast to the conventional transmission of intelligent information over any of the types of metal cables. A dielectric transmission medium which carries the optical signal is referred to as an optical waveguide or optical fiber; it typically consists of either special glass or plastic that satisfies exacting requirements. In optical transmission, a signal of electrical origin is impressed on the optical fiber by an optical source that requires modulation by the signal, and it is reconverted from an optical to an electrical signal at the end of the transmission line by a detector and amplifier.

The ensuing discussion, with few exceptions, is restricted to components and systems that allow the transmission of intelligent information. The term "intelligent" refers to meaningful variations of a transmitted signal either in digital or analog form and is distinguished from random and therefore uncontrolled variations of wavelength, frequency, or amplitude. This therefore excludes unmodulated light beams, such as those used in the course of some medical, industrial, or decorative applications, although in one chapter the application of unmodulated light for medical and other purposes is reviewed. However, optical communications is defined to include all systems and their components that achieve the transmission of data by means of a light beam electronically modulated by electrical signals.

The feasibility of wideband data transmission by guided light has been demonstrated through research worldwide, although the United States, Japan, and Europe have been the main contributors, with some important development also occurring in Australia and Canada. Many of the theoretical predictions have been validated or approached by laboratory demonstrations, and most of the major problems in optical communications have been solved.

The purpose of this chapter is to summarize the topics and provide the reader with an overview of the fiber optics components that form an optical information transfer system. The relationships between system components will be presented, and the important features of each component will be highlighted in the context of system assembly and operation.

COMPARISON OF COMMUNICATIONS SYSTEMS

The progress in the development of practical optical communications systems since the beginning of the 1970s has been staggering. Adequately low optical fiber attenuation and relatively long light source lifetime have been obtained, although in the area of source lifetime, improvements are still needed. The available light output of sources has been steadily increased also but needs development. Considerable progress has been made in the design of practical cables, interconnections, and electronic circuitry, as well as in systems concepts. A few communications links are in field trials or in prototype installations, and several specialized applications for which fiber optics are ideal are in experimental stages (1).

Extension of research and development efforts in the future will depend on the rate and depth of penetration into existing telecommunication networks and, perhaps more significantly, will depend on development of new networks. This growth will be a reflection of technological progress that has been made or that is expected to be made, but the economic aspects are equally important. Thus, a realistic evaluation of the technological potential of optical communications systems recognizes that the advancement of innovative technology and the economy of its implementation are inseparable.

Optical data may be transmitted either through the use of a guiding medium, such as glass or plastic, or it may be transmitted unguided in air or vacuum, without a solid or liquid medium. Correspondingly, modulated light propagation in a restricting medium is referred to as guided data transmission, whereas propagation in the absence of such a medium is classified as unguided data transmission. These classes of data transfer have different systems considerations.

A comparison of guided and unguided transmission indicates that each system may provide advantages for specific applications and conditions but that each may have great disadvantages for other uses.

Guided Transmission. Guided information transfer is always associated with a light-containing medium such as glass. It allows the propagation of light along a nonlinear, often tortuous, path because of the properties of the optical waveguide. It is evident that this transmission mode is insensitive to most environmental disturbances, except thermal, mechanical, and chemical influences. The requirement for a guiding medium, however, tends to restrict its utility and may influence systems costs.

Unguided Transmission. Mediumless transmission is confined to a line-of-sight path (transmitter and receiver must be perfectly aligned). The signal path does not follow the curvature of the earth or the shape of any obstacle, and its transmission properties are susceptible to environmental influences, such as fog, rain, snow, and dirt, as well as atmospheric disturbances. On the other hand, the absence of a guiding medium allows the use of unguided transmission across distances that are difficult or impossible to bridge with a physical transmission medium.

In both cases the signal experiences deterioration along its path. Guided transmission losses are a function of the radiation characteristics of the light source and detection sensitivity of the receiver, the efficiency of the coupling elements of the system, and the transmission properties of the optical waveguide, among others. Unguided transmission losses are principally restricted to the properties of air or vacuum through which the data transfer is taking place, although the interface properties of the emitting and receiving elements with air or vacuum are also of concern.

Unguided light transmission through the atmosphere is restricted to short distances by atmospheric influences; hence, this type of optical communication finds use chiefly in such special applications as arid regions or in the vacuum of outer space. Other potential applications of unguided intelligent data transmission include communication links that do not require high reliability or that can use repeaters spaced so closely that the link operates even during heavy precipitation. For reliable optical communication, however, especially under terrestrial conditions, some type of optical waveguide is desirable; hence, fiber optics communication is confined to uses where a constrictive medium is deployed.

Many of the technological problems regarding components in optical data transmission systems have been solved to the satisfaction of designers and users. A number of problems still exist involving the lifetime and reliability of the major components and the entire system and the need for improved information-carrying capacity of communications systems. Rapid progress is being made, although at present it is characterized more by evolution than by revolution. Integrated optics, however, although still in an experimental stage, will probably revolutionize optical communications in the future.

Furthermore, more economic solutions to the practical implementation of optical systems have to be found, whereby the bidirectional feedback between performance and economics will play a key role. Improvements of the performance–cost ratio will still require major efforts to allow the cost-effective utilization of the potential of fiber optics.

As with most new technical concepts, the ultimate and crucial question facing optical communications systems involves not only the extent and nature of the technological progress made or expected but also the economic advantages of new concepts over currently competitive systems concepts. Only if a new transmission system results in cost savings per unit of data transmitted will it be seriously considered by equipment manufacturers, assuming that the system's technical characteristics are at least comparable to those of existing systems.

Fabrication economy improvements are not possible without technological advances. A better understanding of the optical aspects of optical waveguides and their peripheral components, such as light sources, detectors, and couplers, serves to improve their fabrication economy by the application of physics and technology. Also, a better understanding of their reliability can be obtained through a study of possible failure mechanisms. Consequently, the utility of an optical communication system depends upon three basic questions:

What operational limits are imposed by the transmission medium?

What operational limits are imposed by the terminal devices, such as source, modulator, and detector?

What is the economic viability of an optical communications system compared to more conventional systems?

Answers to these basic questions cannot be found separately because of the interdependence of technology and economy.

Before answering these questions, it is necessary to review the features of existing and competing technologies. First, for comparative purposes there is a brief review of the principal characteristics of communications systems based on twisted wire pairs and coaxial cables.

Twisted Wire Cables. This cable type is one of the oldest transmission media. The original wires were of the open type and have attractive attenuation features which make such a system useful for data transmission over long distances if no amplification is required. However, open wire is susceptible to crosstalk. Twisted wire pairs greatly reduce crosstalk, and many twisted wire pairs can be placed within a single cable with a minimum of electromagnetic coupling. The 22-gauge twisted pair cable has a typical attenuation of 5 dB/km at 0.05 MHz; the attenuation increases with frequency as $f^{1/2}$. Near-end crosstalk varies approximately as $f^{3/2}$ and is independent of the total length of the line. Far-end crosstalk varies as f^2 and is proportional to line length. Hence, the application of twisted wire pairs is restricted to low bandwidths and low data transfer rates [≤ 1 megabit/second (Mb/s)] and consequently they do not compete directly for the same applications as optic fibers. Because of their present economic advantages over other techniques, however, twisted pairs will continue to be used extensively.

Coaxial Cables. Systems made up of coaxial cables, consisting of a center conductor and an outer tubelike conductor, offer many conveniences, but they have limited applications for broadband services. A coaxial cable has all the advantages of a closed waveguide, and its flexibility makes it versatile. The braid used to enhance flexibility, however, represents an incomplete shield, resulting in electromagnetic radiation and interference. Delay distortion, a major cause of signal degradation in most waveguides, is a problem in a coaxial cable only if the losses are excessive. The limiting factor then becomes attenuation. Attenuation in a coaxial cable is due to conductor loss and dielectric loss; conductor loss dominates at the frequencies of major interest. The attenuation has a minimum if the ratio of the radii of the outer to the inner diameter, b/a, equals 3.6 for a fixed value of b. The maximum data transfer rate is generally proportional to the square of the ratio of inner diameter to cable length; a cable-length increase or a cable-diameter decrease by a factor of 3 will result in a decrease in the achievable data rate by an order of magnitude. More specifically, the maximum data transfer rate of a coaxial cable, R_{max}, is proportional to $(r_d a/L)^2$, where a is the diameter of the inner conductor, L is the fiber length, and r_d is the ratio of the distributed inductance and capacitance of the cable. Representative values of R_{max} range from 10 to 100 Mb/s. It is apparent that the dependence of the data transfer rate on L^{-2} and on a^2 is a serious disadvantage. The major advantage of coaxial systems results from their long useful life and their ease of application. In spite of the cost of copper, these systems are relatively inexpensive, at least in comparison to more advanced metal cable systems, but the total systems cost contains a substantial percentage for transmission circuits and switching facilities.

A comparison of the channel capacities of common transmission media serves to illustrate the highly promising technical potential of optical communications. Only the most prolific transmission media are considered here. Although there will be advances in the characteristics of all media in the future, the progress expected for the more mundane types will be relatively insignificant.

Twisted Wires. Pairs of twisted wires provide the standard channels for carrying telephone conversations. With time-division multiplexing, two wire pairs can carry 24 voice channels up to 80 km with the aid of repeaters. Twisted wires form the least expensive and the least developed techniques of intelligent data transmission; consequently, this technique is the most mature now and the one that promises the fewest advances.

Coaxial Cables. The most recent coaxial cables, consisting of a central conductor surrounded by insulated metal conductors, can carry up to 1840 two-way telephone conversations in the submarine version and up to 2700 two-way telephone conversations per pair in the underground version. A submarine cable usually contains 1 center conductor, whereas an underground cable contains 4 to 20 center conductors, together with a number of wire pairs. Repeaters are required every few kilometers.

Microwave Waveguides. Waveguides suitable for telecommunications usually consist of carefully fabricated hollow tubes of 7.5 cm diameter for the transmission of waveguides ranging from 3 to 7 mm. They have a potential capacity for up to 250,000 two-way voice channels. A direct cost–performance comparison of mediumless microwave communications with data transmission systems using metal or dielectric media is meaningless because of the different propagation aspects and the absence of a transmission medium in microwave links.

Optical Waveguides. Dielectric waveguides or optical fibers are most suitable for carrying intelligent information at wavelengths of the order of 1 μm. Hence, they have the potential to carry approximately one billion simultaneous two-way voice channels. At this time this potential has not been explored in practice in the laboratory or in the field, although there are no fundamental technical obstacles to the achievement of this goal. Theoretically, a single optical waveguide is thus capable of carrying the entire present telecommunications load of the United States.

Figure 1.1 highlights the cost and capacity advantages of fiber optics systems in comparison with twisted wire pairs and coaxial cables. The system cost per channel kilometer (channel km) of optical communications systems will display a remarkable decrease in the future, made possible by substantially declining costs at the system level, and will be accompanied by comparable increases in the number of channels per system.

While optical information transfer by fiber cables employs a methodology that permits its straightforward evolution, it may also create additional costs involved with a new and relatively unfamiliar technology. Costs must also be considered in examining and evaluating the advantages of fiber optics. This evaluation must be done in spite of the fact that optical communications will permit extensive use of multiplexing on a point-to-point basis, so that, for example, a data capacity of 100 Mb per fiber will correspond to 100 kb per twisted pair. It can therefore be expected that a very large number of conventional wires will be replaced by a single optical cable, resulting in major re-

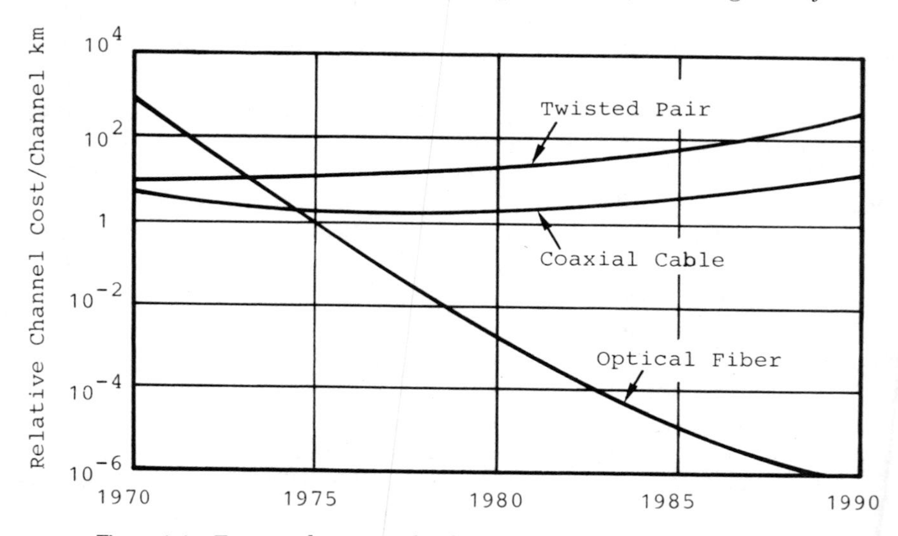

Figure 1.1 Estimated cost trends of major communications cable types.

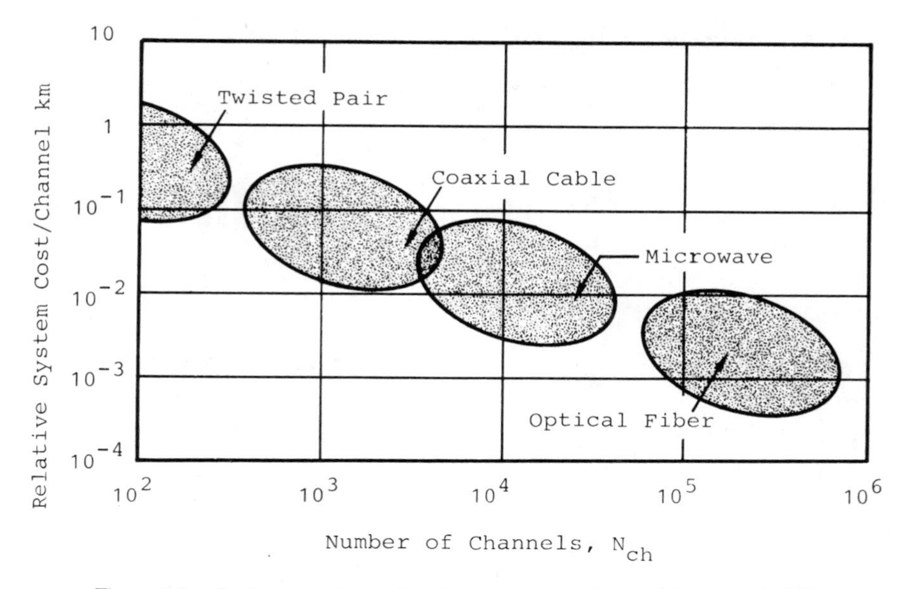

Figure 1.2 Cost comparison of major communications cable types (1978).

ductions in component count, power, size, weight, and cost, as well as reductions on a cable-per-cable basis.

Figure 1.2 relates system cost to system channel capacity for a fixed year in a schematic way. It demonstrates the outstanding superiority of optical communications, particularly in comparison to twisted pairs and coaxial cables. At the beginning of the optical communications age, fiber optics systems could sustain channels at a cost per channel and per unit length that is two to four orders of magnitude lower than that of coaxial cable systems.

A summary comparison of twisted pair, coaxial, and optical fibers is given in Table 1.1. It illustrates that the benefits of fiber optics go far beyond their use as a data carrier and make it attractive for solving problems encountered with conductive cables, such as grounding, crosstalk, electromagnetic noise, and the short-arcing associated with metal wire systems.

A comparison of the major communications transmission media is given in Table 1.2. The table shows the wide divergence of characteristics of the systems and the considerable differences between experimental, state-of-the-art systems and those that are readily available. The outstanding performance of optical fiber systems in almost all respects, however, is evident (2).

REVIEW OF THE POTENTIAL OF FIBER OPTICS

The advantages of fiber optics systems are sufficiently important to many users as to stimulate serious consideration of their installation. The benefits

TABLE 1.1. Comparison of Major Communications Cable Types

Characteristic	Twisted pair	Coaxial cable	Fiber optics cable
Length–bandwidth product (MHz km)	1	20	400
Spacing between repeaters (km)	1–2	1–2	2–10
System cost	Low, slow increase in future	Medium now, slow increase in future	High now, steep decrease in future
System lifetime (years)	20–40	20–40	1–2, 10–40 in future
Crosstalk	High	Low	Negligible
Noise immunity	Low	Medium	High
Electrical input–output insulation	No	No	Complete
Vibration tolerance	Good	Good	Good
Short-circuit loading	Yes	Yes	No
Weight, size	High	High	Low
Cable connections	Soldering, standard connectors	Soldering, standard connectors	Splicing, well-aligned connectors
Fabrication control requirements	Loose	Medium	Precise

fiber optic systems could offer over metal cable systems can be summarized as follows (3):

Larger bandwidth and small loss

Smaller size and weight

Lower material cost

Lower system cost per channel km

Higher system channel capacity

Electrical isolation of input and output of data paths

Immunity to high temperature within reasonable limits

Almost complete immunity to electromagnetic interference (EMI)

Almost complete freedom from signal leakage and crosstalk

Larger distance between repeaters

Almost complete security against detection or interception by unfriendly forces

Most drawbacks that fiber optics systems still possess will disappear on successful completion of additional research and development efforts. The drawbacks are as follows:

Need for more precise control of production parameters to obtain near-ideal fiber dimensions and index profiles

Difficulty of joining individual fiber segments

Limited lifetime of light sources and associated system reliability

Need for fiber protection in order to allow for rough installation and maintenance treatment

Although the potential advantages of optical communications links have been recognized for a long time, significant technological breakthroughs in the fabrication of low-loss, low-dispersion optical fibers have come about only recently. These have made a variety of commercial and military applications feasible and attractive. The development of these vastly superior fibers in turn provides the incentive to develop other improved elements for optical data links; for example, light sources, modulators, multiplexers, couplers, detectors, demodulators, demultiplexers, and receivers. Also, concern about electromagnetic interference, coupled with requirements for higher bandwidth and solutions to crosstalk and ground-loop problems, has increased interest in avionic, naval, and ground-base communications.

The short- and long-term aspects of optical communications systems can be divided into several generations (4).

First-Generation Systems. These systems, broadly, are those installed during the 1975–1980 period. They are systems developed mainly for field trials in a variety of environments and for operation under a number of conditions, they serve to demonstrate and verify manufacturing and installation techniques, performance, and reliability. The field trials are lengthy because they involve establishment of optimum techniques and standards as well as equipment reliability and lifetime. In spite of the economic benefits of very-high-capacity systems, lower-capacity systems (bit rate not in excess of 50 Mb/s) and short-haul video links are being introduced into heavily loaded parts of networks before long-distance, high-capacity systems. Applications are also being made in military equipment, such as airplanes and ships, and in industrial installations, such as high-voltage–generating plants and industrial sites.

Second-Generation Systems. These systems will dominate the 1980–1990 period and will be characterized by partial systems integration, whereby amplification and signal processing will be performed at optical frequencies. The integration will include optical repeaters. Furthermore, such systems will allow a substantial increase in system bandwidth by circumventing the problems of electronic circuit response time that limit the first-generation systems. The introduction of integrated optical components will improve the capabilities of monomode fibers. Optical repeater designs will become simpler than designs of hybrid electronic–optic (E/O) repeaters because of the elimination of E/O and O/E conversion; furthermore, power consumption will be reduced and reliability improved. These advances will improve system economics as well as increase system bandwidth.

TABLE 1.2. Comparison of Characteristics of the Most Prominent Communications Transmission Media

Characteristic	Microwave	Twisted pair	Coaxial cable	Circular waveguide	Optical fiber
State of technology	Well understood No fundamental problems unsolved	Well advanced	Well advanced and predictable	Not yet mature, but few problems remaining Experimental systems in operation	Reaching state of field trials and selected actual uses Few technical problems remaining
Performance	Very good	Lowest of all systems, useful mainly for short-haul use	Good Good for short-haul use	Very good	Excellent and unsurpassed by any other known technology
State of terminal devices	Well advanced and mature	Inexpensive and readily avoidable	Well advanced and mature	Not yet mature Requires special devices with critical dimensions	Still in a state of development Multiplexers, couplers, channel-dropping devices, input/output (I/O) devices not yet matured
Environmental influences	Atmospheric conditions affect performance	Subject to loss increase due to adverse conditions	Immune to environmental conditions	Immune to environmental influences	Little environmental influence except irradiation, in spite of open structure No temperature influence
Installation ease	Difficulty in selecting site and installing in inaccessible remote areas	Requires only small duct space Flexibility and ease of handling strong features	Requires relatively large duct space Easy to install, standardized connectors	Complex installation Requires avoidance of bends and terrain irregularities Installation large percentage of total cost	Very easy Little duct space Unobtrusive Splicing becoming unproblematic

Maintenance ease	Towers are large and expensive Maintenance in remote areas an important problem	Simple because of convenience Easy-to-use connectors	Easy to maintain Simple and standardized connectors	No maintenance required or possible	Generally easy, although not yet proven Convenient and standardized connectors not yet available
Operational convenience	Convenience low because of inaccessibility	High because of flexibility, small size, and ease of installation of cable	Flexible cable Size depends on bandwidth Not well suited for digital signals	Inconvenient Restrictions on installation	Very high because of flexibility, ease of handling Supply of power to repeaters inconvenient because of nonconductive nature of fibers
Electromagnetic interference (EMI) immunity and security	High immunity Proximity to satellite earth stations may introduce EMI Moderate security because of narrow beamwidths	Subject to EMI Not secure because of open structure	Subject to EMI Radiation leakage and limited security	Completely immune to EMI Highly secure except at the terminals	Very high immunity to EMI due to insulating nature of cable Very high security High bandwidth allows application of elaborate coding schemes
Utility	Above average and useful for many applications High bandwidth	Good, but limited due to low bandwidth	Wide range of applications	Useful only in high-density, long-haul applications	Extremely useful and versatile, partially because of high bandwidth Many potential applications not yet conceived
Cost per channel km	Medium	Low for small bandwidth, but high for higher bandwidth	Medium Braiding to reduce EMI and interception increases cost substantially	High channel capacity allows low cost/channel in spite of very high installation cost.	In future will be lowest of all techniques applicable

Third-Generation Systems. The 1990s will see fully integrated optic systems. These systems will fundamentally alter communication economics because of the absence of E/O and O/E conversion elements. Examples include TV cameras that will be able to form an optical image which can be optically scanned, serialized, and then switched and multiplexed for transmission via fiber systems. Another example is the direct conversion of speech to optical signals via acoustooptical transducers. The success of third-generation systems will, at least in part, depend on the experience gained from earlier systems and on the acceptance of these systems by the industry.

The intended application determines the requirements on and the potential problems of optical links. This is shown in more detail in the schematic diagram of Figure 1.3. Applications can be divided into categories depending upon the length of the communications link.

Long-Distance Applications. In this class of applications, including telecommunication and CATV (community antenna television) systems, the major use of optical waveguides is in single-fiber cables operated at high data rates. They utilize the high bandwidth and the low loss of the small fiber core, whereby each individual fiber has its own transmitter and receiver.

Short-Distance Applications. In this class of applications in computer, industrial, ship, and aircraft systems, the use of a multitude of fibers in a cable bundle is often more advantageous. Here each fiber of the bundle usually carries the same information, although this is not a requirement.

Development of fiber bundles is receiving considerable attention from military agencies and a number of commercial enterprises. At present, fiber bundles are available in which the fiber content ranges from single cables of 7 fibers to sophisticated cables of more than 5,000 fibers. The fiber content of the bundles has a significant effect on the choice of the light source, the photodetector, and the coupling techniques.

The purpose of an optical communication link, as with any other communication link, is transmission of information from one point to another without introducing unacceptable degradation of the transmitted signal. In the design of any optical communications system, therefore, there are two major factors to be considered:

Signal-to-Noise Ratio (S/N). The signal-to-noise ratio is a function of many factors, including the source power, source–fiber coupling efficiency, fiber losses, fiber–detector coupling efficiency, and fiber–fiber coupling efficiency. The signal-to-noise ratio is closely related to fiber attenuation.

Signal Distortion. The fidelity of the information transmission is a function of such performance parameters as frequency responses of sources and detectors and the dispersive effects associated with light propagation in optical waveguides. The signal distortion is closely related to fiber dispersion.

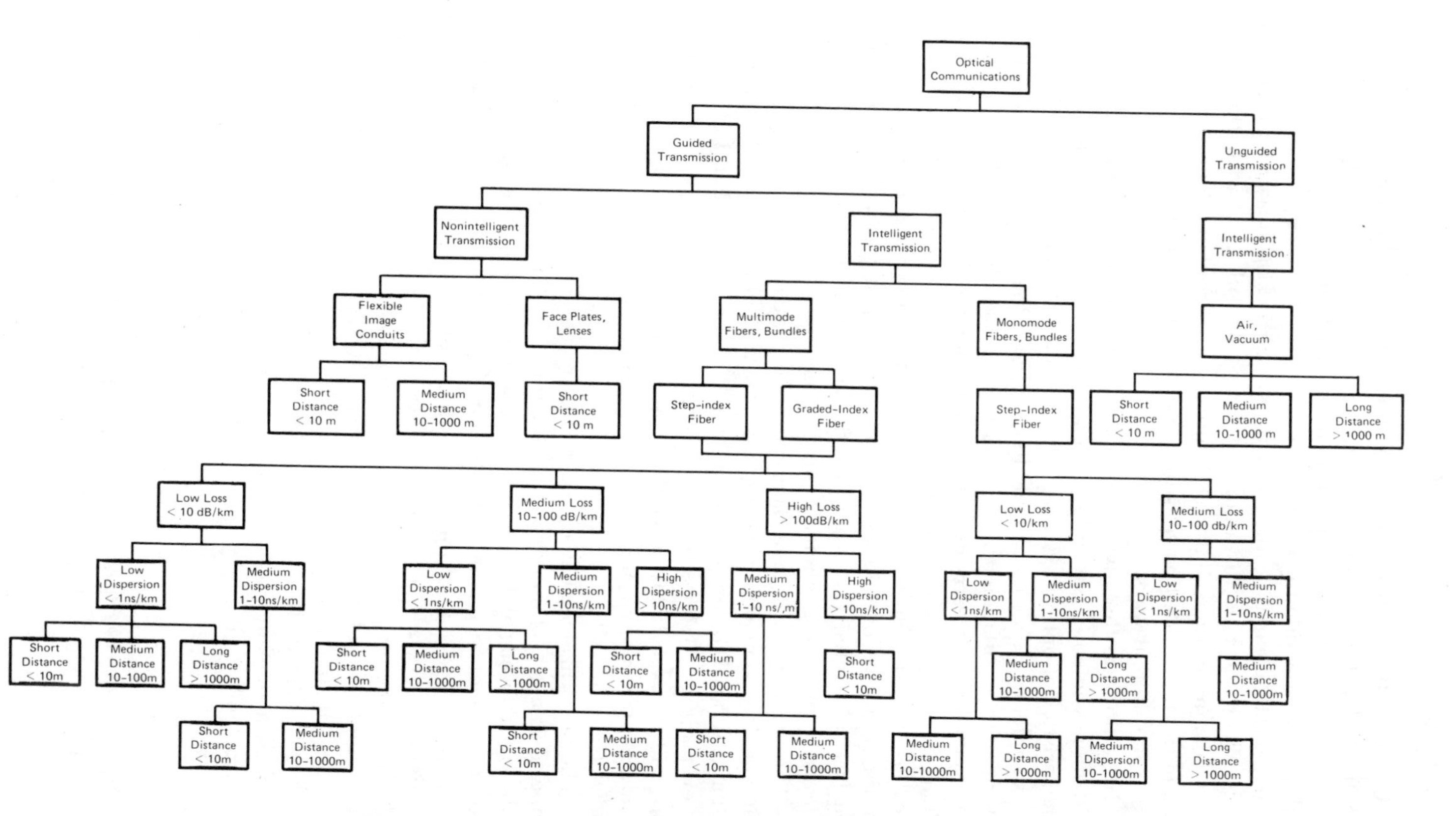

Figure 1.3 Applications tree of principal optical communications systems.

Unacceptable degradation is defined in terms of the way information is transmitted. Because the above factors have a major influence on system performance and cost optimization, it is necessary to understand the sources of the degradation within the entire system.

In a digital communications link the signal consists of a series of discrete signals which must be distinguished from each other on reception. In an analog communications link the information is contained in the detailed shape of the transmitted wave. Noise and distortion affect digital and analog systems in different ways.

In addition to the advantages optical information transmission systems could offer over more conventional metal wire systems, the generally superior characteristics of optical systems will open them to many new applications. For example, while a coaxial cable has a typical loss of about 20 to 30 dB/km, requiring the installation of a regenerative repeater about every 2 km, an optical waveguide cable carrying only a single mode and having a loss of only 2 dB/km allows the extension of the repeater spacing to about 20 km. Because repeaters represent an expensive portion of the overall communication system, the reduction of the number of repeaters by at least an order of magnitude has substantial economic benefits and consequently leads to uses of optical systems where other systems are uneconomical.

The presently available bandwidths of coaxial and optical systems are approximately comparable and are about 20 MHz km per channel for a coaxial system and 50 MHz km per channel for a fiber optics system. In the future, however, optical systems will advance much more rapidly than coaxial systems, with an expected bandwidth improvement of an order of magnitude per decade for optical systems, while only marginal advances are expected in coaxial systems.

The performance of an optical communications system, which is its economic advantage, is limited by the properties of the transmission medium and by the signal format of the information to be transmitted. The choice is usually made on economic grounds, but technical performance limits the range of alternatives. A review of the technological alternatives must thus take into consideration the relevant characteristics of the optical fibers, light sources, and photodetectors. Furthermore, the signal power requirements at the repeater inputs for pulse-coded modulation (PCM) dictate in part the choice of the technical alternatives.

A generalized diagram of an optical communications link useful for guided data transmission is shown in Figure 1.4. In this general form it applies to both digital and analog systems (5). The signal to be transmitted from a system input point to an output point will travel through the following stages:

Signal Shaper–Encoder. The electrical signal is first fed into an encoder–signal shaper. In an analog system this element provides predistortion

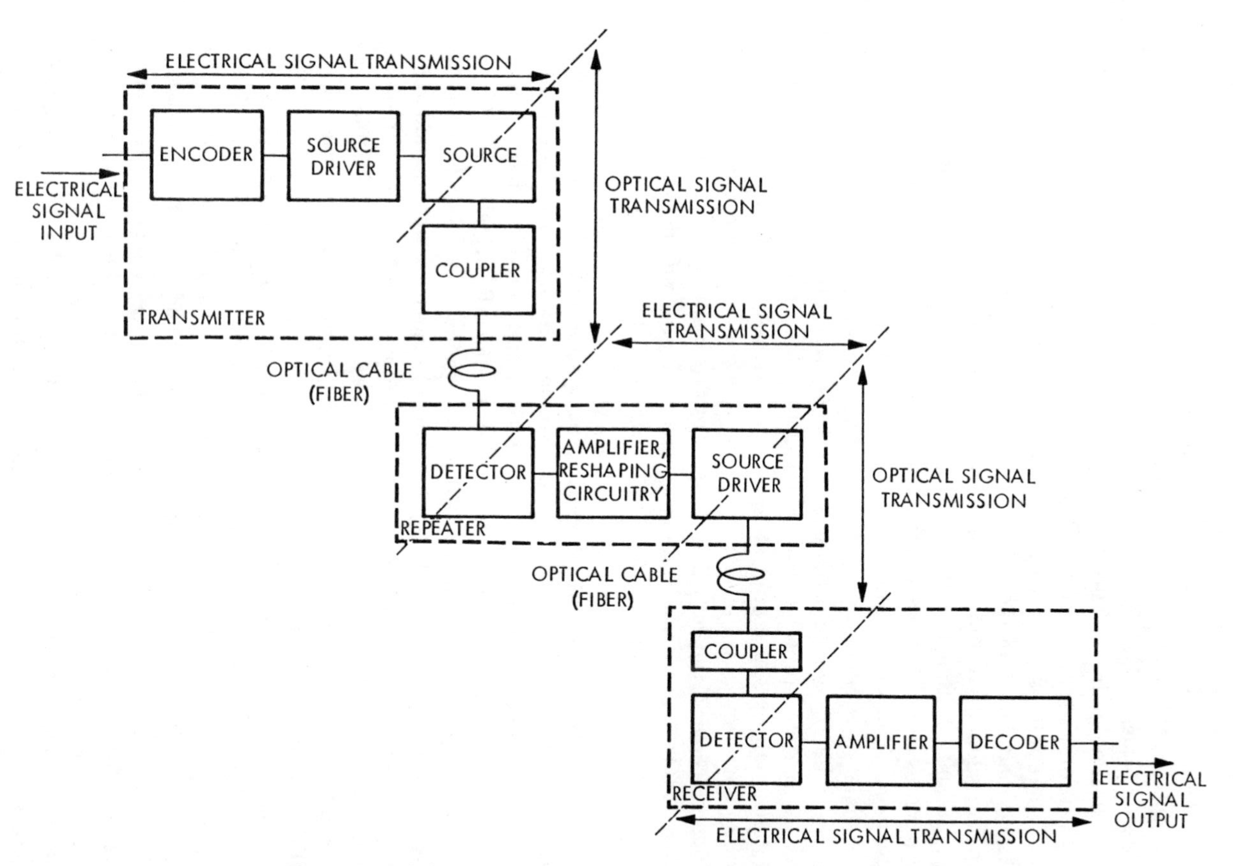

Figure 1.4 Schematic diagram of digital or analog optical communications links.

to compensate for unavoidable distortion introduced later in the system; in a digital system the encoder stage detects the incoming data and regenerates and retimes the symbols appropriately for the optical driver; and in an analog system in which analog signals are transmitted digitally, this stage provides the analog–digital (A/D) conversion and the generation of the appropriate digital symbols.

Source Driver. The signal shaped by the signal shaper–encoder is applied to the source driver. The driver modulates the current flowing through the optical source to produce the desired optical signal. The use of an incoherent light-emitting diode (LED) or semiconductor injection laser allows the direct modulation of the optical source; the use of a coherent Nd:YAG (neodynium-doped yttrium-aluminum-garnet) laser or of a gas laser requires an appropriate additional optical modulator placed between the continuous-wave (cw) laser and the waveguide, which is driven by the source driver.

Source. The source converts the electrical signal into a corresponding optical signal. It is either an incoherent LED, a semicoherent semiconductor laser, or a coherent nonsemiconductor Nd:YAG laser. Principal requirements for the source are faithful reproduction of the electrical signal, monomode excitation, high optical output at low current density, small emitting area, high-frequency response, and long lifetime even with high current density.

Source–Fiber Coupler. The purpose of this coupler is the efficient introduction of the optical power into the waveguide. Its main requirements are a low coupling loss and a perfect match of source and fiber cross-sectional areas.

Optical Cable. The optical cable transmits the optical signal from the transmitter to the receiver, either over a single fiber or over a fiber bundle consisting of either a few or up to several thousand individual fibers which may carry either the same or different information. Principal technical requirements center on low loss and low dispersion, but other criteria are almost equally important, such as dimensional characteristics, degradation, modal interaction, and bending radius, as well as economic considerations. Depending on the fiber, source, and detector characteristics and the total system length, it may be necessary to regenerate the optical signal either electrically or optically by use of repeaters.

Repeater. The repeater acts as a regenerative system element. It is designed to enhance the amplitude and the shape of the signal degraded during transmission over the optical cable. It thus consists of a photodetector, amplifying and reshaping circuitry, and an optical source, and it contains practically all the other circuitry associated with the source, the detector of the transmitter, and the receiver elements. A repeater can therefore be considered a back-to-back receiver–transmitter combination.

Fiber–Detector Coupler. The purpose of this coupler is the efficient detection by the photodetector of the optical signals coming from the fiber. It is designed to provide a match of the respective cross-sectional areas and to minimize reflective losses at the fiber–detector interface.

TABLE 1.3. Design Criteria of Optical Communications Systems Components

a) **Performance requirements**

Transmitter	Transmission medium	Receiver	Component coupling, connection
Overall power consumption	Glass or plastic type	Overall power consumption	Source compatibility
Logic compatibility	Attenuation	Bandwidth	Detector compatibility
Independence of transmission line	Dispersion	Logic compatibility	Connection ease
Optical output power	Dimensions and their variation	Noise characteristics	Source–fiber matching (diameter, NA, contact)
E/O conversion efficiency	Index profile	Signal-to-noise ratio	Detector–fiber matching (diameter, contact)
Emission wavelength	Numerical aperture	Sensitivity	Fiber–fiber matching (diameter, contact, NA, index profile)
Emission area	Dimensional and profile reproducibility	Gain	Attenuation
Spectral width of emission	Jacketing	O/E conversion efficiency	Reliability
Emission directivity	Packing fraction	Bit error rate	Lifetime
Modulation frequency	Splicing–connection ease	Independence of transmission line	Size, weight
Emission mode spectrum	Environmental tolerances	Interference protection	Interference protection
Lifetime	Economy	Circuit–detector technological compatibility	Economy
Interference protection		Economy	
Circuit–source technological compatibility			
Economy			

(continued)

TABLE 1.3. Design Criteria of Optical Communications Systems Components (continued)

b) Principal Units

Components			System	
Transmitter	**Transmission medium**	**Receiver**	**Guided transmission**	**Unguided transmission**
Peak emission power (mW)	Attenuation (dB/km)	Sensitivity (mV/μW)	Data transfer rate (Mb/s)	Data transfer rate (Mb/s)
Emission efficiency (%)	Dispersion (ns/km)	Peak sensitivity wavelength (μm)	System length (km) or maximum tolerable distance between repeaters (km)	Transmission range (km)
Bandwidth (MHz)	Data transfer rate (Mb/s) or bandwidth (MHz km)	Detection efficiency (%)	Transmission efficiency (%)	Transmission efficieney (%)
Peak emission wavelength (μm)	Dimensional properties (core and cladding diameters, spatial variations)	Bandwidth (MHz)	Cost effectiveness ($/km)	Receiver collection area (cm^2)
Emission directivity (degrees)	Optical properties (refractive indices, index profile, spatial variations, numerical aperture)	Noise (mV) or signal-to-noise ratio or noise equivalent power (NEP) (dB or $W/Hz^{1/2}$)		Radiant power at detector (mW)
Lifetime (hours)	Radiation resistance	Detection area (μm^2)		Transmission medium
Emission area (μm^2)		Gain (dB)		Cost effectiveness ($/km)
Thermal dissipation (mW/°C) or thermal resistance (°C/W)				

Detector. The photodetector must be able to follow the signal emerging from the fiber in both amplitude and frequency. At short wavelengths, the achievement of this goal does not present any difficulties, and the detector is able to reproduce the optical signal faithfully as an electrical signal, but at large wavelengths some problems of detection efficiency emerge.

Amplifier and Signal Shaper–Decoder. The amplifier enhances the electrical signal generated by the detector in the optical–electrical conversion process and increases it to a level at which it can be reshaped for proper further use. Again, the amplifier must be distortion-free and its frequency response must match that of the signal. The signal shaper and decoder, finally, converts the raw electrical signal as it is detected into the proper form for use. Its design is a function of the intended application.

In applying fiber optics, a number of criteria must be considered which are largely unique to this technology and which may not exist in comparative technologies, such as twisted pairs or coaxial cables. These criteria are listed in Table 1.3.

The current status and anticipated characteristics of the components that can be combined to form an optical communications system are summarized in Table 1.4. It is understood that considerations similar to those given here apply to repeaters, which are considered to be combinations of receivers and transmitters with associated regenerative circuitry (2).

The electromagnetic spectrum of interest in optical communication is depicted in Figure 1.5, which shows the visible, near-infrared (IR), and far-infrared, wavelength ranges and the available emission and detection ranges of sources and photodetectors, together with the suitable wavelength range of optical fibers and materials applicable to integrated optics.

In analyzing the technological prospects and economic projections for optical systems, it is necessary to define both the present and anticipated operational limits that form the basis for the fundamental questions raised above (5, 6). In this context, the term "operational limit" refers mainly to the trade-offs of distance, data transfer rate, and bit error rate. Furthermore, attention must be given to the component lifetime and degradation.

A discussion of the operational limits of an optical communication system in technological terms leads to the question of the maximum advisable pulse rate or data transfer rate. The rate depends mainly on three factors: (a) waveguide attenuation; (b) material dispersion; and (c) waveguide dispersion.

Basically there are three fundamental limitations that restrict the maximum pulse rate and hence the upper bandwidth of a fiber optics system. These are operations limited by (a) detector noise, (b) pulse dispersion, and (c) delay distortion. Their severity increases with the length of the waveguide (7). In this discussion it is assumed that the power output of the source, the coupling efficiency of the waveguide terminations, and the bit error rate are fixed.

TABLE 1.4. Selection Criteria for Optical Communications Systems Components

Characteristic	Sources			
	LED		Laser	
	GaAs	GaAsAl	Injection	Nonsemiconductor
Advantages	Inexpensive Proven technology Reliable pulsed operation Good match to PIN detector Compatible with ICs	Inexpensive Improved coupling efficiency, power density from small area, spectral match to fiber, frequency response; decreased degradation compared to GaAs LED	Substantial performance advantages compared to LEDs in size, emission angle, coupling, bandwidth and data rate, power, spectral width nuclear radiation resistance	Substantial performance improvements compared to semiconductor injection lasers in spectral width, wavelength, bandwidth, emission angle, coupling Monomode output potential
Disadvantages	Broad emission angle Emission spectrum not optimum match to fiber transmission window; peak too close to OH absorption line Low bandwidth Low output power Peak power and peak emission wavelength inversely proportional to temperature Limited lifetime	Lower thermal conductivity than GaAs, hence thermal dissipation problems and resultant potential performance decrease Relatively low output power	Room-temperature power limitation Duty cycle limited to 10% at limited peak power Substantially more expensive than LEDs	Limited output power Nonlinear input–output Nonsemiconductor technology Immature technology Substantially more expensive than semiconductor lasers Requires external modulator Large emission area

Main applications	Use in first-generation systems Later continued use in short-haul applications	Potential replacement for GaAs LEDs Use in short- and medium-haul systems	Long-haul systems High-performance replacement of LEDs Military applications	Long-haul systems High-bandwidth applications
Key performance parameters	Data rate 100 MHz, efficient rate at 10 MHz Peak wavelength 0.85–0.92 μm Spectral width 100–1000 Å Emission angle 20°–135° Output power 1–300 mW	Data rate 100 MHz Peak wavelength 0.82 μm Spectral width 200–300 Å Emission angle 20–135° Output power 1–100 mW	Data rate 1 Gb/s, potentially 5 Gb/s Peak wavelength 0.80–0.85 μm Spectral width 10–50 Å Emission angle $<$ 15° Output power 250 mW at 10 Mb/s at room temperature, 5–15 W at −200°C	Data rate 1 Gb/s with external modulator, $<$0.01 Mb/s without modulator Peak wavelength 1.06 μm Spectral width $<$1 Å Emission angle $<$5° Output power 0.1–10 mW

(continued)

TABLE 1.4. (continued)

Characteristic	Waveguides		Detectors	
	Low loss	High loss	PN junction photodetector PIN	Avalanche photodetector (APD)
Advantages	Attenuation near theoretical limit (Rayleigh scattering) Relatively good radiation resistance Good mechanical characteristics	Large numerical aperture with resultant high coupling efficiency Lower cost than glass fibers Good mechanical properties Immunity to most environmental influences	Very good performance properties, such as small size, fast and linear response, low dark current, low-voltage operation, ambient temperature operation, good radiation resistance, light weight, ruggedness, low cost	Very good performance properties, such as high internal gain, low noise, small size, fast response, good spectral match with sources Ideal for use with monomode fibers
Disadvantages	Difficulty of producing good monomode fiber that takes advantage of low loss Poor coupling efficiency to LEDs Low packing fraction Low loss difficult to control	High attenuation Low melting point (about 90°C) Spectral incompatibility with efficient infrared (IR) sources Poor immunity to high nuclear radiation exposure	Absence of internal gain requires high-gain low-noise amplifier Response and efficiency degrade at wavelengths over 1.0 μm	High-voltage operation High cost Temperature-compensation requirement Susceptible to saturation, overload burnout by transients Radiation-sensitive

Main applications	Long-haul and short-haul use Land and underwater cables Use in computers and in a variety of industrial applications	Short-haul systems which can tolerate disadvantages, but require low cost and good mechanical characteristics Military potential	Data bus Receiver Coupling to monomode fiber	Receivers in long-haul links Special applications requiring long life and/or maximum bandwidth
Key performance parameters	State-of-the-art loss 2.0 dB/km at 0.85 μm, 1.8 dB/km at 1.06 μm; commercial fibers 5 dB/km State-of-the-art dispersion 0.1 ns/km; commercial fibers 1 ns/km Fiber OD 10–125 μm, bundle OD 5mm Fiber lengths <10 km	Loss in excess of 500–1000 dB/km Dispersion 10–100 ns/km Fiber OD 125–1000 μm Fiber lengths <0.1 km	Response time 10 ns unbiased, 0.1 ns biased Efficiency 90% at 0.9 μm Peak response at 0.85–1.06 μm Noise equivalent power 10^{-14} W/Hz$^{1/2}$	Response time 0.5–5 ns Efficiency 85% at 0.85 μm Peak response at 0.85–1.06 μm Internal gain 30–150 Gain-bandwidth product 20–100 GHz Noise equivalent power 10^{-14} W/Hz$^{1/2}$

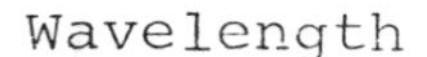

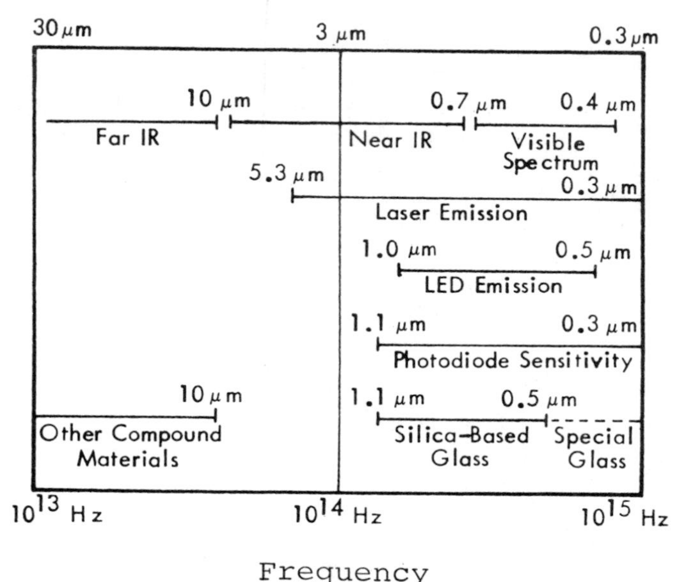

Figure 1.5 Overview of the spectral range applicable to optical communications.

Detector Noise Limitations. Because of waveguide signal attenuation, the amplitude of a light input pulse will suffer diminution as the pulse propagates along the waveguide. Ultimately the amplitude becomes so small that it is indistinguishable from noise, and the receiver is not able to make a zero-or-one decision within the specified probability of error. In this case the amplitude of the pulse, rather than the spread resulting from dispersion, will limit the communication capability of the system.

Waveguide and Material Dispersion Limitations. Waveguide dispersion is a consequence of the apparent change of fiber dimensions (in units of wavelength) with frequency. This results in a frequency-dependent phase and group velocity. Material dispersion is a consequence of the variation of the refractive index of the fiber with frequency. Both waveguide and material dispersion cause a pulse propagating along the waveguide to spread because of the different component velocities. Because of pulse spreading, the receiver eventually will be unable to distinguish between two adjacent pulses that tend to overlap after having experienced dispersion; consequently, the receiver will be unable to decide whether a given time slot contains a zero or a one. In this case, the widening of the pulse, rather than its loss in amplitude, limits the communication capability of the system.

Delay Distortion Limitations. If a waveguide supports several modes with different phase and group velocities, energy in the respective modes will arrive at the detector at different times. Most optical sources, particularly LEDs, excite many modes; if they are able to propagate through the waveguide, distortion will occur. The degree of distortion depends on the amount of energy in the modes arriving at the detector input, which, in turn, depends on the difference in attenuation between the modes and the degree of mode mixing. In this case, the ability of the waveguide to suppress undesirable modes or to convert their energy to a desirable mode is the limiting factor for the communication capability of the system.

OPTICAL FIBERS

The optical waveguide is one of the major components in any viable optical communications system. The dominant portion of all research and development in optical information transfer has been devoted to improvements in fiber characteristics. This is discussed in more detail in Chapter 2.

In applications intended for long-distance transmission of data at high information rates, the most important fiber parameters are signal loss and dispersion. Loss reduces the power margin between the transmitter and receiver, and hence limits the amplitude of the signal. Dispersion widens the transmitted pulses between repeaters and generates interference between successive pulses; hence, it limits the bandwidth of the signal (8).

Losses of optical fibers result from three major causes: material absorption, material scattering, and waveguide imperfections. Absorption is mainly a result of the presence of transition metal ions and hydroxyl (OH^-) radicals in the amorphous structure of the fiber material. Scattering is inversely proportional to the fourth power of the wavelength and is caused by index variations that are small compared to the wavelength. The index variations are produced by inhomogeneities in composition, temperature, etc. and are a function of the glass composition (lowest for fused silica and increasing in glasses with a higher refractive index). Waveguide imperfections are structural defects, such as irregularities along the core–cladding interface and excess cabling loss. In a low-loss optical waveguide, losses due to scattering and fiber imperfections dominate.

Figure 1.6 is an example of the absorption spectrum of a typical low-loss fiber, consisting of an SiO_2–GeO_2 and an SiO_2 cladding. Low-loss regions at wavelengths ranging from about 0.80 to 0.85 μm, from about 1.00 to 1.10 μm, and from about 1.25 to 1.30 μm are apparent, while the strong harmonic of the OH^- absorption line around 0.95 μm prevents the practical use of this wavelength region; operation close to this maximum, however, has been shown to eliminate higher-order modes and with resultant lower pulse dispersion. The 0.85 μm loss minimum can best be exploited by the use of LEDs and GaAsAl injection lasers, whereas only lasers are well suited for the 1.05 μm and

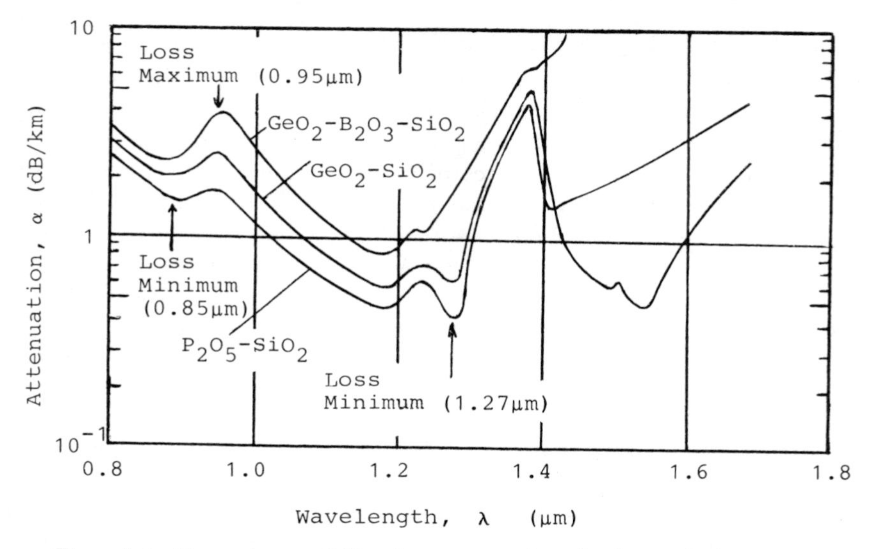

Figure 1.6 Dependence of fiber loss on wavelength of typical glass waveguides.

1.27 μm loss minima. More recent efforts have produced GaAsIn and PAsIn LEDs as well as GaAsIn injection lasers that can be operated at the 1.05 μm wavelength.

The 0.85 and 1.05 μm wavelengths commonly used in fiber optics communications take advantage of the loss minima of the glass at these wavelengths. In principle, there is no reason why the transmission wavelength could not be extended in order to take advantage of the significantly lower scattering loss of the glass at larger wavelengths. Attenuation caused by Rayleigh scattering, which represents the inherent loss limitation, decreases with the fourth power of the wavelength; hence, it is desirable to increase the wavelength of transmission if other considerations are of secondary interest. The successful employment of a large transmission wavelength requires that loss contributions from absorption, particularly those caused by transition metals and the OH^- radical, be reduced below present values.

Thus, if only fiber loss is of concern, an increase in wavelength is advantageous. However, in systems applications other considerations are equally important. For example, detector efficiency and frequency response decrease with wavelength so that improved photodetectors have to be developed.

Only a comparatively small amount of work on detectors suitable for operation at wavelengths in excess of 1 μm is presently being conducted. Also, light sources useful for operation at longer wavelengths have to become available. One of the most promising of optical sources is the GaAsIn LED, which has the potential to extend the LED emission wavelength range from 0.9 to 3.4 μm.

It is also important to note that the radiation resistance of fibers generally increases with wavelength. Therefore, fibers that can be operated at a wavelength of, for example, 1.3 μm are superior to those operating at 0.85 μm.

Dispersion in optical fibers may result from either material, waveguide, or multimode causes. In single-mode fibers, only material and waveguide dispersion is present, with material dispersion being the dominant factor. Material dispersion to a large degree also depends on the spectral width of the light source and the emitted wavelength. Because most fibers today do not allow monomode operation, multimode dispersion is an important consideration, since the pulse spread over a given distance is related to the arrival time difference between modes. This time difference increases with the number of modes the fiber can support, which, in turn, varies as the square of the core diameter.

The dispersion of a signal in a step-index (SI) fiber is usually much larger than that in a graded-index (GI) fiber, as shown in Figure 1.7. Multimode dispersion is significantly less severe in a graded-index fiber because the index profile tends to speed up slower modes and thus to reduce the modal velocity spread.

Furthermore, coupling between modes as a result of structural imperfections, such as directional change, refractive index variations, diameter variation, and strain birefringence, will reduce the pulse spread. Mode coupling is thus desirable from the standpoint of bandwidth. Consequently, the case of negligible mode interaction represents worst-case conditions and is not found

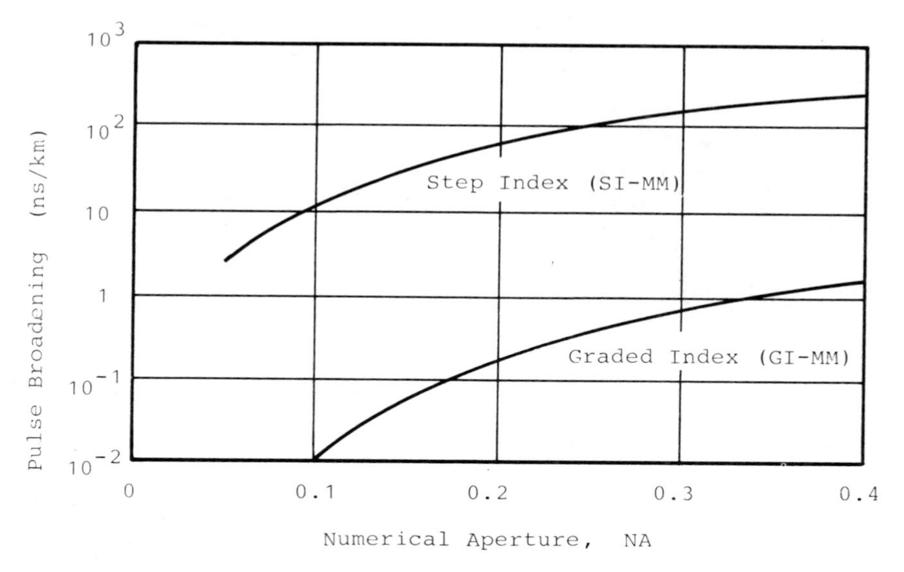

Figure 1.7 Dependence of signal dispersion on numerical aperture of typical glass waveguides. MM, multimode.

in a real waveguide. Coupling may be between guided modes or between guided and radiation modes. If the coupling spectrum is flat, the spread from the mean velocity increases as the square root of the product of fiber length and coupling length. The coupling length is the fiber length where mode coupling reaches an equilibrium state; it usually exceeds 1 km.

Figure 1.8 shows the dependence of the coupling efficiency of the sources of fibers on the numerical aperture (NA) and the packing density of optical fibers. Generally, the coupling efficiency increases with increasing numerical aperture, but at the same time the pulse width received at the detector increases also. In practice, this necessitates a compromise in numerical aperture, which leads to typical NA values ranging from about 0.12 to 0.25. For example, a graded-index fiber with NA of 0.20 yields a pulse spread of about 0.2 ns/km and a coupling efficiency of about 3% if an LED source is used, whereas the combination of a graded-index fiber and an injection laser causes a pulse spread of only 0.05 ns/km and a coupling efficiency of about 10% for a NA of 0.14. The potential coupling efficiency improvements in using a highly directional optical source are obvious.

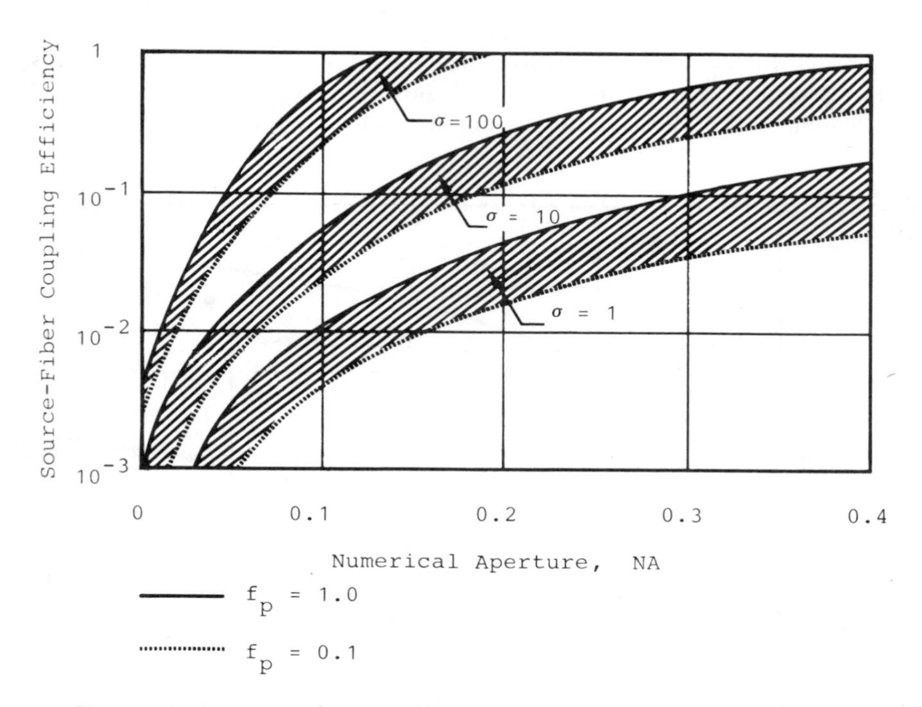

Figure 1.8 Variation of source–fiber coupling efficiency with numerical aperture of fibers; f_p, packing fraction; σ, numerical factor characteristic of a light source.

OPTICAL SOURCES

The availability of fibers of relatively low loss has stimulated the development of improved optical sources. This work, in turn, has led to considerable feedback on fiber research. Although research on LEDs and injection lasers started in the early 1960s and passed through many developmental stages, the requirement for sources for fiber optics systems presented an additional stimulus for further development of these devices. The characteristics of low-loss fibers present a set of unique requirements for signal sources, as discussed in further detail in Chapter 3.

The design, establishment, and performance of optical communication systems depend strongly on the characteristics of the chosen light source. Source requirements and selection, in turn, depend on the specific application. Of principal interest in the selection of an optical source are economic and technical criteria, such as cost, ease of maintenance, brightness, size, spectral properties (such as wavelength and spectral width), efficiency, linearity, modulation capability, and lifetime.

Three major categories of light sources are of practical interest: (a) GaAs and GaAsAl light-emitting diodes (LEDs), (b) GaAsAl semiconductor diode injection lasers (ILDs), and (c) Nd:YAG nonsemiconductor lasers. Historically, LEDs have been the most mature of these; they are available in a variety of shapes, emission spectra, and price ranges. Injection lasers have received substantial developmental effort in recent years, whereas solid-state Nd:YAG ion laser devices are still more or less in the developmental stage.

In general, an optical power source suitable for optical fiber transmission must be reliable, economically viable, and compatible with the fiber. The first two of these requirements are determined by previous experience and economic considerations, whereas the third requirement is intimately related to the properties of the fiber. These include geometry, loss spectrum, group delay distortion, and modal characteristics.

The choice of simple and reliable source–fiber coupling configurations versus high coupling efficiency involves compromises that depend on both source and fiber. Furthermore, the characteristics of the available source have a significant effect on the choice of the fiber; for example, single-mode versus multimode and uniform-index versus nonuniform-index profiles. Fiber design is affected by characteristics such as core diameter and core–cladding index difference. Hence, in selecting an optical source the interplay of these factors with each other and with system requirements must be taken into account.

In optical communications systems, both coherent and incoherent light sources can be used. Whereas either monomode or multimode fibers can be used with coherent sources (lasers), only the multimode fiber can accept enough power from an incoherent source (LED) to be useful.

Both LEDs and injection lasers can be fabricated in the same semiconductor material, GaAsAl, and can have the same basic structure. They differ mainly in their operating mode. In both devices, light emission results from the recombination of carriers injected across a p-n junction, whereby each radiative recombination event generates one photon of energy approximately equal to the energy gap; hence, the wavelength of the emitted light depends on device material composition as well as on dopant type and concentration. In an LED, emission is spontaneous; in an injection laser, it is stimulated. Both LEDs and lasers can be conveniently light-intensity modulated by varying the injected current.

Light-Emitting Diodes (LEDs). In the LED, light is emitted over an angle of 180°, with a spectral width of typically 30 to 50 nm. Coupling efficiency into a fiber is poor because of the incoherent light type and the large-angle emission pattern. The best LEDs have a double heterostructure (DH) and use $Ga_{1-x}AsAl_x$ grown by liquid-phase epitaxy (LPE) as the basic material. These devices have an emission area of about 10^{-5} cm^2; they yield an output power of 10 mW/sr cm^2 at 100 mA, corresponding to a current density of about 10^4 A/cm^2; their external quantum efficiency is about 2% to 3%. Modulation rates in excess of 100 Mb/s have been observed. Mean times to failure (MTTFs, defined as the time at which the current density has to be increased by 20% to realize the same optical power) of up to about 10^4 hours are typical of today's devices, with 10^6 hours anticipated by the next decade. One of the most promising LEDs is the high-radiance Burrus diode, whose characteristics are substantially superior to all other types.

Injection Laser Diodes (ILDs). The semiconductor injection laser diode is almost ideally suited for optical communications. Its light emission pattern exhibits significantly more directivity and its spectral width is more than an order of magnitude narrower than an LED. Furthermore, the emission area is substantially smaller, and, hence, for equal drive current the radiance is considerably higher than that of an LED. However, the input–output relationship is not linear and the device's lifetime is still shorter than that of the LED, although this will not necessarily remain so in the future.

The most promising injection laser is the GaAsAl stripe-geometry double-heterostructure (DH) design. The stripe provides lateral confinement of the injected carriers and hence of the light output. This is important because it allows the emitting area to be matched with the core of the optical waveguide. The spectral width of this laser type is about 2 nm owing to a combination of temperature effects and the existence of a number of longitudinal modes. The emission wavelength depends on the Al concentration, with a typical range of 0.7 to 0.9 μm; metallurgically, the ideal wavelength is 0.8 μm.

Because the radiative recombination time of the carriers is very small (about 1 ns for GaAs and about 5 ns for GaAsAl) and is further reduced by stimulated emission, injection lasers can easily be modulated at high speed by current variations. The modulation speed is limited by a resonance effect due to the phase relationship between the photon and electron densities, which may excite damped oscillations. This leads to a frequency-dependent modulation efficiency which is flat up to a certain limiting frequency (typically ranging from 0.5 to 5 GHz), above which the efficiency drops rather rapidly. The highest useful modulation frequencies observed have been up to 2 Gb/s.

Power coupling from an injection laser into a single optical fiber depends on the matching of fiber and laser area geometries. If exact matching can be achieved, a high coupling efficiency can be obtained, although this means a high fabrication cost, potentially nonoptimized emission characteristics, and reduced lifetime. Substantial improvements in this respect will take place during the coming decade, however, making the injection laser the most useful of optical sources.

Because laser lifetime is related to the drive current, the lifetime can be significantly enhanced over that for continuous wave operation by the employment of pulse modulation at the required rate with a short duty cycle. This allows the additional achievement of an improved performance, mainly a higher permissible optical output power and a higher differential quantum efficiency. However, in the absence of a continuous drive current (using only pulses to turn the laser on), a prepumping time is required before lasing action takes place, thus effectively reducing the laser speed and upper frequencies.

Superluminescent Diodes (SLD). In the SLD both spontaneous light emission (as in an LED) and stimulated light emission (as in a laser) occur simultaneously. Hence, its output usually has a narrower spectral width and a higher radiance than that of the LED, in which only spontaneous emission takes place. The geometry of the simple form of such a device is similar to that of a stripe laser, except for a 10° inclination of the stripe with respect to the normal to the emitting surface to eliminate feedback. A more advanced version of this device suppresses feedback by eliminating one of the mirrors and providing absorption for the backward waves in the cavity. The output of the SLD is incoherent but is about 90% polarized. The major advantage of this structure is that it, unlike LEDs and lasers, which tend to oscillate in many modes at high drive currents, favors the low-order modes for large output powers. However, although the SLD has achieved an output power of up to 50 mW coupled into a fiber with NA of 0.6 and a narrow spectral width of only 5 nm, this device is inefficient compared to lasers and requires rather high drive currents.

Solid-State Ion Lasers. The third important category of optical sources is that of the neodynium-doped yttrium-aluminum garnet (Nd:YAG) ion laser. Although this device is still chiefly in the research stage, it promises to be of increasing importance because of the features it can potentially offer over other sources. The expected advantages of this laser type, as compared with the injection laser, can be summarized as follows.

Its emission wavelength is about 1.05 μm, which coincides with one of the lowest loss minima of glass fibers; this allows operation of the system at a wavelength where Rayleigh scattering in the fiber is significantly less than at the LED or injection-laser wavelength of 0.85 μm. Even more important is the narrower spectral width of the Nd:YAG laser, which results in an even smaller material dispersion than in the injection laser (typically less than about 0.5 nm, and in a number of cases less than 0.1 nm). Another advantage of the Nd:YAG laser is its expected long MTTF, which will eventually exceed that of the injection laser by about an order of magnitude to about 10^7 hours, due to the LED pumping of the Nd:YAG laser. Furthermore, a single-frequency monomode output can be relatively easily obtained from this source type, whereas the LED launches a large number of modes and the injection laser generates a small number. Finally, the coupling efficiency of the Nd:YAG laser is very high, owing in part to the possibility of operating in short lengths (1 to 2 cm) of fiber with Nd-doped cores as end-pumped lasers. For example, an Nd:YAG laser operated at $-6°$C and pumped by an array of GaAsAl LEDs yields an optical output power of about 50 mW in TEM_{00} modes, while at room temperature output powers of 2 to 5 mW are achievable.

Because the fluorescence lifetime of the upper lasing level is about 230 μs, the Nd:YAG laser is not suitable for direct modulation at high pulse rates either by varying the pump or through cavity loss. Therefore, it is always necessary to combine an Nd:YAG laser with an external optical modulator. Because of this additional device, the nonsemiconductor laser material, and the need for pumping devices, this source type will be considerably more expensive than other sources, although as knowledge of the device increases and fabrication techniques improve, its fabrication cost will be reduced significantly. On the other hand, the highly desirable properties of the Nd:YAG laser, mainly its power and emission characteristics, will make this device type attractive.

OPTICAL DETECTORS

The detector of an optical communications system acts as an O/E converter. It demodulates the optical signal received and converts it into a proportional electrical signal, assuming conversion linearity. The characteristics of communications systems based on the transmission of data over optical waveguides therefore depend also on the performance of the receiving or detection components.

Specifically, it is desirable for the receiver to achieve a high incident optical power, a high signal-to-noise ratio, and a low bit error rate, as well as being low in cost. These characteristics depend to a large degree upon the receiver design, particularly the first stages of amplification. Typically, a receiver consists of photodetector [usually either a p-n junction with an intermediate intrinsic region (PIN) or an avalanche photodiode (APD)], front-end amplifier, equalizer, filter, and additional circuitry, depending upon the type of modulation employed. Chapter 4 is a more detailed analysis of the suitable photodetectors.

Unlike electrical signals, optical signals possess an inherent signal-dependent noise. There are several sources of noise associated with the detection and amplification processes of an optical system. These include quantum noise, background noise, surface-leakage current noise, beat noise, and amplifier noise. For pulse-coded modulation (PCM) with a good detector and at a high bit rate, the total receiver noise can be approximated by the sum of quantum and amplifier noise.

In a typical application of a PIN detector, a bipolar front end is about 3 dB better than a resistively loaded front end. The bipolar front end also becomes superior by up to 2 dB to the MOS (metal-oxide–silicon) front end, over 30 Mb/s in conjunction with an avalanche photodetector (APD).

A PIN detector requires only a few volts in order to achieve adequate detection, but it is somewhat limited in frequency response. The APD, on the other hand, allows a higher upper frequency, but its use requires the application of a very high voltage (100 to 250 V), which restricts this type of detector to applications where these voltages are available.

Compared to optical sources, considerably less developmental efforts are being conducted in attempts to advance the state-of-the-art of photodetectors, mainly as a result of the relatively easy availability of detectors with adequate properties, at least at the lower transmission window wavelength. Thus, the development of an improved device or receiver design is not justified. Therefore, in the future comparatively little improvement in photodetector characteristics can be expected, except in cases in which a shift to longer wavelengths is desirable.

PASSIVE COMPONENTS

An important consideration in optical communication is the availability of convenient low-loss connectors and coupling elements. Design and fabrication of these involve extremely accurate mechanical tolerances, optical polishing, and alignment of individual fibers, as analyzed in Chapter 5.

Another important consideration is the present lack of connector standards, which is detrimentally affecting the development of suitable high-performance fibers. In the future, however, some form of standardization can be expected. This will affect at least size and dimensional standards.

In addition to fiber connectors and splices that are applicable to single-fiber connections, there are monofiber–multifiber cable connections and multifiber–multifiber cable connections, as well as fiber–fiber couplers.

Monofiber–Multifiber Connections. Monofiber–multifiber connections are an important concern in systems where fiber bundles carry identical information that is to be mixed in a thick monofiber. The connector must allow efficient coupling to or from system portions. The design of these connectors depends upon standardization of fibers and cables.

Multifiber–Multifiber Connections. This connection concept involves a series of complex problems that center on alignment and low-loss connections. The difficulties of these connectors increase rapidly with the number of fibers per bundle. Again, standardization of the fiber bundles and cables is necessary. Easy, repeatable execution of a connection is a basic requirement for this connector type, which must also compare favorably in its performance to standard metal cable connectors.

Fiber–Fiber Coupling. Coupling into and out of fibers is usually restricted to two main approaches, T couplers and star couplers. The T coupler consists of a glass mixing rod with either an internal mirror or a branching transmission path for injecting, removing, and redistributing signals. The star coupler uses a single dielectric mixing rod to divide an input signal among multiple output fibers or bundles and is the lowest-loss device yet designed for multiterminal distribution. With T couplers, typical throughput losses are about 3 to 4 dB per coupler, and with star couplers a total loss of about 10 dB can be achieved. T couplers, because of their individual losses, allow a maximum of only about ten terminals to be connected per system unless additional repeaters are deployed, whereas no such limitation exists with the star coupler. From a practical standpoint the star coupler is by far the better approach to the coupling concept, and it is the only useful approach for low-loss, low-packing-density fiber bundles. Manufacturing techniques for the T coupler still have to be developed. The T coupler concept is limited to short communications links, such as aircraft and computer multiplexing systems, and the star coupler finds wide use in almost all other applications. Fiber-to-fiber registration couplers are used to provide extremely accurate alignment and registration of cores of all individual fibers in two bundles. Hence, these couplers yield minimum coupling losses, typically 1.5 dB per connection. Several designs of this approach have been developed. Alternate designs will be needed if large bundles or different bundle configurations are to be connected.

SYSTEMS CONSIDERATIONS

In addition to consideration of the characteristics of the individual components of an optical communications system, the establishment of such a system requires attention to its overall performance, cost, reliability, and ease of maintenance. From a technical standpoint, the most important characteristics involve the maximum power that can be coupled into an optical fiber after cabling and the required average power available at the receiver. Thus, system design depends mainly on source–fiber characteristics and receiver requirements and is frequently dictated by cost considerations. Chapters 6, 8, and 9 discuss the system aspects more comprehensively.

To establish optical fiber links as dominant communications systems, attention must be given to efficient, cost-effective optical components and systems. Presently, however, only relatively few systems have been deployed because of system expense, inefficiency, or both. This situation reflects the interdependence of technological and economic factors comprising the present limits in performance and applications (9). A number of optical communications links are, however, in various stages of development and in field trials to determine feasibility, reliability, cost, and operation characteristics, and to discover potential problem areas. To date, almost all experimental systems operated throughout the world have performed better than originally anticipated.

In the future the evolution of optoelectronic technology will lead to extensive use of integrated optics for optical communications systems. Already, substantial research is being done and development is progressing in this field. The potential analogy to the rapid and often revolutionary development of integrated circuits is apparent. The optical frequency equivalent of electronic integrated circuits based on silicon—integrated optic circuits based on nonsemiconductor materials—will provide similarly efficient and economic systems. In spite of the current research on these integrated optic components, numerous problems remain at this time, partly because of the reduction of device dimensions and of tolerances by several orders of magnitude as dictated by optical frequencies. Ultimately, integrated optics systems will have an effect on the field of communications similar to the effect of integrated circuits on computer advances.

Until cost-effective, reproducible, and practical ways have been found to fabricate optical communications systems based on integrated optics, all optic transmission links will consist of discrete components: optical sources, fibers, detectors, couplers, etc.

It is not justifiable to assume that the development of integrated optics will have a significant short-term impact on discrete optic systems. Hence, the immediate concern of the designer of optics communications systems will continue to be directed toward the state-of-the-art of available or expected discrete components. Also, system compatibility has to be considered in the absence of system integration. These are two major reasons for these temporary problems.

Component Development by Divergent Manufacturers. Most components used in optical systems have been developed by a variety of manufacturers specialized in particular components, such as glasses, connectors, circuits, light-emitting devices, etc. Most of these manufacturers never devoted efforts to the assembly of entire systems. The result is that the many specialized talents contributing to the research on component parts for a common system are not inbreeding their developments within a single company; thus, interface incompatibility does result.

Component Development for Dissimilar Applications. Often components originally developed for entirely different applications are put together in a single system. Obviously, these components have usually been optimized with different applications in mind. Thus, cost efficiency results from using mass-produced devices and from reduced development costs, but efficiency is lost because of a lack of performance optimization and systems compatibility.

These deficiencies, although currently of major concern, are only temporary. The low-loss fiber used in conjunction with a wide-angle light-emitting diode is typical of both points above. The fiber was developed for use in low-loss information transfer applications and was not intended for use with such a wide-angle source, whereas the emitter was developed and mass-produced for other applications (mainly for calculators, where a wide-angle emission pattern is desirable) rather than for use with fibers. Thus, research on optical sources presently concentrates on the development of injection and nonsemiconductor lasers rather than on the improvement of LEDs. However, although few available LEDs have been developed specifically for emission into fibers, LEDs are relatively widely employed in systems with less stringent requirements, thereby trading efficiency for reliability and low cost.

The pulse deterioriation along an optical fiber in terms of pulse amplitude and width requires the inclusion of regenerative system parts, commonly referred to as repeaters, whose purpose is to reamplify and reshape signal pulses. The spacing between repeaters is a function of the degree of signal deterioration, which in turn is a function of fiber characteristics and available source and receiver characteristics (degradation of signal pulse amplitude by fiber loss and degradation of signal pulse width by fiber dispersion).

Although system performance may make small spacing between repeaters desirable, system cost requires maximum repeater spacing, so that an optimum distance between repeaters is a compromise between economic and technical considerations. Presently, typical repeater spacings range from 1 to 10 km with 50 km spacing eventually expected, whereas conventional metal wire communications systems require repeater spacings of the order of 1 km.

Repeater spacing is usually determined by the data transfer rate. The spacing is increased until a designated loss limit or dispersion limit is reached. Several limitations can be distinguished (10).

Fiber Loss Limitations. Signal attenuation resulting from relatively high fiber losses can ideally be overcome with a lower-loss fiber but at higher cost. The selection of a fiber having a higher numerical aperture allows the coupling of a higher optical power into the fiber and hence will compensate in part for subsequent fiber losses, but an increase in the numerical aperture tends to increase the mode dispersion in a multimode fiber. Thus, ideally, the maximum repeater spacing in this case will be achieved when the dispersion limit and the loss limit coincide at the desired bit rate. On the other hand, reliability and cost considerations will significantly affect the acceptable repeater spacing.

Multimode Dispersion Limitations. If the repeater spacing is limited by multimode dispersion, a fiber with a smaller numerical aperture may be chosen. Thus repeater spacing can be increased while simultaneously decreasing the loss margin through a decrease in power coupled into the fiber. If, however, the numerical aperture is too small, excess cable loss may become intolerably high.

Material Dispersion Limitations. If the repeater spacing is limited by material dispersion in the fiber, a source with a narrower spectral width should be chosen. For example, if economic or other technical considerations do not preclude its use, an injection laser or, better, an Nd:YAG laser may be substituted for the LED source, thus allowing a repeater spacing increase of at least one order of magnitude.

Pulse Dispersion Limitations. If the repeater spacing is limited by the inability of the optical fiber to support a high data transfer rate due to a nonoptimized fiber refractive index profile, a conventional multimode step-index fiber may be replaced by a graded-index fiber of near-optimum index gradient. A more expensive, but ultimately more desirable, replacement would be the selection of a small-core monomode step-index fiber, which would potentially allow a repeater spacing increase by at least one order of magnitude and eventually, in conjunction with an ultranarrow Nd:YAG laser source, by up to three orders of magnitude.

The system limitations due to signal dispersion and attenuation are summarized in Figure 1.9. It gives the fiber length–bandwidth product as a function of the spectral width of the source, the refractive index differences of the fiber core–cladding, and the fiber index profile. For an ideal index profile, the length–bandwidth product is independent of numerical aperture but inversely proportional to source spectral width, so that the numerical aperture can be selected to allow the highest source–fiber coupling efficiency. For a nonideal index profile, the length–bandwidth product is independent of source spectral width, so that a narrow source is not necessarily required. The length–bandwidth product is a strong function of the index profile of a graded-index fiber; the dependence is stronger the narrower the spectral width of the source.

The maximum allowable fiber length is a function of fiber attenuation. The limitation is a result of the maximum permissible signal-to-noise ratio and is therefore a slight function of the type of detector used. The use of an avalanche photodetector allows a longer fiber or larger repeater spacing than the use of a photodiode.

Typical curves relating the maximum data transfer rate to the optical power at the detector and to the fiber attenuation are shown on Figure 1.10. For dispersion-limited operation the maximum pulse rate is independent of fiber loss and of optical power. This conclusion is reasonable because it is assumed that there is adequate power at the detector, but the pulses have

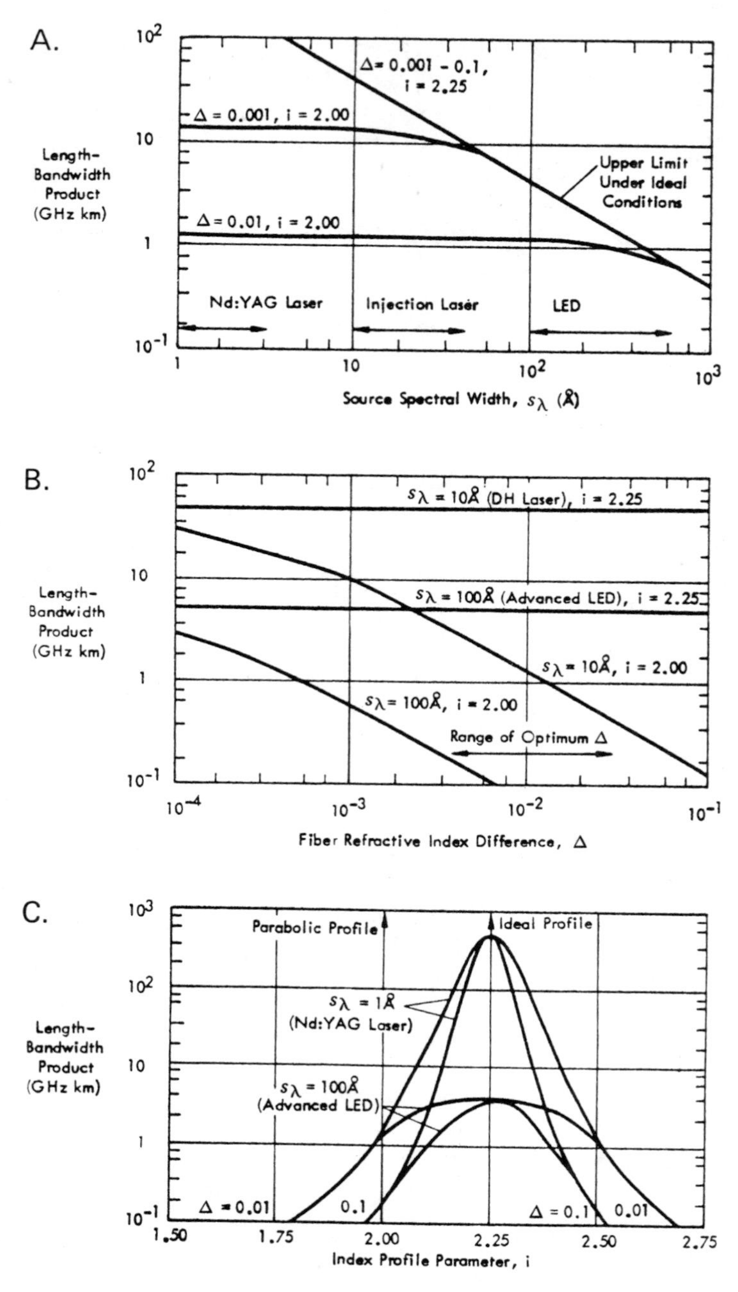

Figure 1.9 Dependence of length–bandwidth products on (A) source spectral bandwidth, (B) fiber refractive index ($\lambda = 0.85$ μm), and (C) index profile parameter.

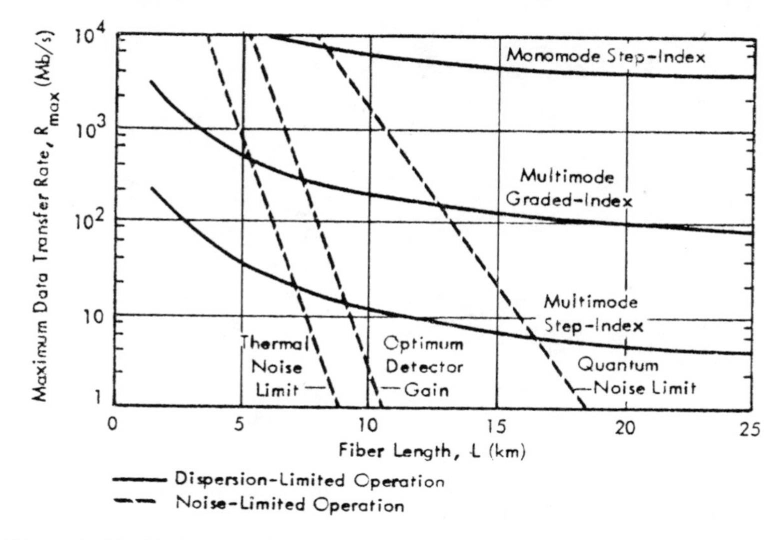

Figure 1.10 Variation of maximum data transfer rate with fiber length (α = 4 dB/km; P_0 = 1 mW); α, attenuation; P_0, initial source strength.

overlapped because of dispersion, precluding an intelligent decision on the presence or absence of a pulse. For noise-limited operation, however, the maximum pulse rate is a strong function of both fiber loss and length.

The slopes of the curves are important. For fiber length L (defined by the distance between adjacent repeaters or by the total system length in the absence of repeaters), for dispersion-limited operation the data rate varies with fiber length as L^{-1} or $L^{-1/2}$. A monomode fibers has the more desirable $L^{-1/2}$ dependence, and a multimode fiber shows an L^{-1} dependence. For noise-limited operation the variation of the maximum data transfer rate with length is exponential, since the pulse rate varies with optical power of its square; an increase in attenuation causes a change in the slope and the location of the curves. For comparison, in a coaxial cable the variation of maximum pulse rate with cable length displays an L^{-2} dependence, which is even less desirable than the L^{-1} or $L^{-1/2}$ variation of dispersion-limited optic systems.

Examination of the curves shows that the intersection of corresponding curves represents the change from dispersion-limited to noise-limited operation. Generally, operation in the noise-limited regime should be avoided. As the data transfer rate demand increases, one must design for reduced spacing between repeaters or between transmitter and receiver.

The reduction of fiber length for increased data rate is gradual in a dispersion-limited operation, but it becomes severe in a noise-limited operation. This has an effect on the economics of the system. Furthermore, it is advantageous to allow for a gradual rather than an abrupt degradation of operating conditions. Taking into account fiber and source aging and other performance

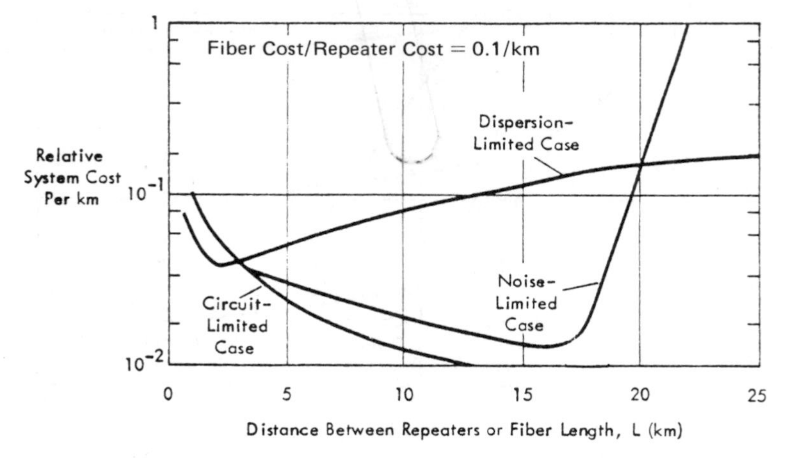

Figure 1.11 Dependence of system cost on fiber length under various system limitations at optimized data rate.

deteriorations, it is advisable to design the system for a fiber length adequately below the dispersion–noise crossover point.

The cost as a function of fiber length is shown in Figure 1.11 for the three limiting cases considered. It is assumed that the data transfer rate is optimized for maximum economy but that the system is not necessarily operating at the maximum transfer rate. In other words, optimum and maximum data transfer rates may not coincide.

For short to medium fiber lengths, a system operating in the dispersion-limited case is usually most expensive; the cost per kilometer increases with fiber length. In contrast, the noise-limited system shows a decreasing cost per kilometer with fiber length up to a point where amplification will require substantial investments and drive the cost steeply above that of dispersion-limited operation. If the system is limited by the speed of the associated electronic circuitry, the system cost per kilometer will display a continuous decrease.

REFERENCES

1. H.F. Wolf & J.D. Montgomery. *Fiber Optics and Laser Communications Forecast.* Gnostic Concepts, Inc., Menlo Park, California, 1976.

2. R.L. Gallawa et al. *Telecommunication Alternatives with Emphasis on Optical Waveguide Systems.* Office of Telecommunications Report 75-72, Washington, D.C., U.S. GPO, 1975.

3. C.P. Sandbank. Fiber Optic Communication Systems. Intelcom 77, Atlanta, Georgia, October 1977.

4. K.C. Kao & M.E. Collier. Fiber Optic Systems in Future Telecommunication Networks. World Telecommunication Forum, Geneva, October 1975.

5. S.E. Miller et al. Research Toward Optical Fiber Transmission Systems. *Proc. IEEE* **61**:1703, 1973.

6. R.L. Gallawa. Component Development in Fiber Waveguide Technology. Intelcom 77, Atlanta, Georgia, October 1977.

7. Special Issue on Optical Communication. *Proc. IEEE* **123** (No. 6), 1976.

8. J.E. Fulenwider. Wideband Signal Transmission Over Optical Fiber Waveguides. Guided Optical Communications Conference, Society of Photo-Optical Instrumentation Engineers, San Diego, California, August 1975.

9. J.E. Goell et al. Long-Distance Repeatered Fiber Optical Communication Systems. Guided Optical Communications Conference, Society of Photo-Optical Instrumentation Engineers, San Diego, California, August 1975.

10. H.F. Wolf. Overview of Optical Fiber Systems Markets. Intelcom 77, Atlanta, Georgia, October 1977.

OPTICAL WAVEGUIDES

Helmut F. Wolf

The future of optical communications depends to a large degree on the characteristics, interface properties, economics, and availability of a suitable transmission medium—optical waveguide or optical fiber. Fibers allow relatively loss-free data transfer over distances that may eventually range from a few meters to hundreds or thousands of kilometers. The fiber transmission properties are almost completely independent of environmental conditions, and at the same time transmitted information cannot be intercepted in an unauthorized way. Most optical fibers are made of glass. Glass offers the lowest achievable loss and dispersion, whereas waveguides made of plastic are characterized by loss and dispersion that are frequently higher by one or more orders of magnitude. Hence, most currently available plastic fibers have only a limited range of potential applications. Fibers have to interface with other elements of a communications system, such as sources and detectors; the interface involves difficulties, compromises, and challenges in addition to those imposed on the fiber itself.

This chapter reviews the important characteristics of fibers suitable for intelligent optical communication. In addition to an analysis of propagation properties, it gives an overview of fabrication techniques, cabling methods, and interface considerations. The applicability of the fiber characteristics to systems requirements is stressed wherever appropriate. A large body of published research has been combined in this review, together with a substantial amount of information derived through private communications and original research. For additional sources, the reader is referred to other published reviews (1–3).

OVERVIEW

Two characteristics of optical waveguides of primary importance are signal attenuation and dispersion. Generally speaking, attenuation (or loss) determines the distance over which a signal can be transmitted without becoming indistinguishable from noise, and dispersion determines the number of bits of information that can be transmitted over a given fiber in a specified time period.

In 1970 the advent of optical fibers with a loss of only 20 dB/km was considered revolutionary. Previously, fiber losses typically ranged from several hundred to over 1000 dB/km (4). Since that time, further significant improvements have been reported in the literature. At the present, minimum losses have been achieved that approach the theoretical limit determined by Rayleigh scattering (about 1 dB/km). Further improvements will be measured in fractions of 1 dB/km, whereas in the past improvements were usually measured in several dB/km.

With actual fiber losses approaching the theoretical limit, the most spectacular advances in the development of fibers have already been achieved, at least in the relatively sterile environment of research laboratories or preproduction facilities. Future efforts will be devoted to the achievement of this performance in large volume in a production environment.

With the attainment of low-loss fibers that approach ideal loss characteristics, attention has shifted to improving fiber bandwidth. Again, substantial improvements in dispersion characteristics have been made in the last few years. At present, the upper limit of the data transfer rate is a few hundred megabits per second (Mb/s), and a few experimental systems are reported as capable of operation at 200 and 400 Mb/s. Thus, while in the early 1970s a bandwidth of a few MHz was considered commendable, presently bandwidths of several hundred MHz have been reported as laboratory achievements, and bandwidths of several GHz will become possible by the early 1980s.

To take advantage of these advances in order to produce large quantities of high-quality fibers with uniform characteristics, improvements in control of the fiber material characteristics are necessary. Thus, attention is again shifting from bandwidth to improved control of the refractive index and fiber dimensions. This, in turn, leads to the need for improved optical sources whose spectral width is reduced and whose frequency response and available light output is increased. At the same time, source and fiber reliability under environmental and aging–life test conditions is receiving considerable attention. Finally, the economic considerations of fiber fabrication, installation, and maintenance are becoming increasingly important and are tending to control the field of optical communications development.

Fibers with a total loss of only a few dB/km at the wavelengths of interest (which are determined by the loss properties of the glass rather than by the properties of available light sources, permitting low-loss operation mainly at two distinct wavelengths of about 0.85 and 1.05 μm) can be fabricated relatively economically today. To take full advantage of the significant potential of these fibers in optical intelligent communications systems, optical waveguides generally have to meet the following requirements at the optical wavelength of interest:

Low transmission loss

High transmission bandwidth and data rate

High mechanical stability

Easy and reproducible fabrication methods

Low optical and mechanical degradation under all anticipated operational conditions

Easy interface with peripheral system components without performance degradation

These and a few other factors can provide substantial advantages for fibers in comparison with more conventional coaxial waveguides and electric transmission media (5).

A general overview of the absorption spectrum of glass with a wavelength range of 0.1 to 100 μm is given in Figure 2.1. This range extends from the ultraviolet (UV) region through the visible and near-IR regions to the far-IR region. The figure can be considered a relative representation of the major loss contributions. The ultraviolet absorption caused by electron transitions is normally absent in the near- and far-IR regions, which are of greatest interest in communications. The Rayleigh scattering loss, because of its λ^{-4} dependence, shows a rapid decline with wavelength, but in the far infrared, vibrations of Si-O molecules are dominant. In the near-IR region losses due to coloring impurities (transition elements and the OH^- radical) clearly dominate the absorption characteristics.

This indicates that the wavelength range of 0.7 to 3.0 μm represents a transmission window which is limited primarily not by fundamental loss phenomena but by the presence of impurities. Most efforts to reduce the absorption loss in this wavelength region have concentrated on reduction of the impurity densities of these elements and molecules. In contrast, electronic transitions and Rayleigh scattering—both dominant in the UV re-

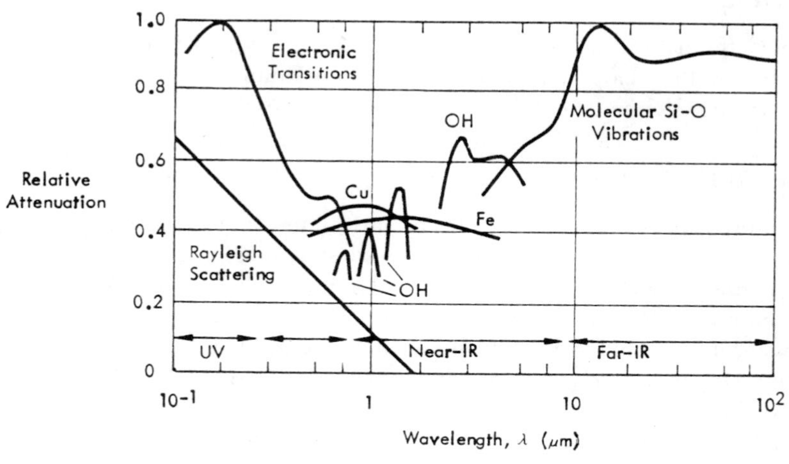

Figure 2.1 Principal contributions to signal attenuation in glass waveguides.

gion—and molecular vibrations—dominant in the IR region—represent fundamental limitations that cannot be overcome by higher material purity.

For the transmission of intelligent signals over an optical path, there are therefore usually two relevant wavelength regions, determined by the impurity content of the fiber. These are the 0.8 to 0.9 μm range and the 1.0 to 1.1 μm range; the 0.9 to 1.0 μm range separating the two useful regions exhibits a high OH-induced loss. The lower range can usually be satisfied with LEDs and semiconductor lasers of the GaAsAl type, whereas the higher range is the almost exclusive domain of semiconductor and nonsemiconductor lasers of the Nd:YAG type. Typical wavelengths chosen for data transmission in glass are approximately 0.85 and 1.05 μm. Both of them are compatible with Si detectors, which provide satisfactory sensitivity at 0.85 μm and marginal performance at 1.05 μm; GaAs detectors are capable of covering the 1.05 μm wavelength adequately.

The strong wavelength dependence of Rayleigh scattering is one of the major reasons for current efforts to raise the wavelength above 1.05 μm, where Rayleigh scattering is further reduced but where molecular SiO vibrations do not yet cause a strong loss increase. In this wavelength region, the only significant attenuation originates from fiber contamination. Thus, it is evident that a near-complete elimination of metal ions and hydroxyl radicals is desirable. The advisability of increasing the transmission wavelength from 0.85 or 1.05 μm to 1.25 μm or higher, as has been suggested (6), has to be weighed against decreases in the spectral sensitivity of most detectors and the difficulties in achieving suitable optical output at this wavelength. None of these difficulties, however, are fundamental, although they apply to present components.

Attenuation Mechanisms

Fiber attenuation mechanisms can be broadly divided into absorption and scattering. Both have comparable practical importance, although their relative significance depends on the wavelength of interest.

ABSORPTION LOSS

Absorption is normally caused by impurity ions, mainly of transition metals and OH radicals. Although high-purity glass is transparent in the visible region, the presence of transition metal ions (such as iron, copper, cobalt, chromium, and nickel) introduces energy levels that can absorb photons. Hence, it is desirable to reduce the metal content of glass to values below about 0.01 ppm. The purity requirement of glasses thus exceeds that of semiconductor devices. Additional absorption results from the harmonics of the vibration of OH radicals that are associated with the presence of water in the glass. The basic frequency of OH vibration corresponds to a vacuum wavelength of about 2.8 μm. The third and fourth harmonics occur at 0.95 and 0.73 μm, respectively, and are thus near the wavelengths of normal interest. Although these absorption maxima are weak by conventional standards, they cause significant attenuation in long waveguides. For example, in fused silica, OH vibrational absorption causes a loss of about 1 dB/km for 1 ppm OH at a wavelength of 0.95 μm.

SCATTERING LOSS

The major scattering losses can be divided into those resulting from material inhomogeneities (fiber surface roughness, fiber curvature, and fiber composition variation) and those caused by the random distribution of individual molecules (Rayleigh scattering). Some inhomogeneities in the glass are unavoidable because of the random arrangement of the molecules in the amorphous material. Because Rayleigh scattering loss decreases with the fourth power of wavelength, very low losses can be obtained only in the infrared spectral range and not in the visible region. For example, Rayleigh scattering at 1.0 μm amounts to about 0.8 dB/km. Scattering losses may also be caused by fiber core–cladding interface roughness. In practice, however, the spatial variations of interface irregularities are considerably less frequent than those of the light wavelength, so that some deviations can be tolerated.

Scattering loss may also result from fiber bending. Because in any optical waveguide a small fraction of the optical energy is traveling outside of its core, at the far side of a bend it would be forced to travel faster than the limiting speed of light; since the evanescent field tail will tend to be dragged along by the field in the core, it will resist this dragging process by radiating away. The amount of radiation loss depends on field intensity and radius of curvature. At small curvature, loss is negligible, but at sharp bends it increases so rapidly that practically all power is lost in a short distance.

PROPAGATION ASPECTS

In the following, the propagation aspects of optical fibers are reviewed in order to present a critical analysis of their future potential. This includes an analysis of the state-of-the-art and of the inherent limitations of fibers. The terms fiber and waveguide are used interchangeably. For more detailed information, the reader is referred to the extensive bibliography on optical communications, with some of the more relevant references given at the end of this chapter. Excellent summaries are provided by Miller et al. (3) and Maurer (7).

Fiber Requirements

Because of the central importance of fibers to the performance of a complete optical communications system, the requirements that optical waveguides have to satisfy are rather stringent. They can be divided into technical and economic demands, both of which tend to constrain the optimization of practical systems. These requirements are outlined below.

Small Loss. The fiber must have a low loss in order to prevent signal amplitude degradation over large distances; the loss minimum or minima must be at a wavelength for which suitable light sources and detectors exist or for which development can be expected.

Small Dispersion. The fiber must have a small signal dispersion in order to prevent signal pulse width degradation over large distances.

High Reproducibility. Fiber geometry as well as dopant and index profiles must be chosen such that loss and dispersion are minimized over the entire length of the fiber. This requirement implies extremely tight control over the parameters to avoid spatial and time variations which lead to performance degradation.

Suppression of Undesirable Modes. Support of only one or a few modes with similar propagation properties is desirable to avoid variations in signal delay caused by different modes arriving at the detector at different times.

Optimized Numerical Apertures. It is critical to select a numerical aperture whose value is compatible with, or at least an acceptable compromise among, the normally conflicting requirements of large bandwidth, large optical acceptance angle, minimum bending loss, and other performance criteria.

Easy Implementation. Easy fabrication, installation, and maintenance of fibers which allow economical, reproducible, and reliable fabrication, field splicing, and other implementation of connectors, as well as easy handling under all anticipated conditions, are major criteria in selecting an appropriate fiber.

Environmental Resistance. Fiber protection against environmental influences such as stress and strain, torsion, bending, humidity, shock, vibration, and irradiation without a significant degradation compared to the characteristics in the naked state are of practical importance in most applications.

Compatibility with Other Components. Compatibility in geometry and characteristics with other components of the system may severely constrain the choice of fibers and other components.

The best fibers fabricated to date have attained loss and dispersion characteristics that approach the theoretical limits. This corresponds to attenuation of approximately 1 dB/km and dispersion of less than 1 ns/km. An attenuation of 1.1 to 1.3 dB/km at 1.05 μm was achieved as early as 1972 on relatively short sample fibers. At that time the basic research was completed, although it took another two to three years to manufacture such a fiber that could be reproduced in a production environment. After these significant attenuation reductions, attention shifted to dispersion improvements. Because pulse dispersion affects the maximum data transfer rate, its minimization was considered imperative. A data transfer rate of 1 Gb/s is available at present on 1 km fibers, and a 2 Gb/s data rate is available in prototype fibers. Theoretically, pulse spreading is well understood, and ways to reduce it are known (although these are largely proprietary). Although a transfer rate of 1 Gb/s km is obtainable, an increase beyond this value may not be worth the additional expense. Similarly, the achievement of 1 dB/km attenuation in cables is of questionable value in systems where a loss of about 3 to 5 dB/km is sufficient for most practical applications.

Fiber Types

Circular or rectangular waveguides consisting of a dielectric of low optical loss can serve as a transmission medium for guided optical waves if the refractive index outside the waveguide is higher than that within it. Such an optical fiber can guide electromagnetic energy of optical frequency along a well-defined path. This energy is carried partly inside the fiber and partly outside.

In general, an optical fiber consists of a central portion, known as the core, made of an inorganic glass or plastic dielectric through which most of the useful information is transmitted. It is surrounded by a cladding material, also of either glass or plastic, whose refractive index is slightly lower than that of the core. Thus, a practical fiber consists of an inner region (core) with a refractive index n_1 and an outer region (cladding) with a refractive index n_2, whereby $n_2 < n_1$. The external field decays rapidly to zero in the direction perpendicular to the direction of propagation, and it decays approximately exponentially in the direction of propagation. The material compositions of core and cladding, as well as their dimensions and index profiles, determine optical attenuation and signal dispersion characteristics of the fiber.

Because the refractive index of the core is slightly higher than that of the surrounding region, total reflection of most optical rays within the core prevents their escape to the outside. Hence, rays will be continuously reflected back and forth during their travel along the fiber, unless sharp bends are encountered; the exception to this is rays that enter the fiber at a large angle. Rays that are reflected often must travel a longer distance within the fiber than rays that remain close to the fiber axis and that encounter no or few reflections; hence, in a fiber whose core has a uniform index of refraction, reflected rays will arrive later at the fiber's end than straight rays. This effect is called modal dispersion. It can be alleviated by inducing nonaxial rays to travel faster than axial rays so that the larger distance is compensated for by a higher ray velocity. Because the velocity of light is inversely proportional to the refractive index of the medium in which it propagates, a fiber whose refractive index decreases radially from the axis allows rays that deviate from the axis to travel faster than rays that deviate less. This situation is schematically illustrated in Figure 2.2.

Most fibers have a nominally circular cross section, although in integrated optics rectangular cross sections are encountered frequently. Several other types of optical waveguides have been suggested, but most of them have

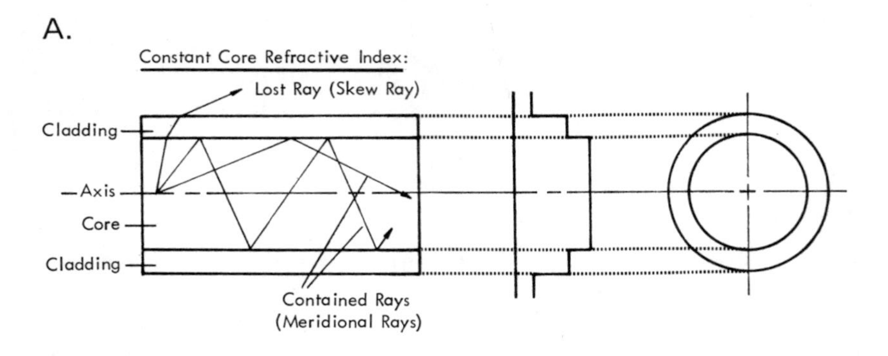

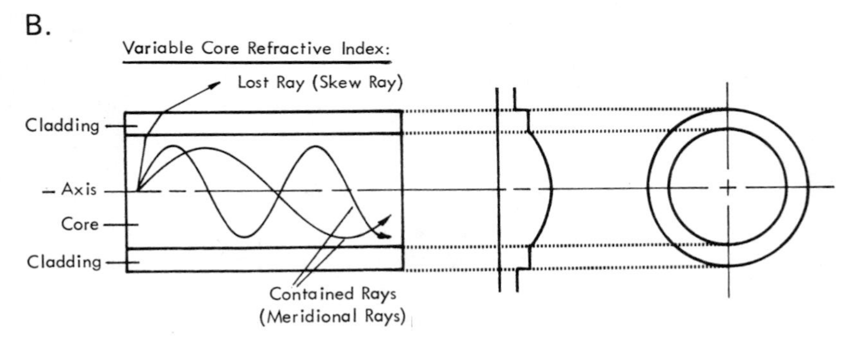

Figure 2.2 Schematic illustration of index profiles and resulting propagation aspects of fibers with different structures.

achieved little practical value. Furthermore, depending on the profile of the refractive index within the core, we can distinguish between step-index fibers, in which the refractive index changes abruptly from core to cladding, and graded-index fibers, in which the refractive index decreases continuously (usually parabolically) from core center to a region within the cladding. These two fiber types have received prominent attention. A further distinction can be made, depending on whether only a single mode or several modes can propagate.

Generally, the refractive index profile in an optical fiber of arbitrary index profile is* (footnote opposite)

$$n(r)/n_o = \left[1-(r/a_1)^i\right]^{1/2} \tag{2.1}$$

where r describes the radial distance from the core center, n_o is the refractive index at the core center, and i is a parameter characteristic of the impurity profile.* For a step-index fiber $i = \infty$ for $r < a_1$, and $i = 0$ for $r > a_1$; for a parabolically graded-index fiber $i = 2$. Usually, however, in a high-bandwidth graded-index waveguide the profile parameter i differs from the value 2 by a small amount. Although often a value of $i = 2.0$ has been accepted, corresponding to a parabolic profile, for most of the highest-bandwidth fibers a value close to $i = 2.25$, representing a near-parabolic profile, yields superior performance, as will be discussed later.

In practice, therefore, there are three major types of circular optical waveguides of interest. Structurally, they differ in their refractive index profiles and in their cross-sectional dimensions. Operationally, they differ in the number of modes and the information capacity they can support. These types are shown schematically in Figure 2.3 and can be described as follows (8–10).

Step-Index Multimode (SI-MM) Fibers. This fiber consists of a homogeneous core of refractive index n_1 surrounded by a cladding of slightly lower refractive index n_2; the relative index difference

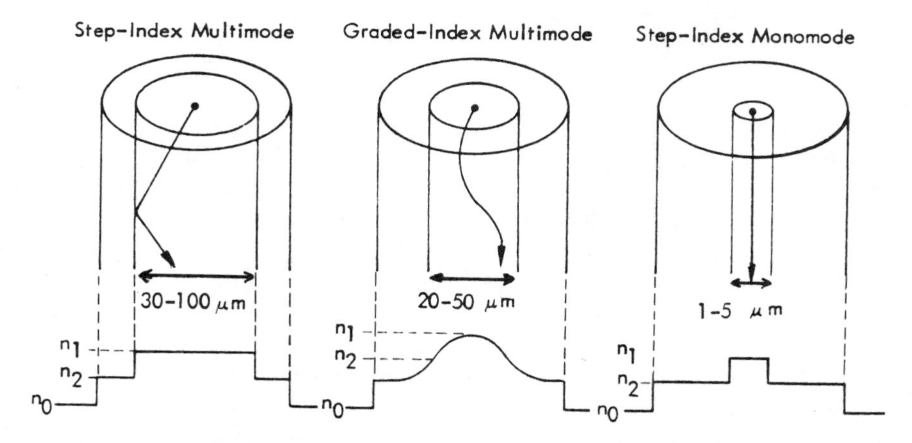

Figure 2.3 Schematic illustration of refractive index profiles of the three dominant optical waveguide types.

---

*Throughout this chapter, subscripts 0, 1, and 2 usually refer to vacuum or air, fiber core, and fiber cladding, respectively. The following symbols are used:

n_o	Refractive index at fiber center
n_0, n_1, n_2	Refractive index of vacuum, fiber core, and cladding
c, λ	Speed, wavelength of light in vacuo
a_1, a_2	Radius of core, cladding
r	Axial distance from core center
β_0, β_1	Angle between light ray and fiber axis outside, inside of fiber
L	Fiber length
NA	Numerical aperture of fiber

$$\Delta = (n_1 - n_2)/n_1 \tag{2.2}$$

is usually of the order of 0.01 (1%) or less for present low-loss fibers based on fused silica. Light is guided by the total reflection within the core. Different partial rays propagate at different angles and therefore exhibit different transit times, resulting in a limited bandwidth of this fiber. Historically, this is the oldest fiber type.

Step-Index Single-Mode (SI-SM) Fibers. In this fiber multipath transmission is almost completely avoided by choosing a core diameter so small that essentially only one axial partial ray can propagate. For this reason single-mode fibers are most attractive if a large bandwidth is desirable. The transmission bandwidth is limited only by material dispersion in connection with the emission bandwidth of the light emitter. The monomode fiber can be considered to be a dielectric surface waveguide in which only the fundamental mode (HE_{11}) can propagate. Because the core radius a_1 has to be below a critical value in order to operate in the single-mode condition,

$$a_1 = \left(\frac{1.202}{\pi}\right)\left(\frac{\lambda}{n_1/(2\Delta)^{1/2}}\right)$$

this condition poses severe fabrication restrictions. For example, if $\Delta = 0.01$, then the core diameter has to be smaller than 3.6 light wavelengths; assuming $\lambda = 0.85\ \mu\text{m}$, then the diameter has to be less than about 3.1 μm. If this condition is not met, the fiber will support more than one mode and exhibit a severe bandwidth degradation. Because SI-SM fibers are the most difficult to produce, their use is limited to special applications.

Graded-Index Multimode (GI-MM) Fiber. In this fiber a reduction of the difference of transit time between partial rays is accomplished by a gradation of the refractive index from the center to the outside. This causes the light to be guided not by total reflection but by distributed diffraction. The radial decrease of the index away from the fiber axis results in an increase in the velocity of the partial rays outside the core center, thus compensating for the larger path length. Therefore, almost the same average velocity can be achieved for all partial rays by a suitable choice of the index profile. The optimum index profile is near-parabolic, but small deviations from this optimum cause significant pulse broadening. Sufficiently accurate control of the index profile is required, but it is difficult to achieve in production. Graded-index fibers can be described by an equation of the general form

$$n(r) = n_o \operatorname{sech} \rho r \tag{2.3}$$

where n_o is the refractive index at the fiber axis, r is the distance from the fiber center, and ρ is the radial variation constant related to the focusing distance.

The GI fiber is the most suitable compromise for a practical fiber that meets most performance and economic requirements.

Typical core diameters range from 1 to 5 μm for monomode step-index fibers, from 30 to 100 μm for multimode step-index fibers, and from 20 to 50 μm for multimode graded-index fibers. In practice, for most commercial fibers the ratio of core to cladding radius is $a_1/a_2 = 0.6$; thus for a 50 μm thick core, the cladding thickness is typically about 15 μm.

OTHER FIBERS

Another interesting fiber type has the index profile shown in Figure 2.4. W-type fibers based on silica are fabricated by chemical vapor deposition (CVD) techniques and exhibit a transmission loss as low as about 3 dB/km at 0.85 μm. The W-type fiber consists of a core and a thin cladding layer surrounded by an external higher-index region; its total diameter is typically about 20 μm. The W-type fiber is effectively a monomode fiber, although it possesses multimode dimensions. Consequently, it permits bandwidths far in excess of those of multimode fibers. However, its fabrication process is considerably more complex and hence less economic. Several possible modifica-

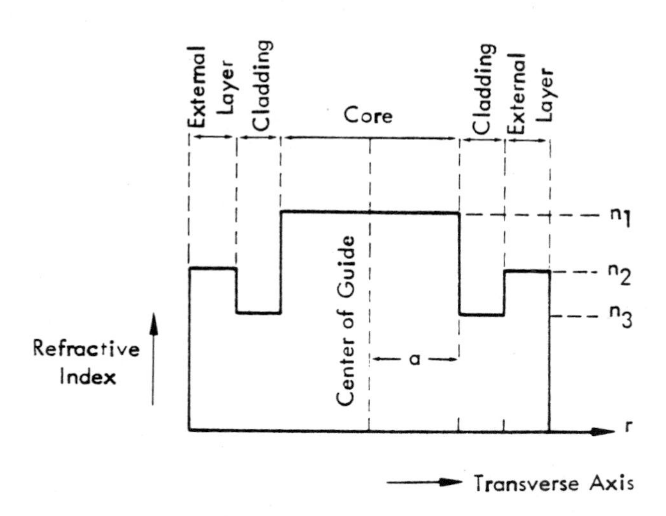

Figure 2.4 Schematic illustration of the refractive index profile of the W-type waveguide.

tions of the simple step-index W-type fiber exist, such as the axial symmetric and the clad-slab nonsymmetric types. All of them exhibit high bandwidth and monomode operational capability in spite of multimode lateral dimensions.

Light Path in Fibers

Assuming that light is contained in the fiber by total reflection within the core, light guidance is determined by reflection and interference. Electromagnetic waves that satisfy both of these transmission requirements are called partial rays or modes.

Propagation Angle. The first condition requires that the angle of propagation within the fiber, β_1, must be larger than the critical angle for total internal reflection, $\beta_1 > \arcsin\ (n_2/n_1)$. For example, for $n_1 = 1.500$ and $n_2 = 1.475$, the critical angle $\beta_1 = 79.5°$. The critical acceptance angle is related to an important property of the fiber, the numerical aperture, which is

$$\mathrm{NA} = \sin\ \beta_1 = (n_1^2 - n_2^2)^{1/2} \tag{2.4}$$

Constructive Interference. The second condition requires that the multitude of rays propagating in the core must superimpose themselves so that constructive interference results. This limits the possible angles β_1 to a discrete set of values.

In a multimode fiber there are two types of electromagnetic waves which propagate simultaneously and which are superimposed. These are discussed below.

Meridional Rays. In a straight waveguide, meridional rays, or contained rays, entering parallel to the axis, propagate along the waveguide with little deflection. If the waveguide is not exactly straight or if the light does not enter exactly parallel to the fiber axis, then the wave will perform a zig-zag or wave-like course because of reflections or light bending at or near the core–cladding interface. All modes corresponding to meridional rays (HE_{1m} modes) have the same group velocity if material dispersion is neglected. All meridional rays focus at successive equivalent positions on the axis of the fiber.

Skew Rays. Skew rays, or lost rays, describe all rays that do not pass through the fiber axis. Waves of this type follow continuous helical trajectories. Skew rays do not focus with the refractive index given by Equation (2.3). Therefore, deviations from the refractive index variation described by Equation (2.3) are difficult to distinguish from skew ray effects due to the light source.

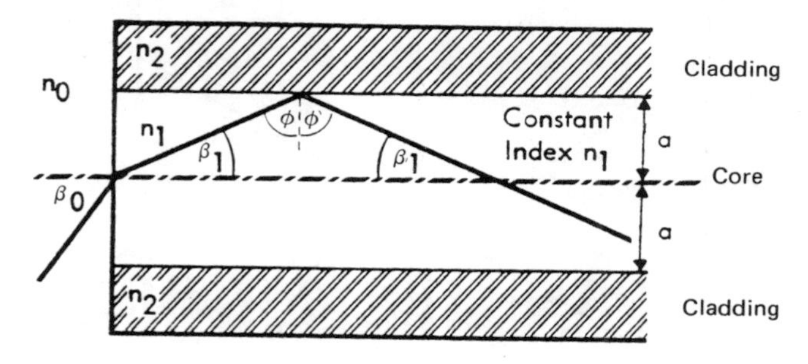

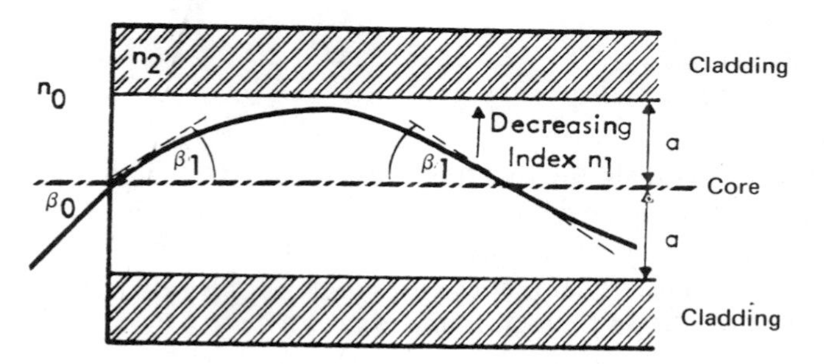

Figure 2.5 Schematic illustration of meridional-wave light paths in step-index and graded-index fibers.

Meridional waves represent the more important wave propagation mechanism, and they are easier to describe. The following discussion will be devoted mainly to this wave category. Meridional waves in step-index and graded-index waveguides are shown schematically in Figure 2.5.

In the following we will use the example of the step-index fiber to describe the propagation of optical power in a fiber in more detail. A step-index fiber consists of a core of diameter $2a_1$ and a homogeneous refractive index n_1 and of a cladding layer of homogeneous refractive index n_2; usually $n_2 < n_1$, whereby the difference $\Delta = 1-n_2/n_1$ is usually about 1% or less, and the core diameter is typically 30 to 100 μm. The relationship between Δ and the ratio n_1/n_2 is shown in Figure 2.6. Light entering the guide at an angle β_0 with respect to the axis will experience diffraction at the air–core interface, resulting in an angle β_1, which is determined by the index ratio according to

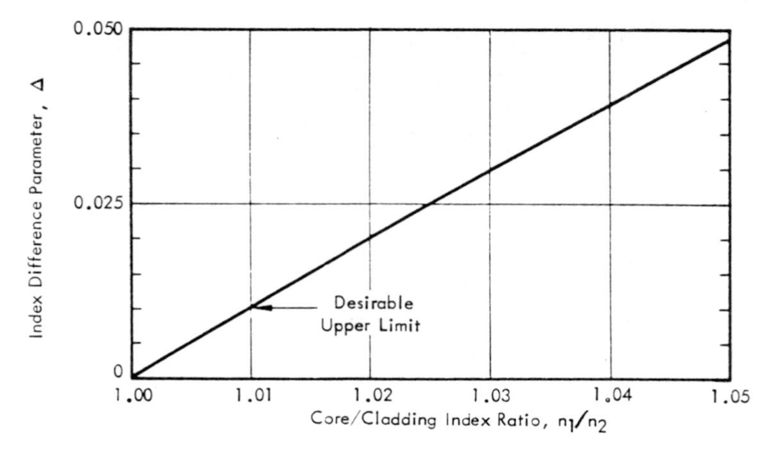

Figure 2.6 Dependence of the refractive-index difference parameter on the core–cladding index ratio for $n_1 = 1.500$.

$$\sin \beta_0/\sin \beta_1 = n_1/n_0 \tag{2.5}$$

If β_1 is sufficiently small, the light ray is totally reflected at the core–cladding interface at a critical angle ϕ, given by $\phi = \pi/2 - \beta_1$ under the assumptions that the interface represents ideal reflective conditions and that the waveguide is not bent. Under these assumptions subsequent reflections occur at identical angles ϕ. The condition for total reflection is given by $\sin \phi_t > n_2/n_1$.

Assuming $n_1 = 1.500$ and $n_2 = 1.485$, then $\Delta = 0.01$, corresponding to an index difference of 1%, and the minimum allowable angle ϕ_t that retains the light ray within the core is $\phi_t = 81.9°$. This corresponds to a maximum allowable entrance angle $\beta_1 = 8.1°$ within the fiber and $\beta_2 = 12.2°$ outside the fiber with respect to the fiber axis, assuming the ambient medium is air or vacuum ($n_0 = 1$). If, on the other hand, $\Delta = 0.02$ (corresponding to $n_1 = 1.500$ and $n_2 = 1.470$), then the extreme, allowable angles are $\phi_t > 78.5°$, $\beta_1 < 11.5°$, and $\beta_0 < 17.4°$. This is illustrated in more detail in Table 2.1. It must be kept in mind that the refractive index is a function of wavelength and normally shows a tendency to decline slightly with increasing wavelength.

Appropriate exit conditions apply at the fiber end at which light is transmitted from the fiber of index n_1 to a medium of index n_0. Usually, however, there is no restriction on the exit angle because most available detectors do not exhibit a significant directivity.

The number of reflections and hence the total length of the light path over a given fiber length L is smaller the smaller the angle β_1 (that is, the more the ray coincides with the fiber axis). Thus, two rays entering the core simultaneously but with different angles will arrive at the fiber end at different times. The time difference between an axial ray and one propagating at angle β_1 may be appreciable. It is given by $dt = Ln_1\Delta/c$, where c is the speed of light. This time delay causes signal dispersion and limits the fiber bandwidth.

Because the time delay between rays of different angles is proportional to the index difference parameter Δ, it is desirable to make n_1 and n_2 as similar as possible. However, this limits the angle of total reflection, ϕ_t, so that a larger amount of optical energy loss will be experienced, particularly in fiber bends. Hence energy loss and dispersion represent a practical trade-off.

It should also be kept in mind that wavelength and light velocity differ in media with different refractive indexes. Thus, the wavelength and speed of light, which in vacuum (or air) are λ and c, respectively, will have decreased to λ/n_1 and c/n_1, while the frequency will have remained unchanged. This means that the wavelength of light entering a fiber of refractive index 1.5 will be about 50% shorter in the waveguide than in air. Likewise, the speed of light within the fiber will be 50% lower than in air.

TABLE 2.1. Properties of a Step-index Waveguide Pertaining to Light Propagation Characteristics as a Function of the Refractive Index of Fiber Core and Cladding, Assuming $n_1 = 1.5$

Cladding index (n_2)	Index ratio (n_1/n_2)	Index difference parameter (Δ)	Numerical aperture (NA)	Angle of total reflection (ϕ_t)	Maximum fiber entrance angle on fiber side β_1	Maximum fiber entrance angle on vacuum side β_0
1.500	1.000	0	0	90.0	0.0	0.0
1.496	1.003	0.003	0.08	85.6	4.4	6.6
1.493	1.005	0.005	0.10	84.3	5.7	8.6
1.485	1.010	0.010	0.14	81.9	8.1	12.2
1.478	1.015	0.015	0.17	80.1	9.9	15.0
1.471	1.020	0.020	0.20	78.6	11.4	17.3
1.463	1.025	0.024	0.22	77.3	12.7	19.3
1.456	1.030	0.029	0.24	76.1	13.9	21.1
1.449	1.035	0.034	0.26	75.1	14.9	22.7
1.442	1.040	0.038	0.28	74.1	15.9	24.3
1.435	1.045	0.043	0.29	73.1	16.9	25.9
1.429	1.050	0.048	0.31	72.2	17.8	27.3

Deviations From Ideal Characteristics

The two most important propagation characteristics of any fiber are its signal attenuation and its signal dispersion over the total fiber length. A more detailed discussion of the loss and dispersion mechanisms will follow below. Loss is mainly a material property, although it is also a function of the diameter, geometric variations, and the refractive index profile. The pulse dispersion also depends on the material properties, but it can be significantly affected by the distribution of the refractive index across the fiber diameter.

In a step-index multimode fiber the difference between the fastest and the slowest possible rays, and hence the total broadening of an input light pulse in the case of equal excitation of all possible modes, is

$$\tau = (Ln_1/c)\Delta \tag{2.6a}$$

As will be discussed later, in actuality the different modes do not carry equal amounts of energy in a practical waveguide because of mode mixing and mode conversion, so that the effective broadening is less than that given here.

Assuming that the transit time of an axial ray (the fastest possible ray within a given core) is

$$t_a = Ln_1/c \tag{2.6b}$$

the pulse broadening due to multipath transmission can be expressed by

$$\tau = t_a\Delta \tag{2.7}$$

which is independent of the core diameter and wavelength. In a near-parabolic graded-index multimode fiber, by contrast, the pulse broadening is

$$\tau = t_a(\Delta^2/2) \tag{2.8}$$

which is again independent of core diameter and wavelength. In a monomode step-index fiber, ideally there is no multipath transmission, so that material dispersion is the only cause of pulse broadening.

The pulse width at the end of a fiber transmission line is also affected by the pulse width of the light source at the emitting end. Hence, the smallest signal dispersion is achieved by the use of a laser beam with a small spectral bandwidth, whereas the emission spectrum of a light-emitting diode is usually too broad for high-bandwidth transmission.

Mode coupling in an optical fiber narrows the pulse response but, at the same time, causes additional loss. Under practical circumstances, this additional loss may limit the usefulness of coupling for the purpose of reducing mode dispersion.

The two types of pulse spreading that limit the data transfer rate and hence the information-carrying capacity of an optic fiber are illustrated in Figure 2.7. These are treated under the term signal or pulse dispersion and can be divided into modal dispersion and waveguide dispersion. In each case, the maximum pulse rate is chosen such that the pulse spreading at the fiber output is half the interval between pulses.

Modal dispersion results from the different arrival times of optical rays that have different angles with respect to the fiber axis. In a multimode step-index fiber of typical dimensions and index difference, the difference in arrival times between the leading edge (produced by rays parallel to the axis) and the trailing edge (produced by rays of maximum propagation angle with respect to the axis, determined by total reflection) of an arbitrarily shaped pulse is approximately 20 ns/km; this corresponds to about 400 cm at the speed that light travels in glass. In order to restrict the pulse rate to values that are less than half the pulse interval, the pulse rate should therefore not exceed approximately 2.5×10^7 pulses per second. In a graded-index fiber of typical dimensions and doping profile, modal dispersion can be reduced by a factor of about 25 compared to a step-index fiber; hence, the arrival time between leading-edge and trailing-edge waves is about 1 ns/km, corresponding to about 20 cm at the speed of light.

Wavelength dispersion results from the different velocities (v) of optical rays propagating in materials of differing refractive indexes and at different wavelengths. The lower the refractive index, the higher the velocity, and the smaller the wavelength, the lower the velocity, assuming a constant frequency. This is illustrated by the relationship between velocity and refractive index

$$v_1 = c/n_1 \tag{2.9}$$

and between velocity and wavelength

$$v_1 = f\,\lambda \tag{2.10}$$

where f is the frequency of the optical wave. Hence, in a graded-index fiber the velocity of an optical ray increases from the center to the outside of the fiber. It also differs for the wavelengths of the lower and upper ends of the source emission spectrum; hence, it is desirable to employ a source of minimum spectral bandwidth for minimized pulse broadening. For example, a typical light-emitting diode has a center emission wavelength of about 0.85 μm and a spectral width of 350 Å (= 0.035 μm). Assuming that there is no modal dispersion, a single pulse from an LED would spread about 65 cm/km, thereby limiting the signal rate to about 1.5×10^8 pulses per second. On the other hand, use of a semiconductor laser source having a spectral width of only 20 Å (= 0.002 μm) results in wavelength dispersion of about 4 cm/km, thus permitting a signal rate of 3×10^9 pulses per second.

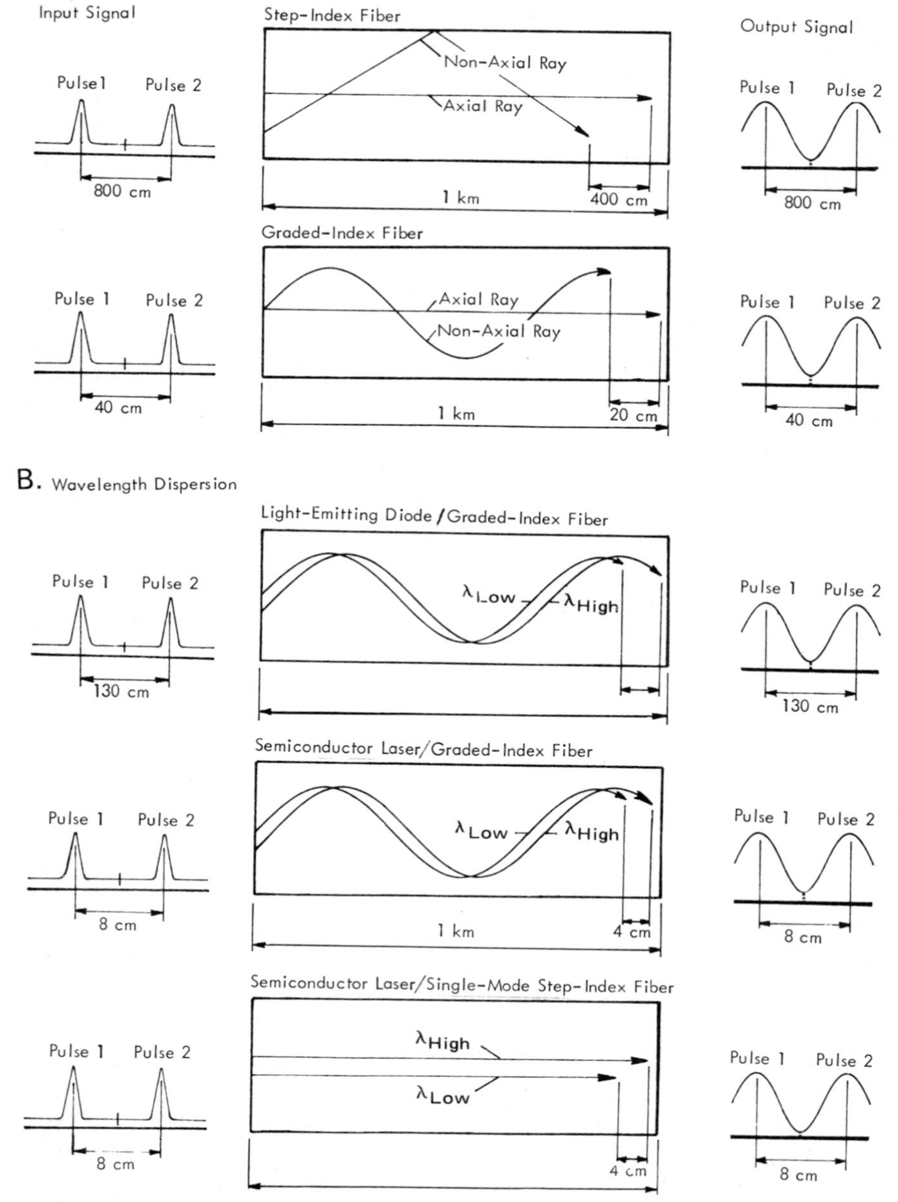

Figure 2.7 Schematic illustration of pulse spreading (dispersion) in optical fibers: A. Modal dispersion. B. Wavelength dispersion.

In spite of the advantages single-mode fibers can potentially offer in regard to signal transmission characteristics, multimode fibers with their significantly larger diameters remain superior to monomode fibers for two important practical reasons: (a) they impose less stringent requirements on the fiber, and they even transmit the incoherent light from an LED; (b) they facilitate splicing because of a larger diameter, or at least they relax the tolerances required for connection.

The usefulness of multimode fibers, often carrying thousands of modes even if the index difference between core and cladding is only a few percent, depends on their dispersion characteristics. Although delay differences among the modes distort the signal and produce a signal response inferior to that of a monomode fiber, overall system economy may place the desired information rate of individual fiber channels in a range where the signal response of multimode fibers is adequate (for example, in systems of less than 100 Mb/s). In such systems, multimode operation is preferable economically over monomode operation.

The pulse dispersion in practical multimode fibers is usually considerably less than that predicted by theory. Two mechanisms are responsible for this loss reduction: (a) the stronger attenuation of higher-order modes, and (b) the exchange of power between modes during propagation. Because most scattering mechanisms, with the exception of Rayleigh scattering, cause higher-order modes to suffer more radiation losses than lower-order modes, the high-order modes do not contribute as much to power transport, so that the delay distortion predicted under the assumption that all modes carry equal amounts of power is improved.

Propagation Modes

The propagation properties of a fiber can be described by the so-called v value, sometimes referred to as the normalized frequency, although it is a structural parameter rather than a frequency. The v value is an important fundamental fiber property, because it contains the most important fiber parameters (11–14). It is defined by

$$v = 2\pi(a_1/\lambda)(n_1^2 - n_2^2)^{1/2} = 2\pi(a_1/\lambda)\mathrm{NA} \tag{2.11}$$

For example, for $a_1 = 25\ \mu$m, $\lambda = 0.85\ \mu$m, and the above given refractive indexes, $v = 50.4$.

Among other factors, the v value determines the number of modes a fiber can support. The concept of modes is important because it determines the bandwidth that is achievable over a certain fiber length. A multimode fiber with a circular cross section can support a finite number of guided modes and an infinite number of unguided (or radiative) modes. The guided modes can be divided into a family of H_{on} E_{on} modes as well as a family of hybrid HE_{mn}

modes. In addition, there are, in analogy to microwave waveguides, TE_{mn} and TM_{mn} modes. Theoretically, a waveguide may be capable of carrying thousands of different modes.

Normally it is desirable to reduce the number of modes carried to a minimum. In a step-index fiber the number of modes (N_m) is related to the v value by

$$N_m = \frac{1}{2} v^2 \tag{2.12}$$

which shows that the number of possible modes increases with the square of the fiber diameter. For instance, using the above numerical example, $N_m = 1270$, meaning that the number of possible modes in this fiber exceeds 1000. Thus, the need to minimize the number of modes requires a reduction of the fiber diameter. This, however, has important economic implications because of fabrication difficulties and because of a reduction in source–fiber coupling efficiency. Conversely, an increase in wavelength tends to decrease the number of possible modes.

The principal advantage of a single-mode fiber is its high bandpass property. Because different modes normally have different velocities, a single mode is equivalent to having only one transmission velocity and hence to an inherently higher data rate capability. The requirement for single-mode operation (counting only modes with one polarization) is $v < 2.405$.

Because the dominant modes differ in their energy propagation characteristics (expressed by their particular field distribution and propagation constants and given, for example, by loss and velocity), it is usually desirable to suppress all modes except one or two in order to obtain the lowest fiber loss and dispersion. For any mode, an equivalent refractive index n_e can be defined such that in a step-index fiber the propagation of a particular mode is limited to the range $n_2 \leqslant n_e \leqslant n_1$, where

$$n_e = b\,\lambda/2\pi \tag{2.13}$$

where b is a term depending on the index variation along the fiber radius and the characteristics of the appropriate modes. Thus for any mode the value of n_e describes the index profile along the fiber radius.

The concepts of equivalent refractive index and v value are useful to describe the mode propagation characteristics. Figure 2.8 shows the dependence of n_e (which can vary from a minimum of n_2 to a maximum of n_1) as a function of the v value for a few of the low-order modes of a step-index fiber. It indicates that all modes except the HE_{11} mode (which is the lowest-order hybrid mode) exhibit a cutoff frequency (15).

The HE_{11} mode propagates with an increasing percentage of energy outside the fiber as the core diameter decreases. This is important because it

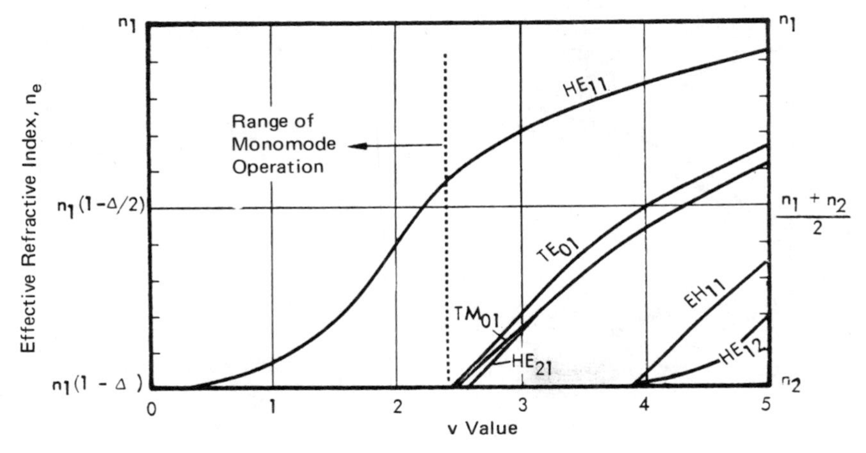

Figure 2.8 Relationship between the refractive index and v value for principal modes in a step-index fiber.

allows single-mode operation of a fiber by a reduction of the core diameter, assuming that the outside medium has a low attenuation. The HE_{11} mode of propagation is the most interesting one. It does not display a cutoff. Hence, single-mode operation requires that $v < 2.405$. This means in practice that within a limited range one can choose fiber parameters such that this condition is met. Thus, in a numerical example, for a wavelength $\lambda = 0.85$ μm, assuming $n_1 = 1.5$ and $\Delta = 0.01$, a maximum fiber radius $a_1 = 1.5$ μm is tolerable for monomode transmission. This small core radius is not feasible in most practical applications, so that one is usually content with a few modes existing simultaneously in a given fiber. On the other hand, an index difference of $\Delta = 0.001$, which is difficult to achieve, would lead to a maximum radius of 5.0 μm or a core diameter of 10 μm.

The number of modes that can be carried depends on the refractive indexes as shown graphically in Figure 2.9. N_m increases very rapidly as the critical v value, corresponding to a specified n_1/n_2 ratio, is exceeded, and it increases much more slowly for higher values of n_1/n_2. It is important to note that the number of modes increases with the square of the core diameter and decreases with the square of the wavelength. Figure 2.9 illustrates the relative inaccuracy of the relationship between N_m and n_1/n_2 around the critical v value. Numerical values for v and of the number of modes are also given in Table 2.2.

The v value can be used to describe the frequency limitations of the fiber through waveguide and material dispersion, as implied above. Figure 2.10 shows the variation of the refractive index as a function of v value and wavelength (16). Although in actuality both n_1 and n_2 are wavelength dependent, in Figure 2.10A it is assumed that they remain constant with frequency. This

A.

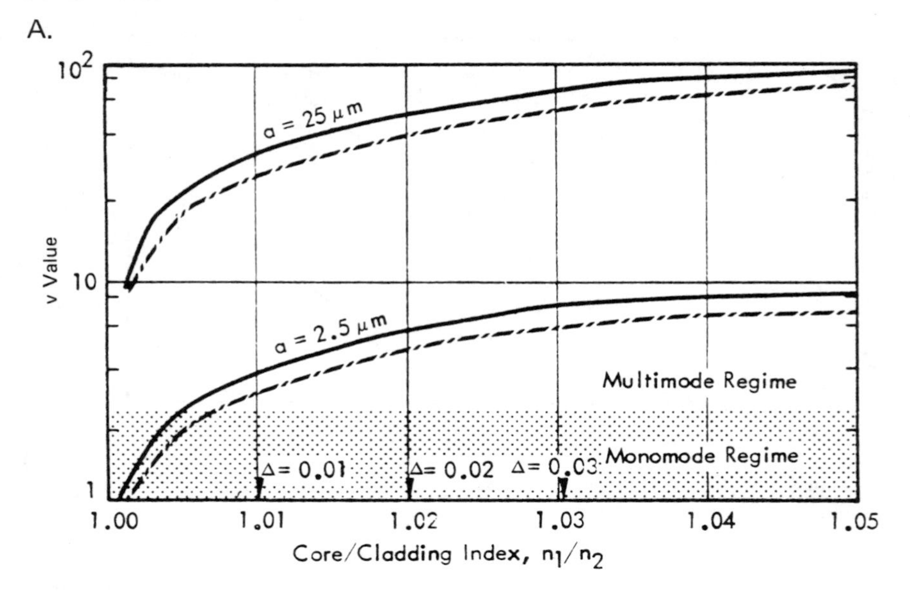

B.

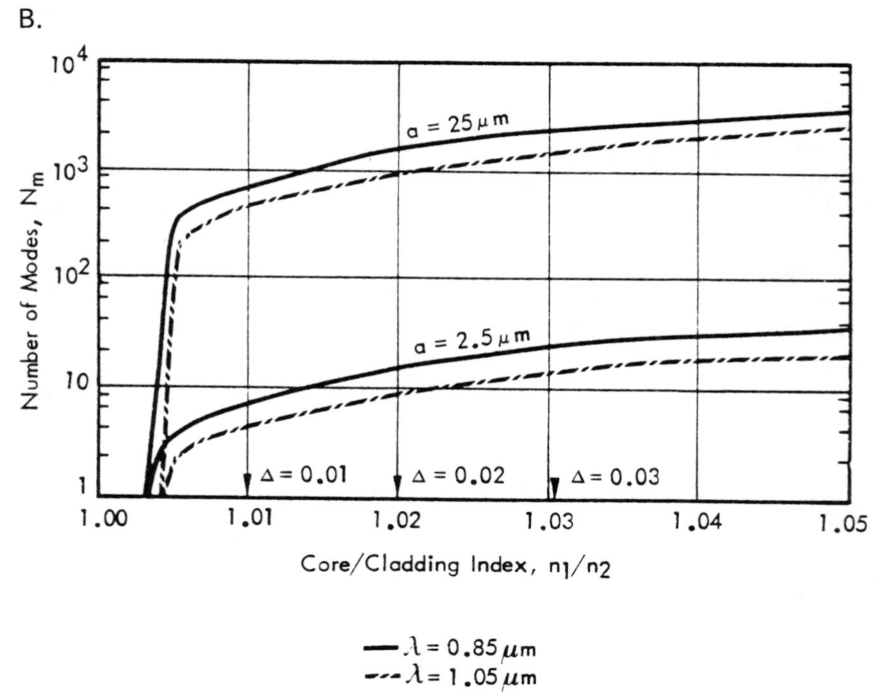

Figure 2.9 Dependence of the v value (A) and mode number (B) of a step-index waveguide on the refractive index, core diameter, and wavelength.

TABLE 2.2. Dependence of the v Value and Mode Number on the Index Ratio, Waveguide Diameter, and Wavelength for a Step-Index Fiber, Assuming $n_1 = 1.5$

		$\lambda = 0.85\ \mu m$				$\lambda = 1.05\ \mu m$			
		$a_1 = 2.5\ \mu m$		$a_1 = 25\ \mu m$		$a_1 = 2.5\ \mu m$		$a_1 = 25\ \mu m$	
n_1/n_2	n_2	v	N_m	v	N_m	v	N_m	v	N_m
1.000	1.500	0	1.0	0	1	0	1.0	0	1
1.005	1.493	2.7	3.6	26.8	358	2.2	2.3	21.7	234
1.010	1.485	3.9	7.7	39.1	764	3.2	5.0	31.7	501
1.015	1.478	4.7	11.0	47.3	1118	3.8	7.3	38.3	733
1.020	1.471	5.4	15.0	54.2	1472	4.4	9.6	43.9	964
1.025	1.463	6.1	19.0	61.2	1872	5.0	12.0	49.5	1227
1.030	1.456	6.7	22.0	66.6	2222	5.4	15.0	54.0	1455
1.035	1.449	7.2	26.0	71.7	2568	5.8	17.0	58.0	1683
1.040	1.442	7.6	29.0	76.3	2914	6.2	19.0	61.8	1909
1.045	1.435	8.1	33.0	80.7	3258	6.5	21.0	65.3	2135
1.050	1.429	8.4	36.0	84.3	3552	6.8	23.0	68.2	2327

N_m, number of modes.

illustration shows, for the lowest-order modes, the normalized modal phase velocity $v_p = c/n_e$ for which, always, $c/n_1 \leq v_p \leq c/n_2$. The effective refractive index n_e depends upon the characteristics of the modes. Below a certain normalized frequency, or above a certain wavelength, determined by a_1, n_1, and n_2 and expressed by $v < 2.4$, only one mode can propagate. Figure 2.10B illustrates the increase of the refractive index with decreasing wavelength, while the equivalent index n_e shifts monotonically from n_1 to n_2 as the wavelength decreases. Note that an increasing wavelength corresponds to a decreasing refractive index and hence an increasing speed of light.

Since in most applications monomode operation is desirable, it can be seen that a small core diameter and a small difference between n_1 and n_2 are required. Assuming, for example, a fiber core diameter of 5 μm, this means that the index ratio should be $n_1/n_2 < 1.004$ for single-mode operation at 0.85 μm. For $n_1 = 1.5$, this corresponds to $n_2 = 1.494$ or an index difference $\Delta = (n_1 - n_2)/n_1$ of 0.4%. This situation is not significantly different at a wavelength of 1.05 μm. It is obvious that this requirement leads to practical fabrication difficulties and is not necessarily a feasible approach.

The simultaneous stringent requirements for both consistently small and uniform core diameter and an extremely small index difference between core and cladding therefore illustrate the extremely tight fabrication control that has to be adhered to in order to achieve monomode operation. Because selec-

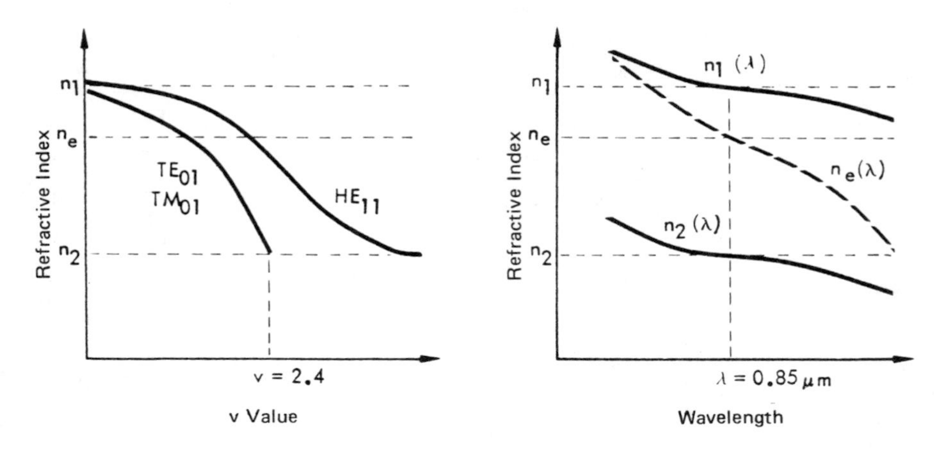

Figure 2.10 Schematic illustration of the relationship between the refractive index and v value for waveguide dispersion and material dispersion.

tion of core radius and index ratio involves a trade-off, it is possible to relax the demanding n_1/n_2 ratio somewhat, but obviously only at the expense of an even smaller core diameter.

Thus, at least in theory, most modes can be suppressed by making the core thin and the index difference between core and cladding small. The few modes that can propagate are weakly guided, but usually the guidance is sufficient for bending radii of a few millimeters. Because at present it is not practical to fabricate single-mode fibers, owing to the difficulties in producing structures of very small diameter and of small index differences, one is usually restricted to waveguides that support several modes, often up to several hundred.

The analysis above was carried out for step-index fibers. It can be used, with appropriate modifications, for other index profiles as well. In a graded-index fiber, the signal delay difference between on-axis and off-axis rays can be minimized by the proper choice of the radial grading of the refractive index. An approximately parabolic profile results in the smallest maximum time delay, because in this case the ray making the larger angle with the axis spends most of its time in the medium having the lower index and hence propagates faster than one close to the axis.

In fibers whose core refractive index displays a graded cross-sectional profile, there is no distinct core and cladding region. Instead, the refractive index changes continuously from its maximum value at the center (fiber axis) to a lower value at the fiber boundary. Outside the fiber it drops abruptly from the index value of the glass to that of the surrounding air. In such a fiber the light is not guided by internal reflection but by the bending of the light rays away from the thinner medium, which is the outer region of the fiber. For all practical purposes, the graded-index fiber can be considered to extend infinitely in the radial direction.

The analysis of a graded-index fiber is considerably more difficult than that of a step-index fiber. This is a result of the complex profile of this fiber type. The number of modes a graded-index fiber can support is

$$N_m = \Delta(2\pi a_1 n_1/\lambda)^2 \; i/(i+2) \tag{2.14}$$

where the index profile parameter is given by Equation (2.1). This relationship implies that the total number of modes that a graded-index waveguide can support is significantly smaller than the corresponding number for a step-index waveguide under equivalent conditions. Thus, a graded-index fiber may be superior to a step-index fiber because of its mode-suppressing feature and, hence, its bandwidth-increasing property. It enhances the velocity of waves that exhibit wide excursions from the center of the fiber, whereas rays parallel and close to the fiber axis are relatively slower.

The number of modes a graded-index fiber can support depends on the fiber properties, as seen in Table 2.3, which shows that as in the case of the step-index fiber the number of modes increases with index difference and with fiber diameter. However, the additional parameter that affects modal number is the index profile. Although from the standpoint of bandwidth an index profile parameter $i = 2.25$ (at $\lambda = 0.85\ \mu$m) is most desirable, the reduction in mode number is not optimized around this value because the number of possible modes decreases with declining parameter i. This leads to the conclusion that if an ideal profile for which $i = 2.25$ cannot be achieved it is more advantageous to fabricate the graded-index fiber with a lower i value than with a higher i value. This, of course, has an influence on other characteristics of the fiber which is not necessarily desirable.

Mode Coupling

The previous discussion of a waveguide that can support a number of independent transmission modes applies only to an ideal optical fiber. In a real fiber, by contrast, there is interaction between the modes. This mode coupling results in the transfer of energy between modes, as might be expected. It takes place if the fiber axis deviates from straightness or if the fiber encounters fluctuations in its diameter or refractive index as a function of distance along the axis. In the event of mode coupling, power is transferred between nearest-neighbor modes, both guided and unguided, until a steady-state power distribution is obtained. The fiber length to achieve this steady-state situation is the coupling length L_c.

TABLE 2.3. Dependence of Mode Number on Index Profile, Waveguide Diameter, and Wavelength for a Graded-Index Fiber, Assuming $n_1 = 1.5$

		Number of Modes, n_m											
		$\lambda = 0.85\ \mu m$						$\lambda = 1.05\ \mu m$					
		$a_1 = 2.5\ \mu m$			$a_1 = 25\ \mu m$			$a_1 = 2.5\ \mu m$			$a_1 = 25\ \mu m$		
n_1/n_2	Δ	i=2.0	2.25	2.5	2.0	2.25	2.5	2.0	2.25	2.5	2.0	2.25	2.5
1.003	0.003	1	1	1	115	122	128	1	1	1	76	79	84
1.005	0.005	2	2	2	192	203	213	1	1	1	126	133	140
1.010	0.010	4	4	4	384	407	427	3	3	3	252	267	280
1.020	0.020	8	8	9	768	814	854	5	5	6	504	533	560
1.030	0.029	11	12	12	1114	1180	1238	7	8	8	730	773	811
1.040	0.038	15	15	16	1460	1546	1622	10	10	11	957	1013	1063
1.050	0.048	18	20	20	1844	1953	2049	12	13	13	1209	1280	1343

Index profile parameter i	Deviation from ideal index profile (%)	Number of Modes, N_m											
		$\Delta = 0.005$				$\Delta = 0.01$				$\Delta = 0.03$			
		$\lambda = 0.85\ \mu m$		$\lambda = 1.05\ \mu m$		$\lambda = 0.85\ \mu m$		$\lambda = 1.05\ \mu m$		$\lambda = 0.85\ \mu m$		$\lambda = 1.05\ \mu m$	
		$a_1 =$ 2.5 μm	$a_1 =$ 25 μm	$a_1 =$ 2.5 μm	$a_1 =$ 25 μm	$a_1 =$ 2.5 μm	$a_1 =$ 25 μm	$a_1 =$ 2.5 μm	$a_1 =$ 25 μm	$a_1 =$ 2.5 μm	$a_1 =$ 25 μm	$a_1 =$ 2.5 μm	$a_1 =$ 25 μm
1.50	−33	2	165	1	108	3	329	2	216	10	988	6	647
1.75	−22	2	179	1	117	4	359	2	235	11	1076	7	705
2.00	−11	2	192	1	126	4	384	3	252	12	1153	8	755
2.25	0	2	203	1	133	4	407	3	267	12	1220	8	800
2.50	+11	2	213	1	140	4	427	3	280	13	1281	8	839
2.75	+22	2	222	1	146	4	445	3	292	13	1335	9	875

Because in an optical waveguide there is a finite number of guided modes and an infinite number of radiative modes, even if the number of guided modes in a given waveguide can be reduced to one, there may still remain a large number of possible radiative modes. Hence, the concept of mode conversion or mode coupling becomes significant. The response of a multimode fiber to intensity-modulated optical signals is influenced by loss, coupling, and delay differences among its various modes. It is important to note that the terms monomode and multimode refer only to guided modes and not to unguided ones.

The coupling of guided modes and the resultant exchange of power between high-speed and low-speed modes causes a mixing of the power among the guided modes, which results in an average velocity for the power transport and a corresponding reduction in pulse broadening. Mode coupling causes a pulse width spread given by

$$\Delta t_{\mathrm{mc}} = (n_1/c)\ (LL_c)^{1/2} \tag{2.15}$$

In the presence of mode coupling, the pulse width is proportional to the square root of length, whereas in the absence of mode coupling it is proportional to the length itself. L_c is a critical coupling length. The total loss of such a mode-coupled fiber is proportional to $(\Delta t_{\mathrm{mc}}/\Delta t)^2 \ \alpha\ L$.

In a monomode fiber there is no signal distortion due to modal velocity differences. Signal distortion is present, however, and is caused by dispersion in the dielectric fiber material and by the fact that the group velocity of the mode depends slightly upon frequency even in the absence of material dispersion. However, for small bandwidth these effects are considerably less significant than is multimode dispersion. For example, for a GaAs light-emitting diode as optical source whose spectral bandwidth is 500 Å (or 50 nm), the width of a narrow pulse increases to about 4 ns/km, whereas for a GaAsAl laser whose spectral width is 2 nm the detected pulse width is only 0.1 ns/km, and for an Nd-doped yttrium-aluminum-garnet (Nd:YAG) laser whose spectral width is less than 0.5 nm the pulse width increases to only 0.01 ns/km. Monomode fibers that are excited by a coherent source are thus theoretically capable of transmitting signals up to a rate of about 10^5 Mb/s over a waveguide whose length is 1 km.

Although signal dispersion requires a minimization of the number of modes in an optical fiber, waveguides that can support a large number of modes may be important in conjunction with the use of incoherent optical sources because the amount of light a fiber can accept increases with the number of modes it is able to support. An incoherent source generally excites every mode with the same amount of power. Also, because of the mode speed equalization feature of graded-index fibers, LEDs permit only half of the coupling efficiency in graded-index fibers compared to that in similar step-index fibers.

Assuming that there is an allowable power flow in both core and cladding, the combined fiber power flow is

$$P = P_1 + P_2 \tag{2.16}$$

and the ratio of optical powers in core and cladding:

$$P_1/P_2 = \tfrac{3}{4}\, N_m^{1/2} - 1 \tag{2.17}$$

where the total number of free-space modes accepted and transmitted by a step-index fiber for small acceptance angle β_0 is

$$N_m \approx 2(\pi a \beta_0/\lambda)^2 \tag{2.18}$$

Thus, the power flow in the cladding decreases at a rate that is approximately inversely proportional to the square root of the number of modes. For example, a fiber with a core radius of 25 μm, a core index of 1.5, and an index difference $\Delta = 0.003$ can transmit 206 modes; about 10% of the power propagates in the cladding, and the time difference between the simultaneously emitted fastest and slowest modes is 13.5 ns after transmission over a distance of 1 km.

Figure 2.11 illustrates the information-carrying capacity of a typical optical fiber system that consists of a GaAs laser source and an Si APD detector.

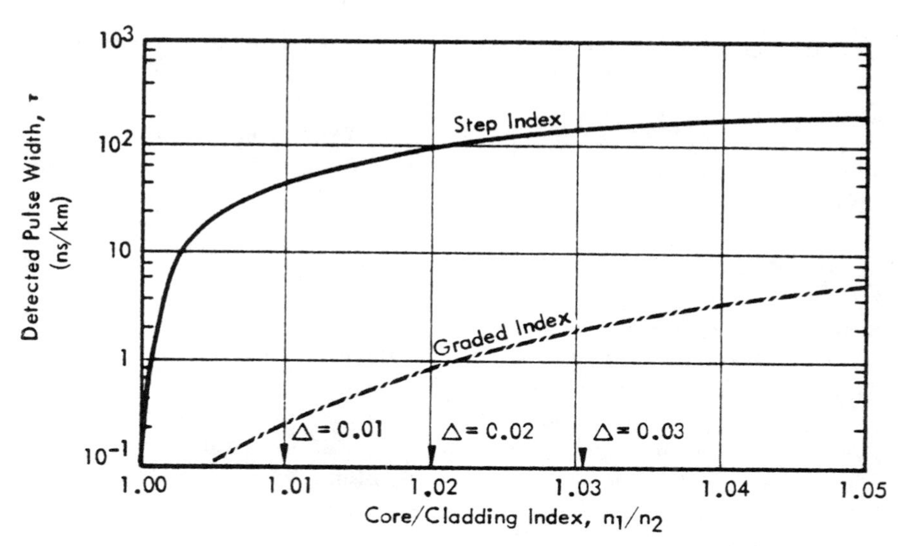

Figure 2.11 Dependence of pulse spreading on the refractive index and fiber type at $\lambda = 0.85$ μm and $n_1 = 1.5$.

The fiber loss limit, indicated by the solid line, corresponds to a parabolic index profile, whereas the material dispersion limit alone, indicated by the dashed line, would result in a higher bit rate. At some fiber length the signal will be limited by the total fiber loss. Thus, in practice the range covering the lower left quarter of the illustration may be used in an operating system. It is evident that the data transfer rate can be enhanced by mode coupling, although at the expense of a mode coupling loss which must be added to the total fiber loss.

Mode coupling also occurs in tapered and bent sections of a fiber. In a tapered section the core diameter continuously increases or decreases from its original value to some other value. Before and after the tapered section, the diameter is assumed to have a constant value. Likewise, a straight fiber connected to a bent section that has a constant radius of curvature and then further connected to a straight section usually shows considerable mode conversion.

Biconical tapered sections of multimode cladded fibers can be used as mode filters and mode analyzers, although at the expense of a higher attenuation. The tapered fiber, shown schematically in Figure 2.12, is useful in filtering out undesirable modes and is thus capable of achieving a relatively high bandwidth, comparable to that obtainable with monomode fibers. Because in the example shown the core diameter at the taper is reduced from 12.5 to 2.5 μm, all modes except the HE_{11} mode are filtered out (the HE_{12} mode has a very small contribution remaining). Obviously, the fabrication of such a filter presents serious production difficulties; therefore, such a filter must be considered to be a laboratory device.

As a consequence of mode coupling, pulse broadening is proportional to $(LL_c)^{1/2}$, rather than to $L^{1/2}$, which holds for an uncoupled fiber. Because L_c is usually considerably smaller than L, the modification of L by L_c tends to reduce pulse broadening. Consequently, after several coupling lengths a benefit will accrue to the information-carrying capability of the fiber. This attenuation–bandwidth trade-off can be described by

$$(s_u/s_c)^2\,\alpha_e = \alpha_l = \text{const} \tag{2.19}$$

where s_u and s_c are the pulse widths in the uncoupled and coupled states, respectively, α_e is the excess loss, and α_l is the loss per coupling length. For a step-index fiber, $\alpha_l \approx 0.5$ dB.

Mode coupling is related to fiber packaging, which is known as cabling. Unless extreme care is taken, external perturbations during fabrication can easily increase variations in fiber dimensions and hence increase mode coupling and, in turn, fiber loss beyond acceptable limits. An example of dimensional distortion is shown in Figure 2.13. In order to avoid uncontrolled permanent mode coupling induced during fabrication and installation, the

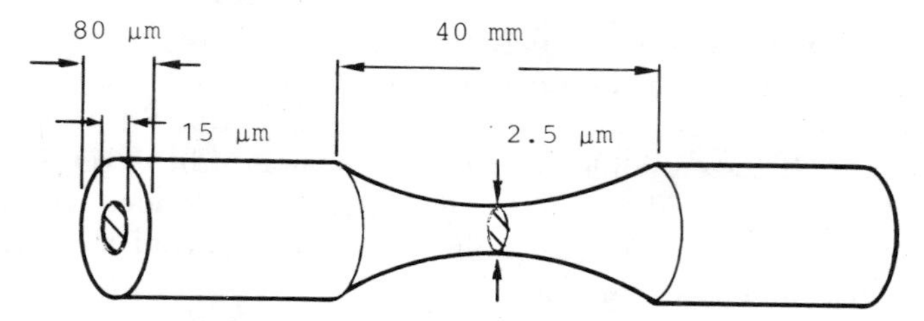

Figure 2.12 Example of a tapered multimode waveguide.

fiber is usually surrounded by a buffer material whose elastic modules are small compared to those of the fiber itself.

Assuming a laser source, the following numerical bandwidth limitations can be expected. The transit time of an axial ray, independent of fiber type, is approximately $t_a/L = 5\ \mu s/km$; of course, as described above, it depends slightly on variations of the refractive index. Hence, pulse broadening in a step-index fiber whose index difference $\Delta = 0.01$ is $\tau = 50$ ns/km, whereas in a graded-index fiber it can be reduced to $\tau = 0.25$ ns/km for the same index difference. This is an improvement by a factor of 200. The resultant maximum bandwidth is approximately given by $B = 1/2\tau$ and the maximum data transfer rate, making the slightly arbitrary assumption $R_{max} = \frac{1}{3}B$, is $R_{max} = \frac{1}{6\tau}$.

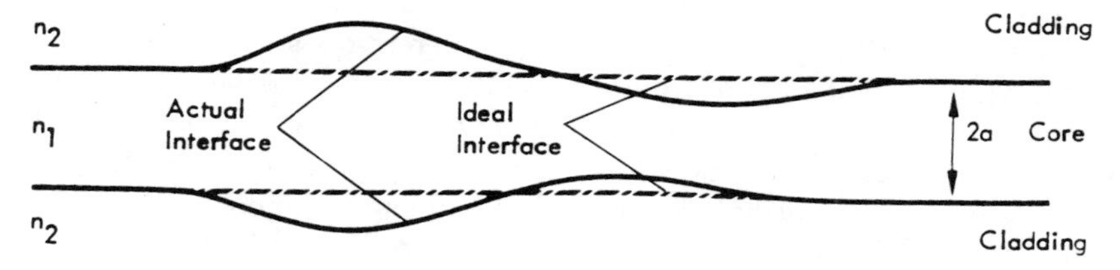

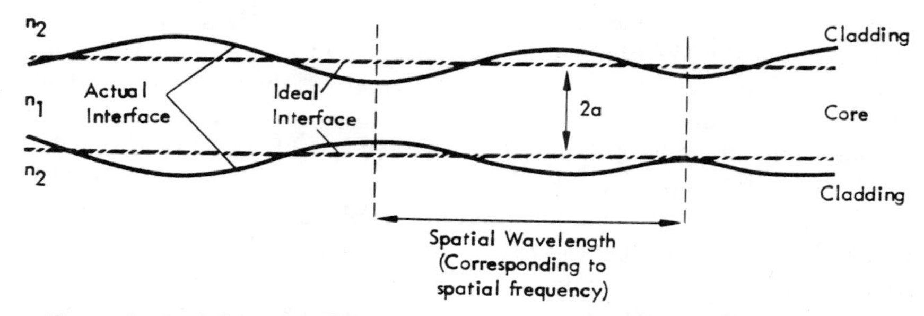

Figure 2.13 Schematic illustration of periodic and nonperiodic core–cladding interface distortions.

Mode coupling between guided modes may sometimes be desirable because of improvements in the delay distortion that results from uncoupled multimode operation, causing a narrower speed distribution of the modes. Mode coupling between guided and unguided modes, however, is normally undesirable unless optical emission from the fiber is intended. In practice, a certain amount of residual mode coupling is usually unavoidable, but it tends to enhance the fiber scattering loss.

Plastic Waveguides

The above discussion was mainly concerned with fibers made of glass. The majority of all optical waveguides are made of this material. Plastic or polymer fibers offer some advantages that make them potentially attractive in some applications, but because of their substantially higher attenuation compared to glass fibers, they find application mainly in short-distance systems. In some of these uses they may be superior to the more expensive glass fibers and may withstand a rougher mechanical treatment, so that they can be deployed in relatively unfriendly environments, such as in industrial areas, without significant degradation in performance.

The transmission spectrum of plastic fibers resembles that of glass fibers, except for an upward shift of the transmission curve. The attenuation of typical plastic fibers shows maxima at about 0.72 and 0.88 μm, with a relatively broad minimum ranging from 0.74 to 0.82 μm. At the attenuation minima the loss is typically in excess of 100 dB/km, although some plastic fibers have losses below 10 dB/km. Typical cladding diameters range from 125 to 1500 μm and typical core diameters from 115 to 1400 μm. Plastic fibers can be used at temperatures up to about 80°C (in contrast to glass fibers, which can withstand temperatures up to 1000°C) and maintain their flexibility at temperatures as low as −40°C. Prior to cabling, plastic fibers weigh only about 40% as much as glass fibers, but they can withstand far greater tension, torsion, and vibration stresses.

Polymer fibers are drawn, for example, with a polysterene core having a refractive index of 1.60 and a methyl methacrylate cladding having a refractive index of 1.49. The high index difference results in a high numerical aperture (NA = 0.6) and in a large angle of acceptance (about 70°), which makes the use of these fibers in conjunction with inexpensive light-emitting diodes economically attractive. Contrary to the situation with glass fibers, mechanical strength considerations do not prohibit plastic fibers from having large diameters. Since plastic fibers are usually employed at low bandwidths, the propagation of a large number of modes is not a serious drawback. Glass fibers have a high tensile strength and hence are quite brittle, so that they require a small diameter in order to minimize problems of fracturing and are grouped in large bundles if a large cross-sectional area is required. Plastic fibers, on the other hand, have a relatively low tensile strength, but possess considerable elasticity. Hence, fiber breakage is not a major concern in bundling or cabling. Individual fibers can be of larger diameter and cables can contain fewer fibers, compared to glass fibers.

Since plastic fibers normally do not have the constraint of small signal dispersion due to their low-bandwidth characteristics and since achievement of monomode operation is not a primary goal, the large core diameter is not necessarily detrimental to satisfactory performance. Hence, it is possible to achieve a further cost saving by fabricating cables of plastic bundles with fewer fibers and with fewer fabrication constraints as a result of the greater ruggedness of the plastic material.

ATTENUATION MECHANISMS

Most of the worldwide efforts in optical waveguide research have been directed toward the achievement of low-loss, low-dispersion fibers. Desirable goals have been an attenuation of less than 5 dB/km and a dispersion of less than 1 ns/km (mainly for applications in telecommunications equipment, where fiber lengths of at least 10 km are needed). Other applications generally can tolerate less stringent fiber characteristics, except for uses within the central processing unit of a computer, where low-loss and low-dispersion fibers are required. The discussion in this section is concerned with an analysis of fiber attenuation and its causes.

Attenuation Classification

Signal attenuation along a fiber is usually expressed in decibels per unit length; it is an important criterion in evaluating the performance of fibers for optical communication systems. This is particularly important for long fibers or when there is large spacing between repeaters.

The fiber loss is defined as the ratio of light output power P_{out} to light input power P_{in} at a given wavelength such that

$$\alpha = 10 \ln(P_{in}/P_{out})/L \tag{2.20}$$

$$P_{out}/P_{in} = e^{-\alpha L/10} \tag{2.21}$$

If α is referred to a standard length L, it is usually expressed in dB/km. Thus a 1 dB loss means that the light intensity has decreased in transversing a fiber by about 10% and a 10 dB loss indicates an intensity decrease of about 63%, whereas a fiber with no loss ($P_{out} = P_{in}$) corresponds to a 0 dB loss. This is illustrated in Figure 2.14. The various loss contributions, if expressed in dB/km, are additive.

There are multiple causes of optical losses in glasses and fibers. Generally, the loss depends on the glass system, glass composition, preparation

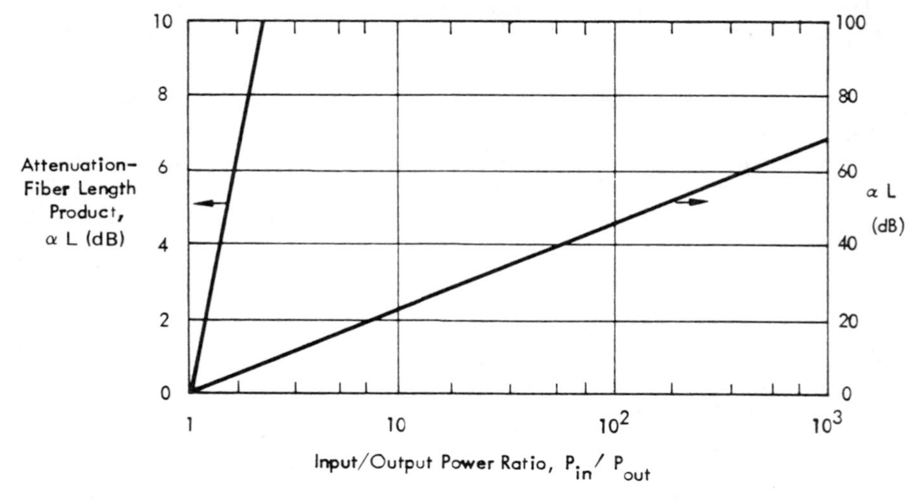

Figure 2.14 Relationship between optical input and output power, attenuation, and fiber length.

and purification method, and waveguide design. It is advantageous to distinguish generally between material-related and dimension-related fiber losses. Both of these types of losses can be further divided into absorptive and radiative components.

MATERIAL-RELATED LOSSES

Material-related losses are associated with the fiber material itself and do not relate to dimensional fiber imperfections. They have the following features:

Absorption Losses. These result mainly from one of three basic phenomena:

Intrinsic absorption. This loss type is related to the basic fiber material. Usually it is negligible in silica glass in the near-infrared range, although in the far infrared it may be significant.

Impurity absorption. This loss mechanism is associated with transition metal ions, such as Cu, Fe, Ni, Co, Cr, etc. The content of these elements must remain substantially under about 0.01 ppm in order not to increase the fiber loss of alkali silicate glasses. In glasses with high silica content, their influence is significantly less dominant. Even in high-silica glasses, however, impurity absorption is strong as a result of the presence of the OH^- radical (frequently referred to in this context as water). It has absorption maxima at 0.725, 0.825, 0.875, and 0.950 μm, with the strongest maxima at the two highest wavelengths. In contrast, Fe and Cu contributions usually occur over the entire visible and near-IR spectrum, without characteristic maxima. The OH maxima are the prime reason for the selection of the 0.85 and 1.05 μm wavelengths for data transmission, since they coincide with loss minima. Although the complete elimination of OH and transition metal content in glasses is not possible, a reduction to acceptable levels can be accomplished by proper process control. A reduction of the impurity-induced loss below that of intrinsic scattering loss is not required, however. The absorption maxima of these transition metal ions are usually very broad, so that it is difficult to identify the contaminating species from the spectral dependence of fiber absorption. The contributions by OH radicals, on the other hand, are sharp and easily identified. The absorption per ion for water does not vary as much from glass to glass as does the transition metal ion absorption.

Atomic defect absorption. Losses caused by imperfections of the atomic structure of the fiber material, either high-density clusters of atom groups or missing atoms, affect the wave propagation properties, depending upon their concentration. Usually, however, this is a second-order effect and can be neglected except in very-low-loss material.

Heat or intensive radiation may cause an increase in absorption losses. Temperature increase may lead to structural variations in the glass and, under certain circumstances and in certain glasses, tend to cause a significant loss increase. The susceptibility of optical waveguides to intensive radiation resulting in temporary or permanent loss increase is a strong function of the composition of the glass and the prior treatment it has received. For example, conventional glass fibers may exhibit a substantial loss increase under strong irradiative influence, whereas a Ge-doped silica fiber may display only a modest loss increase. A sample of such a waveguide showed a 16 dB/km increase at 0.82 μm after exposure to a radiation dosage of 4300 rads compared to more than 2000 dB/km in a conventional glass exposed to an identical radiation dosage.

Radiative or Scattering Losses. Generally, scattering is a result either of a lack of order of the material structure, of structural defects, of particle inclusion, or of random fluctuations (17, 18).

In crystalline materials the first two causes dominate. In amorphous materials (such as those used in fibers), the lack of structural order, particle contamination, and random fluctuations are most important. The inclusion of dust particles is one of the causes for contamination. High-temperature glass formation results in the chemical decomposition of most included particles and subsequently in their appearance as impurity centers. Furthermore, the glassy state retains some of the fundamental behavior of the liquid state; hence, localized material density fluctuations can take place. Scattering losses in optical fibers can be broadly divided into two major classes.

Rayleigh scattering. This is the most important scattering loss mechanism. It is an intrinsic effect and cannot be affected by material treatment. It represents the ultimate lower attenuation limit of optical fibers and is caused by fluctuations in atomic density of the basic material and of intentionally introduced impurities; both result in a variation of the refractive index. Density fluctuations in the basic material cause the primary scattering loss; they amount typically to up to 90% of the total scattering loss in most glasses. Concentration fluctuations, either of the constituents of multicomponent glass or of impurities, cause a proportional scattering loss; typically, the loss induced by concentration fluctuations is 10% to 20% of the total scattering loss and is thus insignificant in comparison to the density-induced loss. Rayleigh scattering is inversely proportional to the fourth power of the wavelength, so that it rapidly decreases with an increase in wavelength. For example, at a wavelength of 0.70μm, scattering may contribute about 5 dB/km to the total fiber loss, but at 0.85 μm, its contribution has decreased to about 2.3 dB/km, and at 1.05 μm it is only about 0.95 dB/km.

Brillouin or Raman scattering. This scattering mechanism is of only secondary significance under normal operating conditions. It is usually observed only at very high light power levels in very long monomode fibers. The threshold power level is proportional to the core cross-sectional area, so that one would not expect an influence by this scattering type in wide multimode fibers. However, it may be observed in very narrow monomode waveguides.

Intrinsic Losses

The intrinsic absorption loss of glasses is a consequence of the random molecular structure of the material, which results in varying local electric fields on a microscopic scale. These cause an intrinsic attenuation in chemically pure materials in a normally transparent region below the fundamental interband absorption edge. The mechanism responsible for this absorption is principally related to local field-induced broadening of the excitonic levels that are generated for energies close to but below the interband edge. The interband absorption edge is proportional to energy according to

$$\alpha_a \sim \exp\left[(E_p - E_G)/E_m\right] \tag{2.22}$$

where E_p is the photon energy, which is assumed to be related to the wavelength by $E_p = 1.237/\lambda$, with E_p given in eV and λ given in μm; E_G is the effective energy gap of the material; and E_m is a material constant ($E_m = 0.3$ eV for soda-lime-silicate and $E_m = 0.5$ eV for fused silica). This mechanism is similar to that observed for amorphous semiconductors ($E_m = 0.4$ eV for GaP, for example).

Equation (2.22) indicates that the absorption coefficient is independent of the detailed nature of the microfields. It also implies that materials with large energy gaps, such as fused silica, should have lower absorption than materials with narrower gaps, such as soda-lime-silicate, provided that the material constants of these glasses are comparable. Absorption loss data for these glasses are shown in Figure 2.15A. The curves represent lower bounds for experimental data. These data exemplify the exponential trend, which extends over at least six decades in attenuation and over 5 eV in photon

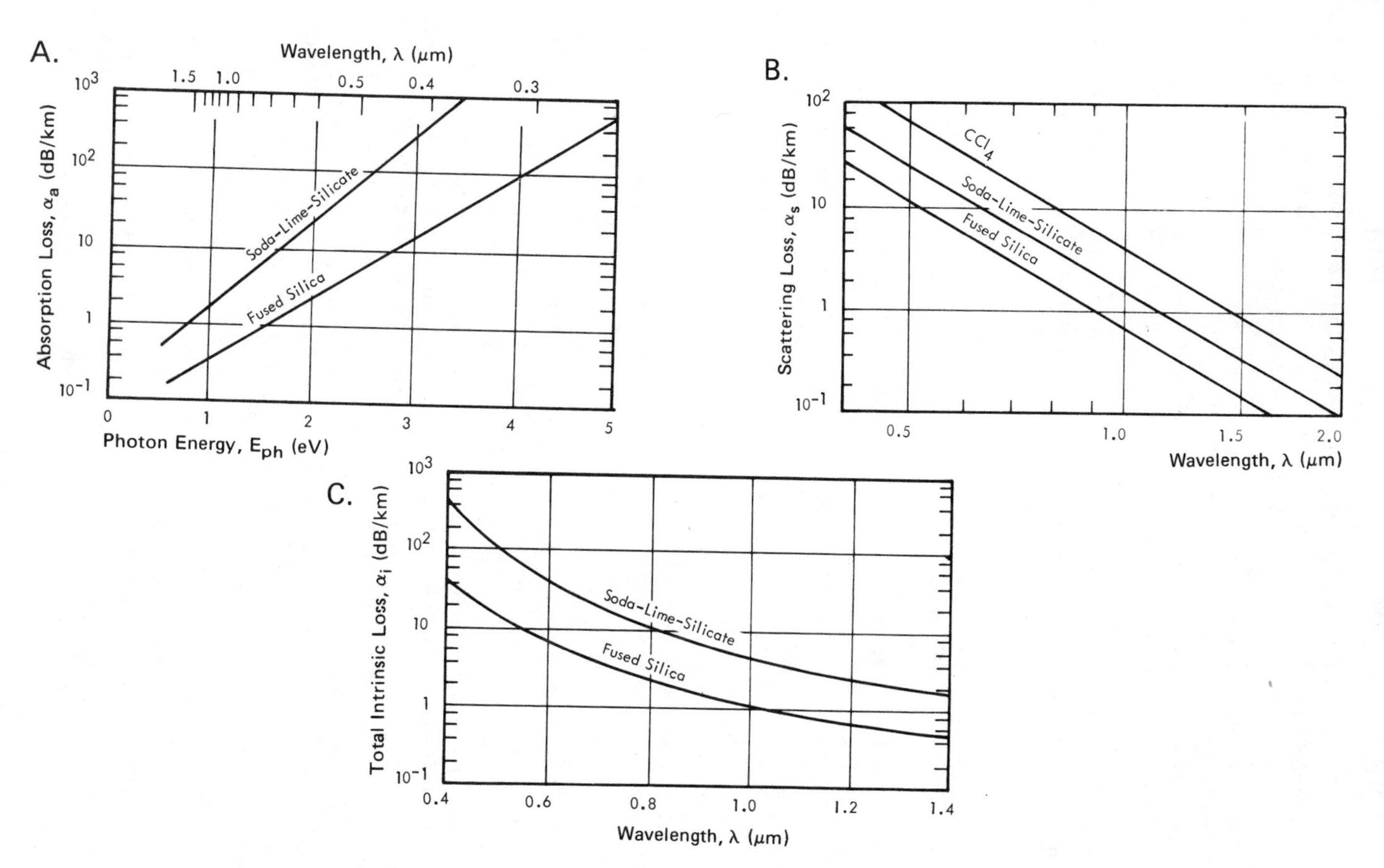

Figure 2.15 Intrinsic optical waveguide loss as a function of wavelength.

energy. This absorption mechanism is intrinsic and not governed by impurity ions. Impurity ions are known to exhibit specific spectral absorption bands rather than the observed exponential dependence. The material constant E_m is directly proportional to the energy kT_f, where T_f is a fictive temperature discussed below. Scattering loss in both liquids and glasses is a result of microscopic variations in the local refractive index (or dielectric constant) associated with their random molecular material structure. In liquids (for example, in CCl_4, which is used for some liquid optical waveguides), the variation of the local refractive index is caused mainly by thermally generated fluctuations in the number of molecules within a region of dimensions less than that of the wavelength. Thus, in liquids the scattering loss is

$$\alpha_{s(d)} = \left(\tfrac{8}{3} \pi^3 \right) \lambda^{-4}(n^8 p^2) kT\kappa_T \tag{2.23}$$

where λ is the wavelength, n is the refractive index, p is the photoelastic coefficient, k is Boltzmann's constant, T is the absolute temperature, and κ_T is the isothermal compressibility of the material. The term n^8p^2 establishes the relationship between fluctuations in density and fluctuations in refractive index. This equation clearly shows the inverse proportionality of the scattering loss to the fourth power of wavelength. It also indicates the strong relationship between loss and refractive index.

Equation (2.23) is based on the assumption of thermal equilibrium. Therefore, it is valid in liquids but not in glasses. Because the random structure of glasses is not determined by the ambient temperature T but by the temperature at which the glass, if heated, would come into thermodynamic equilibrium (fictive temperature concept), the use of this fictive temperature T_f instead of the ambient temperature T allows the determination of scattering losses in single-component glasses. Although glasses comply less readily to density fluctuations than liquids and hence a lower inherent scattering loss would be expected in glasses, the higher fictive temperature of glasses compared to the ambient temperature applicable to liquids partially offsets the lower density fluctuation of glasses. In glasses of lower softening point, a smaller amount of scattering would thus be expected compared to glasses of high softening point. Hence soda-lime-silicate glass (with a softening point of about 800°K) has a lower attenuation due to density fluctuations than fused silica (with a softening point of about 700°K), assuming otherwise identical conditions.

In multicomponent glasses, however, there is an additional mechanism that contributes to their scattering loss. Scattering loss can be attributed to the statistically random distribution of the polarizable components. This mechanism, which in soda-lime-silicate glasses causes losses greater in magnitude than those caused by density fluctuations, leads to an additional contribution to local variations in the refractive index. For a glass of m components, this additional scattering loss alone is

$$\alpha_{s(c)} = \left(\frac{32}{3}\pi^3\right)\lambda^{-4}\left(\frac{n^2}{\rho N_A}\right) \times \sum_{j=1}^{m}\left[\left(\frac{\partial n}{\partial x_j}\right)_{\rho,T,x_{i\neq j}} \left(\frac{\partial n}{\partial \rho}\right)_{T,x_i} \left(\frac{\partial \rho}{\partial x_j}\right)_{T,x_{i\neq j}}\right]^2 M_j x_j \quad (2.24)$$

In this equation, ρ is the density, N_A is Avogadro's number, M_j is the molecular weight, and x_j is the molecular weight fraction, indicating that the additional contribution due to polarizable components is proportional to the square of the refractive index and indirectly proportional to the fourth power of the wavelength. The two scattering terms are additive; hence, $\alpha_s = \alpha_{s(d)} + \alpha_{s(c)}$ (the subscripts d and c refer to variations in density and polarizable components, respectively). A comparison of the total scattering losses of three applicable materials based on these equations is shown in Figure 2.15B.

Figure 2.15C depicts the combination of the inherent scattering and absorption losses of fused silica and soda-lime-silicate glasses as established above. It is assumed that the total intrinsic loss is $\alpha_i = \alpha_s + \alpha_a$. These curves represent the fundamental limit of intrinsic fiber losses. They allow an estimate of the ultimately expected waveguide attenuation as a function of wavelength and hence affect the choice of a suitable optical source. CCl_4 losses are not included in this illustration, since the CCl_4 absorption spectrum is not known, although it has been found that its absorption loss alone exceeds the total loss of fused silica. A comparison of fused silica and soda-lime-silicate indicates that the intrinsic loss of both materials decreases with wavelength, that fused silica has a total loss that is about five times smaller (in dB) than that of soda-lime-silicate, and that for systems limited by intrinsic losses alone the repeater spacing may be twice as wide at 1.05 μm as at 0.85 μm.

Extrinsic Losses

In addition to the intrinsic loss mechanisms discussed above, there are additional extrinsic contributions to loss that are introduced mainly by impurity atoms or ions (19, 20). In the near infrared, the absorption loss of optical fibers arises mainly from the presence of transition metal ions (particularly Fe and Cu) and of water (in the OH radical form) dissolved in the glass. The metal ions can exist in several ionic states, each having absorption bands of different intensity, shape, and wavelength. The relative proportion of the various ions and hence the optical loss of the glass depend upon the temperature and the oxygen potential of the melting environment. Under oxidizing conditions, Cu is the major contributor; only about 0.008 ppm are required to result in a loss of 10 dB/km. Bubbling with reducing gas mixtures normally causes a decrease in the attenuation coefficient resulting from Cu, but this procedure frequently effects a large increase in the loss contribution by Fe. Consequently, a compromise is usually sought.

The OH content of the glass typically arises from three sources: absorption in the starting materials, dissolution in the starting materials, and precipitation from the furnace atmosphere. OH absorption spectra occur at 2.8, 1.4, 0.96, and 0.74 μm, corresponding to the fundamental and the first, second, and third harmonics. In addition, there are two structural combination bands at 2.2 and 2.4 μm. Although the absorption progressively decreases with a decrease in wavelength, the OH radical absorbs significantly in the region between 0.80 and 1.05 μm.

Depending on the wavelength and the relative contributions by impurities, either scattering or absorption may be the dominant loss mechanism. These major contributions to the overall glass loss within wavelength range of 0.6 to 1.1 μm are shown in Figure 2.16, together with the composite characteristic curve. It is assumed that the Fe^{++} content is 0.002 ppm and the OH^- content is 6 ppm. The total loss curve exhibits minima at about 0.83 μm and at about 1.05 μm, separated by a small maximum at about 0.95 μm that is chiefly caused by OH^- groups. Likewise, the very faint maxima at 0.72 and 0.88 μm are a consequence of OH^- contributions. The transmission loss caused by the Fe^{++} content increases with increasing wavelength and displays a maximum within the range of interest at a wavelength of about 1 μm. The transmission loss caused by OH^- content shows several maxima and minima over the useful wavelength range which are chiefly responsible for the loss undulations of the composite curve. Theoretical loss and actually measured loss on typical fibers generally agree well, except between 0.8 and 0.9 μm and above 1.0 μm, where the measured loss exceeds the predicted value. This leads to the conclusion that selected further improvements are feasible. At some other wavelengths the measured values are slightly less than the theoretical values, indicating that the density of absorption centers is less than that assumed in the theory. This leads to the further conclusion that some other impurities, in addition to Fe and OH, are present whose elimination can be expected to result in slight improvements of the material.

In the absence of any absorption centers, the total loss is a function of scattering only. Thus at 0.85 μm a loss minimum of about 2 dB/km can be expected, and at 1.05 μm, one of about 0.8 dB/km. Advances beyond that would require a reduction in the density of scattering centers (which is only marginally achievable) or a decrease in the scattering center polarizability, assuming a given wavelength, which would lead to the use of different materials. Assuming a negligible effect of absorption centers, an increase in wavelength is another possibility. For example, in the absence of absorption, a wavelength shift of 1 to 2 μm would result in a 16-fold loss decrease.

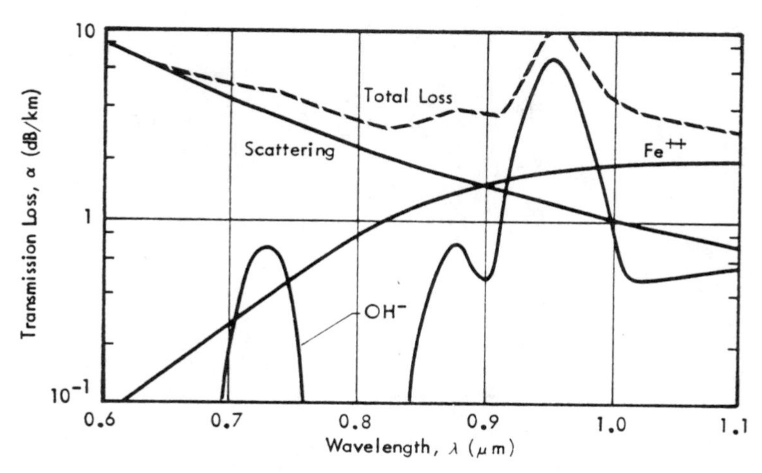

Figure 2.16 Approximate relationship between waveguide attenuation and wavelength for fiber optics glass.

Because the electromagnetic energy resides partly in the core and partly in the cladding, the total fiber attenuation is composed of terms that describe losses in the fiber core and in the cladding,

$$\alpha = p_1\alpha_1 + p_2\alpha_2 \tag{2.25}$$

where the subscripts 1 and 2 refer to core and cladding, respectively, and p describes the fractional power within the core or cladding. When all modes are equally excited, an approximate value for the fractional power within the cladding is described by the term p_2, which is defined as the ratio of cladding power to total power:

$$p_2 = 8/3v \tag{2.26}$$

A different approach to fiber design uses a special jacketed fiber, as illustrated in Figure 2.17. This fiber possesses an external higher-index sheet and consists of a core of radius a_1 and index n_1, a cladding layer of thickness $a_2 - a_1$ and index n_2, and a jacketing layer of arbitrary thickness and index n_1. If the wavelength in free space of the light to be transmitted is λ and if the index difference is expressed by $\Delta = 1 - n_2/n_1$, then the radiation loss of TE modes in this fiber can be expressed by

$$\alpha = \tfrac{2}{3}\left(\frac{\Delta^{1/2}}{a_1}\right) \exp \frac{-4\pi n_1(a_2 - a_1)(2\Delta)^{1/2}}{\lambda} \tag{2.27}$$

Hence, the fiber mode loss decreases exponentially with cladding thickness.

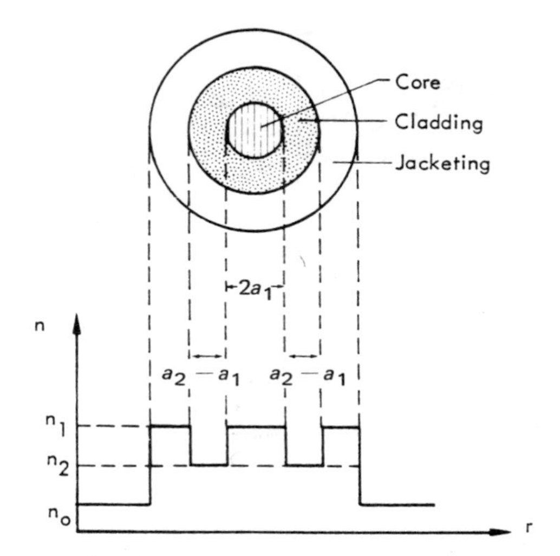

Figure 2.17. Typical refractive index profile of a step-index fiber after jacketing.

Also, a large cladding thickness, as well as a large core diameter, are desirable. Furthermore, this fiber has a higher radiation loss of the higher-order modes. Thus the lower-order modes predominantly are transmitted and a reduction in pulse dispersion can be achieved. Hence, it is possible to choose the geometry and index profile such that an optimum fiber loss can be obtained in practice. Assuming that a loss of less than 1 dB/km is required, cladding thickness and refractive indexes of this fiber type can generally be chosen to result in the desired properties.

In addition to absorption and scattering losses, most practical fibers exhibit attenuation originating from a perturbation of the fiber geometry. Geometry-related losses may have a variety of causes. Normally they are a function of inhomogeneities in the fiber, caused by variations in fiber diameter, by bends, or by other factors. This type of signal attenuation manifests itself dominantly in radiation loss. The fiber cladding may have a further detrimental influence on the fiber loss. Structure-related losses thus have the following features.

Core Bending. In a fiber bend some of the evanescent energy tends to exceed the velocity of light in the cladding and will thus be radiated (21, 22). An outer ray that has to cover the longer distance suffers more of this radiation loss than an inner ray. The critical radius of curvature below which heavy bending losses occur can be estimated from

$$r_c = \frac{(3/4\pi)\lambda n_1^2}{(n_1^2 - n_2^2)^{3/2}} \tag{2.28}$$

This means that from the standpoint of bending loss reduction it is desirable to design a fiber with widely divergent refractive indexes in order to reduce the permissible bending radius as much as possible. Also, the shorter the wavelength, the smaller the critical radius of curvature. Assuming $n_1 = 1.500$ and $n_2 = 1.495$, for $\lambda = 0.85\ \mu$m the minimum allowable radius of curvature is $r_c \approx 62\ \mu$m. If, however, $n_1 = 1.500$ and $n_2 = 1.450$, then $r_c = 8\ \mu$m. In most practical situations, the critical radius of curvature is therefore very small for well-guided modes. For modes close to their cutoff value (where they are no longer guided in the core), however, the allowed radius of curvature may be substantially larger. This has an important consequence in the installation of optical fibers if they are to be twisted around each other or around some other supporting structure, where they cannot be bent abruptly even if their brittleness would allow it. The radiation loss due to fiber bending is approximately inversely proportional to the bending radius.

Core Dimension. Distinction can be made between periodic and nonperiodic waveguide perturbations. If periodic variations of the fiber dimensions correspond to the beat wavelength between two modes, the resultant mode coupling may cause a radiation loss. The loss depends on the extent of the perturbation and the nature of the involved modes. Nonperiodic waveguide perturbations refer to localized variations in fiber diameter. They may cause mode conversion and resultant signal loss.

Cladding Dimension. If the cladding is insufficiently thick, the electromagnetic field may be perturbed by the surrounding jacket of the cable, causing a significant contribution to the total fiber loss. A cladding thickness in excess of 10 to 15 μm is usually considered to be sufficient to prevent cladding-induced losses.

The system loss tolerance limit is a function of the type of the photodetector used in the receiver. Thus a PIN diode is restricted to a maximum of about 45 dB signal loss, whereas an avalanche photodiode (APD) allows a signal loss of about 55 dB. Only a small improvement in system length results from replacement of the PIN diode by an APD. Assuming a maximum tolerable loss of 45 dB, a maximum fiber length up to about 4.5 km can be served with a 10 dB/km fiber, for example, without the use of a regenerative repeater, whereas a 1 dB/km fiber allows a system length extension up to about 45 km.

The choice between cost and performance factors requires a compromise. It is facilitated by the availability of a large variety of fibers that can meet a range of economic and technical requirements. Figure 2.18 is a representation of limitations in the system length as a result of signal attenuation due to fiber loss. In conjunction with the length limitation resulting from

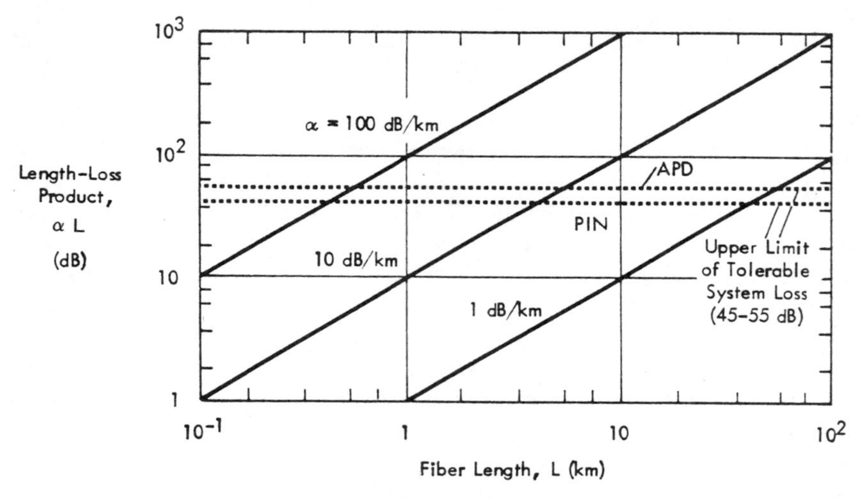

Figure 2.18 Product of fiber length and attenuation.

signal dispersion and noise, Figure 2.18 allows us to estimate the maximum system length before signal regeneration by means of repeaters becomes necessary.

DISPERSION MECHANISMS

Most of the major advances in reducing fiber loss have been achieved during the past decade. Further improvements will be comparatively minute and perhaps not even necessary in view of the economics of removing the last traces of contaminants such as transition metals and OH radicals. The reduction of fiber loss to values below 5 dB/km at a wavelength of 0.85 μm, or below 2 dB/km at 1.05 μm, has made the concept of practical long optical communications links feasible. Hence, pulse dispersion over long fibers has become the focal point of attention. Long-distance optical communications is no longer limited by fiber loss but, more dominantly, by pulse broadening.

Intramodal and Intermodal Dispersion

In weakly guiding fibers (defined as those which have a small refractive index difference), material and waveguide dispersion effects are responsible for the delay distortion that is experienced by optical signals. Both contributions are additive (21).

Figure 2.19 illustrates the concept of pulse spreading for two Gaussian-shaped light pulses entering a fiber at a time interval t_p and having a negligible pulse width s_λ, determined by the spectral width of the source. After propagating through a fiber of length L they will have retained their spacing t_p but will have widened to a width s_o and eventually partially overlap because of dispersion. A measure of overlap is the quantity δ. In a practical system δ is not allowed to exceed a specific value determined by the requirement that the system detector must be able to distinguish between the two pulses. Therefore, the allowable maximum value of δ that corresponds to a certain fiber length is one of the factors that determine regenerative repeater spacing (16). Pulse spreading causes the filling of the spaces between pulses until eventually the receiver is unable to decide whether or not a time slot contains a pulse (one or zero). Thus, pulse spreading limits the communication capability of the system.

An introduction to the significance of pulse broadening was given earlier. In order to determine further the quantitative effect of this important property on the system characteristics, we must evaluate the group delay for each mode of operation. The total power output at distance L from the input is then the sum of the time delays of all modes integrated over the spectral width of the source. This can be discussed in a more practical manner as follows (23, 24). The width of the output pulse can be expressed by its standard deviation, s_o.

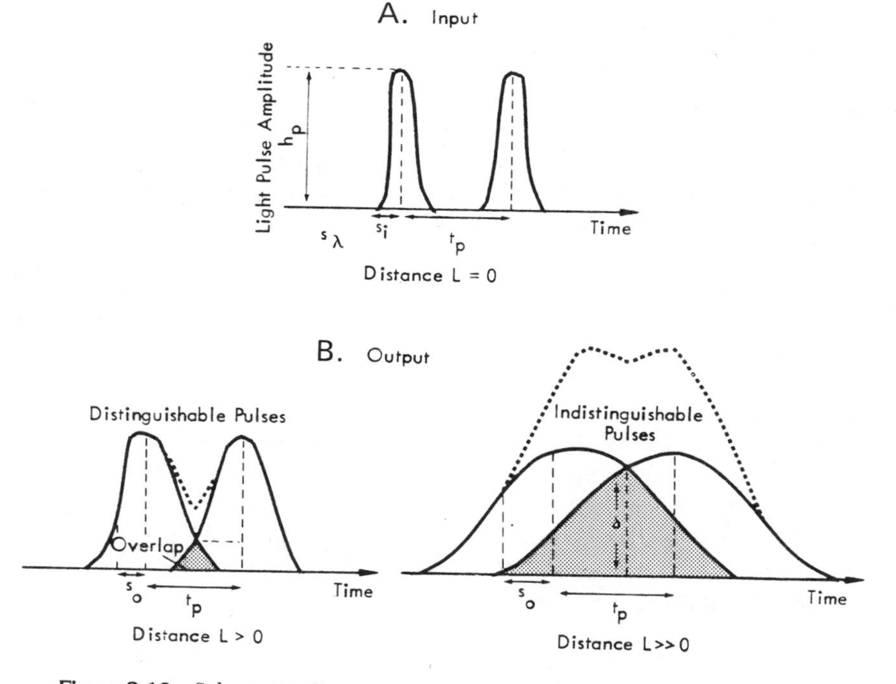

Figure 2.19 Schematic illustration of the principle of signal pulse widening, showing the criterion of pulse distinction. Solid line, individual pulses; dashed line, composite pulses.

It is composed of an intramodal and an intermodal term, s_m and s_n, respectively, such that

$$s_o = (s_m^2 + s_n^2)^{1/2} \tag{2.29}$$

The resultant pulse width after a signal has been transmitted over a fiber of length L is thus a function of a term s_m that describes the characteristics of the source (spectral width and dispersion) and a term s_n that describes the characteristics of the fiber (index profile and index difference).

Intramodal Pulse Broadening. The intramodal term s_m is a function of the spectral bandwidth of the source (s_λ) and the material dispersion ($d^2n/d\lambda^2$); it represents the fundamental limit to pulse broadening and is usually the only component present in a monomode fiber. Quantitatively it can be expressed by

$$s_m = \left(\frac{s_\lambda \lambda L}{c}\right)\left(\frac{d^2n}{d\lambda^2}\right) \tag{2.30}$$

This indicates that pulse spreading as a result of intramodal dispersion increases with the spectral width of the source, with the emitted wavelength, with the amount of material dispersion, and with the fiber length. Although material dispersion is a fundamental material property, the only option a system designer has for reducing intramodal pulse spreading at a given wavelength and system length is the reduction of the spectral width of the source, which explains the major efforts by many laboratories to achieve narrower source pulse width through improved source characteristics. Typically at $\lambda = 0.85\ \mu$m material dispersion is about $d^2n/d\lambda^2 = 7.0 \times 10^{-3}\ \mu m^{-2}$, and at $\lambda = 1.05\ \mu$m it is about $3.5 \times 10^{-3}\ \mu m^{-2}$. Hence at 0.85 μm, using an LED of spectral width $s_v = 500$ Å $= 0.05\ \mu$m, intramodal pulse spreading per unit length of fiber is $s_m/L = 1$ ns/km. For an injection laser source (with $s_\lambda = 20$ Å $= 0.002\ \mu$m) at the same wavelength, the intramodal pulse broadening is $s_m/L = 0.04$ ns/km. This can be translated into an upper limit of the bandwidth–length product of about 1 GHz km for the LED and of 25 GHz km for the laser source. Further detail of this analysis is shown in Figure 2.20. Note that the limitations implied by this illustration apply only to intramodal pulse spreading in the absence of intermodal pulse spreading and attenuation limitations.

Intermodal Pulse Broadening. Intermodal broadening results from delay differences between modes. By index profile shaping through a variation of the profile parameter i, intermodal pulse broadening can be reduced to a value below that of intramodal broadening, so that in theory it can be neglected, providing that the index grading profile represents a near-ideal situation. In reality, however, it is difficult in a production operation to obtain a

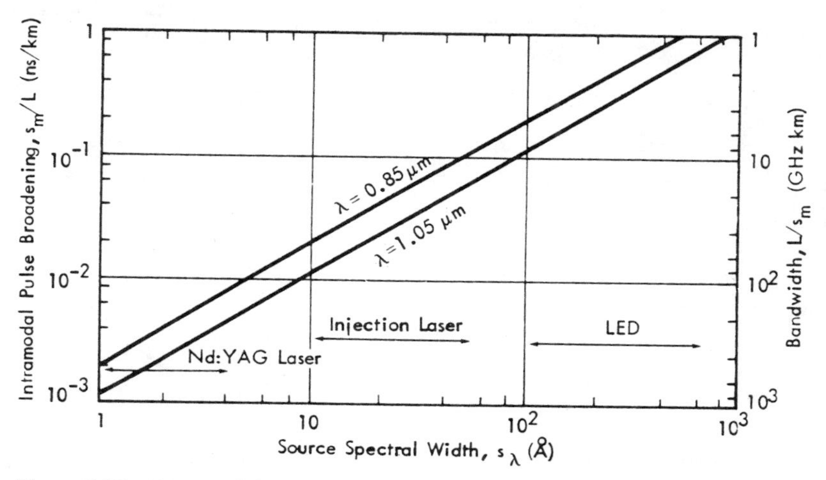

Figure 2.20 Intramodal pulse widening as a function of source bandwidth and wavelength.

fiber index profile that equals the ideal. Intermodal pulse broadening is given by

$$s_n = (Ln_g\Delta/2c)f(i) \tag{2.31}$$

Here n_g is the axial group index and $f(i)$ is an empirical quantity that depends upon the profile parameter i, which, for example, was defined by Equation (2.1). The variation of $f(i)$ with the index parameter shown in Figure 2.21 indicates that $f(i)$ and hence s_n has a minimum at $i = 2.25$. Thus, it is theoretically possible to reduce intermodal pulse broadening to zero by precise control of the index profile, as shown in Figure 2.22. Note again that the illustration refers only to limitations due to intermodal effects in the absence of other limitations. If $f(i) \neq 0$, then for $\Delta = 0.01$, under the assumption $n_g = 1.5$, the intermodal pulse broadening is $s_n = 0.8$ ns/km for $i = 2.0$ and 2.0 ns/km for $i = 3.0$. Figure 2.22 clearly points out the need for very tight control over the index profile. Thus a deviation from the ideal profile by only ±5% causes an intermodal dispersion increase from 0 to more than 0.3 ns/km. Because intermodal pulse width broadening is proportional to the index difference Δ, efforts to improve the source–fiber coupling efficiency by an increase of Δ tend to increase intermodal pulse broadening. Conversely, the desirability of small intermodal pulse broadening tends to result in a reduced coupling efficiency (23).

In actuality, pulse broadening is a function of both the intramodal and the intermodal terms. The effect of the combined influences on the fiber bandwidth in the absence of any attenuation is depicted in Table 2.4, where an

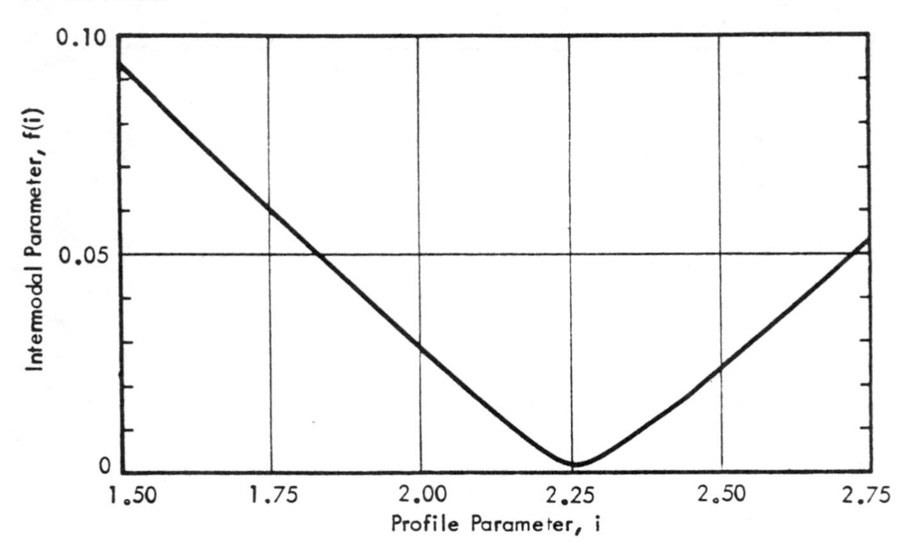

Figure 2.21 Relationship between intermodal parameter and refractive index profile of a graded-index waveguide.

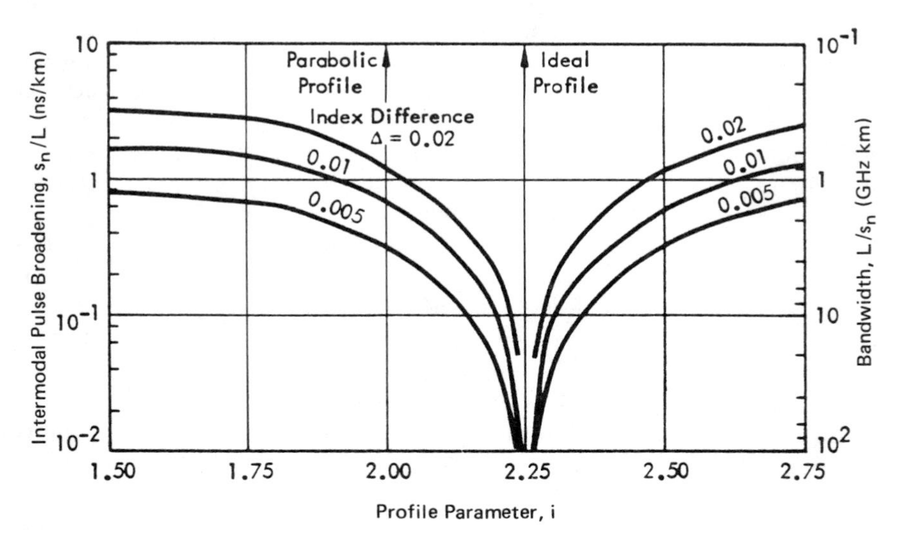

Figure 2.22 Intermodal pulse widening as a function of the fiber index profile (at $\lambda = 0.85\ \mu m$).

TABLE 2.4. Bandwidth Limitation of an Optical Communication System Due to Intramodal and Intermodal Pulse Widening as a Function of Fiber Length, Index Profile, and Source Spectral Width at $\lambda = 0.85\ \mu m$

Fiber length L (km)	Intramodal broadening s_m (ns)		Intermodal broadening s_n (ns)					
	Laser	LED	$i = 2.00$		$i = 2.25$		$i = 2.50$	
	($s_\lambda = 0.002\ \mu m$)	($s_\lambda = 0.050\ \mu m$)	$\Delta = 0.01$	$\Delta = 0.001$	$\Delta = 0.01$	$\Delta = 0.001$	$\Delta = 0.01$	$\Delta = 0.001$
10^{-2}	0.0004	0.01	0.007	0.0007	0	0	0.007	0.0007
10^{-1}	0.004	0.1	0.07	0.007	0	0	0.07	0.007
1	0.04	1.0	0.74	0.074	0	0	0.68	0.068
10	0.4	10.0	7.4	0.74	0	0	6.8	0.68
10^2	4.0	100.0	74.0	7.4	0	0	68.0	6.8

Fiber length L (km)	Maximum bandwidth $1/s_0$ (GHz)							
	$s_\lambda = 0.002\ \mu m$				$s_\lambda = 0.050\ \mu m$			
	$i = 2.00$		$i = 2.25$		$i = 2.00$		$i = 2.25$	
	$\Delta = 0.01$	$\Delta = 0.001$	$\Delta = 0.01$	$\Delta = 0.001$	$\Delta = 0.01$	$\Delta = 0.001$	$\Delta = 0.01$	$\Delta = 0.001$
10^{-2}	135.0	1189.0	2500.0	2500.0	80.0	100.0	100.0	100.0
10^{-1}	13.5	118.9	250.0,	250.0	8.0	10.0	10.0	10.0
1	1.35	11.89	25.00	25.00	0.80	1.00	1.00	1.00
10	0.14	1.19	2.50	2.50	0.08	0.10	0.10	0.10
10^2	0.01	0.12	0.25	0.25	0.008	0.01	0.01	0.01

attempt has been made to show the effect of source spectral width, fiber numerical aperture, and index profile on the total pulse width–limited bandwidth. Obviously, the narrower-width injection laser yields a substantially higher bandwidth, and the variation of the index profile toward the ideal tends to increase the bandwidth by up to about two orders of magnitude. An improvement in bandwidth of almost an order of magnitude can be expected if the index difference of the fiber is reduced from 1% to 0.1%.

Another important conclusion derived from this analysis is that there is only a small difference in bandwidth as a function of the fiber index profile in the case of an LED source, whereas a strong dependence exists in the case of a laser source. Also, in the case of an LED the effect of the fiber index difference between core and cladding is relatively small, whereas in the case of an injection laser it is considerable.

A variation of these conclusions is given in Figure 2.23, which shows the potential data transfer rate of an optical link as a function of source type and fiber refractive index profile. Again, the requirement of a precise index profile is apparent for a laser source, whereas in the case of an LED a much less sensitive dependence of the transfer rate upon the index profile is observed. Thus extremely tight control of the index profile is required for narrow spectral width sources in order to achieve an elimination or reduction of all bandwidth limitations except the inherent material dispersion.

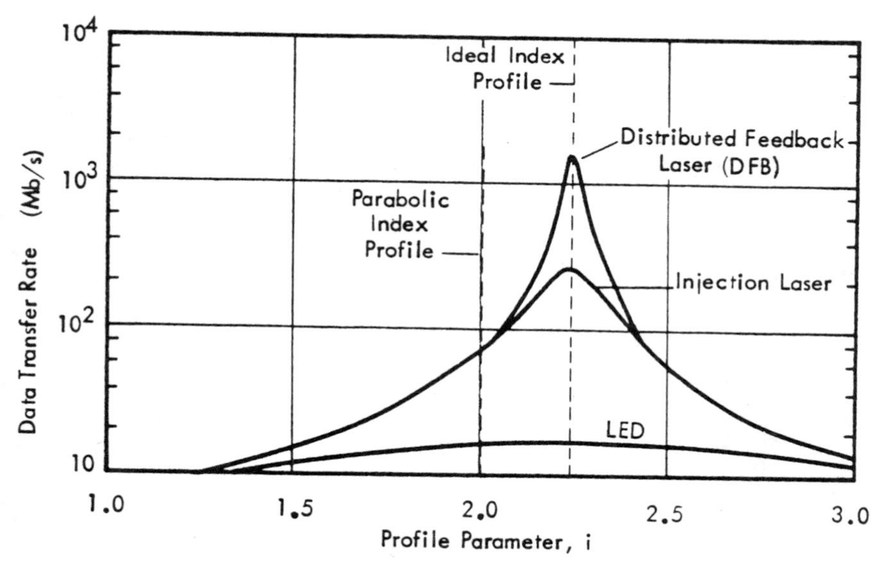

Figure 2.23 Illustration of the influence of the index profile parameter on the data rate capability of graded-index fibers.

Contributions to Dispersion

Because of the importance of the derivation of the pulse dispersion–limited bandwidth of an optical fiber, we want to discuss this subject in more detail. We assume again that the length–bandwidth product of a fiber can be expressed by the general relationship

$$\frac{L}{s_o} = \left[\left(\frac{s_\lambda \lambda (d^2n/d\lambda^2)}{c}\right)^2 + \left(\frac{n_g \Delta f(i)}{2c}\right)^2\right]^{-1/2} \tag{2.32}$$

which is composed of the terms s_m/L and s_n/L. The product of fiber length and bandwidth as a function of spectral width of the source, numerical aperture of the fiber (index difference), and index profile is given in Figure 2.24. These data have been calculated for key numerical values of these parameters in order to allow more general conclusions and to outline the inherent performance limitations.

In the following, we want to summarize the conclusions regarding the behavior of the product of system length and bandwidth, L/s_o, as a function of source spectral width, fiber index difference, and index profile parameter.

Source Spectral Width. At small width of the optical emission spectrum, the length–bandwidth product is independent of the spectral width of the source. It approaches 1.4 GHz km for a fiber of $\Delta = 0.01$ and 13.7 GHz km for a fiber of $\Delta = 0.001$, regardless of the spectral width of the source. At large width of the emission spectrum, the length–bandwidth product declines at a rate inversely proportional to the spectral width, regardless of the numerical aperture of the fiber. Thus, if $s_\lambda = 10$ nm, the highest bandwidth that can be achieved for a fiber 1 km long is about 5 GHz even for the most ideal index profile, and at $s_\lambda = 100$ nm it is about 0.5 GHz. On the other hand, in a fiber whose index difference is $\Delta = 0.01$, the bandwidth will not be increased by using an optical source that has a width of less than about 20 nm.

Fiber Index Difference. In the case of an ideal index profile ($i = 2.25$ at 0.85 μm), the length–bandwidth product is independent of the index difference of core and cladding. If the index profile deviates from the ideal, the length–bandwidth product varies inversely with the index difference, although at unrealistically low values of Δ the product is still independent of Δ. Although in the case of an ideal profile the length–bandwidth product depends strongly on the spectral width of the source, it becomes almost independent of source width at usually encountered values of Δ if the index profile is nonideal. Thus, for combining a reasonably large index difference with a reasonably wide source, it is necessary that a near-ideal index profile be achieved. For example, if a 5 nm wide source is used in conjunction with an ideal profile, an L/s_o value of about 10 GHz km can be expected even for a large value of Δ, such as $\Delta = 0.05$. In order to obtain the same length–

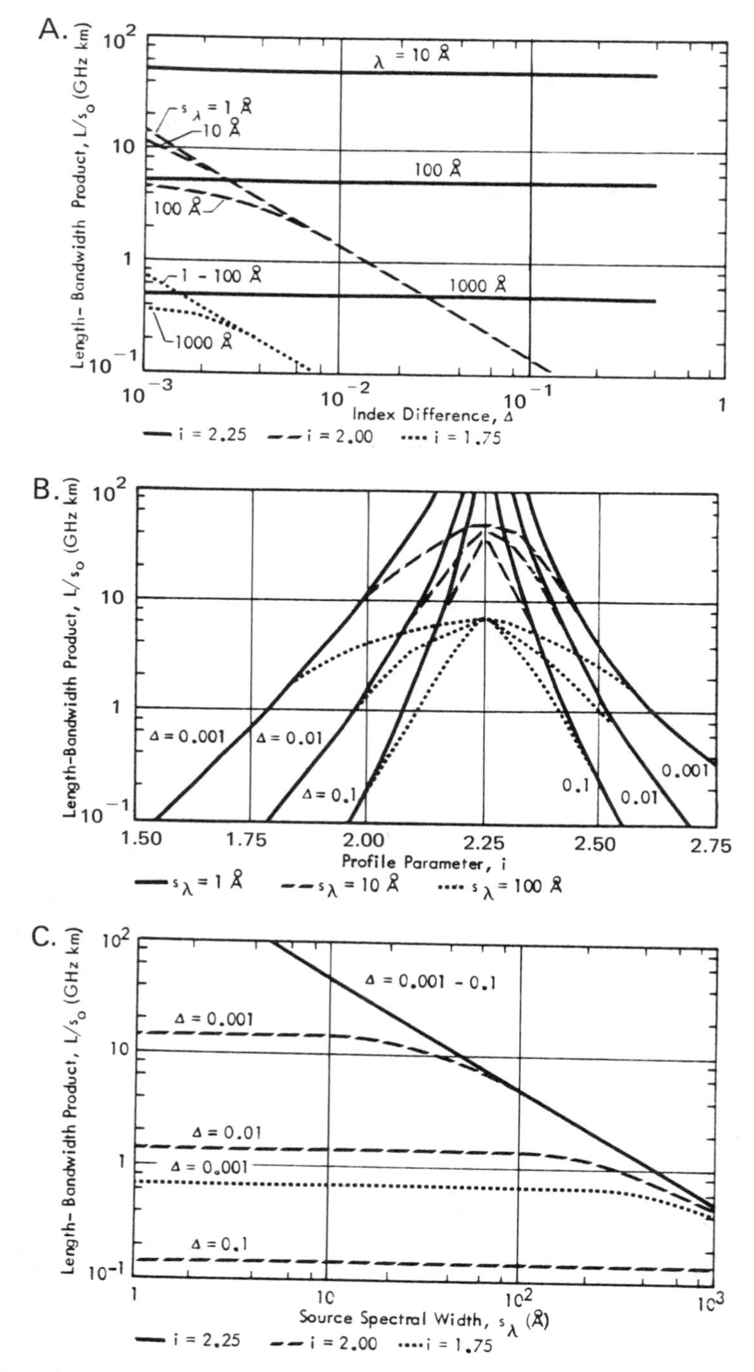

Figure 2.24 Dependence of the length–bandwidth product of graded-index waveguides on fiber and source characteristics as determined by intramodal and intermodal pulse widening.

bandwidth product with a fiber whose profile parameter $i = 2.00$ (rather than the ideal value of 2.25), a source with a spectral width of about 0.1 nm and a fiber index difference of about $\Delta = 0.001$ have to be used; this is not realistically achievable.

Profile Parameter. The length–bandwidth product displays a strong dependence on the index profile of the fiber. The dependence is the stronger the smaller the index difference and the narrower the source emission spectrum. The largest L/s_o value can be achieved if $i = 2.25$ (at 0.85 μm); in this case the length–bandwidth product is independent of index difference. For example, if the spectral width is $s_\lambda = 0.1$ nm, the highest L/s_o value achievable under any circumstances is about 500 GHz km; for $s_\lambda = 1$ nm it is about 50 GHz km; and for $s_\lambda = 10$ nm it is about 5 GHz km. While the peak L/s_o value for a given spectral width is obtained at $i = 2.25$, the dependence of L/s_o on the parameter i is strongest for a large value of Δ; however, for a very small value of Δ the achievable length–bandwidth product is relatively insensitive to index profile variations. The sensitivity increases, however, with decreasing spectral width of the source.

Pulse broadening in both step-index and graded-index fibers can be reduced by a reduction of the index difference, but then the acceptance angle of the fiber is reduced also, causing a significantly lower launching efficiency from a source with a large radiating angle, such as a light-emitting diode.

In a long multimode step-index fiber the bandwidth may be higher than apparent from the above analysis because of mode conversion, whereby optical power is transferred from slower to faster modes and vice versa. Assuming a fiber length of about 10 km between repeaters, a bandwidth of only about 10 MHz is possible in a step-index fiber. In a graded-index fiber, on the other hand, the practical limitation due to fabrication control difficulties is given by a transmission bandwidth of about 1 GHz km. Again assuming a fiber length of 10 km between repeaters, a bandwidth range of about 100 to 300 MHz is possible. Single-mode fibers, however, whose bandwidth is limited only by material dispersion and source bandwidth, allow an upper fiber bandwidth limit of about 10 to 50 GHz km if a sufficiently narrow monomode laser is employed. This brings the transmission bandwidth of a fiber 10 km long into the 1 to 5 GHz range. However, the best present GaAsAl lasers oscillate in several modes, resulting in a minimum source spectral width of 0.002 μm (or 20 Å). This still allows a bandwidth of 3 to 4 GHz km, which can be obtained with single-mode step-index fibers.

If an LED source rather than a laser source is used to launch the rays, material dispersion limits the fiber bandwidth even in a graded-index fiber to significantly lower values. For example, in the typical wavelength range of 0.8 to 0.9 μm of GaAsAl LEDs, the bandwidth is only about 200 to 300 MHz km, or smaller by a factor of about 3 than in the case of a laser source. Monomode step-index fibers are exclusively used in conjunction with laser sources in order to take advantage of the high-bandwidth properties of this fiber.

FIBER FABRICATION

After the description of waveguide properties, in the following we will direct our attention to the manufacturing techniques of step-index and graded-index fibers. These types of fibers require fundamentally different fabrication methods. Further refinements and improvements of manufacturing techniques are expected in the future, which will result in a diversification of the fabrication processes.

Suitable Material Systems

Most present fibers are made of glass, consisting of either silica or silicate. Glass waveguides offer the lowest losses presently achievable, which typically range from less than 2 dB/km to several hundred dB/km. For most long-haul applications an attenuation of less than 10 dB/km is desirable. Fibers made of plastic generally have a much higher loss and frequently exhibit more severe degradation with radiation level and time than fibers made of glass. The feasibility of fabricating glass fibers with losses of less than 20 dB/km was predicted in 1966, when losses exceeding 1000dB/km were typical of available materials. This prediction triggered interest in the use of glass as the transmission medium for long-haul optical communications systems. It was responsible also for new efforts in the development of techniques capable of producing low-loss glass fibers. Since the attainment of fibers exceeding the predicted 20 dB/km value in 1970, there has been steady progress in reducing losses even further by advanced fabrication techniques. In the course of these efforts loss minima approaching 1 dB/km have been achieved, at least in laboratory samples, and adequately low dispersion characteristics are also obtainable.

As a result of these continuing research activities, the material development effort has shifted in emphasis from achieving lower loss and lower dispersion characteristics to achieving improved process control and reproducibility of properties.

In general, there are three major material systems used in the fabrication of optical fibers.

Silica System. This glass is composed mostly of SiO_2, with another metal oxide added so that there is a small index difference between core and cladding. The cladding must have a lower refractive index (for waveguide action) and a higher thermal coefficient of expansion (for good bending properties) than the core. Silica is historically the oldest material system used in optical fibers. Furthermore, it has been studied extensively by the semiconductor industry. Silica is insoluble in water and has the lowest known coefficient of thermal expansion.

Silicate System. Silicate consists of silicon, oxygen, and one or more metals with or without hydrogen. The silicate system is superior to the silica system because of its better handling properties; for example, it has a lower melting point than SiO_2. Because many silicate components are soluble in water, they lend themselves to standard purification techniques of crystallization and ion exchange.

Polymer System. This is the only organic material used to fabricate fibers; silica and silicate-based systems consist of inorganic materials. Polymer or plastic materials involve different chemistry considerations than those applicable to inorganic materials. Generally, plastic fibers have considerably higher losses, caused mainly by enhanced absorption; furthermore, absorption minima normally do not coincide with those found in inorganic materials. Further research on plastic materials, however, is expected to yield substantially better performance than presently achievable, although it is unlikely that plastic fibers will exceed the performance characteristics of glass fibers.

The various materials presently used in the fabrication of optical waveguides made of glass are listed in Table 2.5. The most promising materials are tricomponent oxide systems consisting of soda, lime, and silicate, sodium, boron, and silicate, and sodium, aluminate, and silicate. The addition of lime increases the chemical stability of the silicate and hence its durability, but it causes compositional variations. The sodium-aluminum-silicate system will ultimately yield a lower attenuation and allow a greater ease of manufacturing compared to a pure silica-based system because of the lower intrinsic scattering in this material; however, at present it is still difficult to achieve the suitable fabrication methods and to produce the high-purity chemicals which allow the absorption loss levels of sodium-aluminum-silicate to be reduced to those of the silica glasses.

The majority of all optical waveguides are made of a variety of inorganic materials. The most prominent of these are (a) fused or vitreous silica; (b) sodium borosilicate; (c) germanium borosilicate; (d) soda-lime-silicate; and (e) sodium calcium silicate.

In addition to solid dielectric waveguides, fibers consisting of a liquid core (for example, CCl_4) have been suggested. Although these have yielded

TABLE 2.5. Typical Properties of Optical Waveguides Made of Different Materials and by Different Fabrication Methods

Material	Fabrication technique	Typical core diameter (μm)	Numerical aperture (NA)	Loss at 0.85 μm (dB/km)
Alkali-lead-silicate	Clad rod	50–80	0.45–0.55	35–50
Soda-lime-silicate	Clad rod	20–40	0.15–0.25	50–60
Soda-borosilicate	Double crucible	10–20	0.10–0.20	20–25
Silica–GeO_2	Vapor deposition	5–15	0.10–0.15	3–5
SiO_2–P_2O_5	Vapor deposition	10–50	0.15–0.20	1–3

a loss of only 5 dB/km (at 1.28 μm), they are less suitable for practical applications.

Chemical compounds used as typical starting materials in the fabrication of fibers can be divided into solid substances used in melting operations and liquid chemical vapor deposition (CVD) dopants with high vapor pressure. Some of the more common materials are the following: Solid materials used in melting include SiO_2, $(C_2H_5)_4SiO_4$, Pb_3O_4, B_2O_3, As_2O_3, $BaCO_3$, Li_2CO_3, Na_2CO_3, $NaNO_3$, K_2CO_3, KNO_3, and $CaCO_3$. Liquid materials used in chemical vapor deposition include $POCl_3$, $GeCl_4$, and BBr_3.

The addition of oxides to the fiber material generally causes an increase in the refractive index. B_2O_3, however, causes a slight index reduction. Oxides frequently added to glasses are NaO, CaO, and Al_2O_3, usually in combination with SiO_2. These systems yield relatively inexpensive fibers because of the low materials cost, and they are relatively easy to work with because of their low melting points. However, opposing these desirable advantages is the drawback of a high impurity content in silica, which has hindered the achievement of the very lowest attenuation values determined by scattering alone.

The minimization of optical losses requires the fiber materials to have the following principal properties:

Large energy gap. The effective energy gap of the material must be large in order to minimize optical absorption in the red and near-infrared regions due to the tail of the ultraviolet absorption band.

Low transition temperature. The glass transition temperature must be low in order to minimize density variation scattering and to reduce the broadening of the ultraviolet absorption band. Density fluctuations are usually caused by volatile glass components that are partially and randomly lost at high temperature.

Good component matching. In a compound glass, a perfect matching of the dielectric properties of its various components must be achieved to minimize scattering from compositional fluctuations caused by the segregation of some of the components of the multicomponent glass.

Low impurity concentration. To reduce absorption by transition metal ions and hydroxyl radicals whose fundamental absorption lines or harmonics cause significant loss maxima, it is necessary to reduce the content of these contaminants to values near 1 ppb or below.

Considerable efforts have been expended to develop materials for fibers that possess these characteristics and that are thus capable of transmitting signals with low loss and low dispersion over long distances. Because vitreous silica, in its pure form, has the lowest optical loss of any solid or liquid material known, and because typically it contains only a few parts per million of OH radicals, so that the OH absorption bands are relatively weak, most research efforts are directed toward producing this material in a more economic and more reproducible way.

Fused silica, which in its pure form has an exceptionally low refractive index of only about 1.46, can be doped during formation in order to increase its refractive index to about 1.50. Subsequent doping by the diffusion of boron tends to reduce the refractive index slightly below that of pure SiO_2. The resulting borosilicate glass is suitable as a cladding for a pure fused silica core. Therefore, this material is most suitable for producing the best fibers. Fused silica is one material that combines the advantages of a large energy gap with those resulting from a lack of compositional fluctuations because it is a single component. Unfortunately, fused silica has a relatively high melting point compared to other glasses, which leads to an inconvenient and uneconomic fabrication process.

In contrast, soda-lime-silicate, as the simplest of all materials potentially used for the fabrication of optical fibers, has been investigated to a much smaller degree. It contains approximately 70% to 75% silica, 20% sodium oxide, and up to 5% other compounds. Soda-lime-silicate has a significantly lower softening temperature, which permits easier fiber pulling. Furthermore, soda-lime-silicate allows an easy variation of its refractive index, so that core and cladding can be produced from the same material with slightly different compositions. On the other hand, soda-lime glasses normally have a higher attenuation than fused silica fibers, which is caused mainly by transition metal ions, although there appears to be no fundamental reason why the refractive indexes of soda-lime-silicate and fused silica should differ. Ultimately, both will approach the theoretical lower limit given by Rayleigh scattering.

Borosilicate glass is similar in composition to soda-lime-silicate, with a small amount of boric acid (about 5%) added to lower the viscosity without increasing the thermal expansion coefficient. It frequently has a tendency toward phase separation and devitrification, which can be prevented or reduced to a minimum by the choice of an optimum glass composition. For example, doping of the melt with 1.5 mole percent of thallium oxide, as used in the fabrication of some graded-index fibers, has resulted in the reduction or elimination of these problems.

Material Contamination

As discussed above, the principal source of optical attenuation in the visible and near-infrared parts of the spectrum is the presence of transition elements and OH radicals. Furthermore, the oxidation states of individual ionic impurities affects the attenuation level.

Transition Metals. These elements are mainly Cu, Fe, Ni, Co, and Cr, but Mn and V also cause a slight absorption increase in the wavelength range of interest. It is possible to decrease Cu absorption by producing the fiber under reducing conditions, but this results in an increase in the Fe absorption at 1.1 μm. It is thus necessary to optimize the reduction–oxidation (redox) state of the glass in order to achieve the lowest attenuation level either over a broad wavelength region or at the particular emission wavelength of interest.

In practice, the achievement of a low attenuation level by the redox process is further complicated by the fact that the optimized redox states are often unstable unless a buffering agent (for example, As_2O_3) is used; this, in turn, requires careful attention to avoid additional loss due to the presence of the buffer. Buffering agents are also frequently used to control the redox state of elements in order to give the fiber increased radiation resistance.

OH Radicals. The OH absorption maxima cause attenuation peaks at two wavelength regions of interest (0.7 to 0.8 μm and 0.9 to 1.0 μm). Owing to the hygroscopic nature of many of the silica and silicate systems used in the fabrication of optical fibers, it is almost impossible to specify the OH content of raw materials. However, it is possible to dry the appropriate chemicals before use by heating and bubbling dry gas through the materials. This procedure has generally been found to reduce the OH concentration to acceptable levels.

In order to obtain highly purified materials, three conditions have to be met during fiber fabrication: (1) availability of a reliable purification method; (2) reduction of the effects of the surroundings on the purification process to a tolerable level; and (3) availability of a reliable analytical method to determine minute impurity concentrations. It is usually possible to find among conventional purification methods one that is adequately reliable. Examples of suitable methods are recrystallization or growing of single crystals from solutions or melts, and purification by distillation, sublimation, extraction, coprecipitation, ion exchange, chromatography, filtration, or electrochemical methods. The choice of a particular technique depends upon the individual circumstances.

After purification, subsequent contamination of the material by the environment may reduce the effectiveness of the original purification process. This later contamination may be caused either by impurities already present in the surroundings (such as dust), by particles formed during production by mechanical abrasion and chemical reactions (such as side reactions and corrosion processes), or by contamination from the chemicals used in the fabrication process (even when working with closed, clean equipment).

External contamination can be partially avoided by carrying out all manufacturing, packaging, and analyzing operations under clean room conditions, where the air dust count is usually required to be less than 100 particles per cubic meter (in the range of 0.5 to 5.0μm), although this does not reduce contamination from chemicals and mechanical operations. Unless proper care is taken in avoiding process contamination, the use of expensive high-purity starting materials is not justified, since these materials will be contaminated during subsequent processing. Care should be taken during the cleaning procedures. Particularly important is elimination of surface dust on vessels, glass rods, and tubes. Dust-free solvents must be used.

In general, clean room procedures and chemical purity requirements are similar to those common in the semiconductor (particularly MOS) industry.

Much useful information on this subject has been accumulated by integrated circuit manufacturers, who have similar stringent requirements on environmental cleanliness and purity of all applicable chemicals and other materials.

For the analytical evaluation of the impurity content of the glass materials used in the fabrication of optical fibers, a variety of methods is applicable. These include mass spectrometry, absorption and emission spectroscopy, neutron activation, and several others. No one method is suitable for all compounds and all impurities. Specific methods are usually employed owing to minute quantities involved. They generally include instrumental and specific manual chemical techniques.

Historic Fabrication Techniques

In order to manufacture optical fibers that have the optimum characteristics described above, one can employ a number of processes. Each of these fabrication techniques typically offers some advantages that make it superior in a given situation, but most of them also have drawbacks in either technology, performance, or fabrication economy (25, 26). There are five basic manufacturing processes in practical use. They differ conceptually and economically. Generally, there is a trade-off between fabrication economy and quality of fiber properties. The major methods can be broadly divided into several categories.

Older techniques include the rod-in-tube method, the double-crucible method, and the single-material method.

Modern techniques include the outside vapor phase deposition method and the inside vapor phase deposition method.

A brief description and evaluation of each of these techniques will be given in the following.

ROD-IN-TUBE TECHNIQUE

The rod-in-tube technique, shown schematically in Figure 2.25A, is the oldest and the least advantageous method; hence, its use is limited to a few applications where high fiber quality is of secondary importance. Its main disadvantage is that it does not allow the achievement of a low loss. It has all the disadvantages but few of the advantages of the other methods. It will be discussed here mainly because of its historic significance. In the fabrication of a step-index fiber, the core of the desired refractive index is produced in the form of a rod, while the cladding glass of a slightly lower refractive index is made independently in the form of a tube. The rod is then inserted in the tube and the tube is subsequently collapsed at an elevated temperature corresponding to the softening point of the glasses. The resulting clad core is then pulled down and fastened to a rotating drum on which it is wound. The fiber diameter depends upon the drawing speed and the temperature.

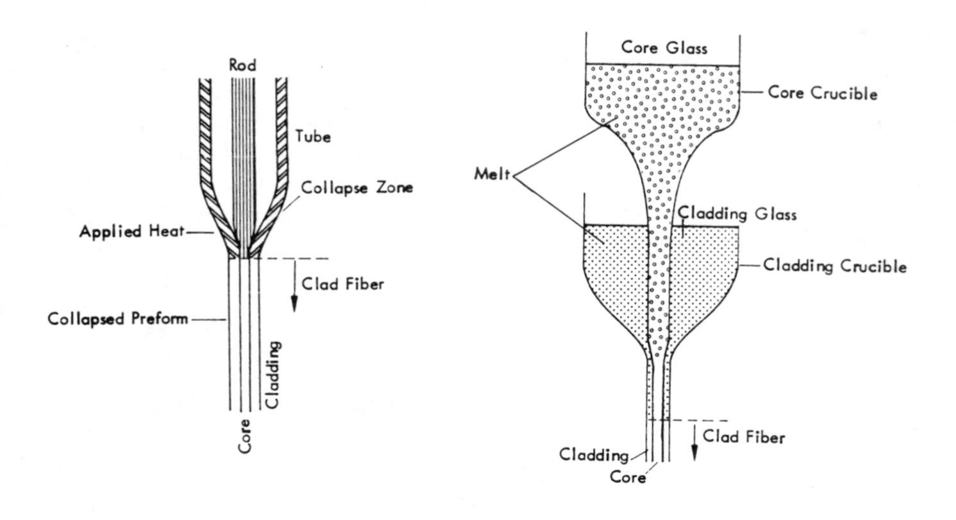

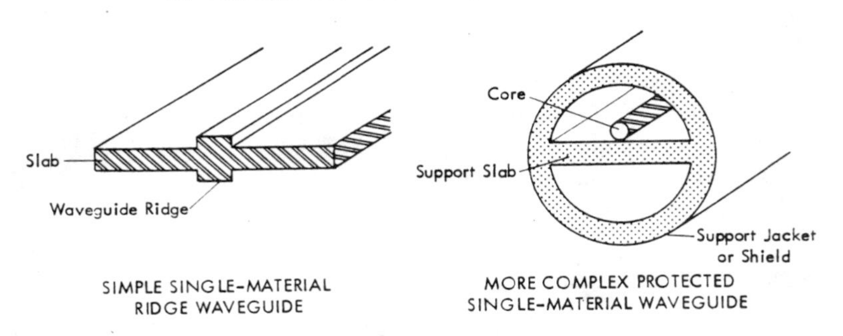

Figure 2.25 Schematic illustration of the dominant historical fiber fabrication processes.

The main disadvantage of the rod-in-tube technique is that imperfections of the inner surface of the tube and the outer surface of the rod result in core–cladding interface irregularities that cause high scattering losses. As a result, rod-in-tube fibers typically have an attenuation in excess of 20 dB/km; this method is not suitable for achieving substantially lower losses. Furthermore, the lack of index control and the problem of glass purity outweigh any economic advantages this method may offer.

DOUBLE-CRUCIBLE TECHNIQUE

For a relatively long time the double-crucible technique has been used to fabricate reasonably satisfactory optical fibers. This method is shown schematically in Figure 2.25B. It avoids several of the drawbacks of the rod-in-tube method—for example, it does not have the core–cladding interface irregularity problem—but it still does not yield low-loss fibers. The lowest losses that have been continuously achieved by this method are about 15 dB/km. In this fabrication technique, the molten core and cladding glasses are held in two concentric crucibles so that the glasses can flow out forming a clad fiber. With core and cladding nozzle diameters of typically 0.5 and 3.0 mm, fiber core diameters ranging from 10 to 20 μm (or $\pm 35\%$ variation) and cladding diameters ranging from 10 to 20 μm (or $\pm 20\%$ variation) have been obtained. The two crucibles are usually made of high-purity platinum; sometimes refractory materials are also employed. In spite of the use of these extremely pure materials, the resulting fiber is usually not sufficiently free of contamination to result in low losses, such as are achievable by the more advanced vapor phase deposition methods.

Furthermore, the simultaneous requirements of purification and maintenance of the purity of the raw materials over periods of time and from fiber to fiber in a large production environment are the most serious drawbacks of fibers produced by the double-crucible technique. Since the shape of the index profile near the core-cladding interface depends upon impurity diffusion, the impurity profiles required for near-parabolic distribution in graded-index fibers are difficult to achieve by this method. On the other hand, the double-crucible technique offers a continuous process by which very long fibers can be pulled. Also, the ratio of possible refractive indexes covers a wide range. This method has the further advantage that it allows wide margins for the core-cladding radius ratios, making it easier to pull monomode fibers with very narrow core and relatively thick cladding.

Frequently, high-temperature operation of the crucibles used in the fabrication of fibers by the double-crucible method causes leaching of trace impurities, mainly Fe and Cu, from the platinum. This effect has been observed particularly in the sodium-calcium-silicate glass system. However, by careful choice of the glass composition in the sodium borosilicate system compatible glass pairs can be found which have viscosities suitable for fiber pulling at a temperature low enough to reduce the leaching rate to a negligible level. Thus the process temperature can be reduced from about 1100°C for sodium-calcium-silicate to about 850°C for sodium borosilicate.

In order to overcome the difficulty of producing the near-parabolic impurity concentration profile of a graded-index fiber by the double-crucible method, the earliest waveguides that did not rely on a step-index profile (for example, the Selfoc [self-focusing] fiber) were manufactured by a modification of the basic process. In this modified double-crucible technique, thermally mobile ions are used to generate a diffusion-controlled parabolic-index fiber with a relatively large index difference between core and cladding. This fiber type has most of the advantages and disadvantages of the more conventional double-crucible fiber, and it is limited to index profiles determined by diffusion processes. It has been fabricated since 1968; a second-generation Selfoc fiber now has a loss of about 10 dB/km at 0.85 μm. This relatively small loss indicates the absence of an appreciable iron content, which in earlier versions of this fiber was largely responsible for a significantly higher loss value. The loss of 10 dB/km is believed to be composed of 3 dB/km scattering loss, 4 dB/km absorption loss due to Cu, 1.5 dB/km due to Fe, and 1.5 dB/km due to Tl, with the ultimately achievable total loss estimated to be about 5 dB/km.

Selfoc fiber is based on sodium borosilicate glass. It has an optimum composition which allows the prevention of the undesirable tendency of the glass toward phase separation and devitrification. The fiber core is doped with 1.5 mole percent of thallium oxide (Tl_2O_3). The glass consists of a combination of thallium nitrate ($TlNO_3$), boric acid (H_3BO_3), sodium nitrate ($NaNO_3$), quartz powder (SiO_2) or synthetic silica (SiO_2), and a small amount of arsenic oxide (As_2O_3). All raw materials other than silica are purified by the combined techniques of ion-exchange separation, solvent extraction, and recrystallization, whereas the silica is made from monosilane by vapor deposition. The refractive index profile of the Selfoc fiber is described by the distribution of the thallium ion concentration. The desired parabolic distribution theoretically can be obtained by the proper selection of the values of the ion-exchange length l_i, the diffusion coefficient D of the thallium ions in the glass at the diffusion temperature, the fiber core radius a_1, and the fiber drawing speed v_d. These properties relate to the profile-determining ion-exchange parameter by

$$K = Dl_i/a_1^2 v_d \tag{2.33}$$

which typically ranges from $K = 0.01$ to 0.1.

SINGLE-MATERIAL METHOD

The single-material waveguide was the first fiber that used only one glass. The attraction of single-material fiber is enhanced by the possibility of using fused silica as the basic material because of its exceptionally low absorption and scattering losses. Since double-material fibers have exceptionally low indexes of refraction (about 1.458), it is difficult in the case of these fibers to find cladding materials that offer both lower index and low absorption coefficient. Hence, the development of single-material waveguides has received some support. The fabrication process of single-material fibers is relatively simple and inexpensive and is based on the assumption that the simplicity of the process should facilitate material purity. In spite of the very-high-purity one-component fused silica glass used in single-material fibers, the best loss values achievable with this method are relatively poor. Furthermore, index profiling is not possible.

Several structures for single-material fibers have been suggested. The simplest single-material optical waveguide consists of a fused silica slab which is enlarged in one cross-sectional area along the entire length of the waveguide (shown in Figure 2.25C). A long, narrow ridge forms the active waveguide and is capable of guiding electromagnetic radiation, preventing it from escaping into the space outside of the slab or even from spreading into the plane of the slab. Although the precise shape of the enlargement is not critical, the number of modes the ridge can support depends on the dimensions of the ridge (thickness and width) in relation to the thickness of the slab. In order to obtain a small number of modes, a small product of ridge thickness and width and a large thickness of the slab are required. If x_r and x_s are the thicknesses of ridge and slab and y_r is the width of the slab, then the number of modes the single-material fiber can support is

$$N_m = \left(\tfrac{1}{2}\,\pi\right) x_r y_r / x_s^2 \tag{2.34}$$

Typical dimensions are a slab thickness of 3 to 4 μm, a ridge thickness of 6 μm for a monomode fiber and 15 μm for a multimode fiber, and a ridge width of about 10 μm. Because the handling of such a waveguide is inconvenient, it is usually enclosed in a tube made of the same material. The outer tube serves only as a shield to prevent contamination of the guiding region by dust or other external influences during handling and to provide a mechanical support of the assembly. In practice, maintaining extreme cleanliness, one inserts a slab of the fused silica into a tube, a narrow rod of the same material is placed on the slab in the center of the tube, and the entire structure is then pulled into a fiber, whereby a high-temperature treatment under proper drawing conditions fuses the different parts together without changing the geometric proportions. This is shown in Figure 2.25C. Because such a structure consists of three separately fabricated elements (core, support plate, and support jacket) it is apparent that the fabrication of a single-material waveguide poses serious difficulties resulting from the need for complex fabrication techniques and the need for maintaining good dimensional and material control during the fabrication process.

VAPOR-PHASE METHOD

The above fiber fabrication techniques meet only some of the requirements of low-loss, high-bandwidth waveguides. Vapor-phase epitaxial (VPE) techniques, however, have the potential to produce fibers of low loss, with index tailoring (arbitrary gradient capability). Also, any desirable ratio of core-to-cladding refractive indexes can be achieved. In addition, the technique is economic (27). Vapor-phase epitaxial techniques are well known in the semiconductor industry and have been developed to a high degree of sophistication. The VPE method is based on the principle of vapor-phase oxidation. In this process volatile gaseous feed materials are delivered to a reaction zone in an epitaxial reactor, where they are heated in the presence of oxygen. The product of this oxidation reaction is a metal oxide soot which is subsequently fused to form a high-purity glass if the materials are carefully selected. The typical reaction taking place is

$$SiCl_4 + O_2 \rightarrow SiO_2 + 2Cl_2$$

Both the outside and the inside vapor deposition process have achieved considerable perfection. These processes allow the fabrication of fibers of the lowest attenuation and of the lowest dispersion (28).

Outside Process. In the outside process, a glass bait is rotated in the flame of a reaction burner while traversing back and forth over the length of the host. Finely divided soots are thus generated by the chemical reaction described above, and a fiber preform is created that is capable of producing several kilometers of fiber. The intended glass composition is controlled by the variation of the various gas flows. Because the gas flows can be adjusted at the end of each traverse, a very fine control of the index gradient can be achieved. The purity of the glass fiber depends mainly on the purity of the feeding materials and is enhanced by the fact that the process is another distillation step. Subsequent steps remove the bait and fire the sample, which sinters the highly reactive soot. This forms a consolidated glass fiber blank from which the final fiber is drawn. Losses ranging from 2 to 10 dB/km and index differences ranging from 0.007 to 0.018 have been achieved by this technique.

Inside Process. The inside process is basically similar to the outside process. Oxidizable vapors are delivered to a reaction zone where they are converted to the corresponding metal oxide soot particles, and the soot is consolidated to form a glass. Also, the gas delivery system is identical to that of the outside process. However, in the inside process the chemical reaction is contained inside a suitable reaction chamber, which ultimately becomes a physical part of the fiber, whereas in the outside process the reaction takes place outside of a reaction chamber. In the inside process the reaction chamber consists of a fused quartz or Vycor tube.

The germania-borosilicate glass for the graded-index core is obtained by simultaneous vapor-phase deposition and fusion onto the inner diameter of a rotating commercial fused quartz tube mounted in a standard glass working lathe; the tube dimensions are typically 12 × 14 × 122 mm. In the inside process the waveguide is thus built from the outside to the inside (hence the name; the cladding glass is deposited first, followed by the deposition of the core glass). With each pass of the O_2–H_2 burner a thin layer of the core glass is deposited along the length of the tube from the vapors of controlled porportions of semiconductor grades of BCl_3, $SiCl_4$, and $GeCl_4$ that flow through the silica tube in an oxygen carrier gas. The silica tubing eventually serves as the cladding. The GeO_2 content of the deposited glass is continuously increased throughout the process in order to obtain a desired graded-index profile. When the deposited layer has reached a predetermined thickness ratio of core and cladding, the temperature is increased to collapse the tubing completely into a solid preform rod. The optical quality and the core–cladding diameter ratio are determined by the deposition step, whereas the circularity of the core depends mainly on the collapsing step. The fiber diameter is usually monitored during pulling by continuous optical inspection and mutual adjustment by varying the speed of the takeup drum while maintaining a fixed preform feed rate and furnace temperature. The soot is generated at the location of the burner flame and collects downstream from it. As the burner passes over the soot, the soot is consolidated to form a clear glass; thus, no separate consolidation step is performed. A preform yields about 3 km of 100 μm diameter fiber.

Compared to the outside technique, in the inside process the collapse step is done at a higher temperature, thereby eliminating inside holes and avoiding thermal shock. In the inside process it is important to have good dimensional control of the starting tube diameter. The inside process offers the significant advantage that almost no humidity (OH) enters the reaction zone; this allows the fabrication of fibers having a loss as low as 2 dB/km. However, the presence of minute amounts of OH still presents a limitation which has to be overcome by an additional expensive and complex process step. In general, the performance of fibers made by the inside process and of those made by the outside process is comparable. The inside process, however, requires a high-silica jacketing. A comparison of the gas flow diagrams of the outside and inside processes is made in Figure 2.26. The structure of the

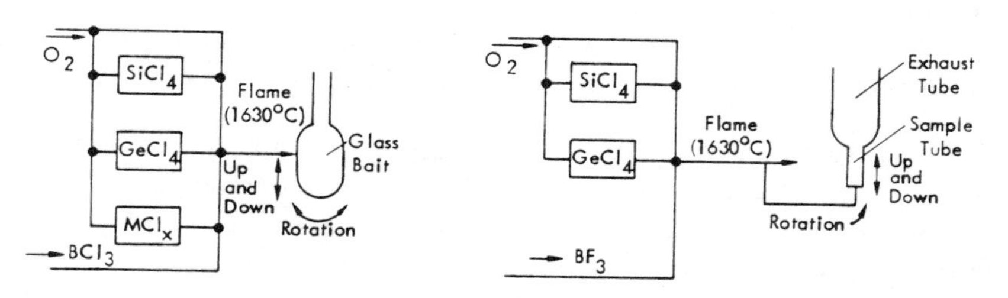

Figure 2.26 Schematic illustration of the present fiber fabrication processes.

resulting fibers is similar. Most current useful waveguides of the step-index and graded-index types use one of the VPE fabrication techniques.

The advantages and disadvantages of the various optical fiber fabrication methods are summarized in Table 2.6. It is evident that the vapor-phase epitaxial process yields the most suitable waveguides; in particular, the inside process represents the most advantageous fabrication method.

FIBER CABLING

Individual glass fibers are generally not thick enough to withstand external forces on their own without suffering axial distortion, mode coupling, and loss. In order to be handled, installed, and used, optical fibers have to be protected to resist the detrimental influences of stress, strain, pressure, torsion, bending, chemical exposure, and other factors. A structure that provides this protection is called a fiber cable. The bare glass fiber is so brittle and its diameter so small (of the order of 100 μm, including cladding) that it has to be surrounded by a series of protective layers. These processes, usually referred to as coating and cabling, frequently include also the addition of one or more strengthening members that consist of either metal wire, plastic, or cord string. The protective jackets must be carefully designed to provide effective protection.

Cabling Requirements

An important requirement for the protective process is that the properties of the bare fibers must not be significantly degraded. With proper care it is possible to achieve a minimum of property variations during the three main stages of the optical fiber fabrication process: bare fiber, coated fiber, and cabled fiber. In order to evaluate the characteristics of coating materials suitable for fiber protection, the mechanical properties of the fiber and their compatibility with those of the plastic have to be investigated.

The fiber is the part of the cable that is most sensitive to both external and internal influences. Hence, proper cable design as well as process improvements and strength-improving coatings must be applied to reduce the amount of cable load that is carried by the fiber.

There are two major inherent problems associated with the packaging of low-loss optical waveguides: fiber breakage and increased attenuation.

Fiber Breakage. Since glass is a brittle material, glass fibers, unlike metal wires, fail without previous noticeable permanent deformation as a result of a tensile component or stress. In theory, the strength of glass is of the order of 700 kg/mm^2, but in practice wide variations are observed owing to the size and frequency distribution of surface flaws. These flaws can cause stress concentrations many times larger than the nominal stress in the same area.

TABLE 2.6. Comparison of Advantages and Disadvantages of Principal Waveguide Fabrication Techniques

Method	Advantages	Disadvantages
Rod-in-tube process	Inexpensive Economical for high-loss-tolerant applications	High scattering loss (≥15 dB/km) Difficult impurity control Difficult index control Difficult dimensional control Problems of potential periodic spatial diameter variations
Double crucible process	Continuous or quasicontinuous process Large index ratio range, although without adequate index profile tuning Allows wide range of core–cladding radius ratios Allows very small core–cladding diameter ratio, thereby offering possibility for small-core monomode fibers	Material purification problem (≥15 dB/km) Raw material composition reproducibility problem Expensive to fabricate Lack of index control Index profile limited by diffusion techniques and lacking versatility
Single-material process	Use of only one material, avoiding the need for separate cladding layer Possibility of using fused silica, the best optical material	High loss due to contamination (≥20 dB/km) Lack of index control Potentially complex cross-sectional structure and hence inherent fabrication difficulties and resulting expenses
Outside LPE/CVD process	Potential for wide range of index profiles, with extremely fine tuning Potential for low attenuation (<10 dB/km) Potential for low pulse dispersion (<1 ns/km) Potential for fibers of large length (≥1 km)	Incomplete elimination of OH radicals in glass Fiber consolidation is a separate fabrication step
Inside LPE/CVD process	Lack of the need for a separate consolidation step Even more economic fabrication than outside price for fibers Same other advantages as outside process	Need for high silica jacket

Furthermore, flaws can grow and propagate, eventually causing failure at stress levels considerably lower than the theoretical breaking stress.

Fiber Attenuation. Increases in the optical losses of a waveguide result from small lateral forces acting on the fiber that cause so-called microbends and that are defined as microscopic perturbations of the fiber geometry. The small-amplitude axial distortions produced by such forces are capable of causing significant radiation losses which produce an excess signal loss in the packaged assembly.

Fiber preparation, the fiber handling techniques employed in the packaging operation, and the design of the package itself determine the extent to which final package performance is affected. Very small external forces exerted on a fiber can cause lateral deformations, mode coupling, and increased fiber attenuation. For example, minute irregularities in the machined surface of a metal drum suffice to cause substantial loss on fibers wound on this drum with only a few grams of tension. This is particularly important in view of the fact that the very lowest fiber losses that have been measured were achieved on fibers in which no mode coupling occurred and which were free of all external forces. Maintaining these low losses in a cable requires better fibers and, more importantly, effective jackets designed to provide optimum shielding against external forces. Three major considerations enter the choice of materials used for jacketing and cabling of fibers: (a) adhesive properties of glass and plastic; (b) Young's modulus of glass and plastic; and (c) coating method.

Fibers coated with plastic having polar or functional groups in its molecules (for example, nylon-12 or phenol resin) have a high tensile stress and large relative elongation, and they possess a small allowable radius of curvature. Fibers coated with a thermosetting plastic during pulling have a high mechanical strength. Also, thermosetting plastic is capable of preventing cracks in the fiber surface from propagating. On the other hand, fibers coated with a thermoplastic material exhibit only marginal mechanical strength, even when the fibers are coated during pulling. The tensile strength of fibers coated with thermosetting plastic increases significantly if the thickness of the plastic is less than 5 μm, while the strength of fibers coated with thermoplastic increases much less. The increase in tensile strength in the case of thermosetting plastic coating is due to the high Young's modulus and the strong adhesion of the plastic to the fiber, which prevents the occurrence of surface cracks.

It is usually advisable to improve the handling capabilities of fibers protected by a thin plastic coating layer (whose thickness is typically of the order of 5 to 10 μm) by the application of a second, much thicker coating layer (typically of a diameter of about 250 μm). Such a second coating frequently consists of nylon or of EVA (ethylene vinyl acetate). Other materials that have been investigated include low- and high-density polyethylene, polyethylene –terephthalate, and polyurethane. Again, mechanical and thermal considerations prevent an injudicious choice of materials. It is desirable for the second coating to have limited adherence to the first coating because frequently the fiber length is reduced due to thermal expansion of the plastic during the jacketing process and after cable installation.

The selection of a suitable plastic material for fiber jacketing normally proceeds as follows. First, attention is paid to the adhesive properties of the plastic under various thermal conditions typically encountered. Second, the interfacial interaction between plastic and fiber is evaluated, since there must be a uniform stress on the fiber. Third, the plastic is required to have a high Young's modulus, which allows a thin coating layer that is still being able to prevent the propagation of cracks. Fourth, the plastic must have good thermal characteristics. It must allow expansion and contraction, and it must possess excellent aging and water-resistant properties. Furthermore, when the coating thickness is very large, the tensile stress produced by thermal expansion of the plastic must not cause the fiber to break.

After the application of the coating layers, fibers are usually covered by an additional thick protective material. This process is designed to provide several functions, of which the major ones are (a) protection against mechanical abrasion; (b) improvement of the chemical resistance; (c) improvement of the tensile strength; and (d) protection against lateral mechanical stress.

Plastic coating usually results in a slight increase in fiber attenuation. This is believed to be caused by microbends introduced during processing in the form of localized stress and pressure. The small lateral deformations of the fiber within the plastic coating lead to coupling of energy into radiative modes and to consequent fiber loss. Such deformations may be caused by local variations in the coating thickness or by particles trapped at the interface. The small increase in loss after cabling is due to a similar compressive effect after sheathing of the coated fibers. Coating typically causes a loss increase by about 1.0 to 2.5 dB/km, and cabling may cause an additional loss increase by 1.5 to 2.5 dB/km over the full spectral range, depending upon the care with which the coating and cabling processes are achieved. Typical results are summarized in Table 2.7.

Examples of a few fiber cables are shown in Figure 2.27 (29). The fibers consist of a strength member of a rod in the cable center to pull pieces a maximum of 0.5 km long through a conduit. They are placed between a few layers of concentric plastic of slippery surface in order to allow an easy fiber move-

TABLE 2.7. Illustration of the Influence of Cabling on Waveguide Attenuation at Different Wavelengths

	$\lambda = 0.85\ \mu m$			$\lambda = 1.05\ \mu m$		
		Increase			Increase	
Fiber condition	Attenuation (dB/km)	dB/km	%	Attenuation (dB/km)	dB/km	%
Bare	7.3	—	—	6.9	—	—
Coated	9.7	2.4	33	9.0	2.1	30
Cabled	11.5	4.2	58	11.1	4.2	61

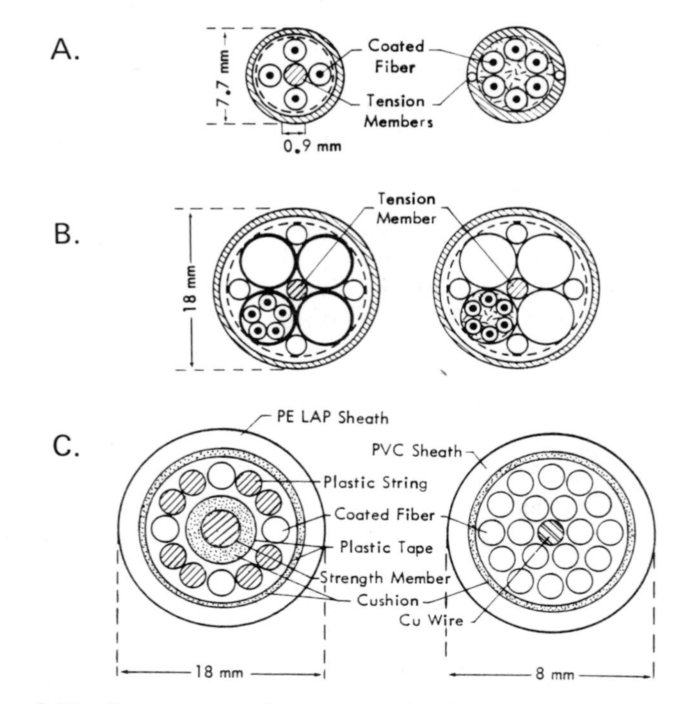

Figure 2.27 Cross section of examples of underground and indoor fiber cables.

ment during cable bending without breakage. The cables have a mechanical strength compatible to that of conventional metal telephone cables.

The mechanical stability of fibers and cables is normally evaluated in stress fatigue and environmental tests (30). For example, mechanical stress tests of cabled fibers encompass bending tests (1000 bends of 180° over rolls of five times the cable diameter); impact tests (200 impacts of 0.5 kg weights from heights of 30 cm, or 0.25 kg from heights of 60 cm); tensile strength tests (up to 100 kg in increments of 12.5 kg for 1 min); twist tests (± 90° from center, 160 cm segment, 1000 twist cycles, at −55°, +25°, and +70°C), and vibration tests. Environmental tests include salt water tests; stagnant water tests (where the water includes various kinds of garbage normally found in a trench); humidity tests; temperature cycling tests (between −55° and +85°C); high-temperature tests (600°C); and high-pressure tests (1.4×10^3 atm), corresponding to the pressure at the ocean bottom. Degradation after combined tests (environmental and stress) may range up to a factor of 5. Fatigue studies (time–stress in high humidity) indicate that extrapolation from short-term strength to allowable stress for long times with very low probability of failure is practical with a derating factor of 3 to 5. This means that a fiber capable of

3.4×10^3 atm for times of less than 1 min should withstand 10×10^3 atm for 10 years with less than 10^{-6} probability of failure.

Environmental Tests

The mechanical properties of optical fibers can be generally and broadly divided into two major categories, which influence the choice of tests to perform to establish reliable operation under expected ambient conditions (31, 32). These are dynamic strength and static strength.

Dynamic strength includes tensile and bending strength and is a function of the operating conditions of the fiber. Static strength includes intrinsic and extrinsic factors that may affect the useful fiber lifetime.

The dynamic and static properties are determined by the characteristics of the fiber itself, by the cable that surrounds the fiber for environmental protection, and by the operational conditions.

The evaluation of bending strength is of importance because during installation and in service the cable will be exposed to bending and will remain in that position for a long time while being exposed to a variety of other influences that also tend to be detrimental to its longterm reliability.

Laboratory tests for measuring fiber and cable strength are carried out on sections of finite length, usually of the order of a few meters, whereas actually installed cables typically have lengths of 1 km or more. Therefore, small-length measurements must be extrapolated to large-length conditions by using Weibull statistics. The accuracy of these extrapolations is a function of the accuracy of the statistical parameters being measured and of the length ratio of actual to tested cable.

Calculations of the tensile strength of optical fibers by Weibull statistics indicate that the short-term tensile stress S_T of an optical fiber is related to its length L and to the stress duration t through the failure probability F by

$$F = 1 - \exp\left[-\left(\frac{S_T}{S_{T0}}\right)^m \left(\frac{t}{t_0}\right)^b \left(\frac{L}{L_0}\right)\right] \tag{2.35}$$

where the subscript 0 refers to normalized terms, m is the Weibull slope of failure probability versus stress at constant time-to-failure (corresponding to the inverse of the standard deviation), and b is the Weibull slope of failure probability versus time-to-failure at constant stress. This equation can be used to predict the strength of long fibers at constant failure probability:

$$S_{T1}/S_{T2} = (L_2/L_1)^{1/m} \tag{2.36}$$

This equation assumes that the strength distribution of both short and long fibers can be characterized by a single slope m. For fibers in excess of a length of about 10 meters, the Weibull slope approaches a constant value and can be approximated by $m = 4.5$.

The measurement of bending strength is simple in principle but difficult to interpret in practice as a consequence of the nonuniform stress distribution through a fiber cross section. The surface area which sees most of the maximum bending stress is typically an order of magnitude smaller than that of pure tension; hence, fewer surface flaws experience maximum stress. Also, unlike the gradual buildup of tensile stress, in bending, the fiber is subjected to the maximum stress at its surface in a step manner, potentially leading to bending fatigue if the stress duration is not controlled during testing. Furthermore, there is a finite possibility of inflicting damage to the fiber surface because the fiber is in intimate contact with a form, a danger not present in tensile testing.

The maximum bending stress S_B on the surface of the fiber of radius a_2 intimately in contact with a form (such as a mandrel) is given by

$$S_B = 2Ya_2/D_m \tag{2.37}$$

where Y is Young's modulus and D_m is the mandrel diameter.

Static fatigue, another parameter of interest to the cable designer, is the slow growth of existing flaws (such as a crack) in the presence of both net tensile stress and moisture. The presence of all three factors—flaw, stress, and moisture—is necessary for fatigue to exist. Flaws are normally present in most fibers, the stress is usually highest at the fiber surface, and exposure to moisture is also usually higher at the surface. These are typical conditions which are inductive to static fatigue. Within a limited range, the crack velocity obeys the relationship

$$v_F = w_x K^M \tag{2.38}$$

where w_x is a constant that will be determined below, and K is defined as a stress intensity factor given by

$$K = w_y \, S_T \sqrt{l_F} \tag{2.39}$$

where w_y is a constant ($w_y = 1.25$ for semicircular flaws normal to the fiber axis) and l_F is the crack length. The exponent M is the stress corrosion susceptibility constant; it is a material property that depends upon fiber composition, geometric perfection, and test conditions. Its magnitude will also be estimated below.

Equation (2.38) indicates that the flaw velocity increases with stress and crack length, depending upon the magnitude of the exponent M. When the crack length reaches a critical value $l_{F(c)}$, the flaw velocity begins to deviate from the above relationship; in this case, unstable crack propagation is observed instead of the previously found slow and stable flaw growth. Above, this critical value given by $K = K_c$, the crack velocity rapidly reaches its terminal value which is one-third of the longitudinal wave velocity and which leads to catastrophic fiber breakage. The fracture strength that corresponds to the critical crack velocity is approximately

$$S_F \approx w_z (10\, l_c)^{1/2} \tag{2.40}$$

as empirically determined. The material constant w_z depends upon the fiber glass type and is as follows:

Glass type	w_z (10^6 N/m$^{3/2}$)
96% silica	1.83
soda-lime	1.84
fused silica	1.89
borosilicate	2.00

The stress corrosion susceptibility constant M can be empirically found from the relationship between strength and its derivative, the stress rate. For aged and abraded fibers of constant length this relationship is given by

$$\left(\frac{S_{T1}}{S_{T2}}\right)^{M+1} = \frac{dS_{T1}}{dt} \bigg/ \frac{dS_{T2}}{dt} \tag{2.41}$$

The slope of this curve of strength versus stress rate plotted on log–log paper yields $M+1$, from which M can be determined. It typically ranges from 20 to 40. It is also related to the exponents m and b of Equation (2.35) by:

$$M = m/b \tag{2.42}$$

Knowledge of the term M allows the determination of the constant w_x from the empirical equation

$$w_x = 10^{-(8.94 + 5.41M)} \tag{2.43}$$

The above analysis indicates that it is possible to determine the lifetime of an optical fiber that is subjected to an applied stress S_{Fa}. Under the assumption that S_{Fa} is less than the fracture strength S_F^* (i.e., $S_{Fa} < S_F^*$), the fiber lifetime can be obtained by

$$t_f = 2\,\frac{1/K_i^{M-2} - 1/K_c^{M-2}}{w_x\, S_{Fa}^2\, w_y^2\, (M-2)} \tag{2.44}$$

where K_i is the initial value of the stress intensity factor. Because $1/K_c^{M-2}$ is small compared to $1/K_i^{M-2}$, the above equation can be simplified to

$$t_f = 2K_i^{2-M}/w_x\ (M-2)w_y^2\ S_{Fa}^2 \tag{2.45}$$

If K_i is given in $N/m^{3/2}$, the above equation yields the fiber lifetime in seconds. It is rather sensitive to the accurate determination of the various terms of which it is composed. Hence, substantial errors may be encountered in practice.

The term M depends upon the flaw size and therefore is not an intrinsic property of the material. Because fiber life estimates based on large values of M are considerably larger (by several orders of magnitude) than those based on a more realistic value obtained from fiber specimens rather than bulk glass, it is necessary to use M values that are obtained from actual fibers.

We can thus summarize the analysis of fiber failure prediction as a function of mechanical stress and time (33). If S is the applied stress, S_0 the nominal stress, and S_{min} the possible lower limit of the detrimental stress, then $S-S_{min}$ is the effective stress. Then the accumulated probability of rupture in fibers is given by

$$P(S,t) = 1 - \exp\left[-S_T^m\ (t/t_0)^n\right] \tag{2.46}$$

where m and n are exponents characteristic of the material, S_T is the stress-related term, and t/t_0 is the time-related term. Hence,

$$S_T = \frac{S-S_{min}}{S_0/S_{min}}$$

Furthermore, t_0 is a reference time. Typically, we can assume $t_0 = 1$ min, $m = 3$, $n = 0.15$, $S_0 = 800\ N/mm^2$, and $S_{min} = 50\ N/mm^2$ (with a usual range of 40 to 80 N/mm^2). If no shear stress is encountered, then the lifetime of fibers can be determined, if in the above equation P is kept constant, from

$$(S-S_{min})^m\ t^n = \text{const} \tag{2.47}$$

These relationships are shown in Figure 2.28.

An extension of this work can be used to arrive at predictions of large sample sizes of fibers and to obtain statistical distributions of fiber characteristics. This, in turn, allows the generation of theoretically predictable life test data and a realistic estimate of the expected fiber lifetime. This is particularly important in communications links that require a nominal lifetime of several decades, in the absence of actual life tests of this duration. The reliability prediction indicated above has still to be modified to include the realities of manufacturing, laying, and operating cables, which introduce mechanical stresses that include not only pulling stresses but also a combination of pulling, pressing, bending, and shear stresses. Regular shear stresses in a cable as a result of torsion, even if low, may increase the breaking probability above the upper

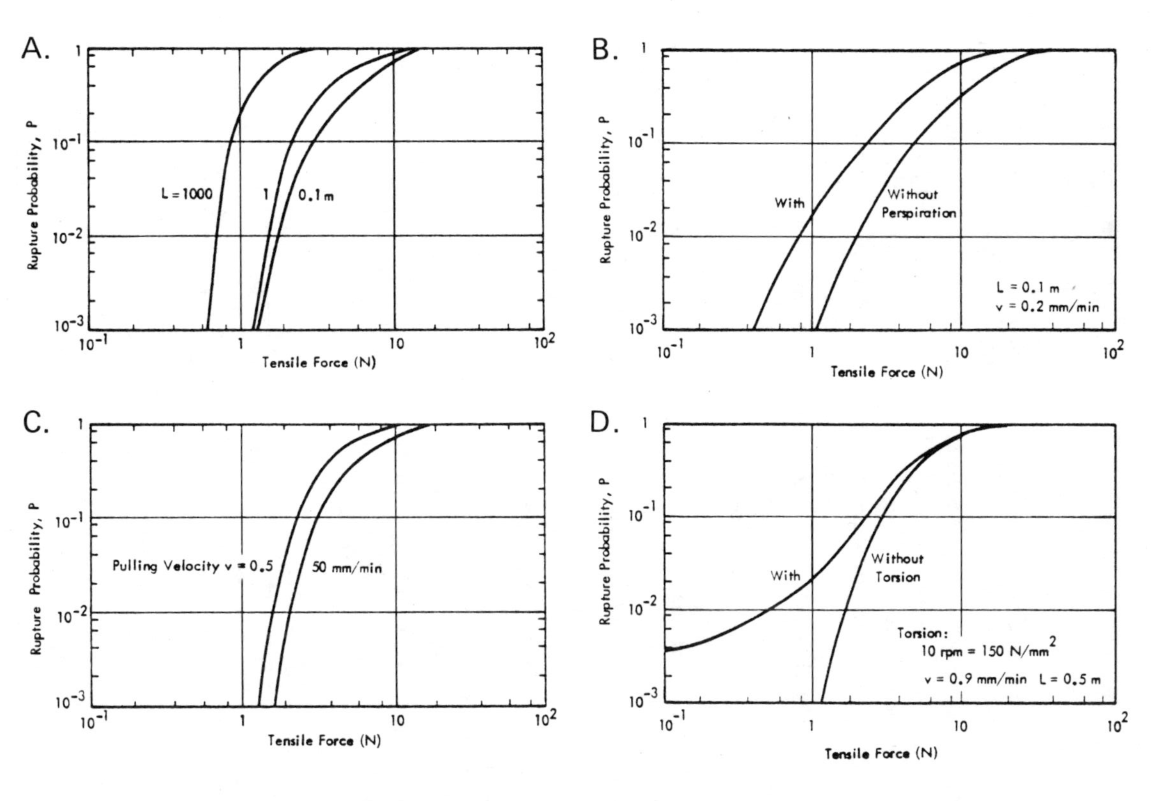

Figure 2.28 Examples of the influence of environmental forces on fiber rupture characteristics.

admissible limit, which for cable plants is of the order of 10^{-6} to 10^{-5} for usual test samples.

In these considerations the minimum stress S_{min} plays an important role, even if the shear stress is negligible. The allowable stresses evident from Figure 2.29 are so low that it is theoretically impossible to keep these values during processing, laying, and operation of the cables. Only a sufficiently high value of S_{min} will permit fibers to withstand the mechanical stresses during these conditions. Hence, all fibers have to be mechanically protected as soon as they have been formed, since the fiber characteristics are usually determined by optical, rather than mechanical, requirements. Mechanical requirements are satisfied mainly by the finish and protection of the fibers. However, the mechanical enclosure can protect only the built-in fiber strength. Hence, the intrinsic permanent axial stresses of a fiber (pressing stresses outside and pulling stresses inside, which are usually considered to be detrimental from an optical standpoint) may be advantageous from a mechanical standpoint and therefore should be optimized.

Other mechanical and environmental test data will now be summarized. The higher the test velocity, the higher the rupture forces and hence the lower the rupture probability. The rupture forces in an atmosphere of 80% humidity are up to 50% lower than in an atmosphere of 35% humidity. Aging of samples with temperature cycling (45 6-hour cycles, +60°C/−30°C) does not affect the rupture probability. A significant increase in rupture probability is observed as a result of touching the fiber by hand. Tensile tests on fibers exposed to torsion show increased rupture probabilities; there is no lower limit for torsion stresses that would result in zero rupture probability.

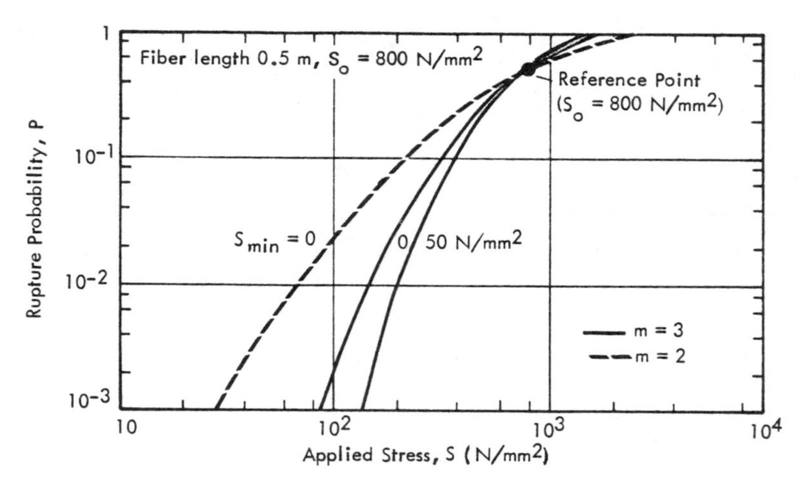

Figure 2.29 Examples of the influence of stress on fiber rupture characteristics.

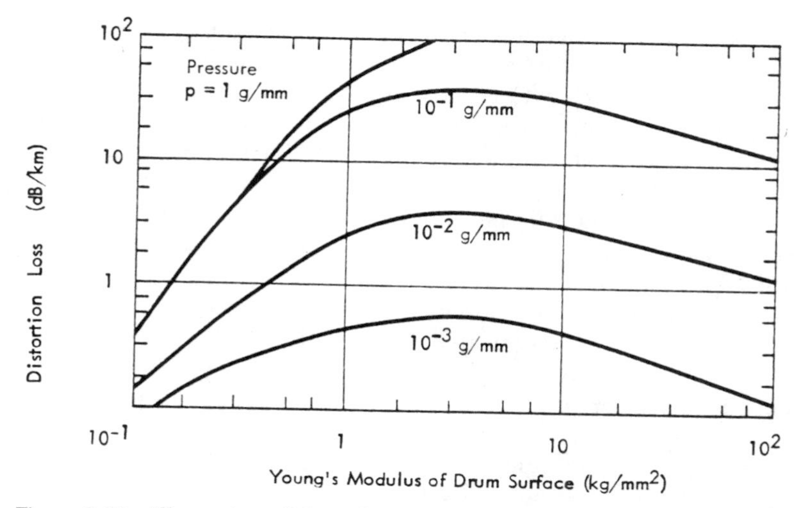

Figure 2.30 Illustration of the relationship between fiber attenuation and Young's modulus of the drum surface for different applied mean lateral forces.

Although the forces to which a fiber is subjected in a cable are difficult to calculate, the sensitivity of the fiber to such forces can be estimated by winding it on a drum with minute surface irregularities. An example of the influence of drum Young's modulus and pressure on the fiber loss is given in Figure 2.30. The fiber has a core diameter of 80 μm, a cladding diameter of 120 μm, a relative index difference $\Delta = 0.005$, and Young's modulus of 7000 kg/mm^2; the drum has a surface roughness (standard deviation) of 1 μm and a correlation length of 1 mm. If, for instance, the tensile force is $F = 10$ g and the diameter of the drum is $D = 20$ cm, corresponding to $p = 2F/D = 1.0$ g/mm, then the distortion loss may be as high as 40 dB/km, depending upon the Young's modulus of the drum. For low pressure, the excess loss decreases with increasing rigidity of the drum surface (large Young's modulus), as the fiber ceases to conform to the irregularities of the surface. For low rigidity of the drum surface (soft drum, small Young's modulus), the loss is also reduced, independent of pressure, since the fiber sinks into the surface and smoothes the irregularities. Thus, both soft and hard surfaces have a tendency to decrease the excess loss for a given pressure; the effect of a surface, however, depends strongly on the pressure applied.

Consequently, since the excess loss should usually be below 0.5 dB/km, an extremely soft drum surface has to be employed whose Young's modulus does not exceed 0.1 kg/mm^2, regardless of the mean lateral force. Typical winding forces caused by the pulling operation itself or applied in rewinding operations range from 10 to 100 g. Assuming a typical drum radius of 10 cm, this causes a mean lateral force ranging from 0.1 to 1.0 g/mm. Loss increases ranging from 1 to 100 dB/km as a result of drum storage are thus common. Therefore, it is necessary to reduce the winding force drastically. This however, has economic consequences because the operation time associated with a slow pulling and winding is longer. It is also necessary to use a very soft drum surface. An increase in the index difference of the fiber (or the numerical aperture) also results in a smaller excess loss.

The plastic jacket design is also influenced by the applied pressure. Usually fibers are organized by bundles and pressed together by binding or sheathing forces, by cable deformations, and by pressure on the cable once it has been placed. Assuming that all fibers in a bundle have plastic jackets and that the fibers have the same characteristics as those given in the previous example, the distortion loss as a function of jacket radius and jacket characteristics can be established.

This is shown in Figure 2.31, where four jacket types are analyzed. These are: a jacket made entirely of a soft material (S); a jacket made entirely of a hard material (H); a jacket made of a hard inside and soft outside material (H/S); and a jacket made of a soft inside and a hard outside material (S/H). From these data it can be concluded that the choice of the material is determined by pressure if the jacket is made from one material alone. The decrease of the loss contribution with increasing jacket radius in the case of the hard jacket is a result of the increase in stiffness. The corresponding increase experienced by the soft jacket is negligible. Generally, a thicker jacketing tends to reduce the excess loss, and a soft shell combined with a hard interior jacket

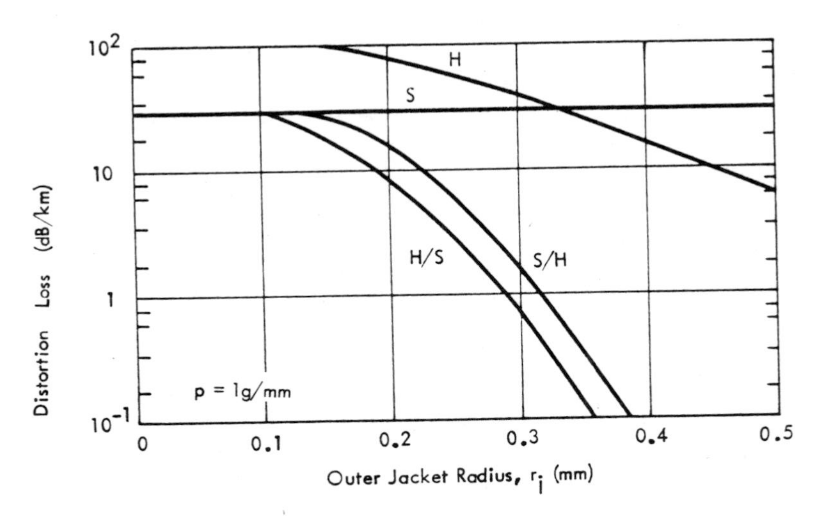

Figure 2.31 Illustration of the relationship between fiber attenuation and the jacket radius for different values of jacket hardness.

results in the lowest excess loss. For example, assuming an outer jacket radius of 250 μm (corresponding to a jacket diameter of 0.5 mm) and a mean lateral force of 0.1 g/mm, a soft jacket causes about 20 dB/km distortion loss and a hard jacket adds about 5 dB/km, but a soft interior and hard exterior jacket combination adds only about 1.5 dB/km and a hard exterior and a soft interior jacket combination adds only 0.7 dB/km. The excess loss of the hard-shell structure becomes negligible when the modulus of the inner jacket is reduced to zero. This implies that the protection provided by a stiff shell that surrounds the fiber in a loose way without any material in between is desirable. However, in practice there are other forces present in addition to those laterally active outside the shell (for example, forces that press the fiber against the inside jacket wall in a cable bend). These forces may increase the distortion loss of a loosely jacketed fiber; they are not analyzed here. Reinforced jackets usually allow the achievement of cable properties similar to those of hybrid jackets (H/S and S/H). The advantage of the reinforced jacket is its anisotropy, which combines stiffness with lateral compressibility. Reinforcements may consist, for example, of strong, fine fibers running parallel or slightly stranded to the optical fiber embedded in a relatively soft jacket material.

The above discussion allows us to draw the following conclusions in regard to the effectiveness of jacketing:

Fiber Properties. The effectiveness of the jacket is a strong function of the characteristics of the fiber to be protected, particularly of the refractive index difference. The distortion losses of both the jacket and of the unprotected fiber are a function of Δ. Hence, the total distortion loss may typically vary as Δ^{-2} for a fiber without jacket and as Δ^{-6} for a fiber with jacket.

Jacket Properties. For equal jacket diameters a soft plastic jacket is usually superior to a hard plastic jacket, except when refractive index difference and lateral forces are small.

External Forces. The fiber jacket must have high flexural rigidity and good lateral compressibility. Below a specific retention length the jacket absorbs irregularities impressed from the outside, whereas longer irregularities deform the fiber and can lead to a distortion loss if they comprise spectral components in the vicinity of the critical wave number of the fiber.

In practice, during fabrication and installation, cable forces are likely to be stronger than those encountered on a storage drum and may necessitate fiber protection by the more expensive hybrid jacketing technology or by reinforcement.

Mechanical stress on optical fibers may lead to increases in bandwidth and attenuation due to mode mixing within the fiber as a result of statistical microbends. Therefore, bandwidth and attenuation involve a tradeoff when mechanical stress is present. From the standpoint of bandwidth, a certain amount of stress and hence mode mixing is advantageous, especially with long fibers (1 to 5 km), because in long fibers the bandwidth increase is larger than in short fibers. The mechanical forces that cause mode mixing, however, should be moderate and equally distributed along the fiber. On the other hand, from the standpoint of fiber loss, mechanical stress is not desirable because the occurrence of microbending may be appreciable. Most of the influence of stress on loss is observed below about 10 kg/mm^2. Fibers with a small numerical aperture are much more susceptible to stress influence than those having a larger numerical aperture. Coating of the fibers by an encapsulant generally results in significantly reduced loss increases due to microbends.

Mode mixing can reduce mode dispersion but not material dispersion. Hence, in good fibers the possible bandwidth increase is limited, whereas in poor fibers, which have a slightly distorted graded-index profile, mode mixing may excite some of the unwanted modes (which are in this case of different velocity and which can also be propagated, contrary to the situation in fibers of perfect graded-index profile) and hence reduce the bandwidth. Thus, step-index fibers generally display more significant stress-induced improvements in bandwidth than graded-index fibers (34).

Refractive index variations of plastic materials are more sensitive to mechanical and thermal stress than those of silica and optical glass (35). In general, it has been found that changes of transmission loss and pulse width are approximately proportional to the square root of the amount of the stress. Typical test results for temperature variation (ΔT) and weight variation (ΔW) are summarized in Table 2.8.

Normally, fibers are arranged in bundles in order to take advantage of the relatively efficient packaging concept; a cable containing only a single fiber is not considered to be an economic packaging approach because of its poor jacketing and cabling utilization. The hexagonal packaging concept is the most popular approach because of its high packing fraction (the ratio of the active core area to the total bundle area).

TABLE 2.8. Examples of Increase in Waveguide Attenuation and Dispersion (per °C or kg) Due to Temperature or Mechanical Stress

Experiment	Fiber length (km)	Test conditions	Transmission loss $\Delta\alpha$ (dB/km)	Dispersion $\Delta\tau$ (ns/km)
Temperature increase		≤120°C		
	1.0	On Al drum	$+0.5\sqrt{\Delta T}$	$+0.5\sqrt{\Delta T}$
	0.7	Free hanging	$-0.2\sqrt{\Delta T}$	$+0.03\sqrt{\Delta T}$
Pressure	0.8	≤12 kg, on drum	$+0.4\sqrt{\Delta W}$	$+0.3\sqrt{\Delta W}$
	0.6	≤2 kg/cm^2, free hanging	$-0.5\sqrt{\Delta W}$	—
Pulling	0.8	≤0.4 kg	$+3.0\sqrt{\Delta P}$	$+2.0\sqrt{\Delta P}$

If bundles are used as parallel conductors excited from a common source (thus carrying identical information), bundle terminations are generally based on a scheme in which the individual fibers are held in a closely packaged array to maximize coupling efficiency. In hexagonal close coupling, the packing fraction is only 42% for a 85 μm diameter core and a 125 μm total fiber diameter. Thus the input loss due to the packing fraction is about 4 dB.

FIBER–FIBER CONNECTIONS

The coupling of individual segments of fibers—either at the source or receiver ends where short sections of fibers are coupled to the long fiber transmission line, or during splicing while being installed, or in the connection of terminals—requires considerable attention in order to retain the relatively low fiber losses. This is particularly important in monomode fibers of extremely small core diameter. Fiber connections in general can be classified as follows: (a) connections between optical cables (fiber–fiber connection); (b) connections between optoelectronic components and the fiber ends as an extremity cable (component–extremity connection); (c) connections between the fiber cable and the extremity cable (fiber–extremity connection). The last two connection types are less problematic than the first type; satisfactory fiber–fiber connections are much more difficult to achieve.

Connection Requirements

In addition to the characteristics of source, detector, and fiber, the efficiency of the coupling mechanisms between source and fiber, between fiber and detector, and between individual fiber segments affects the optical power available at the fiber output. Thus coupling or connecting of optical power from one system component to another is one of the important considerations affecting an optical communications system.

Although generally fibers of any length can be manufactured by a suitable continuous fabrication method, it is usually not desirable to produce fibers whose length exceeds 1 km. This is directly related to the convenience with which cables can be installed in ducts, where installation personnel using metal cables have found the most convenient cable length to be about 0.3 km. Thus a length increase by a factor of 3 for optical cables compared to metal cables is considered to be the maximum tolerable in practice.

Since one of the principal advantages of optical communications compared to electronic communications is the potential for wider repeater spacings (roughly 10 km versus 1 km), a maximum optical cable length of 1 km would obviously not take adequate advantage of the better characteristics of optical fibers. Hence, it is necessary in applications that require cable lengths in excess of 1 km to splice fiber segments or to use connectors. These fiber connections are also required in other applications where only a small fiber length is available in spite of the need for longer sections. In addition, fiber breaks that occur during installation or during maintenance require splicing of individual segments.

Fiber-to-fiber connections may be temporary or permanent. Temporary connections are achieved by coupling and through the use of connectors. Most fibers are fabricated in 1 km sections, although advanced long-haul systems allow fiber lengths far in excess of 1 km; permanent low-loss connections are usually required during installation. The technique used for this type of connection is splicing, which normally results in connection losses far below 1 dB, typically of the order of 0.1 dB.

Splicing and connectors introduce additional losses into an optical system. These losses can be a significant factor in system design (36). If, for example, the splicing or connector loss in a fiber whose loss is 5 dB/km amounts to 0.5 dB, then the connection has an equivalent line length of 0.1 km. Therefore, the effective potential repeater spacing will be decreased due to the connection by 0.1 km for each connection on the line. Hence, the need for low-loss splices and connectors is apparent.

In a numerical example, assuming a repeater spacing of 10 km and a length of individual fiber segments of 1 km, there are nine fiber–fiber connections, two component–extremity connections, and two fiber–extremity connections. Hence, the losses introduced by these connections may become substantial. System considerations usually limit the total loss resulting from connectors to a maximum of 10 dB for a one-repeater system. In the given example, in which 13 connections are used, a maximum loss of 0.75 dB per connection would be tolerable. However, anticipated breakage and subsequent splicing will dictate inclusion of a reasonable safety margin, so that it is advisable in system design not to exceed 0.5 dB per connection. In practice, an average typical extrinsic insertion loss of 0.3 dB per channel and connection of a multifiber can be considered to be routinely achievable.

In monomode fibers whose diameter is typically less than 5 μm, connection losses at a fiber–fiber connection may result from either a separation of the fiber ends, a displacement between the axes of the fibers, or an angle between fiber axes. Assuming two nominally identical fibers, a loss per coupling location of 0.5 dB is experienced if the separation between fiber ends is about $5a_1^2$, if the displacement between axes is $0.5a_1$, or if the angle between axes is $1.5/a_1$, where a_1 is the fiber core radius (in μm). For example, the 0.5 dB loss per coupling is obtained in the joining of two fibers of 2.5 μm diameter if their separation is 8.0 μm, if their misalignment is 0.5 μm, or if the angle between the fiber axes is 1.2°. Thus, extremely accurate fiber joining is required in the case of monomode fibers.

Splicing

Several methods for the efficient splicing of fibers have been suggested. The more promising splicing techniques are characterized by low loss and ease of achieving a permanent and reliable connection in an economic manner. A number of tools have been developed for field use which provide a fast and effective splicing connection.

The Bell Laboratories molded plastic splicing technique, for example, is an improvement over most other known fiber connection methods (37, 38). It allows an average loss of less than 0.1 dB. This is particularly important with fibers whose loss approaches the theoretical minimum, which is close to about 1 dB/km. Whereas most other methods have drawbacks resulting from difficulties associated either with the preparation of the fiber ends or with the mechanical alignment of the previously prepared ends, the molded plastic technique overcomes these difficulties by avoiding the handling of individual fibers, so that no problems of fiber end preparation and mechanical alignment are encountered. The technique is based on dissolving of the plastic coating over a short region to expose the individual fibers and placing the region in a mold, whereby the fibers are held accurately in position by a thin spacer plate. Subsequent molding of a suitable plastic material around the fibers, exposure of a narrow region, fracture of the entire assembly, and application of tension are followed by the splicing operation. Splicing is achieved by placing the two tapes with the prepared terminations in an alignment channel, whereby epoxy is used to index match and to hold the assembly together.

The importance of low-loss connections will become clearer from an analysis of the factors that contribute to connection losses. A general distinction can be made between fiber-related (intrinsic) and technique-related (extrinsic) losses, or between propagation-related mechanisms and the way the connection is made. This distinction is important, first, in order to allow the fiber manufacturer an accurate measurement of the fiber losses and its separation from the hardware losses, and, second, in order to allow the fiber user to obtain a minimum system loss. Both loss types are additive.

INTRINSIC LOSSES

Intrinsic fiber losses include all terms that are related to fiber geometry and profile. In a fiber in which all modes are equally excited, the total intrinsic coupling loss is (39):

$$\alpha_C = -10 \log(N_{m_o}/N_{m_i}) \tag{2.48}$$

where N_{m_o}/N_{m_i} is the ratio of the number of modes in the output fiber to that in the fiber; subscripts i and o refer to the input and output fiber sections, respectively. Because in a graded-index fiber the number of modes is given by

$$N_m = 2\pi^2\left[i/(i+2)\right]\, a_1^2\, \mathrm{NA}^2/\lambda^2 \tag{2.49}$$

the coupling loss in terms of fiber properties is

$$\alpha_C = -10 \log \frac{i_o(i_i+2)}{i_i(i_o+2)} - 10 \log \left[\left(\frac{a_{1_o}}{a_{1_i}}\right)^2\right] - 10 \log \frac{\text{NA}_o}{\text{NA}_i} \tag{2.50}$$

Thus the coupling loss can be divided into three contributions given by profile-related, diameter-related, and numerical aperture–related terms,

$$\alpha_C = \alpha_I + \alpha_A + \alpha_N \tag{2.51}$$

For a perfect joint of two identical fiber sections $\alpha_C = 0$. A joint of two dissimilar sections has an appreciable coupling loss; for example, if $i_i = 1.05i_o$ (5% difference in index profile, where $i_i = 2.25$), $a_{1_i} = 0.9a_{1_o}$ (10% difference in core diameter), and $\text{NA}_i = 0.95\ \text{NA}_o$ (5% difference in numerical aperture), the individual contributions are $\alpha_I = 0.10$ dB, $\alpha_A = 0.83$ dB, and $\alpha_N = 0.42$ dB, with a total loss $\alpha_C = 1.35$ dB. The joint of a step-index fiber ($i_i = \infty$) and a graded-index fiber ($i_o = 2.0$) results in a loss of about $\alpha_I = 3.0$ dB. In order for a fiber joint not to exceed a total coupling loss of 0.5 dB, it is necessary that α_I not exceed 0.20 dB (corresponding to ±20% difference in index profile parameter), α_A not exceed 0.15 dB (+1% diameter difference), and α_N not exceed 0.15 dB (±1% NA difference). A joint between a step-index fiber (for which $i_o = \infty$) and a graded-index fiber (for which $i_i = 2.0$) correspondingly leads to a coupling loss of about 3.0 dB.

The variation of the individual coupling loss contributions with increasing mismatch of the two fiber sections is graphically illustrated in Figure 2.32. From the above theoretical expression it is evident that the coupling loss is unidirectional. This is important because it demonstrates that a fiber joint may have a high coupling loss in one direction and a negligible loss in the opposite direction. The directivity results from the ratios of fiber properties. It indicates that if a fiber section difference has to be tolerated at all in order to achieve a low coupling loss, it is more advantageous for the feeding fiber to have the lower index profile parameter, the larger core diameter, and the higher numerical aperture of the two. In addition to these three coupling loss mechanisms, which are assumed for waveguides in which all modes are excited equally, there is another loss contribution due to mode coupling caused by geometric disturbances of the fibers. It generally depends upon the core–cladding dimensions and refractive indexes as well as the interconnection hardware. Typically it is about 0.2 dB per connection.

EXTRINSIC LOSSES

Technique-related losses may be the result of lateral and angular misalignment of the fiber cores, separation of the fiber ends, mechanical distortion of the fibers, and fiber, finish, and Fresnel reflections. Lateral misalignment arises from hardware inaccuracies as well as variations of the fiber outer diameters. Typically, a displacement of only a few percent can be tolerated. A maximum angular misalignment of about 1° is tolerable. Furthermore, the fiber

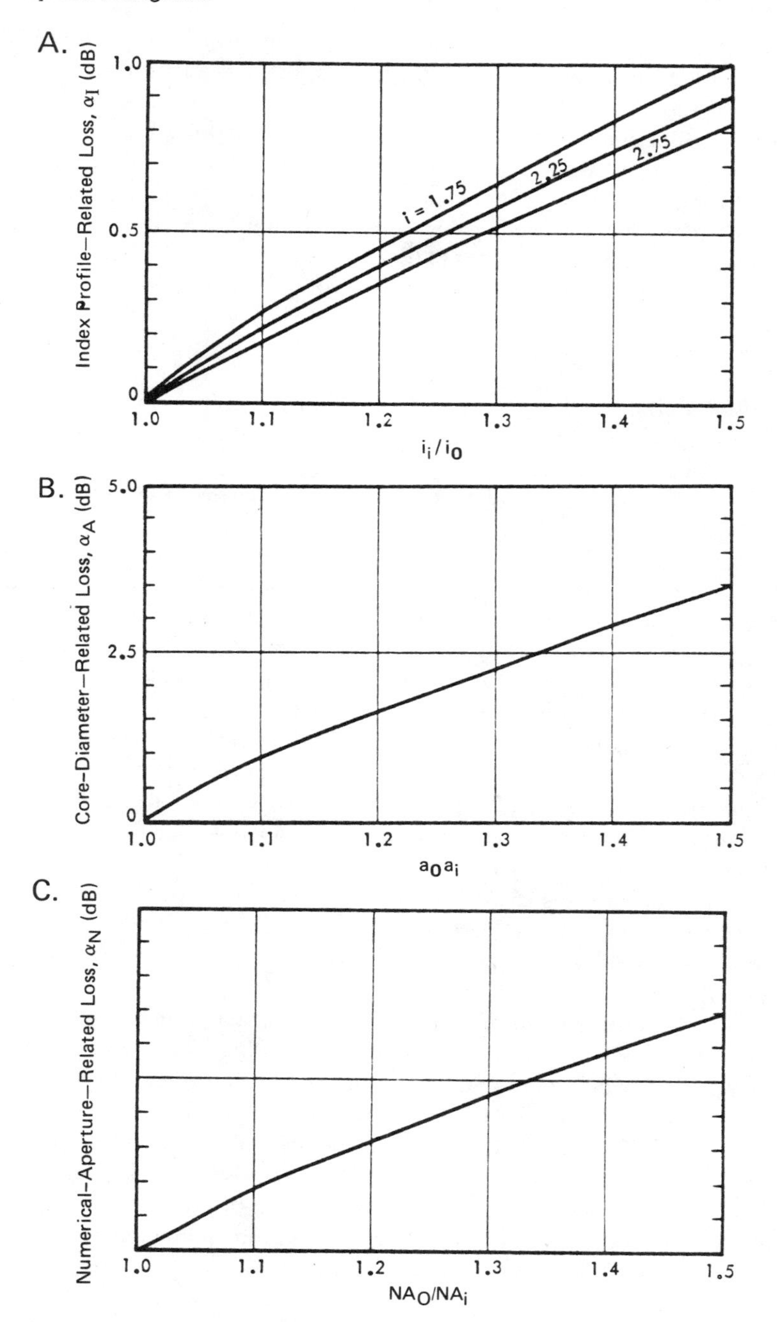

Figure 2.32 Components of the intrinsic waveguide joint attenuation as a function of index profile, core diameter, and numerical aperture of two step-index fiber sections at the joint.

ends must be finished smoothly to avoid scattering losses at the joint. A fiber end separation of a few microns results in an appreciable loss; the loss can be reduced by the use of an index-matching liquid, which also reduces the effect of Fresnel reflections. In the absence of an index-matching liquid, even perfectly smooth ends cause Fresnel reflection losses of about 0.4 dB. Fiber distortion during joining may produce a loss in excess of 1 dB, particularly where adhesives (epoxies) are used in order to hold fibers. The magnitude of the major extrinsic joint losses in relation to the magnitudes of their causes are shown in Figure 2.33 for typical step-index fiber connections.

OTHER CONSIDERATIONS

In addition to the low-loss requirement, a fiber connection must be easy and fast so that the connection can be made in the field. Some of the splicing methods developed so far have very low loss characteristics, but they do not lend themselves easily to practical applications because they require small and delicate parts that are difficult to handle or because they have complex assembly techniques that require accurate manipulation of fibers. Fiber splicing has received considerable attention, particularly under on-site installation conditions, which require fast, easy, and accurate splicing methods (40). These conditions apply mainly to pole and manhole situations in various environments. General requirements for splicing equipment that is to be used in these operations are light weight and compactness, reliability, low-loss connection, ease of operation, satisfactory work efficiency, and economy.

Splicing generally consists of three phases: coating stripping, fiber breaking, and fiber setting. Results on representative samples of splices done in this manner indicate that the average loss per splice is 0.25 dB for step-index and 0.21 dB for graded-index fibers, with a maximum of 0.57 and 0.35 dB, respectively. Mechanical stress tests have confirmed that the breaking strength of splices is comparable to that of unspliced fibers of identical characteristics and that no degradation of splice characteristics takes place over extended periods of time.

From the previous discussion it is apparent that connector design must concentrate on reliable, accurate, and easy fiber–fiber alignment. For this reason, long, hollow cylinders offer one solution to this general difficulty. However, because in such a tube the outer fiber diameters are aligned rather than their centers (as is required for minimization of connection losses), such a connection scheme is marginal in most practical applications. If the inner tube diameter is very large, it is even conceivable that the two fiber sections may be located at opposite sides of the cylinder and thus not achieve optical connection at all. Furthermore, the retention of the fibers usually requires adhesives or epoxies, which can lead to undesirable distortion losses. Also, it is normally difficult to fabricate accurately dimensional small tubes suitable for fiber connections.

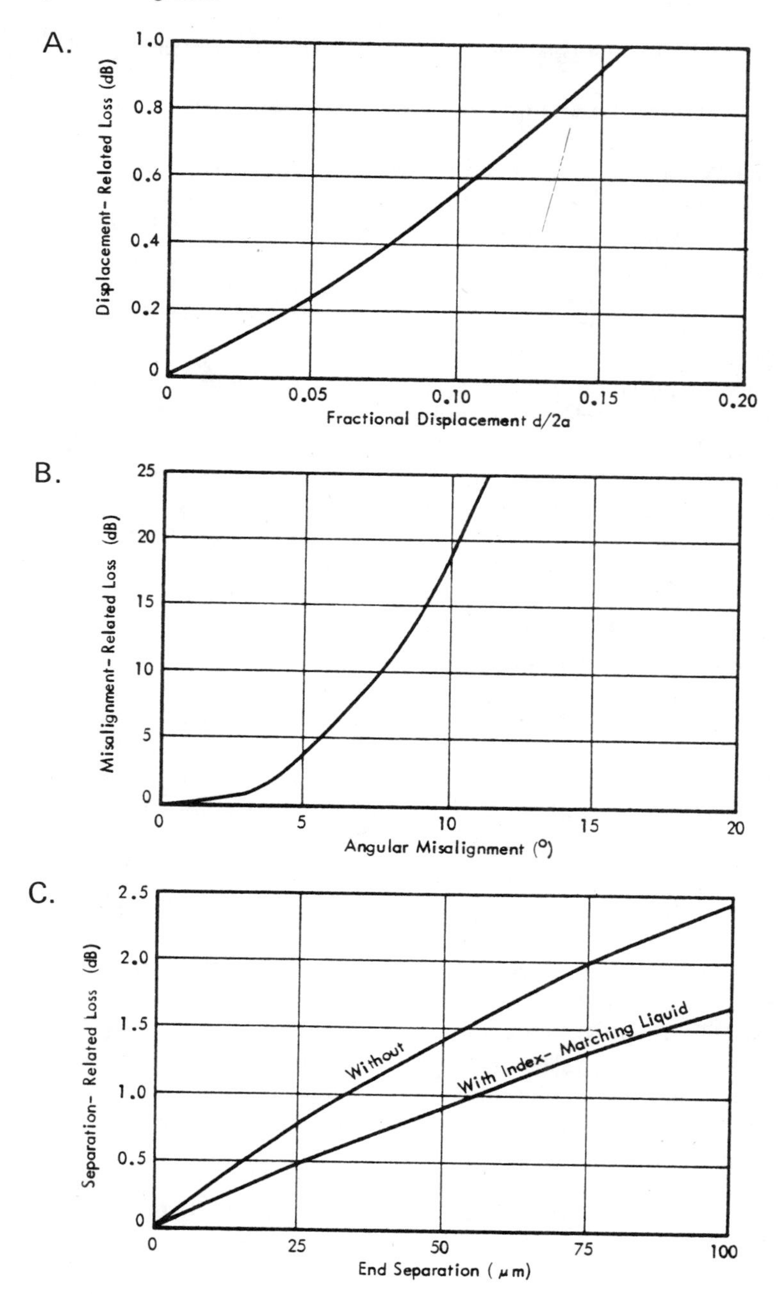

Figure 2.33 Components of the extrinsic waveguide joint attenuation as a function of displacement, angular misalignment, and end separation of two step-index fiber sections at the joint.

Cables that contain several fibers or fiber bundles present considerably more difficult splicing challenges than do single-fiber cables. One possible splicing technique for bundles uses a self-aligning tube that is square in cross section and has inner dimensions that are slightly larger than the optical fiber diameter. Assembly of a splice involves insertion of two fibers with good ends about half-way into each end of the square tube filled with uncured epoxy. Bending of the fibers prior to the insertion results in the rotation of the tube so that a diagonal of the square cross section is in the same plane as the bent fibers. Such a procedure allows the accurate alignment of the fiber ends, with ± 0.1 core radius accuracy, corresponding to about $\pm 3\mu$m; splice loss is about 0.1 dB.

Another possible technique utilizes the arrangement of fibers in cable subgroups (tapes) in glass as a rigid matrix material. In this way the fibers are accurately held in their relative positions so that they are not affected by subsequent splicing in the field. Because linear arrays (tapes) of fibers are suitable as building blocks for optical fiber cables, the method offers advantages over other splicing techniques. By using a glass for the rigid matrix material the problem can be simplified by preparing all optical fiber ends comprising the tape for splicing in a single operation, thereby allowing the efficient, fast, and low-loss splicing of fiber tapes in subsequent operations.

In a three-rod connector the alignment of the core centers rather than that of the outer fiber diameters is achieved. In this technique, three deformable elastic rods of identical diameter are arranged in parallel and in mutual contact; the rod diameters are chosen such that the smallest expected fiber occupying the center of the array is just contacted by the three rods. Thus, the center of each fiber will automatically fall into the center of the rod array regardless of the core diameter, assuming that the fiber has a perfectly circular cross section. Application of a radial force to the array causes alignment by deformation. This connection method results in a low connection loss (less than 0.2 dB per connection with an index-matching liquid). A further development of this concept is a multifiber (bundle) connector which consists of two linear arrays of rods and which provides the simultaneous linear alignment of a number of individual fibers. Because the fiber alignment occurs close to the lateral center of the assembly, each joint is surrounded by continuous alignment rods, and any rod discontinuities are not in the vicinity of the fiber mating plane. Again, the loss of such a connection typically does not exceed 0.2 dB per fiber.

The most critical parameter when joining two fibers is the lateral displacement of their axes. Assuming that there is no gap between the two fiber end faces, that both fibers have the same diameter, and that there is no angular misalignment between the fiber axes, the coupling efficiency of the coupling of two fibers is given by the overlap of the core areas. In the absence of reflection losses (for example, by using immersion liquid between the fiber ends), the coupling efficiency for $d/a_1 < 0.4$ is

$$\eta_c = 1 - (2/\pi)(d/a_1) \tag{2.52}$$

where d is the lateral displacement of the fiber axes and a_1 is the core radius. This means that the practical requirement of a coupling loss not exceeding 0.1 dB necessitates a maximum allowable displacement of less than 1.8% of the core diameter. This translates into a maximum displacement of 1.8 μm for a 100 μm fiber core. Thus the need for precise alignment control is evident.

In the connection of multimode fibers the requirements are substantially relaxed and the losses resulting from the connection are typically reduced by an order of magnitude compared to monomode fibers. Hence, in this case, a precisely engineered support for the fiber ends is usually sufficient, provided that the fiber cores are centrally positioned and their diameters are equal. This support may be a simple capillary tube or a *v* groove. Such simple connection methods are not applicable to monomode fibers, where adjustable connectors are necessary.

Coupling

Couplers applicable to optic fibers serve to provide feeding and tapping of optical energy. The use of couplers is a consequence of the development of multiterminal communications systems and data buses. In such applications, communication between two locations takes place over a common optical path, with each terminal communicating to every other terminal and sending out information on a time-shared basis. The relative effectiveness of the various types of multiterminal systems depends upon the properties of the available fibers as well as upon the properties of the couplers themselves. The performance characteristics of couplers are determined by loss, angular mixing (angular distribution output), and scrambling effectiveness (spatial distribution of output). All three parameters have a major influence on the coupling efficiency and hence on system design.

Basically, two principal types of couplers can be considered for the distribution of data to and from a set of remote terminals; these will be treated in more detail in Chapter 5. One is a serial distribution system that employs T-type access couplers, and the other is a parallel system that employs star-type access couplers.

The T coupler is the more conventional coupler and represents the tapped trunk line approach, where each terminal uses an access coupler to tap into a large main trunk line bundle.

The star coupler represents another approach. An input and output bundle from each terminal is brought to a single, centrally located coupler, where a power division or scrambling function is provided so that the input from each terminal communicates with all the output bundles.

These approaches allow a number of variations in their practical imple mentation, with the optimum design usually dictated by availability and individual system requirements. A comparison of both approaches is given in Table 2.9.

The T-coupler approach, where each terminal uses an access coupler to tap into the main trunk line bundle, is inherently more modular than the star-coupler approach, since an additional terminal may be added just by cutting into the main bundle. However, since each terminal needs a coupler, the total coupling loss of the system is higher than that for one with a star coupler. With the T coupler, on the other hand, less bundle length is needed, but the main trunk line bundle must contain more fibers than the bundles that can be used with a star-coupler system. However, if one T coupler is made to serve several terminals, the coupling loss per terminal can be reduced, allowing a large number of terminals without the use of a repeater.

A star coupler typically consists of a glass mixing rod with a silver-mirrored surface on one end and several fiber bundles attached permanently to the opposite end. Functionally, the star bus design can be considered as shortening the main bus to a single mixing rod and extending the length of the

TABLE 2.9. Comparison of Advantages and Disadvantages of Principal Optical Couplers

Coupler type	Advantages	Disadvantages
T Coupler	Inherent modularity, allowing addition of terminals by cutting into main bundle Flexibility without extensive wiring Need for less bundle length	Coupling loss higher than with star coupler because each terminal needs a coupler Main trunk line bundle must contain more fibers than in star-coupler system Trunk signal level drops when new terminal is added to the system Limitation of system to a few T couplers Receiver must be equipped with a wide dynamic range automatic gain control (AGC) to handle strong signals from adjacent terminals and weak signals from remote terminals
Star coupler (parallel system)	Low coupling loss Insertion loss of the mixer coupler occurs only once, independent of the number of terminals in the system No maximum number of couplers in system No AGC requirement Less stringent design requirements	Does not lend itself easily to modularity Requires more extensive fiber network to wire the system

terminal stubs. Either two bundles, or a single bundle split at the terminal end into emitter and receiver subbundles, connect the star coupler to each terminal; light from one terminal source is transmitted through the bundle into the mixer–coupler, where the light rays spread out and are reflected back to all bundles connected to the mixer and to all terminal receivers. Typical star-coupler loss is about 7 dB per coupler (41).

The main advantage of the series T-coupler system compared to the parallel star-coupler system is the greater flexibility of the tapped trunk line bus. On the other hand, the main disadvantage of the T-coupler system is the signal level drop with each new terminal added, which is caused by the insertion loss of the additional coupler. This results in a limitation of the total number of terminals that can be added and the need for wide-range automatic gain control of the receiver in order to accommodate weak signals from remote terminals and strong signals from near terminals. The parallel star-coupler system has only one mixer and thus does not have this limitation; hence, it offers less stringent design requirements on both receiver and transmitter. However, there is a significantly larger amount of fiber cable necessary to wire the system. In other words, in the parallel system the main bus is shortened to a single-point mixer, while the length of each terminal arm is extended.

Because all couplers are based on certain bifurcation and scrambling principles, they consist of a bifurcation section and a scrambling section. The bifurcation section provides an input port and couples a portion of the radiation from the main throughput bundle to the output port. The scrambling section ensures that input radiation is evenly distributed over the fibers of the main trunk line; it usually consists of a dielectric rod. Because of the multimode nature of these structures, the entire optical energy from the small input port is coupled onto the larger main bundle, whereas only a fraction is coupled off the main bundle. Examples of optical interconnection configurations are shown in Figure 2.34. They illustrate the most common approaches.

A multiterminal communications system typically uses a large number of couplers. Hence, the total throughput loss characteristics are an important consideration. In order to provide a minimum loss contribution, it is necessary that the main trunk line bundle butt directly against the scrambling rod.

FIBER–COMPONENT CONNECTIONS

The connection between the fiber and the terminal components (source and detector) is one of the critical aspects of any discrete optical communications system. Particularly, the source–fiber interface is of overriding importance. This is so because of the usual mismatch of the normally large emission angle of the source and the normally small acceptance angle of the fiber. The total coupling efficiency of the source-fiber connection depends upon source radiance, source radiation pattern, fiber numerical aperture, fiber packing density,

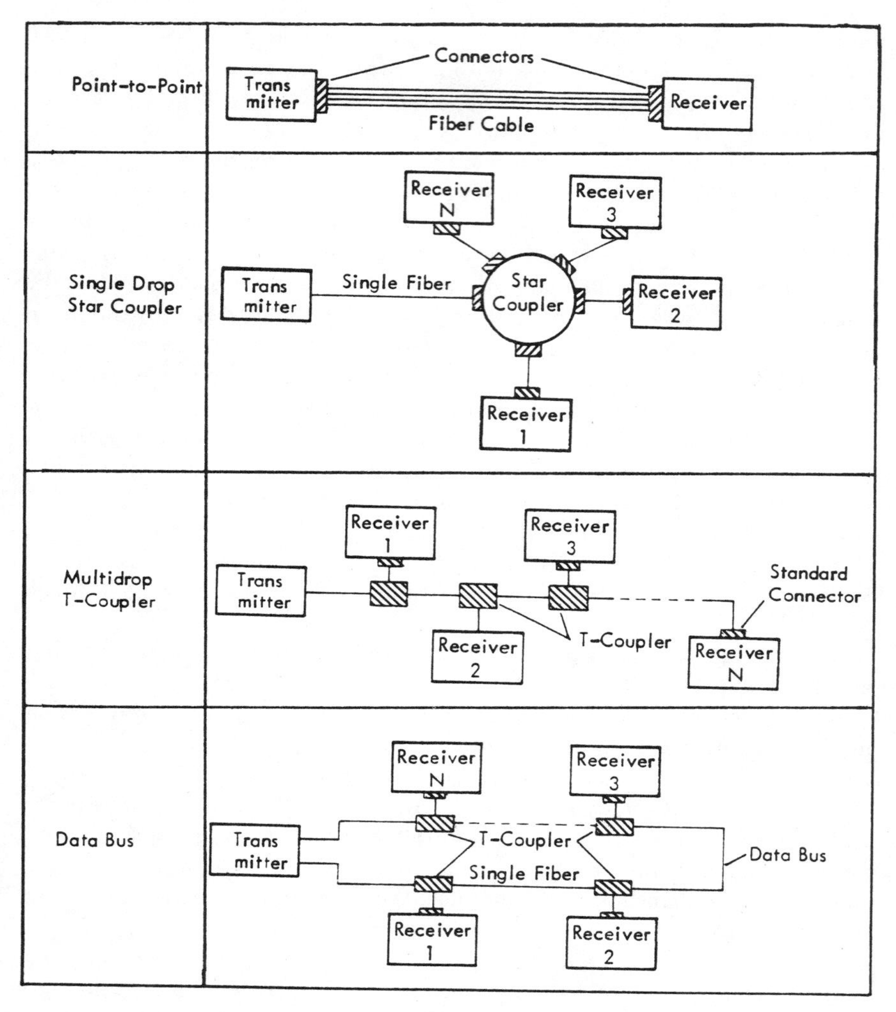

Figure 2.34 Schematic illustration of the most prominent fiber optics connection techniques.

fiber acceptance properties, Fresnel reflections at the source–fiber interface, and source–fiber separation.

Source Connection

If source and fiber areas are mismatched or misaligned, considerably reduced coupling efficiencies can be expected (42). For example, if the output area of the light source is substantially larger than the core area of the fiber, then the coupling efficiency is reduced in a manner corresponding to the ratio of the two areas. Similarly, misalignment of two identical areas results in a coupling efficiency loss proportional to the reduction of the overlapping areas.

Hence, most efforts are directed toward the matching of light source output area and fiber core area. Therefore, for single fibers, one attempts to reduce the diameter of the light source to that of the fiber. Consequently, a laser is superior to a light-emitting diode, especially when small fibers are considered. For fiber bundles the problems are not as severe, since LED sources usually have active area dimensions comparable to those of the bundle.

The coupling efficiency of a source–fiber connection is defined by the ratio of the power accepted by the fiber to the power emitted by the source, P_f/P_s. The power a fiber can accept is

$$P_f = (\pi\beta_1 a_1)^2 R_f \tag{2.53}$$

where β_1 is the acceptance angle of the fiber, a_1 is the core radius, and R_f is the radiance of the fiber. For a light-emitting diode the emitted power is

$$P_s = R_s A_s \tag{2.54}$$

where R_s is the source radiance and A_s is the emission area. Hence, the source–fiber coupling efficiency is given by

$$P_f/P_s \leq \beta_1^2 (A_f/A_s) \tag{2.55}$$

where A_f is the effective fiber core cross-sectional area. Assuming $R_f \leq R_s$, a high coupling efficiency therefore requires the fiber area to be as large as possible or the source area to be as small as possible. The acceptance angle of the fiber also strongly affects the coupling efficiency, but its choice is rather limited by other considerations, as discussed earlier in this chapter. Again, bandwidth requirements usually do not allow the fiber diameter to exceed rather small values, so that any efficiency increase is limited to decreases in the source active area.

Hence, major efforts by most laboratories engaged in the development of optical sources are devoted to the reduction of device dimensions. Obviously, laser diodes are much better suited for this purpose than light-emitting diodes, partly because of their edge-emitting characteristics, which confine emission to a small area. As a result, most major achievements in decreasing device size have been concentrated on semiconductor lasers.

For fiber bundles, the packing density is another factor of concern. The coupling efficiency between an optical source and a fiber bundle is proportional to the packing fraction. The optical power coupled into a fiber bundle is

$$P_i = C_f f_p (\mathrm{NA})^2 \tag{2.56}$$

where the proportionality factor C_f depends upon the source output power, source–bundle mismatch, and Fresnel reflection loss. The dimensionless packing fraction (f_p) is the ratio of the sum of all core areas to the total light acceptance area. The definition of the packing fraction is derived from the fact that only light incident on the core area is useful, whereas light incident on the cladding or the space between fibers is lost. For hexagonal cables, which is the most common packing scheme used for fiber bundles, the packing fraction is

$$f_p = \frac{\pi}{2\sqrt{3}}\left(\frac{a_1}{a_2}\right)^2 \approx 0.91\left(\frac{a_1}{a_2}\right)^2 \tag{2.57}$$

where a_1 and a_2 are the radii of core and cladding, respectively. The dependence of the packing fraction on the core–cladding diameter ratio is shown in Figure 2.35. Thus a naked fiber, for which $a_1 = a_2$, has the highest achievable packing fraction (about 91%) for a hexagonally packaged bundle. For a fiber bundle whose individual core diameters are each 80 μm and whose cladding diameters are 100 μm, the packing fraction is $f_p = 0.58$. For a bundle whose cores are 10 μm thick and whose cladding layers are 10 μm thick, it is only $f_p = 0.10$. For a single fiber for which fiber area and source area are matched, the packing fraction is simply given by the ratio of the areas of cladding and core. Thus the packing fraction $f_p = (a_1/a_2)^2$, so that for a fiber whose core diameter is 80 μm and whose cladding diameter is 100 μm, the packing fraction is 0.64; for a 10 μm thick core with a 10 μm thick cladding layer (radius 15 μm), the packing fraction becomes $f_p = 0.11$.

For both single fibers and bundles, the coupling efficiency increases with the square of the numerical aperture, so that it would appear desirable to increase the numerical aperture as much as possible. However, a large numerical aperture is equivalent to a large difference between core and cladding refractive indexes, which is not advantageous because of the detrimental effect on the data transfer rate, which requires a small index difference. Furthermore, a large numerical aperture also results in a large number of modes that can be supported by a fiber, which again affects the data transfer rate in an undesirable way.

In summary, the optical power coupled into the fiber is a function of mainly the radiance of the source, the numerical aperture of the fiber, and, in case the optical consists of a fiber bundle, the packing fraction of the bundle. The radiance of the source depends upon the characteristics of either laser or light-emitting diode. The dimensionless numerical aperture, as defined earlier, is a measure of the light cone that a fiber can accept and still give total reflection.

The coupling of lasers to the major types of fibers can be accomplished simply by butt joining. Again, the tolerances required for monomode fiber coupling are much tighter than those applicable to multimode fibers. This,

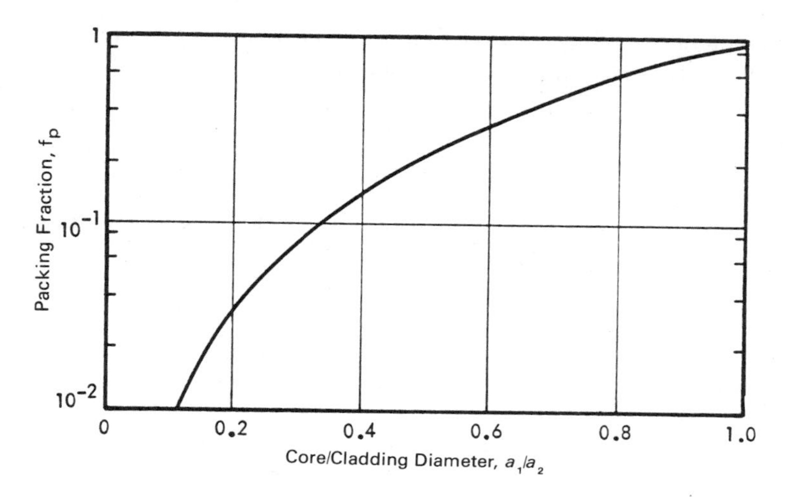

Figure 2.35 Dependence of the packing fraction of fiber bundles on core and cladding diameters for a hexagonal waveguide arrangement.

however, is not necessarily a severe drawback if the laser diode, together with a short piece of fiber and a fiber–fiber connector, is fabricated as a unit in order to avoid laser–source coupling in the field. The coupling efficiency of butt-joined lasers to fibers ranges typically from 15% to 30%. The coupling efficiency is a function of the near-field mode matching in the case of monomode fibers and of the adaptation between the laser far-field radiation lobe and the fiber acceptance angle in the case of multimode fibers. The coupling efficiency of laser–fiber joints can be considerably enhanced by adapting the laser output to the fiber by using small lenses. Efficiencies ranging from 60% to 80% have been achieved. However, this is not a practical method of achieving high coupling efficiency.

The coupling of light-emitting diodes to single fibers is considerably more problematic. One of the reasons for this is the substantially larger emitting area of these optical sources compared to those of lasers. Another reason is the radiation pattern of LEDs, which approximates a Lambertian source. The larger area of LEDs in comparison to laser diodes is not one of principle but one dictated by the significantly lower brightness of light-emitting diodes. Although theoretically a larger brightness of LEDs can be obtained by an increase in current density, lifetime and reliability considerations do not make such an operation advisable. Thus, a large emitting area is needed to obtain a relatively reasonable light output power. An important consequence of the large device area is that such devices cannot be used to couple into monomode fibers. Even with multimode fibers the coupling efficiency of light-emitting diodes butt-joined to fibers does not exceed about 5%. Furthermore, the launching efficiency of LEDs cannot be increased by lenses. Consequently, the optical power that can be launched from an LED into a multimode fiber is typically about 10 to 30 dB lower than that from a laser source. In addition, the overall efficiency is considerably lower. In spite of these significant technological deficiencies, LEDs are considered to be useful optical sources at present, at least in systems of lower bandwidth, because of their better availability and lower cost.

Because of the strong dependence of the coupled power upon the numerical aperture of the fiber, there is a trade-off between source–fiber coupling efficiency and maximum pulse rate. It can be alleviated only by increasing the optical output power of the source (thereby increasing the factor C_f), by reducing interface reflections through better fiber terminations, and by improving the packing fraction. An increase in optical emission power of the source, on the other hand, leads to serious consequences in regard to lifetime.

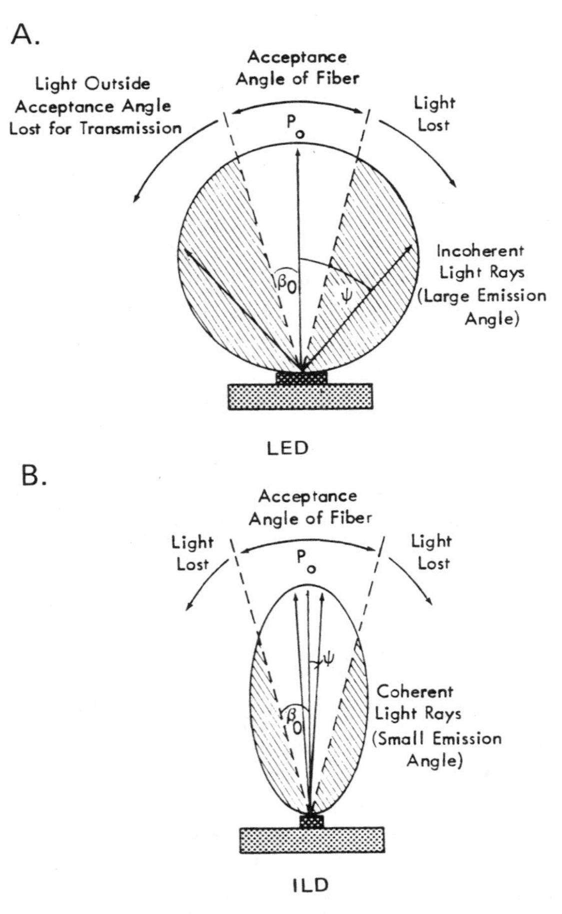

Figure 2.36 Comparison of the radiation patterns of light-emitting diodes (LEDs) and injection laser diodes (ILDs).

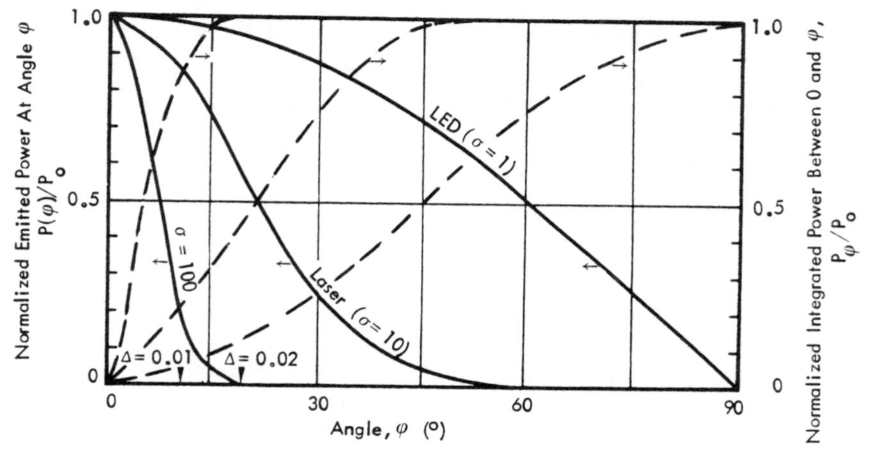

Figure 2.37 Characteristics of radiation patterns of real and hypothetical optical sources.

Source Radiation Pattern

Of additional concern is the mismatch of the radiation pattern of the source and the numerical aperture of the fiber. Light-emitting diodes not only emit incoherent light and over a larger active device area, but they also emit light in more than one direction, following the characteristics of a Lambertian source. If the light intensity of such a source is plotted as a function of the emission angle, whereby light intensity is represented by the lengths of arrows, then the arrow tips follow a circular pattern, as shown in Figure 2.36. In contrast, the light output power of a laser source is unidirectional, coherent, and usually originating from a much smaller device area. Consequently, a laser can potentially cause an enhanced coupling efficiency compared to a light-emitting diode because of the improved match between numerical aperture of the fiber and the more suitable radiation pattern of the source.

Generally, the radiance of a light source can be expressed by

$$P(\phi) = P_o\,(\cos^{\sigma}\phi) \qquad (2.58)$$

where P_o is the maximum optical output power of the source in any direction (this direction is normally chosen as the direction parallel to the fiber axis), ϕ is the angle of emission that describes the angular departure from the maximum intensity, and σ is a numerical factor characteristic of the light source. For an ideal light source the exponent should approach infinity, meaning that all light is emitted in only one direction. In practice, for typical laser diodes $\sigma \approx 10$, and for typical light-emitting diodes $\sigma \approx 1$. The normalized radiation pattern $P(\phi)/P_o$ of typical laser and LED devices is shown in Figure 2.37, together with the emission pattern of a hypothetical, more ideal source for which $\sigma = 100$.

In addition, the total optical power contained within a cone of half-angle ϕ is also given in Figure 2.37; it can be represented by the integrated light output over all angles between 0 and ϕ,

$$P = \int_0^\phi P(\phi)\, d\phi \tag{2.59}$$

which is usually normalized by the maximum power P_o. In this illustration, the values of the minimum acceptance angle of the fiber (β_0) that correspond to $\Delta = 0.01$ and 0.02 have been included. The dependence of the emission angle corresponding to half the maximum output power upon the directivity exponent is shown in Figure 2.38.

The angular radiation pattern of most surface-emitting LEDs is Lambertian, corresponding to a light power $P = P_o \cos\phi$. A more directional beam pattern can be obtained from an edge-emitting double heterojunction (DH) LED. Its directivity coefficient is about $\sigma = 2$, corresponding to $P = P_o \cos^2\phi$. In this device the active junction region of the device is sandwiched between two layers of semiconductor material with a refractive index less than that of the active region. The emission patterns of both edge-emitting DH LED and DH lasers are elliptical in cross section. In the case of the DH LED, the approximate half-power beam divergence angles are $\pm 50°$ and $\pm 30°$ and in the case of the DH laser $\pm 20°$ and $\pm 5°$ perpendicular and parallel to the junction plane, respectively.

Because a fiber whose index difference is $\Delta = 0.01$ has a numerical aperture of 0.21 and an acceptance angle of only 12.2°, only about 5% of the total light output of an LED whose directivity coefficient $\sigma = 1$ is within the light cone acceptance of the fiber. About 95% of the output is lost, even if all other

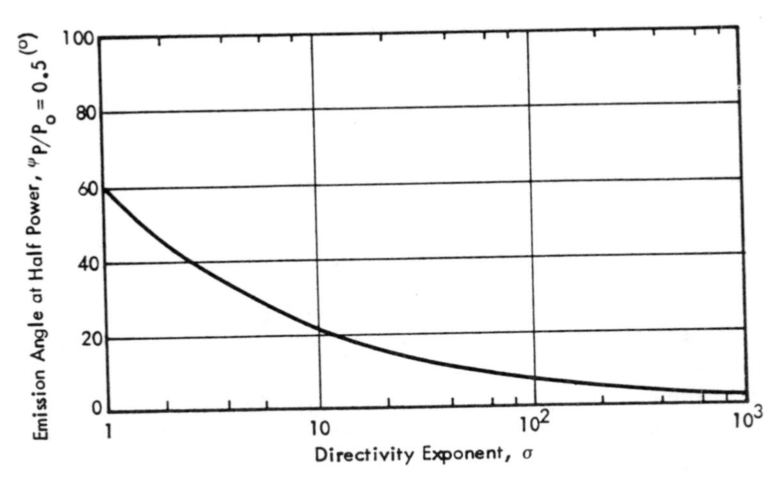

Figure 2.38 Variation of the source emission angle with source directivity.

criteria, such as negligible source–fiber spacing and misalignment, have been met. The use of a laser in conjunction with the same fiber results in a 22% utilization of the emitted light. Finally, the hypothetical source having $\sigma = 100$ allows the utilization of about 96% of its optical output. If the fiber whose $\Delta = 0.01$ is replaced by one whose $\Delta = 0.02$ (corresponding to a NA of 0.30), the optical output utilization will increase from 5% to 11% for an LED and from 22% to 45% for a semiconductor laser.

On the other hand, the use of an LED offers one advantage over a laser source. Owing to its larger emission angle, angular misalignment is less critical for the LED–fiber combination. Because of the higher emission directivity, a laser has to be aligned quite accurately with the fiber into which it feeds. For example, an angular misalignment of 10° leads to a coupling loss increase of only 0.25 dB for an LED but to almost 1.0 dB for a laser. This is illustrated for typical LED and laser sources in Figure 2.39A. The previously mentioned near-ideal source with its even higher directivity requires substantially more critical source–fiber angular alignment. The coupling efficiency of an optical source that is directly butted against the fiber or fiber bundle and whose angular emission distribution is symmetric and varies as $\cos^{\sigma}\phi$ is approximately

$$P_f/P_s = \tfrac{1}{2}\, f_p(\mathrm{NA})^2(\sigma + 1) \tag{2.60}$$

where P_s is the optical power emitted by the source, P_f the optical power the fiber is capable of accepting, and f_p the packing fraction of the bundle (for a single fiber $f_p = 1.0$, for a hexagonal bundle of typical multimode fibers $f_p = 0.6$, and for a hexagonal bundle of monomode step-index fibers $f_p = 0.9$). A ratio of $P_f/P_s = 1$ means that fiber and source are perfectly matched. If $P_f/P_s < 1$, the directivity of the source in relation to the numerical aperture of the fiber is too small; if $P_f/P_s > 1$, the source directivity is excessive for the fiber under consideration.

The coupling efficiency decreases also with increasing separation of source and fiber. This is particularly apparent for a light-emitting diode because of the characteristic radiation pattern which results in increased illumination area as a function of distance. This is shown in Figure 2.39B for LED and laser in conjunction with a typical multimode fiber.

Assuming an NA of 0.2 and perfect area matching of fiber and source, a numerical evaluation of typical source–fiber combinations predicts the following values: For the combination of a single monomode fiber and a semiconductor injection laser ($\sigma = 10$), the coupling efficiency $P_f/P_s = 0.22$; for a multimode bundle in conjunction with an LED ($\sigma = 1$), the coupling efficiency $P_f/P_s = 0.024$. This is shown in more detail in Figure 2.40.

It is evident that the source–fiber coupling loss can be appreciable, particularly if a fiber with a small numerical aperture is chosen, since efficiency increases with the square of NA. In addition, the coupling efficiency increase is approximately proportional to the directivity parameter of the source.

In summary, in present fiber bundles an optical power loss of 3 to 4 dB can be expected per source–fiber coupling location. Furthermore, an additional loss of about 5 to 10 dB results from the low numerical aperture (0.13–0.16) of the fiber. The relatively large size of present light-emitting diodes adds another loss contribution of 5 to 10 dB, which can be almost completely eliminated by the use of smaller lasers.

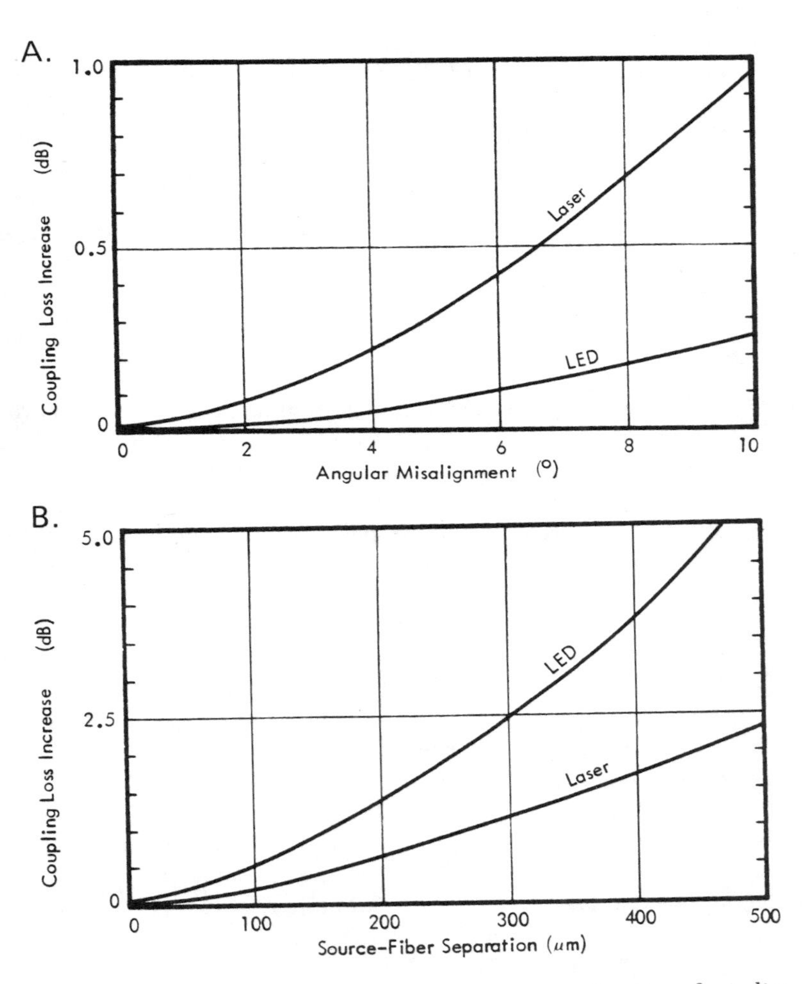

Figure 2.39 Source–waveguide joint attenuation as a function of misalignment and separation.

For a fiber bundle it is possible to reduce the coupling loss by the insertion of an appropriate optical lens between source and fibers. This is schematically shown in Figure 2.41, making the assumption that an intervening lens with a magnification M causes an angular translation $\Theta_1/\Theta_2 = M$. In this way, the coupling loss can be reduced to the loss caused by the packing fraction, if the angular source emission (corresponding to an angle Θ_1) can be sufficiently collimated so that the angle Θ_2 is always less than the angle corresponding to the fiber numerical aperture (β_0). This allows simultaneous maintenance of

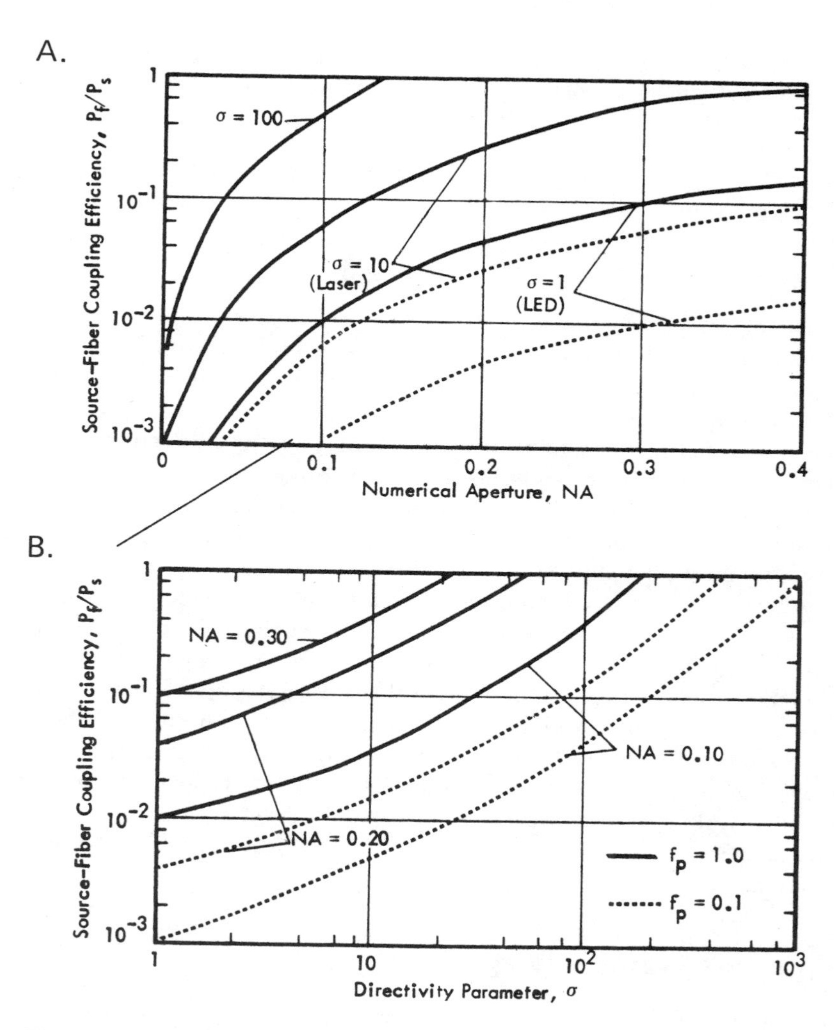

Figure 2.40 Influence of fiber and source characteristics on source–fiber coupling efficiency.

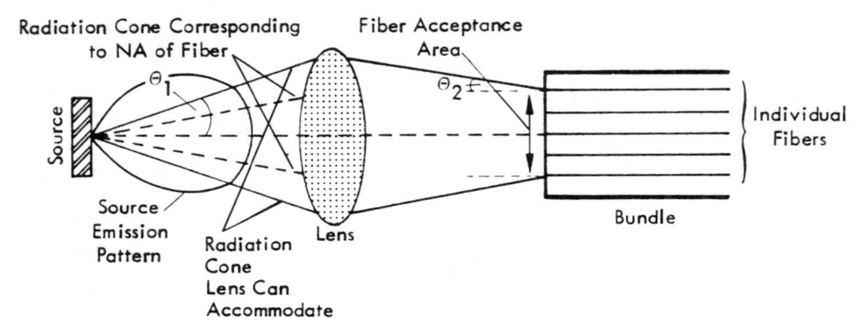

Figure 2.41 Illustration of the improvement of source–fiber coupling efficiency with the use of a lens between the source and the fiber.

the area of the radiation pattern at the bundle input at a level equal to that of the bundle area itself.

Detector Connection

In principle, considerations similar to those of the source–fiber connection in regard to the coupling efficiency apply to the waveguide output at the receiving end, where the fiber couples into a photodetector. However, usually there is only a small angular divergence of the light leaving the waveguide, requiring no special features to improve the emerging light directivity. Furthermore, most photodiodes and phototransistors do not exhibit the strong directivity characteristics of laser sources, so that accurate angular alignment requirements are normally less stringent. Also, there is no critical acceptance angle for most photodetectors.

In spite of the easier connection of the fiber to the detector compared to the fiber–source connection, it is necessary to match fiber cross-sectional area and detection area as closely as possible to increase receiver sensitivity and signal-to-noise ratio to their optimum values. Also, it is desirable to reduce the spacing between fiber and detector as much as possible because of considerations similar to those applying to the transmitting end, although the less stringent directivity requirements aid in the achievement of a low-loss connection. Generally, it is possible to obtain a loss of only about 1 dB per fiber–detector connection, in contrast to the more than 4 to 5 dB per source–fiber connection.

FIBER MEASUREMENTS

The application of fiber optic components in practical systems requires relatively simple yet accurate measurements of the most critical parameters. Measurements routinely performed center typically on fiber and source and depend on the establishment of sophisticated test laboratories. Laboratory development of fiber optics components requires much more precise and versatile instruments compared to those needed for factory or field measurements. Common measurements for research and development on fibers include fiber attenuation, fiber dispersion (bandwidth) and index profile, and fiber geometry (numerical aperture, core diameter and concentricity, and outside diameter).

Attenuation Measurements. Loss measurements are made typically at only one fixed wavelength in order to monitor the characteristics of fibers as they come from production (43). There is substantial interest, however, in the entire 0.6 to 1.1 μm range, with rapidly developing significance of the 1.1 to 1.6 μm range. Thus, the preferred spectral loss instrument includes numerous spot wavelength emitters to cover the 0.6 to 1.6 μm range. Sources are especially required at 0.82, 0.90, and 1.05 μm. Measurement accuracy required is $\pm$0.1 dB/km for research and $\pm$0.5 dB/km for routine production checks. A typical spectral loss measurement test setup is depicted in Figure 2.42. Kao and coworkers (44, 45) have given one of the first detailed descriptions of attenuation coefficients using the single-beam and double-beam methods.

Dispersion Measurements. The bandwidth of a fiber is a function of the dispersion of a signal along its length. Thus, bandwidth is determined indirectly through measurement of dispersion. This measurement is usually made at a fixed wavelength, such as 0.85 μm, but future measurements will also be needed at several other wavelengths in the 0.6 to 1.6 μm range. A typical laboratory measurement setup is illustrated in Figure 2.43. Dispersion measurements use a pulsed light source operating in the time domain and converting to the frequency domain by fast Fourier transform. An alternative method uses a swept continuous-wave (cw) source, with a spectrum analyzer after the detector. Because the spectrum analyzer provides only amplitude information but no phase information, a pulsed source method is frequently preferred. The achievable accuracy of $\pm$5% is usually adequate for most present measurements, but an accuracy of $\pm$1% will be required as the data transfer rate of fibers is increased beyond 1 Gb/s. The transfer rate is also related to the index of refraction of a graded-index fiber, as measured across its cross-section. This index profile is evaluated by analyzing the near-field pattern of radiation from the fiber. Alternately, an interferometer pattern may be projected through a microscope. A programmable calculator or minicomputer is used for calculations that involve the automatic plotting of the output.

Geometry Measurements. Dimensional control of optical fibers is critical, since a variation of dimensions between one fiber and a mating fiber through either a connector or a splice tends to decrease the amount of light coupled between the two portions. Typical dimensional variations are schematically illustrated in Figure 2.44. Owing to its importance, fiber dimensional variation is extensively measured in the laboratory. Continuous on-line

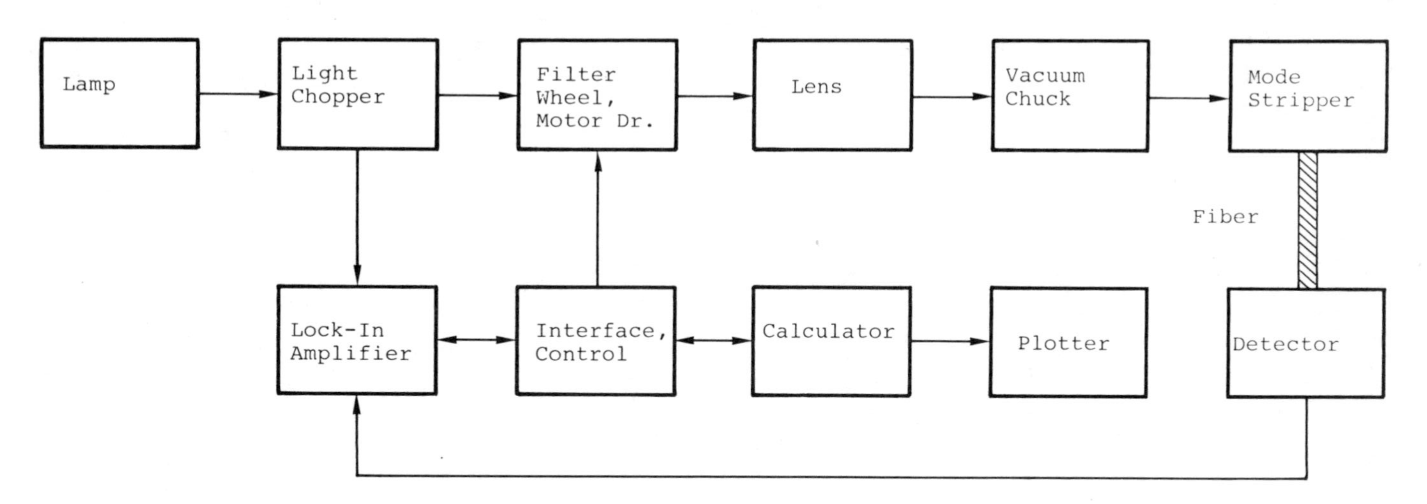

Figure 2.42 Schematic diagram of a fiber attenuation test arrangement.

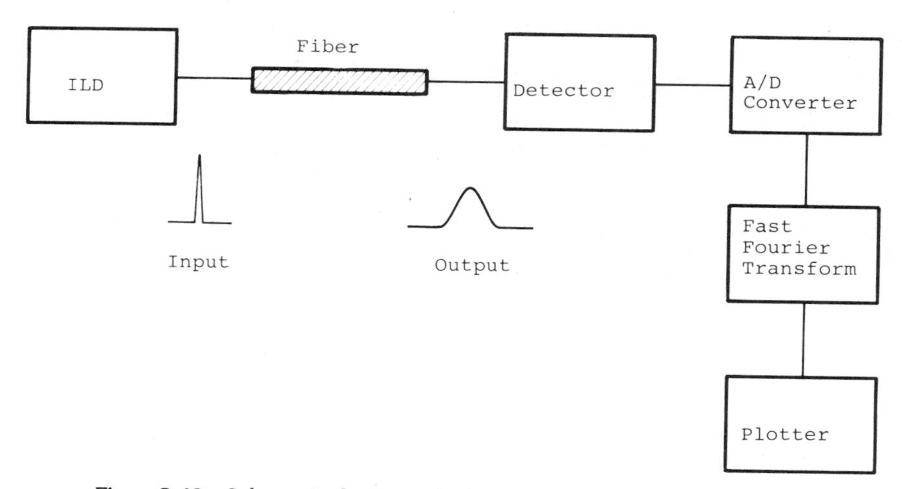

Figure 2.43 Schematic diagram of a fiber dispersion test arrangement.

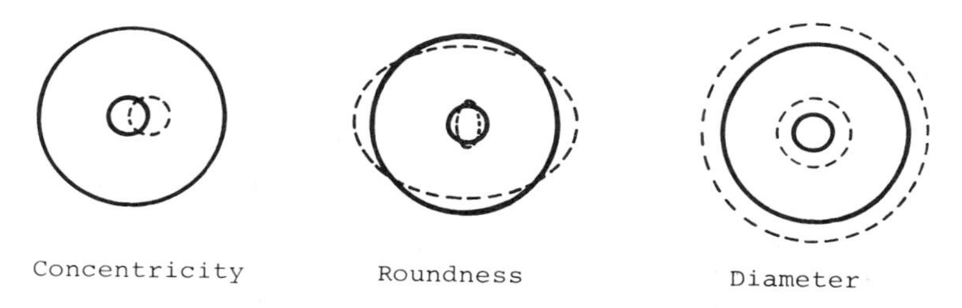

Figure 2.44 Common variations of fiber dimensions.

monitoring will be required in the future. A representative arrangement used for routine monitoring of the fiber outside diameter is shown in Figure 2.45.

Other Measurements. In addition to spectral loss, dispersion–bandwidth, index profile, and geometry–dimension measurements, various diagnostic measurements are frequently made, particularly for fibers of new design. Optical time domain reflectometry (TDR, also called photo probe or radar) is useful in detecting and analyzing variations of dispersion or dimensions along the fiber length. A calorimetric test may be used to separate absorption loss from scattering loss and thus help isolate any causes of unusually high loss. The light emitted through the fiber wall may also be measured; this is a measurement of scattering as a function of angle of incidence.

The measurement of fiber optic components and assemblies falls mainly into three categories:

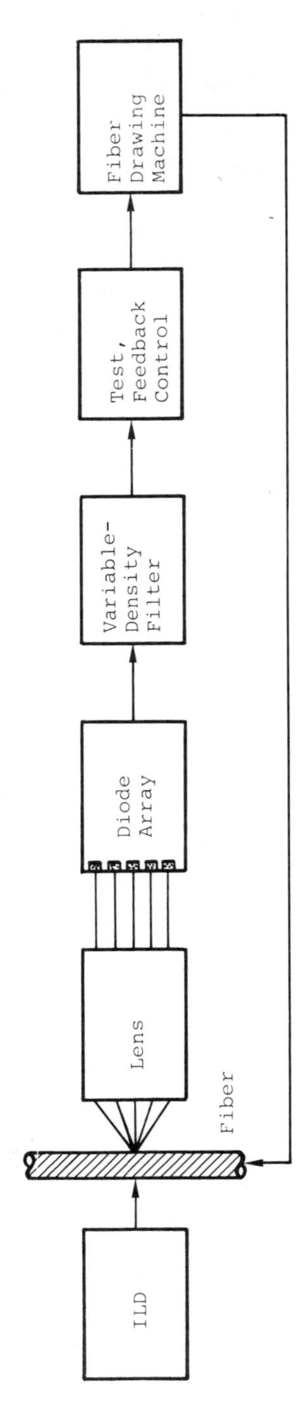

Figure 2.45 Schematic diagram of a fiber diameter test arrangement.

1. In-process monitoring of fiber and cable production.
2. Final inspection of cable harnesses, transmitter and receiver modules, emitters, detectors, and other components.
3. Incoming inspection of components: fiber inspection by cable producers, cable inspection by cable system assemblers, and active element inspection by transmitter and receiver manufacturers.

Fiber geometry measurement is becoming increasingly important. Fiber outside diameter and core diameter are the only parameters now being regularly measured in production. These dimensions are continuously monitored and their dimensional variances fed into a closed-loop control process. Fiber loss, index profile, and core concentricity are also frequently monitored in the factory. In addition, there is a requirement for fault (break) location. The optical time domain reflectometer (TDR) is a useful instrument for this purpose. In addition to locating fiber breaks, it can provide useful information about the extent of dispersion in the fiber and its variation along the fiber length. The laboratory TDR must resolve a break location within ±5 cm, must display the measurement results on an oscilloscope or readout, must be capable of measuring end-to-end loss (up to 50 dB) with accuracy of ±1 dB, and must be free from drift. The field TDR must be capable of locating a fiber break with an accuracy of ±50 cm, through up to about 50 dB of attenuation. A digital readout of distance to the break is preferable to an oscilloscope display. This instrument must be rugged to withstand normal transportation and field handling without damage, it must be free from drift, and its size must be small.

In graded-index fibers it is of primary importance to obtain an accurate and economic measurement of the profile of the refractive index along the core radius (46, 47). Ideally, such a measurement should be made on a continuous basis along the entire fiber length for every fiber fabricated. In practice, one measurement is usually made for every fiber. A sufficient amount of statistical information has been accumulated to predict the index profile along a given fiber if its profile is measured at only one point. There is generally adequate confidence in the fact that the index profile does not change appreciably along the fiber if reasonable process control is maintained.

Most modern index profile measurements utilize the convenient feature that in a fiber in which all modes are equally excited, a close resemblance exists between the near-field intensity distribution and the refractive index profile. This scanning technique can be used on short lengths of a fiber in any circularly symmetric index profile, provided a correction factor is applied. Larger errors occur, however, for fibers with near–step-index profiles owing to the high leakage of some rays.

Although the near-field scanning technique has already been used to obtain a qualitative indication of the index profiles of several different fibers, it is necessary to take into account the propagation characteristics of practical fibers in order to apply this technique to accurate quantitative measurements.

This includes the consideration of mode conversion, differential mode attenuation, and the existence of leaky rays which can cause significant deviations from the predicted near-field distribution. The first two of these factors can be eliminated by the analysis of only short fiber sections, but the third factor, leaky rays, has to be taken care of by its inclusion in the profile-determining equations by provision for a corrective factor applicable to short fiber sections. This factor becomes larger as one progresses from core center toward the core–cladding interface. Since the application of the near-field scanning technique is valid only if all modes are equally excited, an incoherent light source such as an LED has to be used.

REFERENCES

1. H.F. Wolf & J.D. Montgomery. *Fiber Optics and Laser Communications Forecast.* Menlo Park, Gnostic Concepts, Inc., 1976.

2. D. Gloge (Ed.). *Optical Fiber Technology.* IEEE Reprint Series, IEEE Press, New York, 1976.

3. S.E. Miller et al. Research Toward Optical-Fiber Transmission Systems. *Proc. IEEE* **61**:1703, 1973.

4. K.C. Kao & G.A. Hockham. Dielectric-Fibre Surface Waveguides for Optical Frequencies. *Proc. IEEE* **13**:1151, 1966.

5. D. Marcuse. Optical Fibers for Communications. *Radio Electron. Eng.* **43**(11):655, 1973.

6. R.L. Gallawa. Component Development in Fiber Waveguide Technology. Intelcom 77, Atlanta, Ga., October 1977.

7. R.D. Maurer. Glass Fibers for Optical Communication. *Proc. IEEE* **61**:452, 1973.

8. T. Uchida et al. Optical Characteristics of a Light-focusing Fiber Guide and its Applications. *IEEE J. Quantum Electron.* **QE-6**:606, 1970.

9. S. Maslowski. Optical Fibre Transmission—Single Mode Versus Multimode. World Telecommunication Forum, p. 1.3.3.1, Geneva, October 1975.

10. R. Ishikawa et al. Transmission Characteristics of Graded-Index and Pseudo-Step-Index Borosilicate Compound Glass Fibers. International Conference on Integrated Optics and Optical Fiber Communications (IOOC 77), p. 301, Tokyo, July 1977.

11. D. Gloge. Optical Power Flow in Multimode Fibers. *Bell Sys. Tech. J.* **51**:1767, 1972.

12. D. Gloge. Weakly Guiding Fibers. *Appl. Opt.* **10**:2252, 1971.

13. D. Gloge & E.A.J. Marcatili. Multimode Theory of Graded-Core Fibers. *Bell Sys. Tech. J.* **52**:1563, 1973.

14. S. Kawakami. Radiation and Mode Conversion Effects in Optical Fibers. IOOC 77, p. 407, Tokyo, July 1977.

15. P.J.B. Clarricoats et al. Propagation Behaviour of Cylindrical-Dielectric-Rod Waveguides. *Proc. IEEE* **120**(11):1371, 1973.

16. R.L. Gallawa. Optical Waveguide Technology. World Telecommunication Forum, p. 1.3.2.1, Geneva, October 1975.

17. S. Kobayashi et al. Low-Loss Optical Glass Fibre with Al_2O_3 Core. *Electron. Lett.* **10**:410, 1974.

18. D.A. Pinnow et al. Fundamental Optical Attenuation Limits in the Liquid and Glassy State with Application to Fiber Optical Waveguide Materials, *Appl. Phys. Lett.* **22**:527, 1973.

19. J. Schroeder et al. Rayleigh and Brillouin Scattering in K_2O-SiO_2 Glasses. *J. Am. Ceram. Soc.* **56**:510, 1973.

20. M. DiDomenico, Jr. Material Dispersion in Optical Fiber Waveguides. *Appl. Opt.* **11**:652, 1972.

21. A.W. Snyder & D.J. Mitchell. Bending Losses of Multimode Optical Fibers. *Electron. Lett.* **10**(1):11, 1974.

22. A.Q. Howard, Jr. Bend Radiation in Optical Fibers. Society of Photo-Optical Instrumentation Engineers *(SPIE)* **77**:57, 1976.

23. D.B. Keck. Transmission Properties of Optical Fiber Waveguides. *SPIE* **63**:3, 1975.

24. S. Geckeler. Optimization of Graded-Index Fibers with Nonlinear Profile Dispersion. IOOC 77, p. 435, Tokyo, July 1977.

25. D. Kuppers et al. Preparation Methods for Optical Fibres Applied in Philips Research. IOOC 77, p. 319, Tokyo, July 1977.

26. T. Izawa et al. Continuous Fabrication of High Silica Fiber Preform. IOOC 77, p. 375, Tokyo, July 1977.

27. W.G. French & L.J. Pace. Chemical Kinetics of the Modified Chemical Vapor Deposition Process. IOOC 77, p. 379, Tokyo, July 1977.

28. E.N. Randall. Current Practices for Glass Optical Waveguide Fabrication. *SPIE* **63**:9, 1975.

29. T. Nakahara et al. Design and Performance of Optical Fiber Cables. IEEE Conference on Optical Fiber Communication, p. 81, London, September 1975.

30. R.L. Lebduska. Fiber Optic Cable Test Evaluation. *Opt. Eng.* **13**:49, 1974.

31. S.T. Gulati. Strength and Fatigue of Optical Fibers. Intelcom 77, p. 702, Atlanta, Ga., October 1977.

32. R.E. Love. Strength of Optical Waveguide Fibers. *SPIE* **77**:69, 1976.

33. U.H.P. Oestreich. Applicable of Weibull Distribution to the Mechanical Reliability of Optical Fibers for Cables. IEEE Conference on Optical Fibre Communication, p. 73, London, September 1975.

34. K. Inada et al. Transmission Characteristics of a Low-Loss Silicone-clad Fused Silica-Core Fibre. IEEE Conference on Optical Fibre Communication, p. 57, London, September 1975.

35. D. Gloge et al. Optical Fiber End Preparation for Low-Loss Splices. *Bell Sys. Tech. J.* **52**:1579, 1973.

36. D.L. Bisbee. Optical Fiber Joining Technique. *Bell Sys. Tech. J.* **50**:3153, 1971.

37. C.G. Someda. Simple, Low-Loss Joints Between Single-Mode Optical Fibers. *Bell Sys. Tech. J.* **52**:583, 1973.

38. R.M. Hawk & F.L. Thiel. Low-Loss Splicing and Connection of Optical Waveguide Cables. *SPIE* **63**:109, 1975.

39. H. Murata et al. Splicing of Optical Fibre Cable on Site. IEEE Conference on Optical Fibre Communication, p. 93, London, September 1975.

40. A.F. Milton & A.B. Lee. Access Couplers for Multiterminal Fiber Communication. *SPIE* **63**:114, 1975.

41. T. Yamada et al. Launching Dependence of Transmission Losses of Graded-Index Optical Fiber. IOOC 77, p. 263, Tokyo, July 1977.

42. P. Kaiser. Numerical-Aperture-dependent Spectral-Loss Measurements of Optical Fibers. IOOC 77, p. 267, Tokyo, July 1977.

43. K.C. Kao & T.W. Davies. Spectrophotometric Studies of Ultra-Low-Loss Optical Glasses. I. *J. Sci. Instrum. (Ser. 2)* **1**:1063, 1968.

44. M.W. Jones & K.C. Kao. Spectrophotometric Studies of Ultra-Low-Loss Optical Glasses. II. *J. Sci. Instrum. (Ser. 2)* **2**:331, 1969.

45. W.J. Stewart. A New Technique for Measuring the Refractive Index Progress of Graded Optical Fibers. IOOC 77, p. 395, Tokyo, July 1977.

46. K. Hotate & T. Okoshi. Semiautomated Measurement of Refractive-Index Profiles of Single-Mode Fibers by Scattering-Pattern Method. IOOC 77, p. 399, Tokyo, July 1977.

47. K. Iga & Y. Kokubun. Precise Measurement of the Refractive Index Profile of Optical Fibers by Nondestructive Interference Method. IOOC 77, p. 403, Tokyo, July 1977.

3

OPTICAL SOURCES

Helmut F. Wolf

Optical sources represent the direct link between the electrical and optical portions of the communications system input. They provide the optical power to the waveguide by modulation of their input current by the signal delivered from the electronic circuitry. The characteristics of the generated light usually must be such that it has a high output power at an appropriate wavelength, high directivity, linear electrical–optical conversion, long lifetime, and dimensional and other properties that are compatible with those of the fiber, in addition to the very important requirement of a very narrow spectral width (near-ideal monochromatic output spectrum). For optical communications systems there are two types of light sources of principal interest: semiconductor light-emitting diodes and semiconductor injection lasers. Because optical sources fabricated in silicon are not feasible due to the nonradiative recombination process in silicon, the various GaAsAl alloys are the most useful semiconductor source materials. In the future, nonsemiconductor lasers will also become increasingly important, partially because of their narrow spectral bandwidths and their substantially better coherence and narrow-angle radiation patterns, in spite of their fabrication technology, which is not directly compatible with integrated circuits.

INTRODUCTION

The development of optical sources has traditionally concentrated on improvements of optical output power, emission efficiency, emission beam pattern, emission monochromism, modulation frequency, and lifetime. Economic considerations also affect the usefulness of a particular source. The source acts as an electrooptic converter, which is intended to modulate an emitted light beam through variations in the drive current. In order to function properly and to find practical use, the E/O converter must satisfy at least the following requirements in an economic manner:

1. *Output wavelength.* The emission peak wavelength must coincide with one of the loss minima of the interacting waveguide.

2. *Output power.* The optically emitted power must be as high as possible and must be achievable with the highest conversion efficiency or the

minimum drive current to prevent junction heating and subsequent gradual or abrupt (catastrophic) performance degradation.

3. *Output directivity.* The optical output must be highly directional to achieve a high source–fiber coupling efficiency, and it must provide coherent radiation.

4. *Output bandwidth.* The optical output must be as monochromatic as possible to reduce material dispersion as a result of different wavelengths propagating in the fiber at different velocities.

5. *Output degradation.* The device lifetime must be very high because the very high current densities and the resultant localized thermal increases affect device performance detrimentally.

6. *Output distortion.* Easy modulation and linear input–output characteristics must be achievable, particularly for analog operation.

Two principal source types can generally be distinguished, depending on the direction of the photon radiation relative to the electron current flow. Coincidentally, these can be broadly identified as light-emitting diodes and injection laser diodes, respectively, although there is some overlap of these categories.

> *Surface-emitting source.* In surface-emitting devices the major portion of the optical radiation is emitted perpendicular to the device surface and parallel to the direction of the current flow. Examples are planar, dome (hemispherical), and truncated light-emitting diodes (LEDs).
>
> *Edge-emitting source.* In edge-emitting devices the major portion of the optical radiation is emitted parallel to the surface and perpendicular to the current flow. Examples are the edge-emitting LED and the semiconductor laser (ILD).

The properties of the major types of optical sources suitable for optical communications systems are compared in Table 3.1. These data are approximate and will improve as the state-of-the-art advances.

Because waveguides of very small attenuation typically have a small numerical aperture (usually less than 0.3), it is difficult to couple light efficiently from wide-angle-emitting sources into these fibers, so that narrow-angle sources must be employed with low-loss fibers. The edge emission in such highly directional structures is appreciably narrowed in the direction perpendicular to the junction plane owing to the internal waveguiding mechanism in devices that employ heterojunctions. Heterojunctions and the radiation mechanisms will be discussed below.

Materials that allow the generation and emission of photons, i.e., of optical power, are semiconductors in which a p-n junction is able to emit light of a certain wavelength as a result of the application of an electric field. Generally, any direct-gap semiconductor in which a p-n junction can be formed by either thermal diffusion, ion implantation, or epitaxial deposition potentially is a useful optoelectronic material. In order to be an efficient light emitter, how-

TABLE 3.1. Characteristics of the Three Principal Optical Source Categories

Characteristic	Light-emitting diode	Semiconductor laser	Nonsemiconductor laser
Material	GaAs, GaAsAl	GaAsAl, GaAsIn	Nd:YAG
Light type	Incoherent	Semicoherent	Coherent
Wavelength (μm)	0.7–1.0	0.8–1.2	1.0–3.0
Spectral width (nm)	20–60	1.5–3.0	<0.5
Directivity coefficient	1–2	8–12	10–25
Radiance at 100 mA (W/sr cm^2)	0.1–100	1–1000	0.01–10
Emitting area (mm^2)	0.01–10	0.0001–0.01	0.1–1.0
Bandwidth (MHz)	10–150	200–1000	500–2000
Lifetime (hr)	10^4	10^5	10^4
Linearity	High	Medium	Low
Harmonic distortion	High	Medium	Low
Suitability for monomode fiber	No	Yes	Yes

ever, a useful optoelectronic material must meet three principal requirements:

1. *Photon-emitting transition.* It must either be a direct-gap semiconductor, allowing band-to-band transitions, or an indirect-gap semiconductor, allowing impurity level-to-band transitions.

2. *P-n junction formation.* It must be a material in which p-n junctions of suitable characteristics can be formed.

3. *Energy gap variation.* It must allow the variation of its energy gap by variations in its stoichiometry.

Few semiconductor materials meet all three of these requirements simultaneously. The most significant materials that satisfy all of the requirements are the alloy families $GaAs_{1-x}P_x$ and $GaAs_{1-x}Al_x$, with the two extremes GaAs/GaP and GaAs/GaAl, respectively, where the alloy composition is determined by the atomic fraction $0 \leq x \leq 1$. Some of these alloys are capable of light emission in the visible and near-infrared ranges, as shown in Figure 3.1. Semiconductor materials that are able to emit light in the near infrared are GaAsAl, InPAs, PbS, InSb, PbTe, PbSeTe, and several others, although most of them have only marginal interest for fiber optics because their properties do not match the requirements imposed by these systems. For fiber optics applications, the most important optoelectronic materials are semiconductors

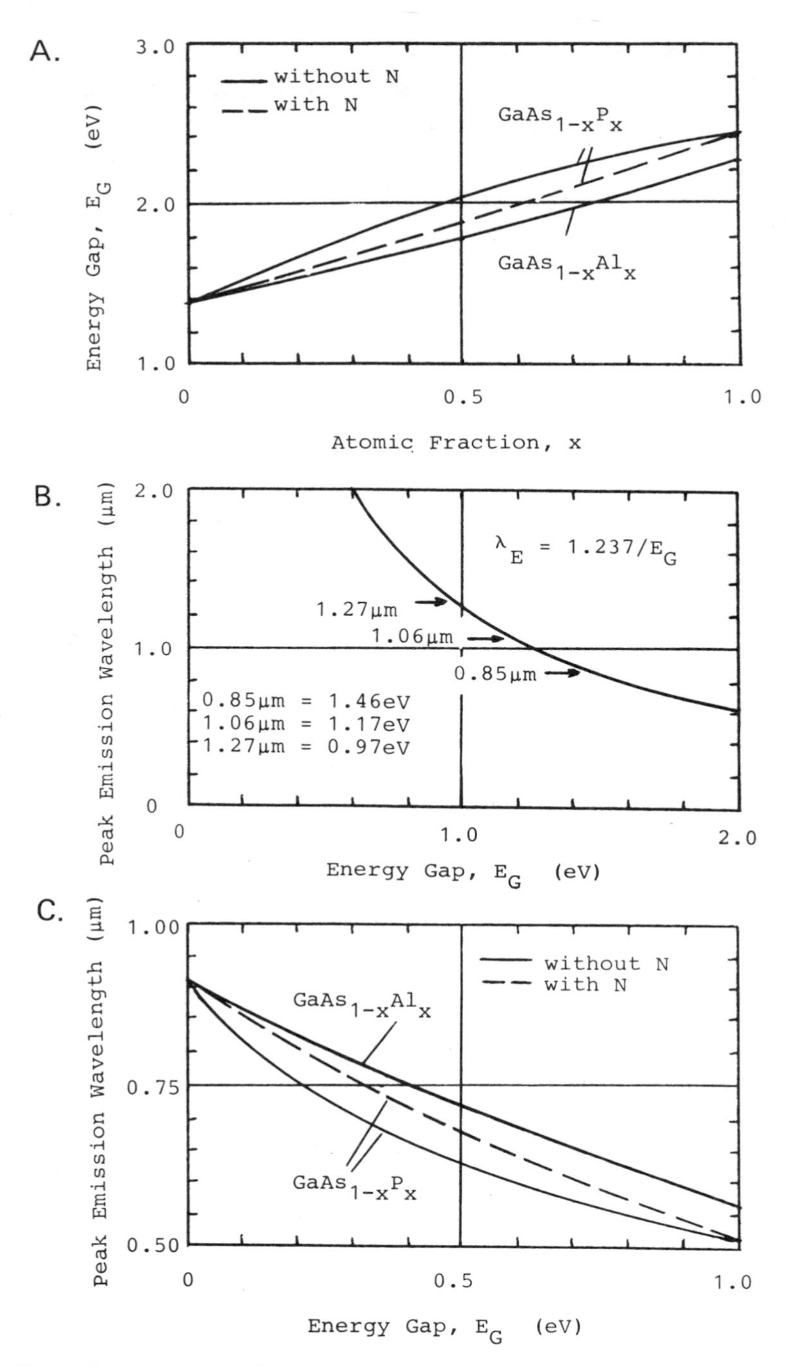

Figure 3.1 Variation of peak emission wavelength with atomic fraction and semiconductor energy gap for selected materials suitable for fiber optics source applications.

that produce light in the near infrared, i.e., those materials whose energy gaps are in the range $1.0 \leq E_G \leq 1.5$ eV. The energy gaps of some of the interesting materials are listed in Table 3.2.

The emission of photons from a semiconductor is best described by the energy band model of solid materials. This model assumes a conduction band mostly empty of electrons and a valence band mostly filled with electrons, separated by the energy gap. The wavelength of light emitted from a direct semiconductor (capable of band-to-band electron transitions), such as GaAsAl, corresponds to the energy gap, while that emitted from an indirect semiconductor (capable of impurity level-to-band electron transition) corresponds to the energy difference between the impurity level and the appropriate band edge. Hence, the wavelength of light emission of an indirect semiconductor differs slightly from that of an otherwise similar direct semiconductor. For direct-energy-gap materials the relationship between the wavelength of dominant (peak) emission and the semiconductor energy gap is given by

$$\nu h = E_G \tag{3.1a}$$

from which follows

$$\lambda_E = ch/E_G = 1.237/E_G \tag{3.1b}$$

TABLE 3.2. Characteristics of Selected Semiconductor Materials Suitable for Sources in Fiber Optics Applications

Material	Transition type*	Energy gap† E_G (eV)	Emission range† (μm)	Comments
GaAsAl	d, i	1.37–2.26	0.55–0.90	Film–substrate lattice mismatch
GaAsP	d, i	1.37–2.39	0.52–0.90	Most advanced technology
GaAlP	d, i	2.26–2.39	0.52–0.55	Stoichiometry control due to high Al reactivity
InGaP	d, i	1.34–2.39	0.52–0.92	Film–substrate lattice mismatch
InAlP	d, i	1.34–2.45	0.50–0.92	Film–substrate lattice mismatch
ZnSeTe	d	1.82–2.69	0.46–0.68	Large operating voltage, traps in material
SiC	i	2.80–3.20	0.39–0.44	High-temperature process, traps in material

*d, direct transition; i, indirect transition.

†Depends upon mole fraction x.

where λ_E and E_G are expressed in μm and eV, respectively; h is Planck's constant ($h \approx 4.14 \times 10^{-15}$ eV s), c is the speed of light in vacuo ($c \approx 3.0 \times 10^{10}$ cm/s), and ν is the frequency of the light.

In an indirect semiconductor, such as Si, transitions and hence optical emission are much less likely because the transitions are accompanied by the scattering of phonons for conservation of momentum, resulting in the consumption of energy that is not available for photon emission. Hence, in an indirect-gap semiconductor, light is emitted only when a suitable impurity is incorporated which is able to create energy levels that have a broad distribution of momenta. This will effectively convert an indirect-gap material to a direct-gap material. An example of a material using this process is GaP with nitrogen (N) incorporation. For silicon there are no impurities known to produce such a conversion. Thus, a useful optoelectronic material requires a direct-gap semiconductor, or an indirect-gap semiconductor with a suitable impurity to allow direct transitions.

The incorporation of N in $GaAs_{1-x}P_x$ and in a few other compounds is one of the important discoveries made during the development of light-emitting devices. Nitrogen acts as an isoelectronic trap because it has the same number of valence electrons as P or As. The most important property of the N isoelectronic trap is the short range (a few atomic distances) of its electric potential, causing the N trap levels to have a large spread in momentum. Therefore, an indirect-gap semiconductor (such as GaP) may become an efficient light emitter when N is incorporated. If the concentration of the nitrogen atoms is sufficiently high, enough N trap levels can be created to allow direct transitions. Another isoelectronic trap is the Zn–O complex in GaP, which results also in the efficient generation of light.

In a material in thermal equilibrium, no transitions across the energy gap occur in which an electron in the conduction band can recombine with a hole in the valence band; consequently, no light is emitted. This is so because the electronic states in the conduction band are mostly empty and those in the valence band are mostly full. However, if a nonequilibrium concentration of excited electrons in the conduction band and holes in the valence band can be generated simultaneously, then the probability of recombination between electrons and holes is high. If the material is a direct-gap semiconductor, light will then be emitted. This can best be achieved by the formation and forward biasing of a p-n junction. Thus, the second important criterion in selecting a useful optoelectronic material is that a suitable p-n junction can be formed in it.

The energy gap of some semiconductors can be altered by variation of the mole fraction x. For example, the combination of GaAs with another suitable binary compound, such as GaP, results in the ternary alloy $GaAs_{1-x}P_x$. Depending on the alloy composition, the energy gap of $GaAs_{1-x}P_x$ can be altered from about 1.4 eV for GaAs (direct energy gap) to about 2.2 eV for GaP (indirect energy gap). Because fiber optics applications require the exact matching of the wavelength of maximum emission of the source and the wavelength of minimum loss of the waveguide, the adjustment of the emission wavelength peak to achieve near-perfect matching is the third important requirement for a useful optoelectronic material. It must be remembered, however, that at a certain critical value of x the ternary alloy will change from a direct to an indirect material if one of the initial binary compounds is direct and the other is indirect. For $GaAs_{1-x}P_x$ the critical atomic fraction $x_c = 0.46$, with direct transitions occurring below x_c and indirect transitions above x_c.

The various $GaAs_{1-x}P_x$ and $GaAs_{1-x}Al_x$ alloys are fabricated by epitaxial growth on GaAs, GaP, or GaAl substrates. Since GaP substrates are transparent, they permit the collection of light from either side of the junction. Hence, higher quantum efficiencies are obtained with GaP substrates. Their transparency, however, allows light transmission to adjacent devices when several devices are arranged as an array. Two technologies are normally used in the fabrication of optoelectronic materials. These are vapor-phase epitaxy (VPE) and liquid-phase epitaxy (LPE). VPE uses the various constituents of an alloy which are introduced in vapor form, whereas LPE is the growth of epitaxial layers from a liquid which consists of molten Ga and other materials such as GaAsAl or GaAsP and dopants such as Zn (for p type) or Te (for n type). The simplicity of the LPE equipment and the fact that molten Ga gathers impurities during the epitaxial growth make the LPE technique superior to the VPE technique. However, it is incapable of handling a large number of wafers, and the stoichiometry of the epitaxy cannot be easily varied in LPE.

Thus, useful optical sources are exemplified mainly by GaAs and GaAsAl semiconductor lasers and light-emitting diodes; other semiconductor alloys have also been investigated and, as will be reviewed below, have characteristics that are in some cases superior to those of GaAsAl-based structures. The $GaAs_{1-x}Al_x$ and $Ga_{1-x}AsAl_x$ alloys used in both LEDs and ILDs have the attractive feature that a variation of the mole fraction x allows the emission peak wavelength to shift over a relatively wide range, from about 0.78 to 0.90 μm, without significantly affecting the emission efficiency. With present GaAsAl devices, at a drive current of 100 mA the optical power launched into a fiber is typically +5 dBm for an LED and +10 dBm for an ILD. However, the power launched depends also on the numerical aperture of the fiber and is less for low-NA waveguides (1). This is illustrated in Figure 3.2.

Although for many fiber applications demonstrated power levels and extrapolated reliabilities are acceptable, output stabilities are frequently not adequate. To couple an appreciable fraction of output power from a source into an optical waveguide, particularly one of small numerical aperture, requires a laser source that is invariant with both time and drive current. Hence, lateral mode stability is an important requirement for high-performance lasers. Lasers that support higher-order lateral modes that result in multiple beams also have limited power collectable from any single lobe. Most injec-

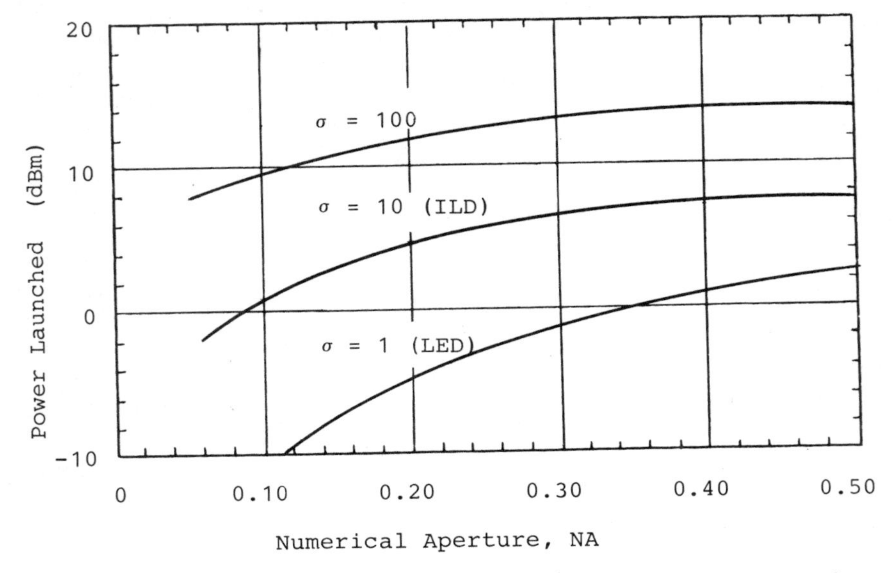

Figure 3.2. Typical relationship between the power launched by optical sources and the numerical aperture of the connecting waveguide.

tion lasers exhibit some degree of modal instability in the lateral direction parallel to the junction, owing primarily to the lack of abrupt refractive-index steps in this direction. This causes nonlinearities (kinks) in the input–output characteristics that correspond to changes in the lateral modes. Hence, a laser optimally aligned at one drive level may lose alignment as it traverses a kink.

LIGHT-EMITTING DIODES

Traditionally, light-emitting diodes represent the oldest technique applicable to fiber optics that is able to provide incoherent light by carrier injection. Although the promising potential of optical information transfer is partially derived from the laser, a source of coherent light is not necessarily required for guided transmission, and relatively-low-capacity fibers can operate by using an incoherent light emitter. Hence, the light-emitting diode represents an inexpensive, relatively reliable source of incoherent optical power. Radiation is generated by the recombination of carriers injected across the p-n junction of the diode. The recombination process determines the speed or frequency response of the device. It typically allows a pulse rate less than about 100 Mb/s, whereas present semiconductor injection lasers are capable of emitting semicoherent light at a maximum pulse rate of several hundred Mb/s.

Early light-emitting diodes consisted of a single p-n junction formed in GaAs. Because these structures contained only one junction, they were referred to as homojunction devices. Advances in liquid-phase epitaxial (LPE) growth techniques that allowed the precise formation of several deposited layers and hence p-n junctions in GaAsAl led to the subsequent fabrication of devices that contained multilayer structures. Because of the existence of several junctions in these diodes they were referred to as heterojunction devices. At present, heterojunction light-emitting diodes and injection lasers can be reproducibly fabricated in any $GaAs_{1-x}Al_x$ alloy, ranging from GaAs (where $x = 0$) to GaAl (where $x = 1$) (2). The internal quantum efficiency (defined as the number ratio of generated photons to injected carrier pairs) of typical LEDs is as high as 0.8 (i.e., 80%), but the external quantum efficiency (defined as the number ratio of emitted photons to injected carrier pairs) of a single p-n junction device is rather low and normally does not exceed about 0.1 (i.e., 10%). The external quantum efficiency of the basic device can be considerably enhanced by reflectors and dome-shaped structures that tend to increase the directivity of the emitted radiation.

The total external transfer efficiency of light-emitting diodes, defined as the ratio of optical output power to electrical input power (given in lm/W), can be approximated by

$$\eta_e = P_{opt}/P_{el} = 6.8 \times 10^2 V(\lambda)\eta_a \qquad (3.2)$$

where $V(\lambda)$ describes the sensitivity of the human eye as a function of wavelength λ, and η_a is the external quantum efficiency (3, 4). Light generation is largely confined to semiconductors of direct energy gap. In these materials carrier transition across the energy gap is accomplished by the generation of photons but not of phonons. The optical power emitted from semiconductors of small energy gap is also a function of the distance of the location of light generation from the semiconductor surface. These semiconductors (which typically have energy gaps less than about 2 eV) are characterized by a small carrier lifetime and hence a small diffusion length L_D but also a high optical absorption. Therefore, it is necessary to generate the light, on the one hand, relatively close to the surface so that it will not be absorbed by the semiconductor and, on the other, sufficiently far from it to avoid the recombination of a large percentage of the carriers without light emission. The optimum depth at which the light is generated is given by

$$x_{jo} = [L_D/(1 - \alpha_c L_D)] \ln \alpha_c L_D \qquad (3.3)$$

Hence, in an example and assuming $\alpha_c = 10^4$ cm^{-1} and $L_D = 1$ μm, the optimum junction depth is about $x_{jo} = 1$ μm. This means that the p-n junction should lie about 1 μm under the semiconductor surface in order to maximize the external quantum efficiency. The relationship between x_{jo} and L_D is depicted in Figure 3.3. In addition, the surrounding material should have a high refractive index, so that most of the light can be emitted at the first encounter with the surface. The fraction of the light that can leave the semiconductor can be expressed by

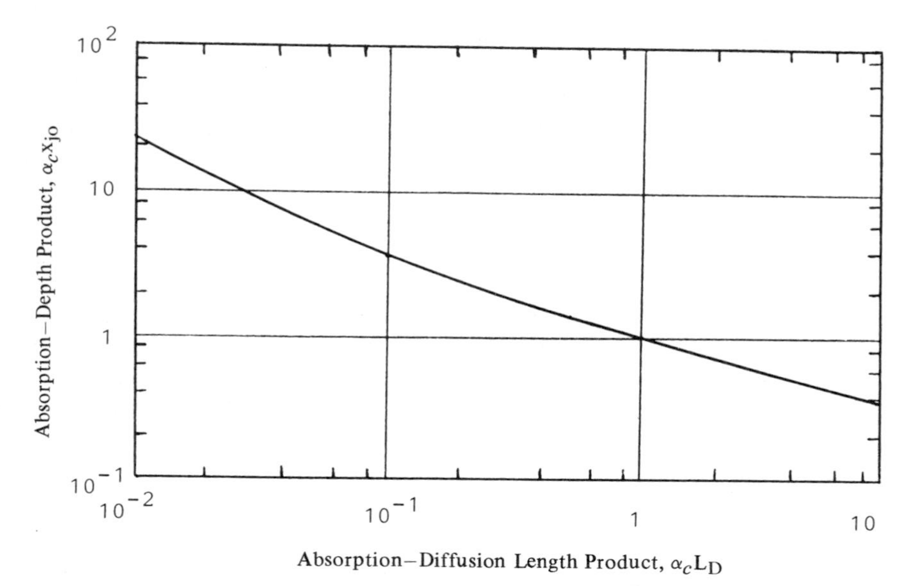

Figure 3.3 Relationship between optimum source junction depth and the product of carrier diffusion length and photon absorption coefficient.

$$\eta_{\text{opt}} = 1/(1 + \alpha_c V_c/TA_c) \tag{3.4}$$

where V_c and A_c are chip volume and surface area, T is the average transmission (for example, 0.02 for GaP), and α_c is the absorption coefficient. This means that an LED consisting of a cube whose side length is 1 mm yields an emission efficiency of 0.7 (i.e., 70%) as a ratio of the externally emitted light to the internally generated light. Contacts to the junction tend to reduce this value, so that a maximum efficiency of about 0.5 can be expected.

In direct materials, primary emphasis of research efforts is normally directed to high emission efficiency rather than to high generated light intensity. In indirect materials the reverse is usually the case; here, one is interested mainly in achieving a high internal light intensity, together with a small absorption coefficient. Consequently, the addition of Si as a dopant to GaAs has been used to reduce the absorption coefficient.

A distinction can be made between small-area high-brightness and large-area low-brightness devices. The high-brightness devices achieve their larger optical output per unit area (radiance) either by a higher current density or by a more efficient conversion of electrical energy to optical energy. However, a higher drive current density is usually associated with a shorter device lifetime. Large-area LEDs, on the other hand, are not well suited for coupling into single fibers because owing to the larger source area compared to the core cross-sectional area a high coupling loss is experienced.

Types of LEDs

Although light-emitting diodes have severe disadvantages if they are to be used in optical communication systems of high sophistication, they offer a practical solution in inexpensive systems of relatively low data transmission capabilities. There are several types of LEDs available which differ in structure and packaging techniques. The more important of these are illustrated in Figures 3.4 and 3.5 and are described in the following.

Planar LED. This diode is the simplest of the various structures that are available and is formed by techniques generally known to the semiconductor industry. It is fabricated by the diffusion of p-type impurities into an n-type GaAs substrate. Current flow through the junction causes spontaneous isotropic emission which impinges on the sidewalls and horizontal surfaces of the device. Because of the refractive index difference between GaAs ($n = 3.6$) and the surrounding air ($n = 1.0$) or epoxy ($n = 1.6$), only the internal emission not experiencing internal total reflection is emitted. This corresponds to an internal maximum angle relative to the normal of 16° for air and 26° for epoxy. It results in a limited amount of available or external emission.

Dome LED. This diode has a higher emission efficiency than the planar LED because of its chip geometry. The n-type GaAs forms a hemisphere around the diffused junction, which has a smaller diameter. The dome diameter is chosen such that all of the internal emission arrives at the boundary within the critical angle given by the refractive indexes of diode material and surrounding material. Thus, the intensity distribution of the external emission is uniform. However, because of geometric considerations the hemisphere must be larger than the radiating junction area. A further improvement in

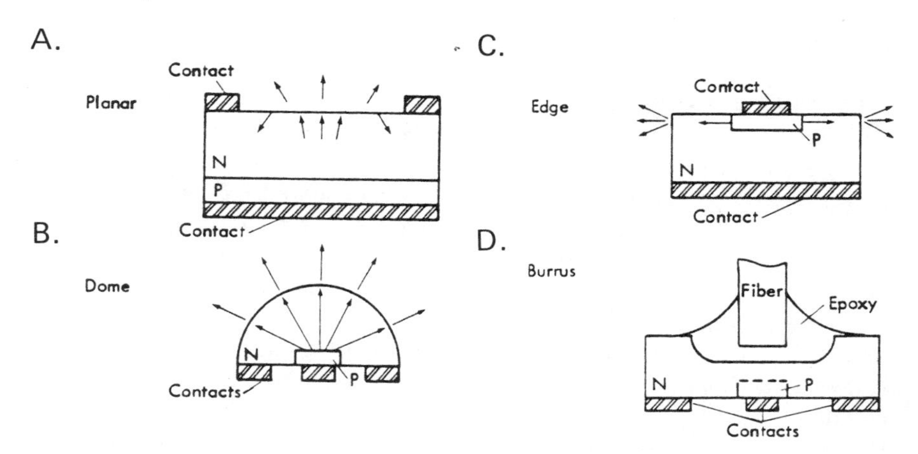

Figure 3.4 Schematic illustration of the structure of the most prominent light-emitting diode classes.

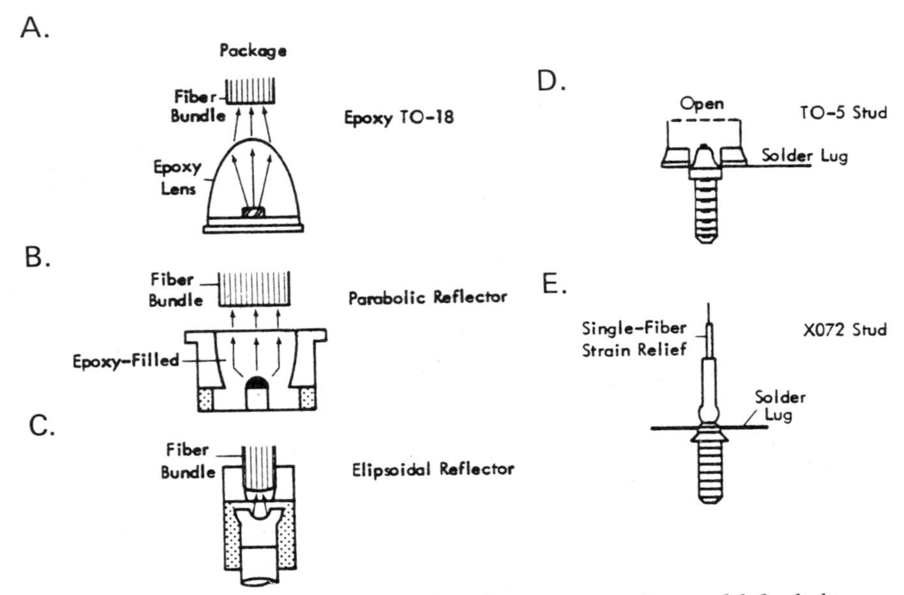

Figure 3.5 Examples of suggested packaging approaches useful for light-emitting devices.

angular radiation characteristics can be realized by using a truncated sphere rather than a hemisphere. The truncated sphere acts as a lens, similar to the immersion lens in a microscope objective, so that much better collimation can be achieved. Because both hemisphere and truncated sphere devices have large effective emission areas and hence relatively small radiance, they are useful as sources mainly for coupling to large-diameter fiber bundles.

Edge LED. The maximum emission of this diode is generated from the sidewalls. This device allows easier collection and focusing of the emission within a smaller beam angle for a given size reflector aperture than the dome LED. Furthermore, its fabrication is simpler. Although the total radiated power is usually slightly smaller than that of the dome LED, its on-axis radiance is higher because of a smaller emitting area. A typical square-edge LED has a side length of about 100 μm; the light is emitted along four sides of a small square-shaped emitter which has a total external efficiency of about 3%. However, only one of the four sides can be used to couple optical power into a single fiber whose core is butted against the emitting edge. Although the optical power is about 1 W at a drive current of 100 mA, only a tenth of this can be coupled into a fiber with NA of 0.3 because of the Lambertian emission pattern, which allows only a small portion of the light to be collected. For coupling to large fiber bundles, a circular or square LED chip is mounted at the focal point of a small ellipsoidal reflector that collimates the total emitted radiation. The radiance of such a device, if used to couple into fiber bundles, is low in spite of a large total output power owing to the relatively large total emission area.

B-LED. This high-radiance diode (frequently referred to as a Burrus diode, after its inventor) is a very useful source, although its principal application is in conjunction with single fibers to which it is directly attached (5). In its simplest form it is similar to a planar LED, with the diameter of the emitting area defined by the bottom p-type diffusion and/or bottom contact. This design produces a near-perfect match of emitting and core areas and allows the smallest emitting area of all LED types. It achieves the highest current density and hence the highest radiance. A well etched in the top n region permits close butting of the single fiber with the emitting area for maximum emission collection and minimum absorption. Representative types have either a simple homojunction structure or a complex heterojunction structure. Its main advantage is its high radiance and hence its large coupled power to a single fiber.

Properties of Various LEDs

The light-emitting diodes described above are compared in Table 3.3. It is apparent that the device with the highest radiance and the smallest active area is the Burrus diode. The planar LED, on the other hand, yields the lowest light intensity per unit of emission area, and its emitting area is largest, but the achievable bandwidth and hence potential data transfer rate is highest; it also has economic advantages. Dome and edge LEDs in reflector packages typically provide the smallest packaged aperture diameters for the low-

TABLE 3.3. Comparison of the Present Characteristics of the Major Classes of Light-emitting Diodes

	LED type			
Characteristic	**Planar**	**Dome**	**Edge**	**Burrus**
Active area (mm^2)	8–15	6–12	2–4	0.01–0.05
On-axis radiance at 100 mA (W/sr cm^2)	0.2–0.5	0.2–0.5	0.3–1.0	6–60
Total radiated power at 100 mA (mW)	1–3	2–4	1–3	0.5–10
Power coupled into single fiber (μW)	0.1–1.0	0.5–5	1–10	30–300
Half-power emission angle (degrees)	±5	±10	±15	±45
External quantum efficiency (%)	1–5	30–40	1–5	45–55

radiance types, the highest power output, and adequately collimated emission. These devices are able to couple the largest power into fiber bundles. Based on information in Table 3.3 and on other considerations, the advantages and disadvantages of these diode structures in practical use are analyzed in Table 3.4. Again, the benefits of the planar LED and of the B-LED are evident and depend on the selection criterion applied.

The comparison of the various light-emitting structures implies that small-area, high-radiance, etched-well devices are optimal among light-emitting diodes for coupling into single fibers. However, for coupling into fiber bundles the optimum source design is the large-area, high-radiance, intensity-shaped emitter. The Burrus diode is most suited for coupling optical power into small-diameter monomode step-index fibers, and it is not useful in conjunction with fiber bundles. This is shown by the fact that one of the limiting factors in system performance is the relatively small coupling efficiency of source into fiber. With the trend toward smaller numerical fiber apertures, techniques that use lenses or reflectors become less advantageous. Therefore, butting the fiber directly to the source is an increasingly desirable approach.

TABLE 3.4. Comparison of the Advantages and Disadvantages of the Major Classes of Light-emitting Diodes

LED type	Advantages	Disadvantages
Planar	Potentially highest bandwidth and data transfer rate Easy fabrication and packaging	Largest emitting area Lowest emission radiance Smallest coupled power into fiber
Dome	High radiated power Well suited for coupling into bundles Easy collection and focusing within small beam angle Small packed aperture diameter	Difficult fabrication Requires parabolic reflector for improved radiance
Edge	Very large radiance Well suited for coupling into bundles Smallest packed aperture diameter of the low-radiance types	Emission from side rather than top of device; hence, difficult packaging approach Requires elliptical reflector for improved radiance
Burrus	Smallest emission area Largest half-power emission angle High current density No need for collimating optics or reflectors Allows mounting in stud package	Unsuitable for fiber bundles DH structure requires complex fabrication process

The Burrus diode, in a numerical example, has a radiance on the order of 100 W sr^{-1} cm^{-2} and allows the coupling of about 1 mW of optical power into a fiber whose numerical aperture is 0.30 and of about 0.2 mW into a fiber whose numerical aperture is 0.15. The conventional Burrus LED frequently uses the double-heterojunction structure and is fabricated by the liquid-phase epitaxial growth process together with etching to form a well for the fiber. In a modification of the basic process for the B-LED, many microlenses are created by photographically defined etching. Epitaxial regrowth in the emitting structures increases the net incident angle for light to be emitted and facilitates the collimation of the emitted light.

The GaAsAl heterojunction LED in which the active area is limited by the edge of the chip yields a relatively high quantum efficiency. It represents a highly directional source. This device, known also as the large optical cavity (LOC or OC) laser, can be considered to be either a LED or an ILD depending upon whether it is operated below or above a certain threshold current level. The cavity, which forms the active region of this device, is defined by two heterojunctions. Its emission power decreases with cavity length because of an increase in internal light absorption. Thus, it is desirable to restrict the active area as much as possible to the radiant edge. The variation of the output power of the OC device with cavity length is shown in Figure 3.6 for a normal edge-emitting heterojunction diode and one in which the emitted radiation is enhanced by a restrictive groove that provides partial reflection at the oppo-

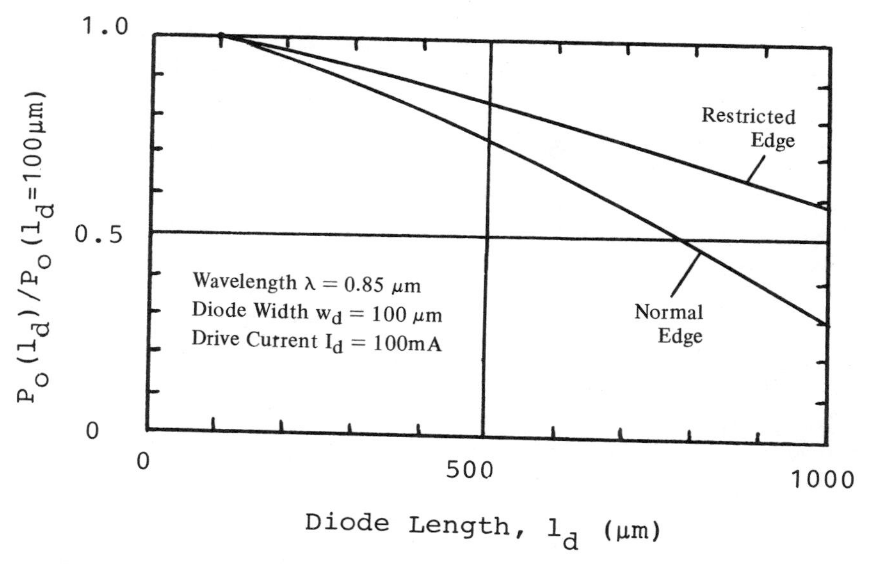

Figure 3.6 Relationship between optical output and length of a GaAsAl heterojunction light-emitting diode with and without reflective back surface (diode length l_d = 100 μm).

site radiant edge. A further discussion of this device is found below in the treatment of injection lasers.

The near-linear relationship between electrical input and optical output of all LEDs, contrary to the input–output relationship of lasers, allows the use of LEDs in both analog and digital applications, although the driver designs differ in both types of uses. The linear input–output relationship of LEDs, however, is largely restricted to low current levels, whereas at high current levels saturation may set in.

Two of the most important properties of light-emitting diodes are radiance and frequency response. Operation of LEDs at high current level tends to increase the optical output power, but it also has a detrimental effect on device lifetime because of an increase in junction temperature. A heat sink is therefore important at high operating current. Frequently, the chip is attached to a copper, BeO, or Al_2O_3 heat sink. Another useful technique is the use of a thin layer of In solder for bonding, which results in a thermal resistance on the order of 10°C/W from junction to ambient. Generally, the methods employed to reduce the junction temperature are similar to those developed by the semiconductor industry for high-power devices. To use a numerical example, a current level of 300 mA typically yields an output power of about 5 mW in the pulsed mode and of about 3 mW in the continuous mode of operation.

The modulation frequency of LEDs, as of other semiconductor diodes, is also of considerable importance. In part, it determines the achievable data transfer rate. As in other devices, the response time of LEDs to a variation of an input signal depends upon the delay associated with the charging of the junction capacitance and upon the minority carrier lifetime. In small junction areas the capacitance-related delay time is usually less than about 1 ns, whereas the minority carrier-related lifetime is typically of the order of 1 to 10 ns at medium injection levels, with decreasing lifetime at higher injection levels, depending on the recombination mechanism. Radiative carrier lifetime can be affected by controlled doping with impurities. Si-doped GaAsAl LEDs have a poor frequency response compared to Zn- and Te-doped devices, with a difference in modulation frequency of about an order of magnitude for comparable doping levels. Typically, maximum modulation frequencies up to 50 MHz have been observed, although some experimental Ge- and Te-doped GaAsAl LEDs have been modulated at frequencies in excess of 200 MHz. However, the increase in frequency response is usually achieved at the expense of emission efficiency. Hence, the choice of the dopant type and concentration must take into account the tradeoff between speed and efficiency.

The structure of a high-radiance LED, fabricated by the LPE process, is shown in schematic form in Figure 3.7. In order to minimize internal absorption, all layers contain more aluminum than the active layer. Current confinement is obtained by means of a hole etched in an insulating phosphosilicate glass, which results in an active area of 60 to 100 μm diameter. A further

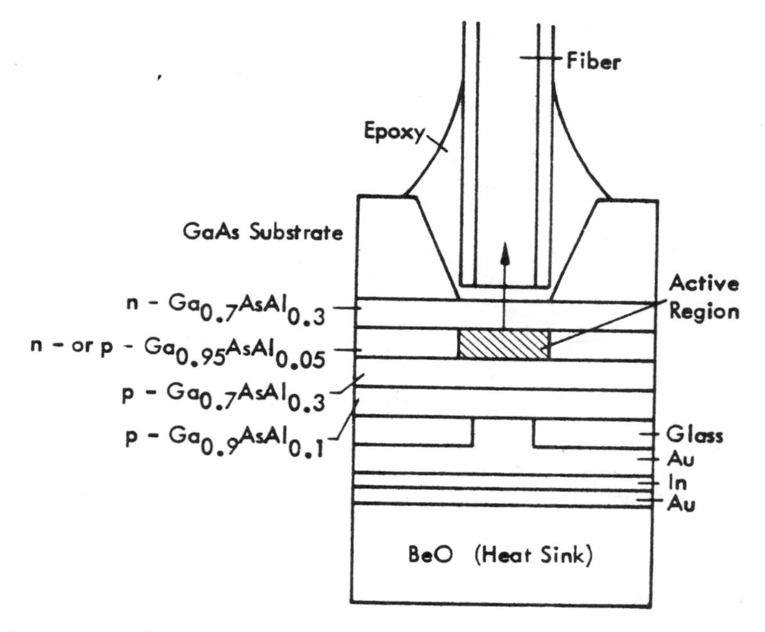

Figure 3.7 Schematic diagram of a high-radiance double-heterojunction GaAsAl light-emitting diode.

means of achieving high-current operation is facilitated by the Au–In–Au layers and the BeO heat sink. Such a device is capable of emitting optical power near a wavelength of 0.85 μm. Its spectral bandwidth can be reduced to about 30 nm (300 Å) by Sn doping rather than Ge doping (which yields about 50 nm bandwidth). Near-linear operation of this device can be achieved under pulsed conditions up to about 6000 A/cm^2 and under cw conditions up to about 3000 A/cm^2. An example of the input–output characteristic of such a device is shown in Figure 3.8.

As mentioned before, there is a trade-off between frequency response and emission efficiency of an LED. This is shown in Figure 3.9 for a typical high-radiance source, in which efficiency and bandwidth are given as a function of Ge doping concentration for otherwise identical structural detail. Similar trends are observed for other light-emitting diodes. An increase in Ge weight content above about 50 mg/g GaAsAl, corresponding to 5%, leads to a significant decline in quantum efficiency, although the bandwidth continues to increase with increasing doping level. An optimum is achieved at about 5% Ge content; this yields a quantum efficiency of about 4% and a bandwidth of about 200 MHz. In the doping range close to optimum, the linearity of the input–output relationship is insensitive to the doping concentration. The smallest distortion, however, about 0.06 dB/°C at about room temperature, is achieved when the Ge content is about 10%.

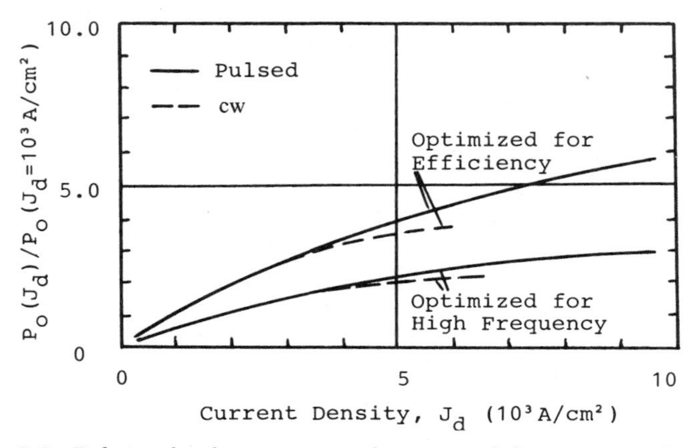

Figure 3.8 Relationship between optical output and drive current of a high-radiance light-emitting diode under pulsed and continuous wave conditions (current density 10^3 A/cm^2).

Whereas GaAsAl LEDs are useful optical sources in the 0.8 to 0.9 μm wavelength range, GaAsIn devices can be used in the 1.0 to 1.1 μm range. These LEDs are usually fabricated by VPE or LPE growth of the GaAsIn layer on a GaAs substrate. Because of the slightly different crystal structures of substrate and film and the resulting strain near their interface, usually a composition grading is employed, whereby a change in the mole fraction x from 0 in the GaAs substrate to a maximum 0.15 in the GaAsIn layer (resulting in an alloy of composition $Ga_{0.85}AsIn_{0.15}$) is achieved. As in other semicon-

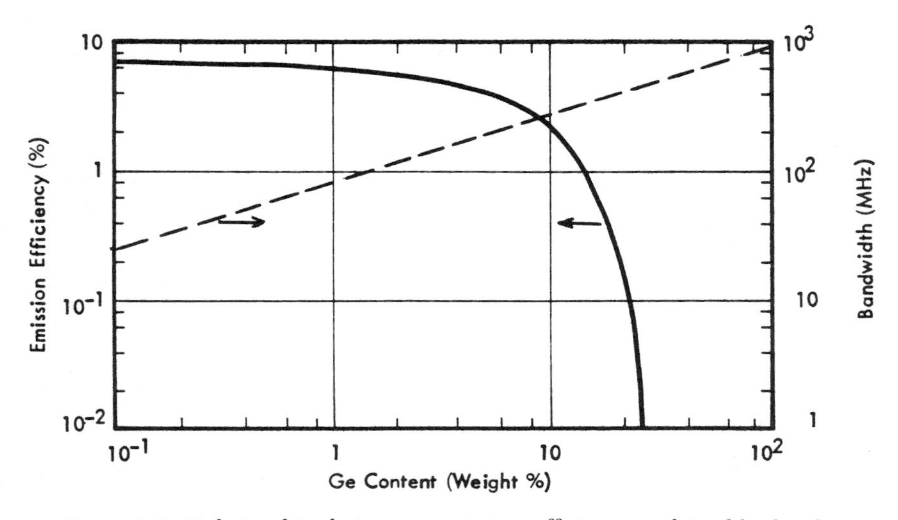

Figure 3.9 Relationship between emission efficiency, achievable bandwidth, and Ge content of a high-radiance light-emitting diode.

ductor alloys, a variation of the mole fraction causes a shift in the peak emission wavelength, so that a decrease in In content results in a decrease of the emitted wavelength. The GaAsAl and GaAsIn layers may contain either a homo- or a heterojunction. For GaAsIn heterojunction LEDs in which $x = 0.15$, an external quantum efficiency of about 0.01 is obtained; in devices in which $x = 0.05$, the external quantum efficiency is about 0.002 at room temperature.

The conversion of the electrical input power into the undistorted optical output power is accomplished by the use of appropriate driver circuits. These circuits establish the correct signal and prebias levels and normally feed directly into the light-emitting devices. In general, the demands on driver circuits for use in analog systems are considerably more stringent than those on digital systems; hence, digital circuits tend to be much less sophisticated and therefore less expensive. As a result, digital techniques are being used increasingly in practical applications, since only two signal states (0 and 1) have to be distinguished, whereas in analog operation faithful reproduction of the input signal by the output signal has to be accomplished over the entire current range (6).

INJECTION LASERS

Although nonsemiconductor lasers are receiving considerable attention for potential optical communications use, the majority of the recent research on lasers has been devoted to semiconductor lasers, partly because of their compatibility with other semiconductor devices. The edge-emitting semiconductor injection laser diode (ILD) offers several distinct advantages over the simple surface-emitting LED. For this reason it is of major interest in optical communications. Among the principal advantages of semiconductor lasers are the smaller emission area, higher radiance, narrower spectral width, and higher frequency response. Another advantage, higher directivity, is of substantial benefit in high-bandwidth systems, but it requires much better compatibility of source and fiber compared to the use of light-emitting diodes.

The development of semiconductor injection lasers began in the early 1960s. The earliest injection lasers had the form of a Fabry-Perot cavity. They consisted of rectangular GaAs chips and contained a p-n junction perpendicular to two polished or cleaved ends that acted as partial mirrors. Light in these devices is generated by the injection of electrons into the p-region, with subsequent radiative carrier recombination in the active region (which can be considered as a cavity). The basic structure of this device is illustrated in Figure 3.10. Such a laser, referred to as a homojunction structure because it consists of a single material, typically has a high threshold current density, far in excess of 10^4 A/cm^2. This is a consequence of the lack of control possible over the thickness of the active recombination region and because of the poor

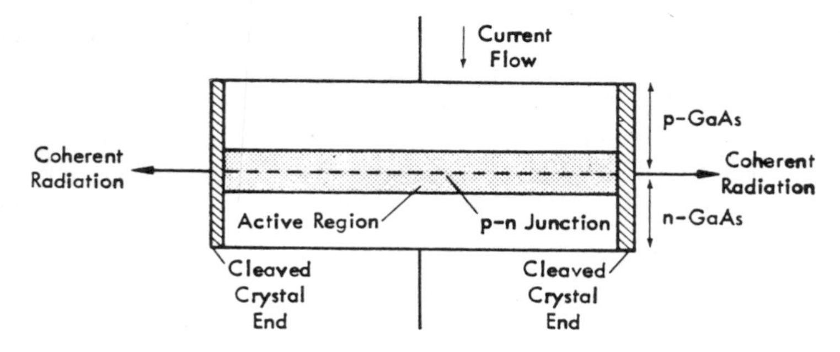

Figure 3.10 Schematic illustration of a Fabry-Perot cavity homojunction injection laser.

characteristics of the active region as an efficient waveguide, with a substantial portion of the optical power lost to the adjacent n-type and p-type regions.

Major Laser Types

Initially, all semiconductor lasers were simple homojunction structures made by Zn diffusion into n-type GaAs. Advances in the controlled liquid epitaxial growth of GaAs and GaAsAl layers led to significant improvements in the reduction of the threshold current density and the enhancement of the external emission efficiency.

The historic development of the injection laser leads from the homostructure 20 years ago to the double heterostructure and the striped and distributed feedback lasers in use today. The development of devices of this class has not yet been completed, and further advances are expected. A brief description of the major laser types follows.

Homojunction Laser. In the earliest demonstration of semiconductor injection lasers in the early 1960s the homostructure was used, implying that only one material, GaAs, was employed. In these devices the electrically injecting p-n junction was fabricated by diffusing a p-type dopant (Zn) into a previously homogeneous GaAs crystal. In these early structures waveguiding did not yet exist, so that the lack of optical confinement caused high emission efficiency losses. Although the threshold current was high, it was lower than predicted, which was correctly interpreted as being caused by small variations in the material refractive index due to variations in the carrier densities. Thus, a small amount of optical confinement was observed in the gain region, although it was not originally intended.

Single-Heterojunction Laser. The realization that improved confinement could be achieved in heterojunction structures of wider-energy-gap and lower-refractive-index material on either side of the gain region led to further research on these devices. In the heterojunction structure the active region is defined by adjacent regions of a slightly differing material, whereby optical confinement within the active region is caused by the lower refractive index of the bounding regions. It was noted that these devices produced much higher injected minority carrier concentrations and optical field densities at lower current densities than homojunction devices. In the late 1960s the first single-heterojunction laser was demonstrated, where a p-type GaAsAl region in close proximity to a p-n junction in GaAs caused a large change in both energy gap and refractive index, but only on the p side of the junction. The threshold current of this device was about one-tenth that of the homojunction laser, but it required a relatively large pumping current because carriers and photons were not well confined to the n side of the junction.

Double-Heterojunction Laser. The double-heterostructure laser, in which GaAsAl layers were fabricated on both sides of the gain region, achieved further reduction in the threshold current level. This device was capable of strong optical waveguiding and current confinement on both sides of the p-n junction, and the threshold current was reduced by another order of magnitude. Today, threshold current densities of about 10^3 A/cm^2 and laser currents of approximately 1 A at room temperature are typical. These devices, if used with an appropriate heat sink, can be operated continuously. As a result, heterojunction lasers constitute the most important laser class. In addition, these devices form the basis of integrated optical sources.

Stripe Laser. Further progress has been made by the use of two-dimensional optical waveguides and current-confining techniques, of which the stripe geometry configuration is the most common (7–14). In such a structure, the heterojunction layers guide the light in the vertical direction, while in the lateral direction the light is guided by the variation of the refractive index resulting from gain in the region underneath the stripe. The weak light confinement in the plane of the p-n junction is produced by the carriers injected in the region under the stripe (15).

Other Considerations of Injection Lasers

Although the two-dimensional stripe laser is characterized by low threshold current (on the order of 100 mA), the optical output frequently does not increase linearly with pumping current. Such a nonlinearity in output, called kink, frequently observed at an output power of a few milliwatts per facet, severely restricts the utility of this device because it is difficult to predict the output power based on the injected current. Attempts to eliminate the kinks have centered on the use of apertures, whereby the beam is laterally stabilized, or by the use of a physical waveguide structure along the plane of the p-n junction to replace the gain-induced waveguide. Such a device is known as a buried-heterostructure laser. In it the light is guided in the center of a moon-shaped active region which is completely enclosed by wider-energy-gap, lower-index GaAsAl; thereby a two-dimensional waveguide structure is achieved in which no kinking occurs at up to about four times the threshold current.

The use of a dichroic filter deposited onto one of the laser facets can further reduce the threshold current significantly (16). The reflector consists of multiple layers of Si and Al_2O_3 and provides reflectivity in excess of 99% over the 0.7 to 1.0 μm range. With these, lasers can operate at powers as low as about 25 mA at an external quantum efficiency of approximately 15%.

A semiconductor laser consists, as we have seen, in its simplest form of a p-n junction whose lateral dimensions are limited by the semiconductor boundaries. The semiconductor dimensions also define two reflecting surfaces, perpendicular to the junction area, between which standing waves of stimulated light occur. Laser action in a semiconductor depends upon two principal conditions: establishment of an inverted population, and excess of stimulated emission gain over radiation loss.

The population inversion condition is related to the stimulated emission gain. This is a consequence of the proportionalities of recombination probability and carrier density product and of electron-hole pair generation probability and carrier deficiency product. The concept of an inverted carrier population is of fundamental importance for an understanding of the operation of the injection laser. Thus, laser action requires a situation in which the product of electron and hole densities, np, during dynamic excitation of electrons across the energy gap exceeds the carrier deficiency product $(N_v - p)(N_c - n)$. The condition for light amplification or photon stimulation is therefore

$$n/N_c \ll 1 - p/N_v \tag{3.5}$$

In order to obtain stimulated gain, the product of the density of electrons at the bottom of the conduction band and of holes at the top of the valence band must exceed the product of the density of electrons at the top of the valence band and of vacancies at the bottom of the conduction band. When a semiconductor is stimulated by light of energy equal to the energy gap, radiative recombination is more probable than pair generation. Stimulated light emission occurs only if the forward current across the junction is of sufficient magnitude to generate an inverted population. This forward current must inject an adequate number of electrons from the n region into the p region. Otherwise there will be no population inversion in the p region adjacent to the junction.

The thickness l_{pi} of the region in which population inversion occurs is the active region of the laser. In this region, band-edge radiation is reflected between the two mirror surfaces. As the band-edge radiation traverses the active region, it stimulates radiative recombination in the inverted population and is thus amplified. Power is extracted through the reflective surfaces. To achieve high efficiency, these surfaces must be planar, parallel, and highly reflective.

In the following we determine the influence of the frequency of light on the photon stimulation process. We irradiate a semiconductor with light of a frequency first greater than and then equal to that corresponding to the energy gap. If the material is irradiated with light of an energy greater than the energy gap, then both the electron density at the bottom of the conduction band and the hole density at the top of the valence band are increased by equal amounts. Thus, a light energy greater than the energy gap of the semiconductor will not result in stimulated emission gain. If, however, the semiconductor is irradiated with light of energy equal to the energy gap ($h\nu_0 = E_G$), transition of excess electrons back into the valence band may be stimulated. As a result, light of frequency ν_0 will be stimulated, since each emitted photon will trigger another conduction band electron to recombine. As these photons transverse the lattice, avalanche photon emission will take place.

The electron population inversion ensures that there will be only insignificant photon loss by the excitation of electrons back into the conduction band at the beginning of the process. As the light burst builds up, however, it will return large numbers of electrons to the valence band by normal recombination, so that photons will begin to be lost. The avalanche then diminishes. A finite time interval is required to obtain another avalanche. Continuous-wave operation would be possible only if the pumping rate exceeded the avalanche process.

The density of the forward current that must flow across the p-n junction in order to result in population inversion is

$$J_i = 2qn_i\,(D_n^*/\eta)L_n\ \exp(l_{\text{pi}}/L_n + h\nu_0/2kT) \tag{3.6}$$

where D_n^* is the effective electron diffusion coefficient as modified by high-level injection and carrier scattering, $L_n = (D_n^*\tau_n)^{1/2}$ is the electron diffusion length, τ_n is the electron lifetime, η is the internal quantum efficiency, and n_i is the intrinsic carrier density. If we assume that radiation losses from the active region are negligible, that the carrier injection efficiency is 0.5, and that the quantum efficiency is unity, then the laser threshold current density at the onset of laser action is

$$J_i = 2qn_i \left(\frac{D_n^*}{\tau_n}\right)^{1/2} \exp\left(\frac{20\lambda}{(D_n^*\tau_n)^{1/2}} + \frac{h\nu_0}{2kT}\right) \tag{3.7}$$

where λ is the wavelength of the emitted light.

Photon losses in a semiconductor laser result in a reduction of emission efficiency. They are the result mainly of the following causes.

Absorption Loss. This loss mechanism is due to electron–hole pair generation. Although it usually dominates, it is taken into account by the inversion condition, which requires that more photons be generated than lost by pair generation; otherwise no emission will take place.

Diffraction Loss. This is the second most important loss mechanism in a laser. It can be minimized by making l_{pi} large compared to λ in order to prevent the diffraction spreading of the laser beam; equal diffraction and absorption losses are found if $l_{pi} \approx 13\lambda$.

Reflection Loss. This loss takes place at the two reflecting surfaces and is small relative to absorption and diffraction losses.

Scattering Loss. This loss contribution is caused mainly by interaction of photons with lattice impurities and is normally negligible.

Increasingly, there is a trend toward raising the wavelength of the emitted power from below 1 μm to the range of 1.1 to 1.3 μm. For this higher-wavelength region the use of GaAsAl lasers is no longer possible, so that other compounds, such as InGaAsP/InP, have to be employed. These materials can be prepared by vapor-phase epitaxial deposition. Presently, current densities below 2000 A/cm^2 can be achieved with these devices. However, there is an undesirable leakage current from the InGaAsP layers into adjacent InP layers because of an energy gap difference. This leakage current is proportional to $e^{-\Delta E_G/kT}$, where E_G represents the energy band gap and T is the temperature. The leakage current thus increases rapidly at small energy gap difference and at high temperature, resulting in higher threshold current and lower conversion efficiency, as will be discussed in further detail below.

The important stages in the development of injection lasers are shown in Figure 3.11, which illustrates the basic structures of the four major laser

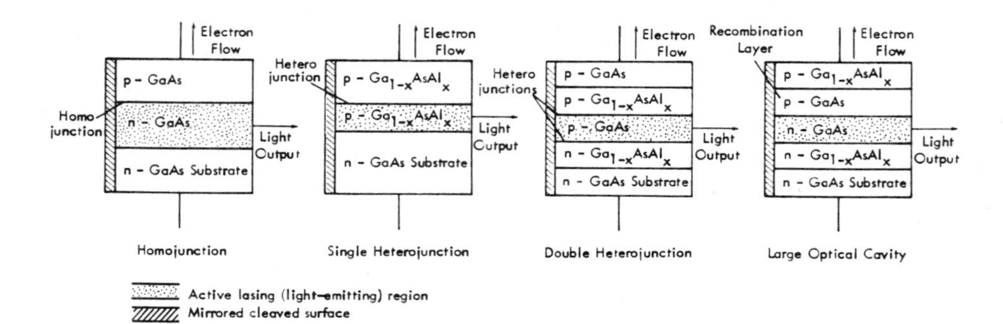

Figure 3.11 Schematic illustration of the structure of the major classes of semiconductor injection lasers.

types: homojunction (HJ), a single-heterojunction (SH), double-heterojunction (DH), and large optical cavity (LOC or OC) lasers (17–19). Generally, these laser types differ depending upon whether carrier injection regions and light generation regions coincide or not and upon the nature of the light confinement. To summarize, in the SH laser the injected carriers and the generated photons are confined by a heterojunction only at one boundary of the recombination region, so that because of the simple structure the emission efficiency is not optimized. In the DH laser the injected carriers and the light contained in the waveguide have the same boundaries on both sides of the recombination region; hence, emission efficiency is considerably improved compared to the SH laser. In the separate-confinement (SCH) laser the injected carriers are confined to a region within the waveguide zone so that carriers and light do not have exactly the same boundaries.

A further comparison of the major injection laser types, together with their optical power and refractive-index distributions, is made in Figure 3.12, with structural dimensions and index variations of typical devices given where applicable (20, 21). Some of the more important parameters of heterojunction lasers are summarized in Table 3.5. These data indicate the general superiority of the DH laser, although the threshold current density and efficiency of the SCH laser are most advantageous. The advantages and disadvantages of these principal semiconductor injection laser types are compared in Table 3.6.

A variation in drive current level causes intensity modulation of an injection laser. Above the threshold current level, linearity is usually adequate, even if the relationship between the electrical input and the optical output is not linear below threshold. Because of the effect of stimulated emission in injection lasers (in contrast to LEDs, where the emission is spontaneous), relatively high modulation frequencies (in excess of 1 GHz) can be achieved. Hence, semiconductor lasers are well suited for pulse-type communication systems. Drive pulses can be applied from zero level or superimposed on a constant bias current set at threshold (22).

Compared to that of an LED, the emitted beam spread of an injection laser is much narrower owing to the higher radiation directivity coefficient and the resulting non-Lambertian radiation pattern. The beam spread (to half power) of stripe lasers is about 40° in the plane perpendicular to the junction and about 15° in the plane parallel to the junction. The coupling loss from an injection laser butted to a multimode fiber with even a low numerical aperture (about 0.15) is much lower than with an LED because of the narrow radiation beam cone. Typically, the laser–fiber coupling loss ranges from 2 to 5 dB.

Although they are in other respects superior to light-emitting diodes, injection lasers emit several modes of radiation and they display a tendency toward filamentary operation. These undesirable properties are normally not known to occur in light-emitting devices.

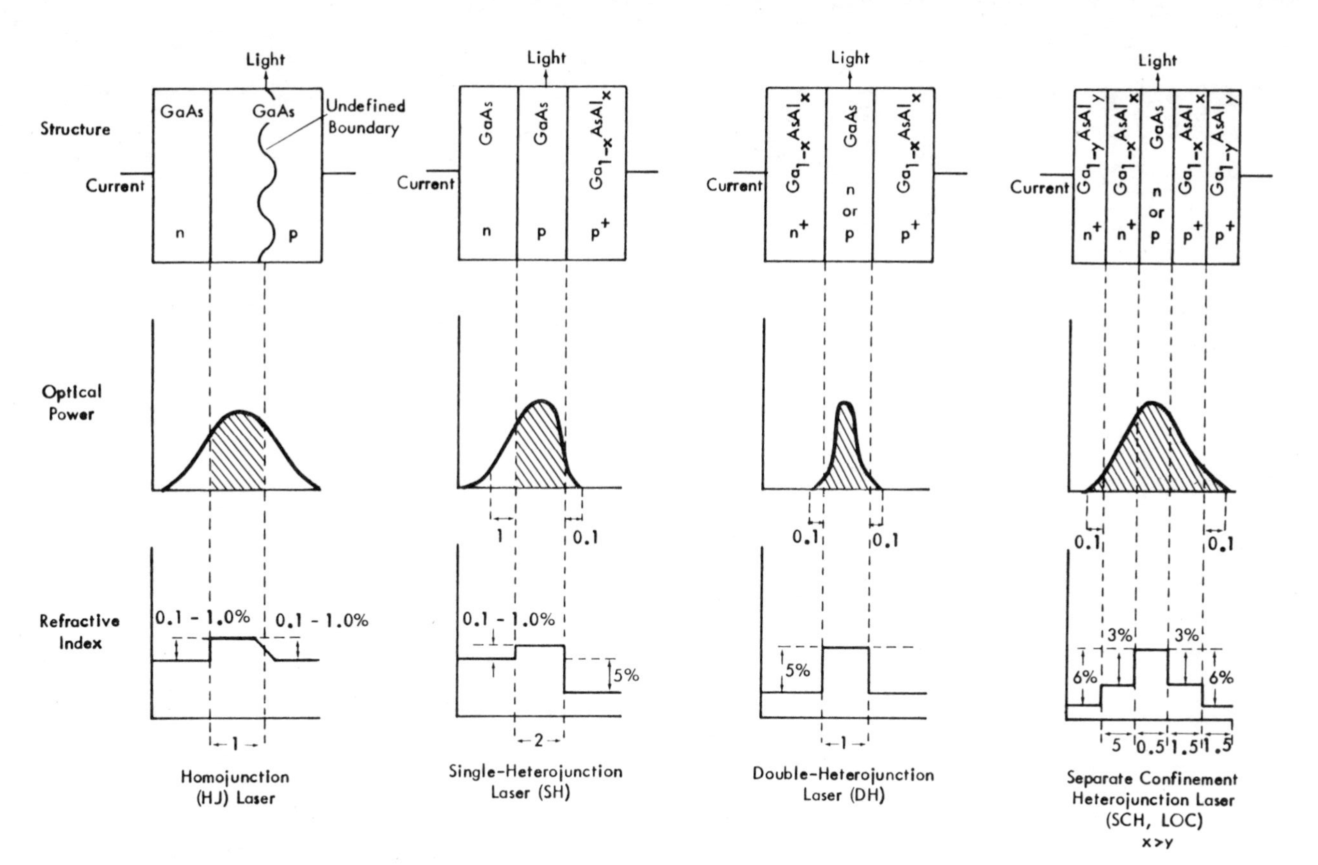

Figure 3.12 Schematic illustration of the structure, location of optical power generation, and refractive-index profile of the major classes of semiconductor injection lasers.

TABLE 3.5. Comparison of the Present Characteristics of the Major Classes of Heterojunction Injection Lasers

Characteristic	ILD type			
	SH	**DH**	**DFB***	**SCH**
Active area (mm^2)	10^{-5}–10^{-4}	10^{-6}–10^{-4}	10^{-6}–10^{-5}	10^{-6}–10^{-5}
Threshold current density (A/cm^2)	5,000–15,000	1000–5000	500–1000	500–1000
On-axis radiance at 100 mA (W/sr cm^2)	10–100	100–1000	500–5000	250–2500
Total radiated power at 100 mA (mW)	1–10	5–50	10–100	5–50
Power coupled into single fiber (μW)	0.1–0.5	1–5	1–10	0.5–5
External quantum efficiency (%)	1–15	25–40	25–50	20–45

*Distributed feedback.

Multimode Operation. Stripe geometry lasers exhibit mode structures parallel to the junction plane. The order of the mode increases with the width of the stripe. Also, stripe lasers that employ doping profile control display these parallel modes. The fundamental mode is obtained for stripes of a width in excess of about 10 μm. Mode confinement parallel to the junction is related to the presence of gain within the stripe confines. The requirements for fundamental-mode operation and low threshold current density usually coincide. Efforts to reduce multimode operation concentrate generally on a reduction of stripe width to less than 10 μm.

Filamentary Operation. Sometimes relatively narrow regions of an injection laser begin to display laser action before neighboring regions reach the laser threshold. These regions are assumed to be inhomogeneities within the active region that possess differing threshold values than their environment. These narrow lower-threshold regions, referred to as filaments, tend to be unstable and to cause noise. This problem can be reduced by confining the stripe width to values less than 10 μm.

The criteria for fundamental-mode operation and single-filament operation are similar. Efforts to reduce filamentary operation are normally devoted to improvements in material homogeneity and better process control, in addition to attempts to reduce stripe width, which also aids in the prevention of multimode operation.

TABLE 3.6. Comparison of the Advantages and Disadvantages of the Major Classes of Injection Lasers

Laser type	Advantages	Disadvantages
Homojunction (HJ)	Simple structure Inexpensive fabrication	High threshold current Relatively large active area Absence of light-guiding feature Inability to adjust wavelength by variation of mole fraction Medium radiance Low quantum efficiency
Single hetero-junction (SH)	Simple structure Single-mode operation Small active area Confinement of light to active region	Medium threshold current Medium quantum efficiency
Striped double heterojunction (SDH)	Single-mode operation Low threshold current Low thermal resistance Small active area Confinement of light to active region	Complex fabrication concept Accurate control of layer composition and thicknesses
Large optical cavity (LOC)	Capability of high data rate modulation Capability to suppress pulse broadening and relaxation oscillation Separation of recombination and radiative layers Small active area	Accurate control of layer composition and thicknesses
Distributed feedback (DFB)	Convenient use in integrated optics Wavelength determined by corrugation spacing Small active area Use of integrated circuit large-volume fabrication techniques High radiance	Combinating of silicon IC techniques and nonsilicon materials Very tight dimensional control

In their basic structure, semiconductor injection lasers, because of their operation at high current densities, are normally not well suited for cw operation, since device lifetime is drastically reduced in such a mode. Hence, they are best applied in pulsed operation. Although their bandwidth is appreciably higher than that of LEDs, they lag in their introduction in practical systems because their mean time to failure (device lifetime) under similar operating conditions is still somewhat lower. Hence, methods to increase device lifetime of lasers have been investigated. A useful lifetime-enhancing technique is the introduction of the previously mentioned stripe geometry. In such a device the lasing region is defined by stripes perpendicular to the radiating direction. The contact metal has access to the semiconductor material only in narrow, well-defined stripes, contrary to the more conventional DH laser, where the metal is allowed to make contact with the entire semiconductor layer. In such a device only a limited number of stable low-order modes can oscillate (23–26). As stripe width and stripe thickness increase, a significant reduction in threshold current density is observed, as indicated in Figure 3.13.

One of the important advantages of an injection laser over a light-emitting diode in practical data communications systems is the enhanced frequency response of the laser. Lateral carrier diffusion within the active region is considered to be among the frequency-limiting factors in DH lasers. It affects the laser relaxation oscillation and the tailing effect (turn-off delay) of the modulated light pulse. Because in a DH laser the oscillation region is usually smaller than the population inversion region, a fraction of the carriers from the inversion region (in which the field intensity is low) diffuses easily into the oscillation region (in which the field intensity is high). Subsequently, the carriers that have diffused from the inversion region are dissipated by the strong field of the oscillation region. Thus, it is desirable to restrict lateral car-

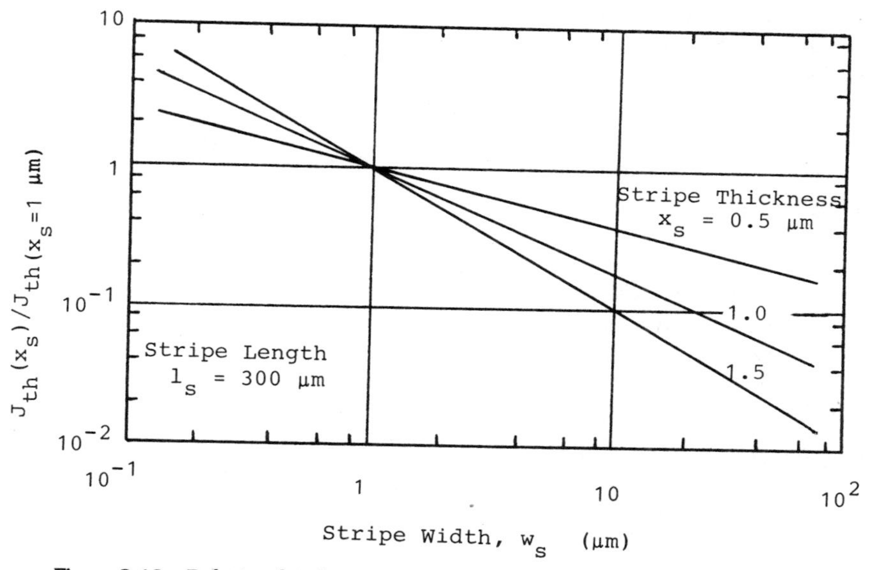

Figure 3.13 Relationship between threshold corrent and stripe width of an injection laser for different stripe thicknesses (stripe with w_s = 1 μm).

rier diffusion into the active region through lateral carrier confinement. Carrier diffusion is a basic property of semiconductors, and as such it affects the transit behavior of injection lasers. Its existence has two major consequences: (a) it reduces the damping-time constant K of the relaxation oscillation, and (b) it causes the tailing effect of the light pulse after pulse turn-off. The damping time is related to the carrier diffusion length in the active region, L_a, and the diameter of the light spot, d_s, approximately by

$$K = (L_a/d_s)^2 \qquad (3.8)$$

It is a direct result of the fact that after the pulse has been turned off the laser continues to oscillate until all carriers stored in regions of a weak field have been dissipated by the field. This fall time is proportional to the bias current density. Therefore, it is advantageous to operate the laser at as low a current density as is compatible with other operating considerations.

The frequency response of an optical communications system is also affected by the spectral width of the optical output, i.e., by the entire wavelength range that is emitted by a single optical source. Furthermore, source spectral widening observed in lasers that are modulated at a high pulse rate tends to reduce the frequency response because of a frequent increase in the number of oscillating longitudinal modes. Because spectral broadening determines the material dispersion limit of the fiber, longitudinal mode selection is required when the laser is to be operated at a high pulse repetition rate. The decrease of the spectral broadening with time for a typical DH laser is illustrated in Figure 3.14, which shows that a light pulse which initially had a width of about 2.8 nm narrows to a width of about 0.8 nm after approximately 3 ns.

The detrimental effect of lateral carrier diffusion in a DH laser can be reduced by operating the device in the fundamental transverse mode and by choosing nearly identical volumes for the oscillating and population inversion regions. Thus, little performance degradation is observed in operating systems in which these measures have been implemented.

The attenuation characteristics of optical fibers require for low-loss operation a wavelength larger than 0.85 μm. Hence, the development of lasers that can be used at wavelengths of 1.05 and 1.27 μm is desirable. Various new materials and structures have been proposed for this purpose.

Whereas the GaAsAl alloy system used in the fabrication of LEDs and ILDs is adequate for many applications around a wavelength of 1.0 μm, at larger wavelengths, particularly in the region 1.1 to 1.3 μm, other materials are required. Among the semiconductors investigated for laser use in this range are InGaAsP–InP and AlGaAsSb–GaAsAl, fabricated either by liquid-phase epitaxy (LPE) or vapor-phase epitaxy (VPE). Encouraging room-temperature lifetimes approaching 10^4 hr have been obtained with some of these

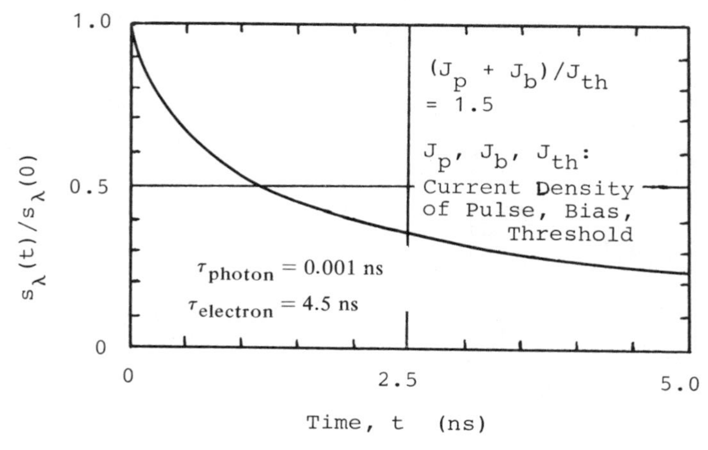

Figure 3.14 Variation of the spectral bandwidth of an injection laser with time elapsed after the onset of a light pulse (spectral width taken at $t = 0$).

materials, especially with InGaAsP lasers produced by LPE growth. Nuese (19) has described an InGaAsP–InP heterojunction laser prepared by VPE that has a threshold current density below about 2×10^3 A/cm^2 and a differential quantum efficiency of 0.49. The emission wavelength of this laser type can be adjusted from 1.1 to 1.6 μm by variation of the mole fraction. However, carrier leakage occurs from the InGaAsP layer into the adjoining InP layers proportional to $e^{-\Delta E_G/kT}$, where ΔE_G is the energy gap difference between the layers. It causes both higher laser threshold current and lower conversion efficiency, particularly at high temperature and at small band gap difference. Data of threshold current density and efficiency of GaAsAl and InGaAsP double-heterojunction lasers emitting optical power over the range 1.1 to 1.4 μm are shown in Figure 3.15. Little difference between the characteristics of lasers made by these materials is normally observed for the same mole fraction in regard to threshold current density and efficiency variations with temperature.

One of the more recent innovations in the development of small sources is the distributed-feedback laser (DFB), which is a device whose wavelength of the radiative longitudinal-mode oscillations is determined by the corrugation period of the structure and not by its material composition. This is in contrast to the other laser types, where emission is determined by stimulation in a single region and where the wavelength of the emitted light is solely a function of the material properties. Although in the DFB laser the wavelength is given by a different mechanism, the device is still fabricated in GaAsAl–GaAs material and employs the DH structure. However, it requires electrical pumping and operation below room temperature, at least in most present devices.

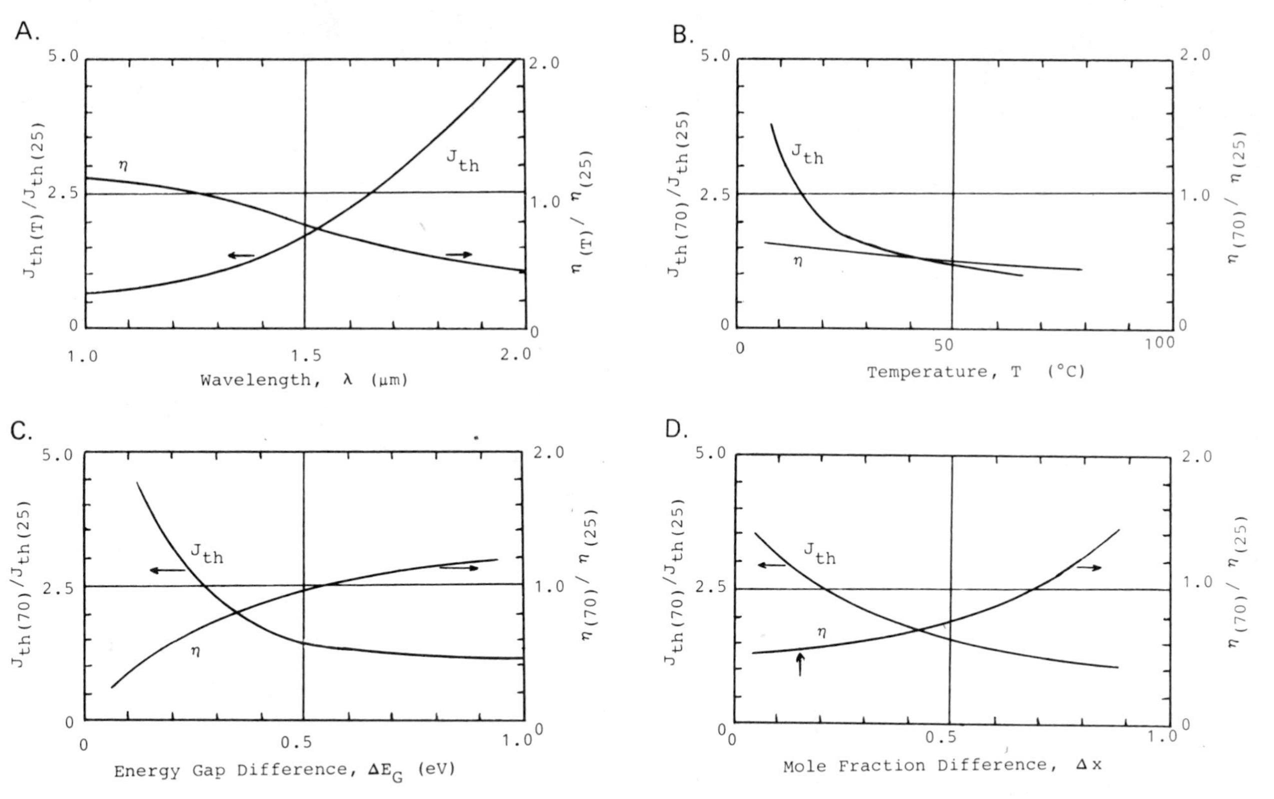

Figure 3.15 Threshold current density and external quantum efficiency of GaAsAl and InGaAsP DH lasers relative to room-temperature values as functions of temperature, emission wavelength, energy gap difference, and mole fraction difference between adjacent layers of GaAsAl—GaAs or InGaAsP—InP semiconductors.

In the DFB laser the feedback is provided by the corrugated interface between the GaAs and GaAsAl layers. Amplitude and wavelength of the laser can be controlled, within certain limits imposed by refractive-index and photolithographic restrictions, by changing the refractive index of the film or substrate layers and by variation of the corrugation period. The wavelength of the stimulated emission is related to the corrugation period d_c and the refractive index of the waveguide, n_g (n_g = 3.6 for GaAs), by

$$\lambda = \tfrac{2}{3}\, d_c n_g \tag{3.9}$$

Typically, the corrugation period is on the order of 0.3 to 0.4 μm. If, for example, n_g = 3.56 and d_c = 0.34 μm, the wavelength of the emitted light λ = 0.81 μm. If, however, a wavelength of 1.1 μm is required and if the refractive index of the waveguide can be reduced to 1.5, then the corrugation period can be relaxed to approximately 1.1 μm, since $d_c = 1.5\lambda/n_g$. The structure of the DFB laser is illustrated in Figure 3.16. The DFB laser, because of its design, provides several attractive features in comparison to the single-region double-heterojunction laser. The more important of these are a better control of the emission wavelength, enhanced mode control, lower emission threshold, more collimated output beam, and polarized output beam (27–29).

Frequency response and spectral width of the DFB laser are also affected by the corrugation period; they depend upon the square of d_s. Thus, it is desirable to reduce d_s to a minimum, although this leads to practical fabrication difficulties. Because the very fine corrugations require precise high-resolution techniques, the major limitation of the DFB laser is the needed photolithography. Although the ultimate limit of photoresist resolution is about 5 nm, corresponding to 50 Å, the smallest features that have been achieved in actuality are of the order of 0.1 μm, obtained by the use of holographic interference between laser beams. In general, integrated circuit techniques, which currently allow lines of about 1.0 μm in production environments and which will ultimately be capable of producing 0.1 μm lines, will aid the development of DFB lasers. Currently, the 0.3 μm lines required for useful DFB lasers are formed mainly by ion milling through a photoresist produced by holography.

One of the attractions of the distributed-feedback laser is its much narrower emission spectrum compared to the single-region DH injection laser. Whereas a spectral width of approximately 2 nm is typical for the single-region DH laser, a width of only 0.5 nm is typical for the DFB laser. In addition, the DFB laser requires a lower drive current to produce the same optical power, i.e., its conversion efficiency is higher.

The emission wavelength of the DFB laser depends upon temperature. It results from the thermal variation of the refractive index, whereby dimensional changes over the temperature range of interest are assumed to be negligible. The stimulated wavelength varies with temperature as 0.05 nm/°C;

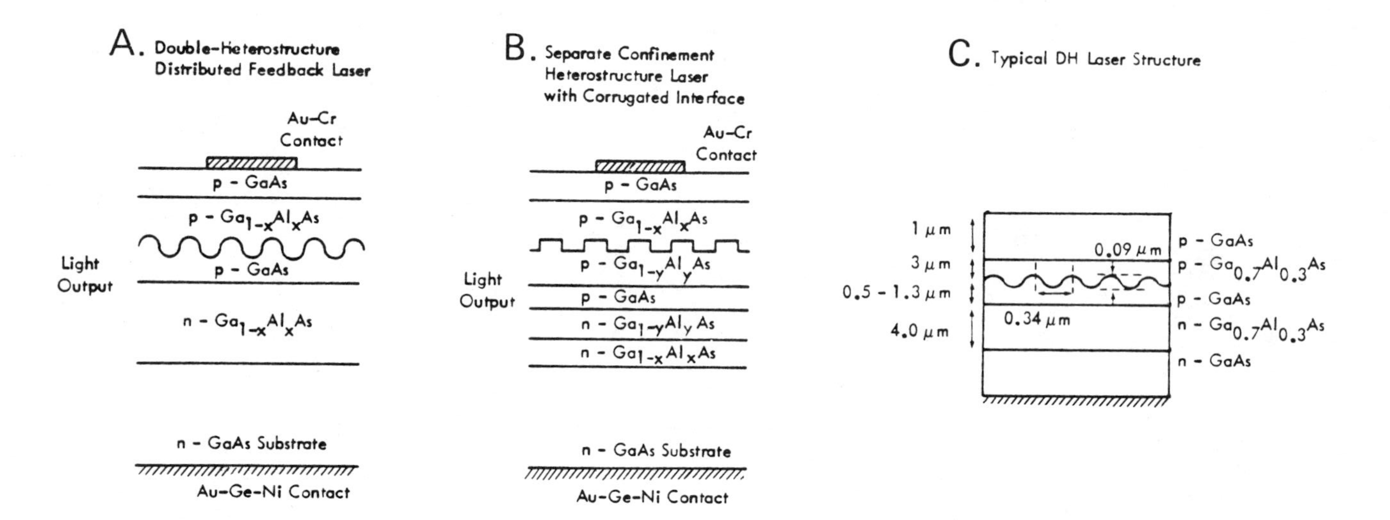

Figure 3.16 Schematic cross section of several advanced semiconductor injection lasers.

hence, for a temperature variation from 0° to 100°C a wavelength change by about 5 nm can be expected. This is insignificant in most practical applications. On the other hand, the wavelength of the spontaneous emission peak shifts by about 0.15 nm/°C; thus, the wavelength increase with temperature over the range 0° to 100°C amounts to about 15 nm.

A typical DFB laser is fabricated as follows: An n-type $Ga_{0.7}AsAl_{0.2}$ layer doped with Sn and a p-type GaAs layer doped with Ge are formed on an n-type GaAs substrate by LPE growth. Then surface corrugations are made on the p GaAs layer by ion milling through a photoresist mask, followed by LPE growth of p-type $Ga_{0.7}AsAl_{0.3}$ and GaAs layers, both doped with Ge. To avoid meltback of the corrugated surface these layers are applied at low temperature (700°C). The final structure has a mesh geometry which limits the injection mechanism to a rectangular stripe. The surface damage caused by the ion milling results in a substantial reduction of the recombination efficiency and hence in a higher threshold current density. This effect, however, will not be observed when improved photolithographic techniques become available.

The separate-confinement heterostructure (SCH) laser, on the other hand, although also containing a corrugated region, is not affected by ion milling because the corrugated region is separated from the gain region. Hence, from the standpoint of performance it has a low threshold current density, since nonradiative recombination due to damage is minimal, but from the standpoint of fabrication it is more complex.

Both devices, the DFB laser and the SCH laser, allow pulsed and continuous wave operation. In the pulsed mode, 50 ns pulses typically are used to stimulate emission.

Source radiance is defined as the radiated power emitted per unit source area per unit angle within a specified small interval centered at wavelength λ (30). Whereas light weight, small size, and appropriate emission wavelength are obtained from most lasers, the requirement of high spectral radiance in a cw room-temperature mode, corresponding to a relatively large amount of optical power (on the order of 1 W), has been difficult to achieve. The conventional GaAsAl–GaAs injection laser, although emitting within a desirable spectral range, possesses an insufficient radiance as a result of its cavity configuration and its comparatively low cw room-temperature output power. Therefore, considerable research efforts have been devoted recently to enhancing laser output power and radiance.

Large radiance is expected in the laser output power is maximized, its spectral linewidth is minimized, and its far-field radiation pattern consists of a single well-defined lobe. Stoll (31) has described useful approaches to achieving injection laser sources of high radiance by first phase and frequency locking a large number of relatively low-powered single-mode devices of a laser array and then merging their outputs into a well-collimated and apodized radiative beam useful for feeding optical power into a fiber. The following is a discussion of the three principal requirements for achieving large injection laser radiance.

LARGE OUTPUT POWER

In order to induce laser arrays to operate synchronously and to generate a spacially and temporally coherent integrated beam, all elements of the array must be locked in frequency and phase to each other. This can be accomplished by using either a master–slave injection locking or a mode filtering technique. The locking technique involves the injection of a small fraction of the optical power generated by an external master laser into each of the cavities of the slave laser array. The mode filtering technique uses a cascade of distributed Bragg deflectors as the mode filtering element. It allows only Bragg-scattered radiation to undergo stimulated emission. In both techniques, phase and frequency locking require the unlocked frequencies of all array lasers to fall within a prescribed frequency bandwidth. For the master–slave technique this condition is given by the expression for the maximum difference in unlocked frequencies which may exist between the master laser and a given slave laser:

$$\Delta\nu_{\max} = \frac{\nu_s}{2Q_s}\left(\frac{P_s}{P_m}\right)^{1/2}\left[1 - \left(\frac{P_s}{P_m}\right)^2\right]^{-1} \tag{3.10}$$

where subscripts m and s refer to master and slave, Q_s is the Q value of the slave laser, P_m is the fraction of master laser power injected into the slave laser cavity, and P_s is the circulating slave laser power at the injection port. For the mode filtering technique the condition for phase and frequency locking requires maintaining nearly equal laser junction temperatures for all devices of the array, since a 1°C change in temperature causes a 0.08 nm shift in laser wavelength.

SMALL SPECTRAL WIDTH

The second requirement for large laser radiance, minimization of the spectral linewidth, can be achieved by suppressing the simultaneous emission of all except one of the multitude of longitudinal modes of a laser (which are common to GaAsAl–GaAs Fabry-Perot lasers). The multimode emission causes a large linewidth of the laser (about 2 nm), which is significantly larger than the more desirable 0.01 to 0.05 nm linewidth of a single longitudinal mode. Hence, the replacement of the large-width cleaved facet reflectors of the conventional laser with distributed Bragg reflectors has been suggested, so that only one longitudinal mode is emitted. The reflectivity of only one longitudinal mode is emitted. The reflectivity of an individual distributed Bragg reflector laser can be described by the full width (in frequency) between first nulls, given by

$$\Delta\nu = (v_l/2\pi l)\left[\pi^2 - (\tanh^{-1}\sqrt{r})^2\right]^{1/2} \tag{3.11}$$

where v_l is the phase velocity of light within the laser, l is the laser length, and r is its reflectivity (peak). This equation implies that single-mode operation can be achieved by adjustment of r and l so that $\Delta\nu$ is equal to or less than the longitudinal mode spacing. In addition to longitudinal mode control, transverse mode control is required to implement power maximization and beam apodization. This can be achieved, for example, by passively confining the laser mode through substrate channeling.

EFFICIENT BEAM MERGING

The third requirement for large laser radiance is that all laser array outputs must be merged together and then extracted from the structure. In the case of the injection-locked array this can be accomplished by cascading the beam splitters and providing an output grating coupler. In the case of the self-locked array only an output coupler is required, since beam combination and apodization within the plane of the source is internally effected by the cavity mode filter.

NONSEMICONDUCTOR LASERS

In the quest for high-radiance sources of small spectral width, nonsemiconductor lasers have also been investigated. Work on these structures has lately progressed such that they may be considered as practical devices for optical communication systems. Much further research is still required before they will be able to achieve the same degree of prominence that semiconductor lasers have reached. Lasers based on garnet materials, such as the Nd:YAG laser, for example, although far from being comparable in developmental state to semiconductor lasers, have exhibited very promising characteristics. Other nonsemiconductor lasers, such as ultraphosphate devices, are also being investigated, but they also require substantial improvements in their operational properties.

One of the most useful nonsemiconductor lasers is the Nd-doped yttrium-aluminum-garnet (Nd:YAG) laser. Potentially, it will be able to offer at least the following major advantages over existing semiconductor lasers:

Emission Wavelength. The emitted wavelength of 1.064 μm coincides almost exactly with one of the low-loss minima of most glass-based fibers and is therefore ideally suited for optical communications systems applications. Most semiconductor lasers require a precise metallurgical composition that involves compromises to achieve the same emission wavelength.

Spectral Width. The Nd:YAG laser has a substantially smaller spectral bandwidth compared to other sources. Therefore, it is capable of reducing the fiber material dispersion and thus makes either longer cables or higher data transfer rates possible.

Monomode Output. The optical output of the Nd:YAG laser usually contains only a single mode. Therefore, it is ideally suited for high-bandwidth operation in conjunction with monomode operation of waveguides if the cross-sectional area of the laser can be reduced to match the dimensions of the monomode step-index fiber core (which is about 1 to 5 μm in diameter).

Lifetime. The requirement of an external modulator which shifts the burden of a high current level from source to modulator results in a potentially long lifetime for the Nd:YAG laser, although at this time only a comparatively small amount of reliability information is available.

A typical Nd:YAG laser, illustrated in Figure 3.17, is pumped by a GaAsAl light-emitting diode; the laser rod may have a cross-sectional area of about 5.0×0.5 mm^2. If operated in a pulsed mode it is capable of yielding a single-frequency output power of a few mW at 1.06 μm. In another approach, a short length (about 1 cm) of an optical waveguide with an Nd-doped core can be operated as an end-pumped laser (32).

One of the major benefits of an Nd:YAG laser is that it may eventually be used directly as an integral part of a fiber end, particularly if fiber and laser are compatible in dimensions and composition. If this can be accomplished in an economic manner, the narrow spectral width of this laser type (about 0.1 nm), which is more than an order of magnitude better than that of semiconductor lasers and more than two orders better than that of LEDs, will allow high–data rate transmission in integrated optics. On the other hand, the Nd:YAG laser has several disadvantages compared to other sources. First, it requires an external modulator, which results in considerable complexity and cost increase; second, it is based on a nonsemiconductor material, which precludes taking advantage of the well-developed integrated circuit tech-

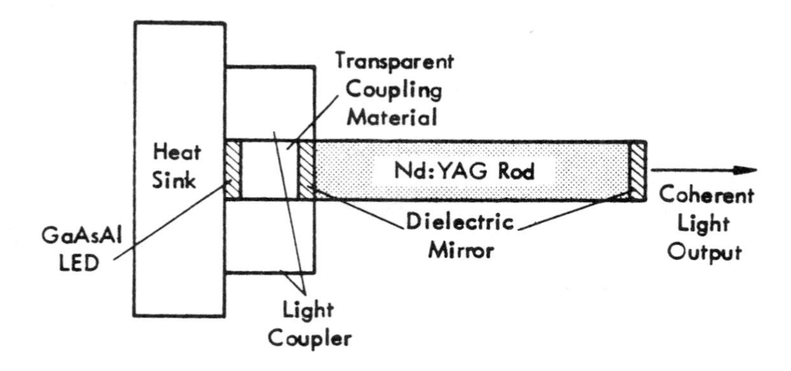

Figure 3.17 Schematic illustration of a nonsemiconductor Nd:YAG laser.

nology with its mass production concepts; and third, its optical output power is still lower than that of the semiconductor sources.

One of the most severe drawbacks of the Nd:YAG laser is related to the inability of using direct modulation, even at very low data rates (such as 1 Mb/s). This is a consequence of the long fluorescence lifetime of the upper laser level, which is about 230 μs, corresponding to a modulation frequency of 4 kHz. An external modulator avoids this difficulty. At data rates below about 100 Mb/s an acoustooptic modulator is a suitable choice, whereas at data rates above 100 Mb/s an electrooptic modulator is required. An electrooptic modulator allows efficient operation potentially in excess of 1 Gb/s, although in practice this rate has not yet been achieved. Magnetooptic modulators have also been suggested in conjunction with nonsemiconductor lasers, mainly because they are normally made of garnet materials, which offers obvious advantages of compatibility and potential integration of both devices on the same chip or in the same structure. However, magnetooptic materials have too much loss to be of practical use in this application.

Other nonsemiconductor lasers of interest belong to the Nb_xLa_{1-x} and Nb_xLi_{1-x} ultraphosphate families. Their main advantage is their ability to provide high Nb content. These laser types require side pumping with an argon laser pump in order to achieve a coherent output at a wavelength of 1.05 μm or above. The growth of the $Nb_xLa_{1-x}P_5O_{14}$ or $Nb_xLi_{1-x}P_4O_{12}$ crystalline materials requires the choice of the optimum mole fraction x in order to obtain the minimum amount of fluorescence quenching. Fluorescence decay times typically are about 120 μs for the $x = 1$ composition, 240 μs for $x = 0.1$, and 325 μs for $x = 0.01$. With an argon laser pump, a threshold of only 2 mW of absorbed pump power can be obtained for end pumping and about 200 mW for side pumping. In end pumping, a nearly concentric cavity with mirrors of about 1 cm radius of curvature and a crystal length of about 1 mm is used. In side pumping, the pump beam is cylindrically focused to a thickness equal to the cavity mode diameter. A typical LNP laser consists of a 300 μm thick crystal and uses an etalon effect of crystal surfaces without a frequency-selective device, resulting in single-frequency operation.

In the past, single-frequency operation of Nd:YAG lasers has been difficult to achieve because of the spatial hole burning effect. This term describes the spatial modulation of the population inversion; it is caused by a standing-wave field which favors oscillations of other modes and which results in an output spectrum of multilongitudinal modes whose intensities fluctuate at random. Several alternate methods have been suggested to avoid this difficulty, such as the addition of a frequency-selective device (for example, tilted etalons or thin-film absorbers), to suppress undesired modes. This technique, however, reduces the available light output power because of the loss of undesired modes. Another method suggested to obtain a single-mode output prevents the modulation of the population by generating a mode for which the field intensity is spatially uniform along the rod axis. However, frequency instability and the short duration (a few seconds) of the desirable mode are serious drawbacks. The LNP laser avoids these disadvantages by using the frequency-selective characteristics of the crystal itself; hence, it does not need an external frequency-selective device. For example, for a pump power of about 10 mW at a wavelength of 0.5 μm the output power of the desired single mode in a 0.3 mm thick crystal is about 3 mW. The crystal thickness must be restricted, however, to a value less than about 1 mm because the spatial hole burning effect in such a structure would still cause a multilongitudinal mode spectrum. This requirement has the benefit of leading to small device dimensions.

SOURCE MODULATION

The modulation of the optical output of any source by the electrical signal is of major concern to the system designer. It represents one of the source limitations. The choice of the modulation technique depends on the optical source to be used. The semiconductor sources (light-emitting diode, injection laser) can be directly modulated up to a maximum of several hundred MHz (with the actual upper frequency being determined by the source under consideration) by variation of the drive current. The nonsemiconductor sources (Nd:YAG laser, ultraphosphate laser), on the other hand, require external modulators and can be operated up to several hundred MHz and above. An optical modulator can also serve as switch or time-division multiplexer.

In agreement with our previous analysis, direct and indirect modulation techniques can be distinguished because of the different operating mechanisms of the various optical sources, as follows (33):

Direct Modulation

This technique applies to all semiconductor light-emitting diodes and injection lasers. It is the simplest, most convenient, and least expensive method used to impress the characteristics of an electrical signal on the optical beam transmitted through the optical waveguide.

Light-emitting Diode. The properties of the LED make it suitable for direct modulation by either analog or digital signals. The bandwidth of the modulated signal may be in excess of 100 MHz, although about 50 MHz is presently considered a practical maximum. The emission response time of the LED is theoretically of the order of 1 ns. By driving the device with short pulses of high peak current and low duty cycle, it is possible to obtain a high peak power; however, in this case the pulse repetition rate must be below 0.1 MHz in order to avoid heating.

Injection Laser Diode. The ILD is suitable for both analog and digital direct modulation. Its input–output characteristic is almost linear above a threshold level. The spontaneous radiative recombination time is on the order of 1 ns and is reduced further by the laser action of stimulated emission. Large modulation bandwidths are thus possible with direct modulation. Laser resonance occurring above 1 GHz results in a modulation efficiency peak at a specific modulation frequency, displaying either peaked high-frequency noise, spikes in the output, damped relaxation oscillation, or regular intensity pulsation at a fixed repetition rate. It is caused by interaction of the optical field in the cavity and the injected electron density distribution and results in oscillations in both optical output and electron density. Modulation with narrowband signals near the resonance frequency is possible, but modulation with wide-band signals is restricted to frequencies below resonance. Once the buildup of the population inversion has been overcome during turn-on, the laser can be modulated directly with signals up to the resonance frequency.

Indirect Modulation

This technique applies only to nonsemiconductor lasers. Optical modulators suitable for this application are of either the reactive or the absorptive types. Both may be either of the bulk form, where the optical beam is freely propagating, or of the waveguide form, where the optical beam is confined to a guiding thin film. The reactive type has received most of the attention in the past because of the large variety of materials that can be used. The waveguide form offers more future potential because it requires less modulating power and is more suitable in conjunction with integrated optics. Three major types of reactive modulators have been investigated: electrooptic, acoustooptic, and magnetooptic, but only the electrooptic type shows promise for the future. The magnetooptic modulator has the further drawback of small optical transparency. Electrooptic modulators are frequently made of $LiNbO_3$ or $LiTaO_3$. They allow operation over the wavelength range 0.4 to 4.0 μm. Typically, the required modulation power in a bulk modulator ranges from 1 to 20 mW/MHz for 100% modulation at a given wavelength. The highest modulation frequencies obtained so far in bulk modulators are in excess of 500 MHz. In thin-film modulators, made of either of these materials, or in GaAsAl–GaAs, bandwidths in excess of 4 GHz can be achieved with a modulation power of 0.1 mW/MHz. GaAsAl–GaAs modulators hold promise for future applications because of their power consumption, bandwidth, and compatibility with other GaAsAl–GaAs devices.

Distortion-free high-speed modulation is one of the important goals in the operation of optical systems at high data rate. Waveform distortion due to relaxation oscillations or output spiking is a major cause of limitations in laser high-speed pulse modulation. At low data rates this waveform distortion can be eliminated by electronic band reduction in the detection system if the modulation frequency is less than the relaxation oscillation frequency. At high data rates, however, this method is not suitable. Some of the feasible methods that can be used to overcome these difficulties and to obtain faithful electronic–optic waveform conversion employ either internal or external light injection. Both methods involve optical rather than electronic mechanisms to reduce the waveform distortion at high modulation rates, where the band reduction method is not applicable. The upper limit of the modulation rate to which the band reduction method can be applied is 0.5 Gb/s for return-to-zero (RZ) signals and 1.0 Gb/s for non-return-to-zero (NRZ) signals.

Internal Injection. In the internal light injection method, optical feedback of the laser output is utilized. A feedback delay of less than that corresponding to the oscillation frequency (typically about one-third of the oscillation frequency) is desirable. This results in an increase in the width of the initial spike before the occurrence of the next spike and the reduction of the pattern effect through the depletion of the excess carrier density (which causes the second spike in the absence of feedback).

External Injection. In the external light injection method a small fraction of the laser output is constantly injected into resonant modes of the laser cavity, whereby the injected radiation suppresses the relaxation oscillation by reducing the height of the initial spike and attenuating the residual oscillation. In addition, the light injection reduces the delay for the lasing onset.

The internal injection method is more advantageous because it allows a larger light output amplitude owing to the utilization rather than suppression of the initial spike and because it allows a more efficient reduction of the pattern effect. Its applicability, however, is restricted to RZ signals, because the peak light intensity cannot be sustained for successive pulses for NRZ signals. The external injection method, however, can be used for both RZ and NRZ signals. The internal injection method can be employed up to about 1 Gb/s. For data rates in excess of this value the external injection method has to be employed because it not only suppresses the relaxation oscillation but also reduces the delay for the onset of lasing. Thus, the internal light injection method allows the extension of semiconductor laser applications in pulse modulation schemes to the Gb/s range.

FAILURE MECHANISMS

Because of the high drive currents, all optical sources tend to fail after prolonged operation, even if used at room temperature. Failure may either be a gradual degradation of device characteristics or an abrupt termination of useful operation. Source failure after a limited operational period represents the most serious problem with all present optical sources. Typically, the lifetime of most sources ranges from about a thousand hours (a few months) to about a million hours (more than one hundred years) at useful current and power levels, extrapolated from accelerated high-temperature measurements. The analysis and prediction of source failures is particularly important for optical sources used in communications, military, and space applications.

In identifying and eliminating the failure mechanisms that occur in light-emitting diodes and in injection lasers, significant progress has been made. It is now known that the major failure modes are caused by metallurgical defects in the vicinity of the p-n junction or by facet damage in cw laser diodes. As a result of intensive investigations of the causes of source degradation, a model has evolved which relates the internal damage mechanism to laser electron–hole recombination (16). It is based on the recognition that the energy released in nonradiative recombination enhances semiconductor point-defect displacement and thus leads to the accumulation of crystal defects in the recombination region. In order to eliminate or reduce these material defects, it is necessary to use starting wafers of the lowest available defect density (on the order of 100 cm^{-2}, (preferably much lower), to use GaAsAl rather than GaAs in the active region, to coat the facets by an oxide or other protective film to reduce surface-related effects, to avoid external strain by mounting the devices with soft solder (such as In), to avoid internal strain by using substrate and film materials of matching lattice properties, and to define the stripe contacts by oxide so that all steps subsequent to crystal growth are restricted to the passivated surface.

Current densities in optical sources usually exceed 10^3 A/cm^2, whereas in devices used for display purposes current densities normally do not exceed about 10^{-2} A/cm^2. Device failures caused by the high current density can be divided into two major categories, similar to those of other semiconductor devices: bulk-related and surface-related failures. Nonradiative recombination centers can form in the recombination region. They tend to reduce the internal quantum efficiency and to increase absorption (34, 35). Because carrier recombination is necessary for these defects to be introduced, both LEDs and ILDs degrade by similar mechanisms and at similar rates. Differences between the two types of sources arise from the different current densities at which they are operated and from the absence of facets in LEDs.

The defects that degrade device performance are introduced during device fabrication and are not intrinsic to the material. Hence, reliable sources require careful attention to maintaining a high degree of metallurgical perfection during all fabrication steps. Defects frequently introduced during material preparation or device fabrication include dislocations, contaminants (such as Cu and Ni), and exposed edge regions. As in other semiconductor devices, complete surface passivation is therefore necessary. Aluminum present in the recombination region tends to prevent or reduce degradation. This is fortunate because the variation of the aluminum content in GaAsAl devices is used also to affect the emission wavelength peak. The MTTF of sources is a strong function not only of the Al content of the $GaAs_{1-x}Al_x$ alloy but also of the dopant type used in the recombination region. A typical relationship between the LED lifetime and Al and dopant content is shown in Figure 3.18. The lifetime given in this illustration refers to the estimated half-power life at a drive current density of 1000 A/cm^2. A lower drive current results in a longer life-

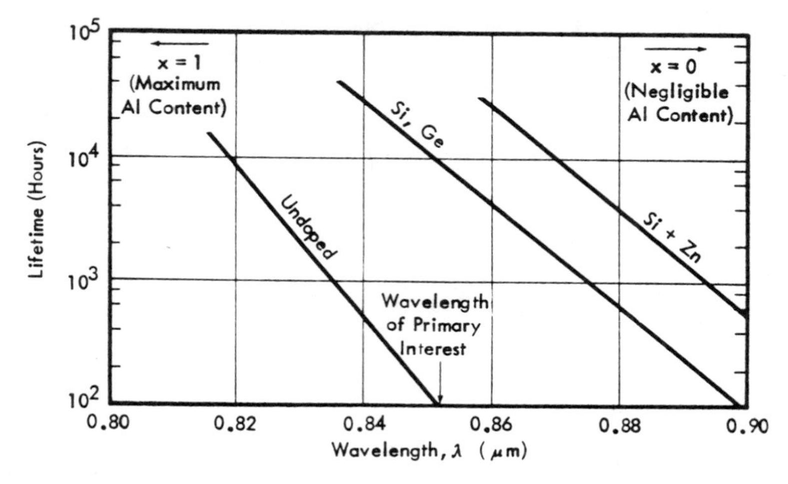

Figure 3.18 Empirical relationship between lifetime of a $Ga_{1-x}AsAl_x$ light-emitting diode and Al, Si, and Ge content.

time at the expense of light output. For a given dopant an increase in lifetime is observed if the Al content is increased, thereby shifting the emission wavelength to smaller values (although this is usually not a feasible approach because of waveguide constraints). Generally, a decrease in wavelength by about 2% due to Al content variation causes a lifetime increase by an order of magnitude.

In principle, defect-caused failure mechanisms in LEDs and ILDs are similar. Injection lasers, however, have an additional failure mechanism associated with facet damage which tends to decrease facet reflectivity. Thus, overall lifetimes of lasers are typically shorter than those of light-emitting diodes operated at the same current density levels. Laser facet damage may be catastrophic or gradual. Whereas catastrophic failures usually occur as a result of a high current density, gradual failures extend over many hours of operation and are caused by a slow erosion of the facets even at low current levels (35). Catastrophic damage, in contrast, is often accompanied by melting or cracking of the structure.

Catastrophic Facet Damage. This failure mechanism depends mainly on the drive current density and the drive current pulse length. It typically occurs at power densities at the mirrors in excess of about 10^6 to 10^7 W/cm^2. The maximum damage limit decreases as an inverse function of the square of the pulse width. Thus, it is desirable to limit the pulse width to very small values. This is depicted in Figure 3.19, where the tolerable power limit (in watts per emitting facet length) is shown as a function of the pulse width. Figure 3.19 indicates that in spite of careful elimination of material and device defects, lasers may suffer substantial degradation as a result of facet erosion. The maximum power per facet length (which is the laser stripe width) has been experimentally shown to follow the relationship

$$P/l = k_l/t_p^{1/2} \tag{3.12}$$

where k_l is a constant characteristic of the laser and t_p is the pulse width. Assuming a pulse length of 100 ns and a facet length of 10 μm, for example, the ratio P/l of a SH laser is about 220 W/cm, corresponding to the maximum allowable optical power that can be used before catastrophic facet damage takes place. Because the maximum power is inversely proportional to the square of the pulse width, a reduction in t_p from 100 to 1 ns will increase the maximum optical power limit before facet damage will take place by an order of magnitude, to about 2.9 W. Likewise, from the standpoint of reliability, it is desirable to increase the facet length as much as possible (as long as it remains compatible with the associated fiber dimensions). Because monomode fibers require a core diameter less than about 5 μm, the compatibility requirement of laser and waveguide demands a laser stripe width that is less than 5 μm. Thus, a 5 μm long facet can have a maximum power of 110 mW at 100 ns. Facet protection typically results in a lifetime improvement of more than an order of magnitude. For example, in a LOC laser with 5 μm wide stripes and operated at a pulse width of 10 ns, facet protection by coating would allow a maximum power increase from about 1.5×10^3 W/cm (corresponding to 0.8 W) to 12×10^3 W/cm (corresponding to 6.0 W) under otherwise identical conditions. Lasers with thin optical cavities (such as DH lasers with very low

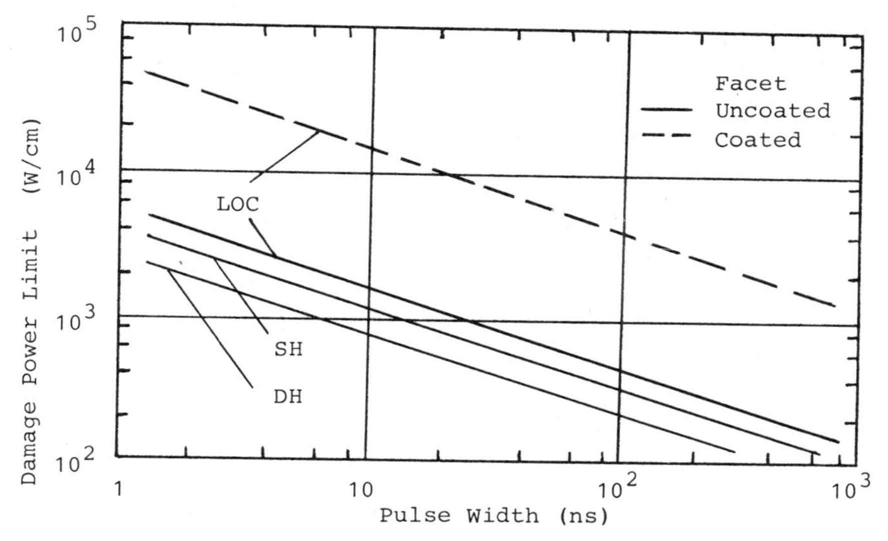

Figure 3.19 Dependence of electrical power level above which catastrophic facet damage occurs in injection lasers upon pulse width and device type.

threshold) fail catastrophically at lower pulsed power than do lasers with wide optical cavities or lasers in which the field can spread out of the active region, as in the case of homojunction or SH lasers.

Gradual Facet Damage. This failure mechanism normally occurs at current densities much below those typically associated with catastrophic damage, although it may also be observed at high current densities. Gradual damage, however, is also related to various ambient influences, such as temperature and humidity. Furthermore, luminescent areas, called dark line defects (DLD), that cross the active region during lasing are responsible for gradual laser degradation. This failure mode is the most prominent of all gradual laser failures and is identified with a three-dimensional dislocation network which originates at a dislocation that crosses the interface between the active region and the adjacent layers (for example, the GaAsAl–GaAs interface) and which grows during the operation of the laser. The dark line defect is generally related to various types of crystal imperfections, such as dislocations, stacking faults, and others caused by high-intensity light during the lasing operation. More specifically, the localized temperature increase during light emission is the main cause of the dark line defect. The DLD is usually restricted to the active region, which is typically only a few micrometers wide. Another cause of this failure mode is the propagation of screw dislocations; this mechanism does not require vacancy collection. Also, in another failure mode the lasing lifetime is related to strain, so that a high laser MTTF requires the elimination of all strain at the device interfaces. $Ga_{1-x}AsAl_x$ layers on GaAs substrates, for example, are compressively stressed when brought from the high epitaxial growth temperature to room temperature; stress levels typically range from 10^8 to 10^9 dyne/cm^2, depending upon the value of x. At epitaxial growth temperature both materials have approximately identical lattice parameters, but they differ considerably at room temperature, leading to gradual degradation. Such stress can be re-

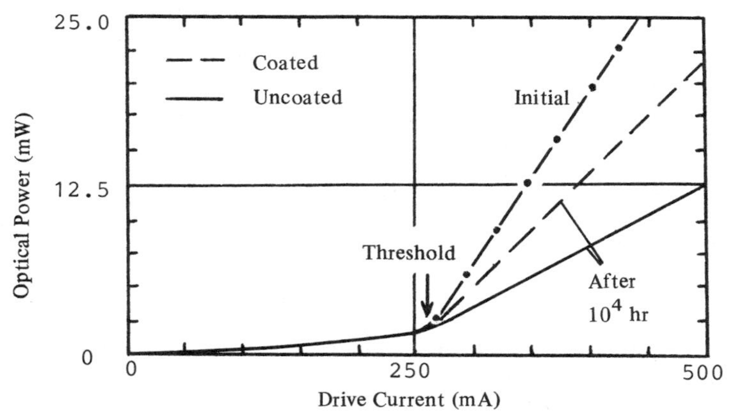

Figure 3.20 Dependence of optical output power of coated and uncoated injection lasers upon drive current and operating time.

moved by the use of $Ga_{1-x}As_{1-y}Al_xP_y$ material (whereby usually $y = 0.015$) for the region of the widest energy gap of the laser.

Generally, gradual changes in optical power level can be compensated by increasing the drive current level, which, however, increases the current density and hence the probability of an eventual catastrophic failure. This is illustrated in Figure 3.20, where the drive current of an injection laser and its resultant optical output power are shown for different times of operation. Assuming first an uncoated device and a required optical output of 20 mW, initially a drive current of about 400 mA is necessary. After 5000 hr of operation the optical power at the same current has decreased to about 10 mW, and after about 10,000 hr it has decreased to about 7 mW. In order to compensate for this device degradation, it is necessary to raise the current level to 510 mA after 5000 hr and to 740 mA after 10,000 hr; this increases the current density to about twice its original value, accompanied by even more drastic degradation and eventual catastrophic termination. On the other hand, if the facet is protectively coated, a substantial improvement in device characteristics is observed.

Lifetime data of optical sources are usually extrapolated from accelerated tests at elevated temperature. This is feasible because lifetime (mean time to failure, MTTF, defined as the time at which the current density has to be increased by 20% to realize the same optical power) is proportional to $e^{E_G/kT}$, where E_G is the semiconductor energy gap and kT is the thermal energy. This is illustrated in Figure 3.21, where it is assumed that MTTF at room temperature T_L can be derived from measurements at high temperature T_H from

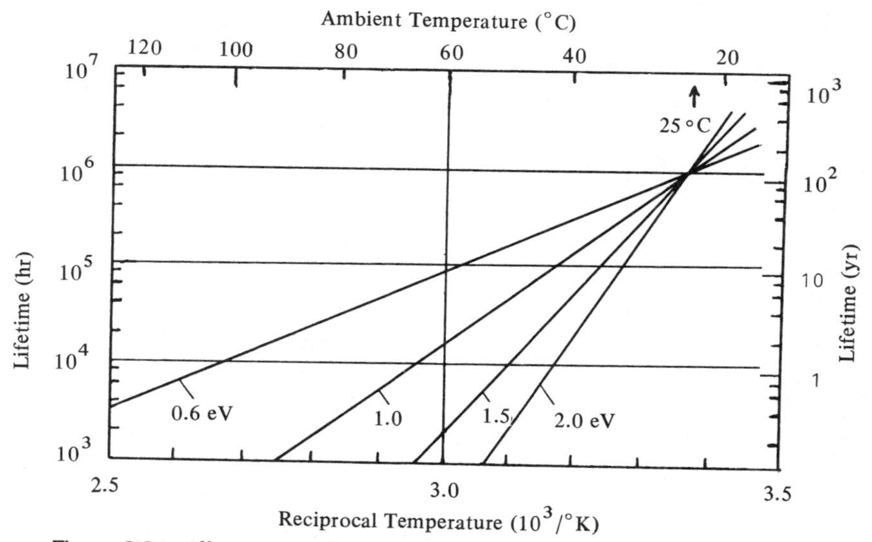

Figure 3.21 Illustration of the acceleration method used to extrapolate room-temperature laser lifetime from elevated-temperature measurements.

$$\text{MTTF}(T_L) = \text{MTTF}(T_H)\ \exp\left[E_G(1/kT_L - 1/kT_H)\right] \quad (3.13)$$

A larger semiconductor energy gap is thus equivalent to a faster laser output degradation with temperature. This can be utilized to determine the anticipated probable mean time to failure. For example, because it is reasonable to expect a room-temperature laser lifetime of 10^6 hr (114 yr), a device made in a material whose energy gap is 1.0 eV must have a lifetime in excess of 10^4 hr at 60°C or of almost 10^3 hr at 100°C. On the other hand, the room-temperature requirement of MTTF of 10^6 hr in a semiconductor whose $E_G = 0.65$ eV is equivalent to a lifetime of 7×10^4 hr at 60°C. Hence, extrapolations on devices made from material of large energy gap allow significantly shorter test times than those whose energy gap is small. For example, the energy gap of GaAsAl is about 1.6 eV, that of GaP is 2.3 eV, and that of GaSb is 0.7 eV.

REFERENCES

1. E. Weidel. Light Coupling From a Junction Laser into a Monomode Fibre With a Glass Cylindrical Lens on the Fibre End. *Opt. Commun.* **12**:93, 1974.

2. H. Kressel & M. Ettenberg. A New Edge-emitting Heterojunction LED for Fiber Optic Communications, *Proc. IEEE* **63**:1360, 1975.

3. H. Strack. Optoelektronik. *E and M, Wien,* **95**:301, 1978.

4. A. Bergh and P.I. Dean. Light-emitting Diodes. *Proc. IEEE* **60**:156, 1972.

5. C.A. Burrus, et al. Direct Modulation Efficiency of LEDs for Optical Fiber Transmission Applications. *Proc. IEEE* **63**:329, 1975.

6. Y.S. Liu & D.A. Smith. The Frequency Response of an Amplitude-modulated GaAs Luminescence Diode. *Proc. IEEE* **63**:542, 1975.

7. K. Saito et al. Buried Heterostructure Lasers as Light Sources in Fiber Optic Communications. IOOC 77, p. 65, Tokyo, July 1977.

8. I. Hayashi. Semiconductor Light Sources. IOOC 77, p. 81, Tokyo, July 1977.

9. M. Nakamura et al., Single-Transverse-Mode Low Noise (GaAl)As Lasers with Channeled Substrate Planar Structure. IOOC 77, p. 83, Tokyo, July 1977.

10. J.C. Carballès et al. Linear Light-Output Characteristics in DH GaAlAs Lasers. IOOC 77, p. 87, Tokyo, July 1977.

11. M. Yamanishi et al. Optically Pumped GaAs Lasers with Two-dimensional Bragg Reflectors. IOOC 77, p. 93, Tokyo, July 1977.

12. I. Melngailis. Lasers at 1.0–1.3 μm for Optical Fiber Communications. IOOC 77, p. 197, Tokyo, July 1977.

13. H. Nagai & Y. Noguchi. $InP/Ga_xIn_{1-x}As_yP_{1-y}$ Double Heterostructure LEDs in the 1.5 μm Wavelength Region. IOOC 77, p. 201, Tokyo, July 1977.

14. T. Yamamoto et al., Carrier Lifetime Measurement of InGaAsP/InP Double Heterostructure Lasers, IOOC 77, p. 205, Tokyo, July 1977.

15. D.R. Scifres et al. Integrated Optical Components in Semiconductor Injection Lasers. *Society of Photo-Optical Instrumentation Engineers (SPIE)* **139**:117, 1978.

16. C.J. Nuese. Advances in Heterojunction Lasers for Fiber Optics Applications. *SPIE* **139**:123, 1978.

17. B.C. DeLoach, Jr. On the Way: Lasers for Telecommunications. *Bell Lab. Rec.* **53**:203, 1975.

18. F.H. Doerbeck et al. Monolithic $Ga_{1-x}In_xAs$ Diode Lasers. *IEEE J. Quant. Electron.* **QE-11**:464, 1975.

19. L. Lewin. A Method for the Calculation of the Radiation Pattern and Mode

Conversion Properties of a Solid-State Heterojunction Laser. *IEEE Trans. Microwave Theory Tech.* **MTT-23**:576, 1975.

20. S. Iida et al. Spectral Behavior and Line Width of (GaAl)As–GaAs Double Heterostructure Lasers at Room Temperature With Stripe Geometry Configuration. *IEEE J. Quant. Electron.* **QE-9**:362, 1973.

21. F.A. Blum et al. Monolithic GaAs Injection Mesa Lasers With Grown Optical Facets. *Appl. Phys. Lett.* **25**:620, 1974.

22. H. Kressel. The Application of Heterojunction Structures to Optical Devices. *J. Electron. Mater.* **4**:1081, 1975.

23. T. Ozeki & T. Ito. A New Method for Reducing Pattern Effect in PCM Current Modulation of DH-GaAlAs Lasers. *IEEE J. Quant. Electron.* **QE-9**:1098, 1973.

24. M. Takusagawa et al. An Internally Striped Planar Laser With 3 μm Stripe Width Oscillating in Transverse Single Mode. *Proc. IEEE* **61**:1758, 1973.

25. M. Takusagawa et al. A New Stripe-Geometry Double Heterojunction Laser With Internally Striped Planar (ISP) Structure. *IEDM Tech. Digest* 327, 1973.

26. F.A. Blum et al. Optically Pumped Grown GaAs Mesa Surface Laser. *Appl. Phys. Lett.* **24**:430, 1974.

27. Y. Suematsu et al. A Multi-Hetero-AlGaAs Laser With Integrated Twin Guide. *Proc. IEEE* **63**:208, 1975.

28. M. Nakamura. Monolithic Integration of Distributed-Feedback Semiconductor Lasers. IOOC 77, p. 227, Tokyo, July 1977.

29. D.B. Anderson & R.R. August. Progress in Waveguide Lenses for Integrated Optics. IOOC 77, p. 231, Tokyo, July 1977.

30. P.M. Kruse et al. *Elements of Infrared Technology.* New York, John Wiley & Sons, 1962.

31. H.M. Stoll. High-Brightness Lasers Using Integrated Optics. *SPIE* **139**:113, 1978.

32. K. Otsuka & T. Yamada. Single-longitudinal-Mode $LiNdP_4O_{12}$ Laser. *Proc. IEEE* **63**:1621, 1975.

33. J.M. Hammer et al. Efficient Modulation and Coupling of CW Junction Laser Light Using Electrooptic Waveguides. *Proc. IEEE* **63**:325, 1975.

34. F.A. Blum et al. Monolithic $Ga_{1-x}In_xAs$ Mesa Lasers With Grown Optical Facets. *J. Appl. Phys.* **46**:2605, 1975.

35. H. Kressel & I. Ladany. Reliability Aspects and Facet Damage in High-Power Emission (AlGa)As CW Laser Diodes at Room Temperature. *RCA Rev.* **36**:230, 1975.

4

OPTICAL DETECTORS

Arjun N. Saxena and Helmut F. Wolf

Optical communications using optic fibers has become a practical reality primarily because of major developments in solid-state light sources and low-loss optic fibers. The basic concepts of the detectors in optical communications have been fairly simple to date. They are two-terminal devices; i.e., they are photodiodes. With improvements in their design and fabrication, these photodiodes have been quite satisfactory for use in optical communications.

The majority of the detectors in use today are made from silicon and germanium. The technology used for their fabrication is much simpler than that used for very-large-scale–integration (VLSI) integrated circuits (ICs) in silicon. Because of the developments in photodiodes and avalanche photodiodes and their acceptable performance as discrete detectors in optical communications, the incentives to develop monolithic ICs containing photodetectors have been rather small. However, with the advent of high-speed microprocessors and various types of memories, it is expected that VLSI chips containing photodetectors (Saxena, unpublished data) will be developed shortly. These will enhance the performance and applications of optical communications. Also, new materials are expected to be developed for use in photodetectors which will have several unique properties, one of them being that the ionization coefficient of one type of carrier (electron or hole) will dominate the ionization coefficient.

INTRODUCTION

Photodetectors convert the information transmitted via optical signals to corresponding electrical signals, which are then processed further to perform the desired functions. The wavelength range of the photons used in optical communications with optic fibers is 0.8 to 1.4 μm. This spectral range is subdivided into three wavelength regions. The wavelength region is determined by the LED and the laser materials used for the light sources. The three regions are 0.8–0.9 μm, 0.9–1.1 μm, and 1.1–1.4 μm.

The type of photodetector which would be preferable for each of the above spectral ranges varies in its construction and in the materials used for its fabrication. As mentioned above, the photodetectors currently available commercially are quite satisfactory for all the spectral ranges listed above. However, further improvements in materials and fabrication are being made continually to enhance performance, in particular at the long wavelengths. For background purposes, the reader is referred to the excellent reviews by Melchior (1), Conradi et al. (2), Stillman and Wolfe (3), and Elion and Elion (4), as well as Chapter 12 of the excellent textbook by Sze (5).

The basic principle of photodetectors is that they absorb incident photons, which produce electrical charge carriers, i.e., electrons and/or holes. The flow of charge carriers produces the electrical signals.

The diameters of the cores of the optic fibers used in communications are between 10 and 50 μm, and their attenuation ranges between 5 and 20 dB/km. The light sources which can couple with these optic fibers are light-emitting diodes (LEDs), injection laser diodes (ILDs), and neodymium-doped yttrium-aluminum-garnet (Nd:YAG) crystals pumped by LEDs.

The amount of light which can be coupled into the fibers at the source end ranges from a few microwatts to a few milliwatts. The light emerging at the detector end after traversing perhaps several kilometers of the optic fiber is attenuated. To detect such weak optical signals, the photodetectors have to meet the following criteria:

High Sensitivity. The photodetectors must be able to detect weak optical signals; i.e., the quantum efficiency should be high.

Wavelength Compatibility. The peak efficiency of the photodetectors should fall in the wavelength region of the light sources used in optical communications.

High speed. The response of the photodetectors to optical signals should be fast. They should have sufficient bandwidth to handle the information rate. Also, this response of the photodetectors should be linear with respect to the optical signals over a wide range of amplitudes.

Large signal-to-noise ratio. The detectors should also have low dark currents.

Room-temperature operation.

High reliability and Long-term Stability.

Suitable package. The packaging of the photodetectors should be reliable and lend itself to good coupling with the optic fibers. Detectors should not be affected adversely by changes in ambient conditions.

Material compatibility. The material and the technology used to fabricate photodetectors should be compatible with monolithic IC technology and should interface with microprocessors.

Compatibility requirements involve considerations of the coupling to the fiber, power supply, electronic circuitry needed to amplify and process the signal, physical dimensions, etc.

A variety of photodetectors are potentially applicable to optical communications systems, but only a few of these (for example, only silicon and some compound semiconductor materials) can be seriously considered for this application. Various types of photodetectors which have been used for the detection of photons in a variety of applications, including optical communications, can be classified as follows:

Semiconductor Detectors

Examples of semiconductor detectors are photoconductive devices, conventional and avalanche photodiodes, phototransistors, and charge-coupled devices (CCDs). These devices can detect impinging photons and contain a depleted semiconductor region with a high electric field that serves to separate photon-excited electron–hole pairs. Such a depletion region can be obtained, for example, by either reverse-biasing of a p-n junction, reverse-biasing of a Schottky or surface barrier, or biasing of an MOS capacitor in a mode suitable for depletion of carriers in the semiconductor. The speed with which the photon–carrier interaction takes place determines in part the frequency response of the devices. Hence, CCDs, for example, cannot be considered for high-data-rate operation. The most important semiconductor detectors

(1) Photoconductor
(2) Depletion layer photodiode
    (a) PN photodiode
    (b) PIN photodiode
    (c) Schottky barrier photodiode
(3) P-n junction avalanche photodiode (APD)
(4) Schottky barrier APD
(5) Electroabsorption APD
(6) New materials for APDs

Nonsemiconductor Detectors

Examples of vacuum photodetectors are photomultipliers and image intensifiers. The photocathode is the key element in a photomultiplier, which detects photons through the process of photoelectric emission of an electron from the photocathode. The most useful materials for vacuum devices are the alloys Ag–O–Cs, Cs–Te, K–Cs–Sb, GaAs–Cs_2O, and Na_2KSb–Cs. The last is the most commonly used photocathode material because of its broad spectral response, relatively high sensitivity, and low dark current. Most of these devices, however, lack the frequency response required of high-data-rate detectors. The most important nonsemiconductor detectors are photomultipliers and microchannel plates.

Detector Characteristics

The semiconductor detectors outlined above will be described in the following sections. Of those listed above, the most commonly used photodetectors in optical communications are the silicon PIN and avalanche photodiodes. Other semiconductor materials are under development and investigation, and should preferably meet the following two criteria:

1. The energy band gap of, and/or the Schottky barrier height of, the semiconductor should be slightly smaller than the energy of the photons at the longest wavelength of the light used for communication.

2. Under ideal conditions, only one type of carrier, i.e., either electrons or holes, should undergo ionizing collisions in the semiconductor. If this cannot be achieved, the ionization coefficient of one type of carrier in the semiconductor should dominate the other.

The most useful optical detectors are those made of semiconductor materials, partly because of their superior characteristics and partly because of their compatibility with other semiconductor devices and the resulting easier fabrication. Although optical detectors can be produced in any of the various semiconductor materials, considerations of transparency, carrier mobility, and fabrication economy must also be considered. Contrary to the situation in optical sources, however, detectors made of silicon are usable, at least at short wavelengths, because of the different mechanism that governs the detection process.

Silicon is used as the semiconductor for the PIN and avalanche photodiodes because its properties (electrical, optical, fabrication technology, packaging, reliability) are well known. These properties have been studied and evaluated over the past two decades, first in the fabrication of discrete diodes and transistors and later in conventional IC and VLSI technologies. It is somewhat surprising that integrated photodiode and phototransistor ICs with microprocessors have not yet found major applications in optical communications. This may be due to acceptable performance of discrete silicon PIN and avalanche photodiodes so far. Also, the major advances in light sources and low-loss optic fibers have taken place only in the past five years, and the microprocessor development is only about three years old. Therefore, the integration of photodetectors with VLSI circuits has lagged somewhat. However, it is expected that in the next few years such integrated structures will be developed and will be used extensively in optical communications. Also, it is anticipated that new semiconductor materials will be developed in which the ionization coefficients of a given type of carrier can be chosen to be dominant over that of the other carrier type. Photodetectors fabricated from such materials will be superior to those fabricated from silicon, although they are not likely to lend themselves to monolithic integration.

The next most popular material used is germanium. Other materials are under development.

DETECTOR TYPES

Photodetectors can be classified, as outlined in the previous section, into two major categories, i.e., those made from semiconductor materials and those made from other materials. Various types of photodetectors will be discussed, although those used predominantly in optical communications will be emphasized and their details given. To evaluate the performance of a photodetector, the following questions need to be answered:

1. How efficient is the detection of the photons in the photodetector? How many charged carriers are produced in the photodetector by the incident photons?
2. Does the photodetector increase the number of charged particles (produced initially on photon absorption) as they are transported out of the photodetector to the external circuitry? What is the mechanism of this gain in, or multiplication of, the number of carriers?
3. When the above charged carriers flow out of the photodetector to the external circuitry to produce the output signals, how well do the output electrical signals correspond to, or faithfully reproduce, the input signals? How clean are these output signals? How much higher are these as compared to the noise characteristic of a given type of photodetector and the external circuitry? Also, how fast is the response of the photodetector?

The physical mechanisms of a few types of photodetector will be discussed. However, because of the limited scope here, their noise characteristics will not be discussed in detail. The general expression for the total photocurrent I_t produced by a photodetector when the incident optical power is P_t can be written

$$I_t = M(q\eta P_t/h\nu + I_d) \tag{4.1}$$

where q is the unit electronic charge, η the quantum efficiency, M the multiplication (gain) factor, P_t the incident optical power, h Planck's constant, ν the frequency of the photons, and I_d the dark current of the photodetector.

The quantum efficiency η, the multiplication factor M, the dark current I_d, and consequently the total photocurrent I_t depend strongly on the material and the design of the photodetector and on the wavelength of light used. The choice of the material and the design depends on the wavelength range and the application of the photodetector.

The parameters defined in the following sections are used to evaluate the performance of the photodetectors.

Quantum Efficiency

The quantum efficiency η is the percentage of incident photons absorbed by the photodetector to liberate charged carriers. In general, η depends on several factors and can be written

$$\eta = f(\lambda, \omega, \alpha_c, E_G, \alpha_n, \alpha_p, t, G_d) \tag{4.2}$$

where λ is the wavelength of light, ω the modulation frequency of incident light, α_c the absorption coefficient of light (which is also a function of λ) in the detector material, E_G the forbidden energy band gap of the detector material, α_n the ionization coefficient of electrons in the detector material, α_p the ionization coefficient of holes in the detector material, T the temperature of operation of the detector, and G_d the geometry of the detector. The geometry determines the total path length of light in the detector, its location with respect to the depletion layer, and the electric field seen by the carriers (which affect α_n and α_p).

Multiplication Factor

The multiplication factor M is a measure of the gain provided by the photodetector. It is defined as the ratio

$$M = I/I_p \tag{4.3}$$

where I is the total output current at the operating voltage (where carrier multiplication occurs) and I_p the total primary current at low voltage (where carrier multiplication does not occur).

Total Noise Equivalent Power

The total noise equivalent power (TNEP) of a photodetector is a measure of the minimum detectable signal in the presence of the noise inherent in the detection and multiplication processes. It is defined as the amount of light (root mean square, rms, of the sinusoidally modulated radiant power) incident on the active area of the photodetector which will produce an output signal equal to the noise output (rms noise voltage). It is usually stated in nanowatts.

Noise Equivalent Power

The noise equivalent power (NEP) is defined as the TNEP normalized to a 1 Hz bandwidth, stated as W/Hz.

Responsivity

The responsivity of a photodetector (R_p) is the average output current divided by the average incident radiant power,

$$R_p = I/P \tag{4.4}$$

Signal-to-Noise Ratio

As the name implies, the signal-to-noise ratio (S/N) of the photodetector is the ratio of the signal produced by the incident light under the given operating conditions to the electrical noise inherently present in the detector under those operating conditions. It is one of the most important criteria used to evaluate detectors.

Response Time

The response time (τ_R) of the photodetector is the transit time of the charged carriers to the output terminals. It is dependent on the characteristics of the photodetector as well as on the external circuitry. In a simplified model, the response time τ_R is the rms sum of the total charge collection time and the RC (resistance–capacitance) time constant due to the series and the load resistances and detector and stray capacitances. The τ_R is a measure of how fast the photodetector responds to the variations in the input optical signals.

A few types of photodetectors made from semiconductor materials will now be described.

Photoconductors

Photoconductors are one of the earliest types of photodetectors (6). Their basic principle of detection is the increase in the conductivity of a semiconductor due to the absorption of light. Figure 4.1 shows a cross section of a photoconductor. It consists of a slab of a semiconductor with ohmic contacts at each end. Figure 4.2 shows the construction of a photoconductor detector using the thin-film and recent photolithographic techniques to give interdigitated ohmic contacts. Other variations of these structures are possible. However, their basic operation remains the same. On light absorption, the in-

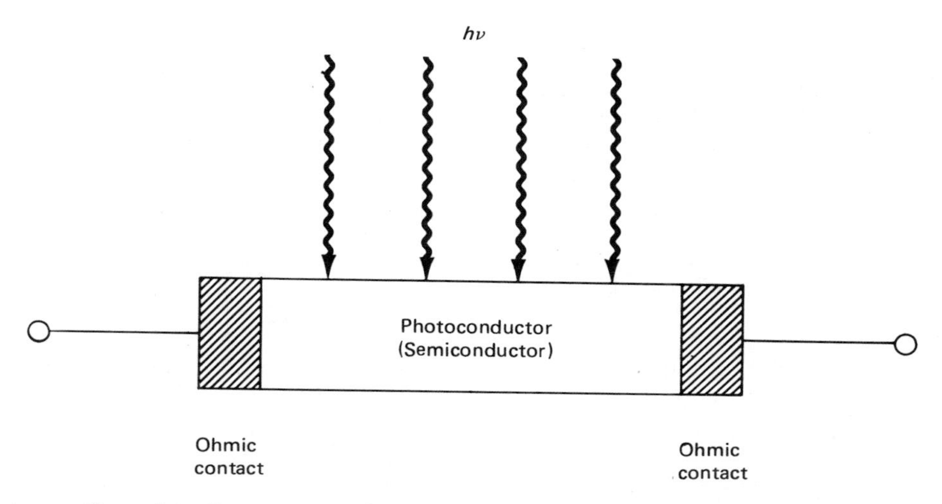

Figure 4.1 Cross section of a photoconductor photodetector. It consists of a slab of a semiconductor with ohmic contacts at each end.

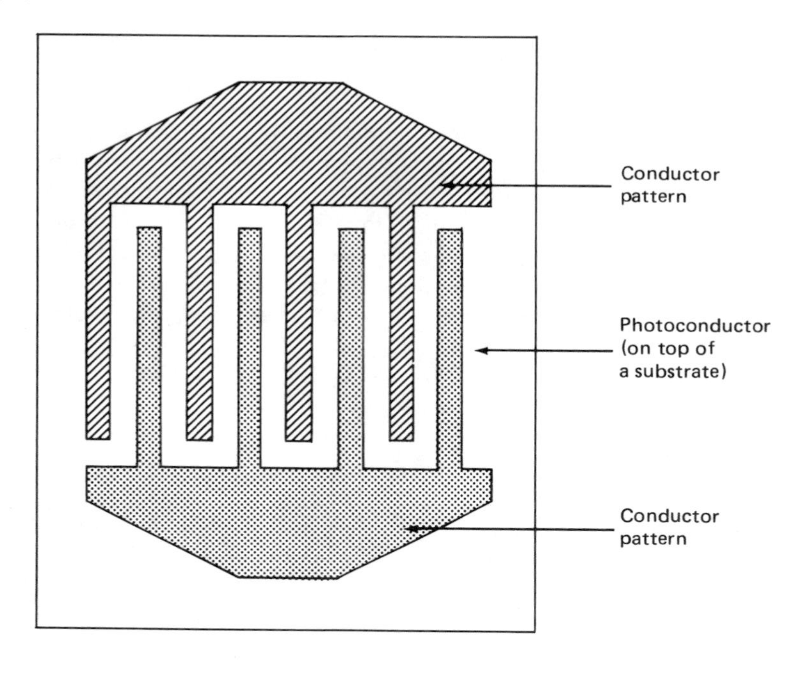

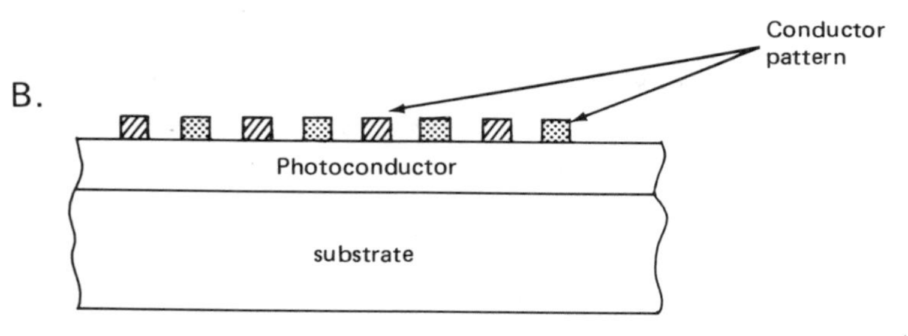

Figure 4.2 (A) Top view of a photoconductor detector with interdigitated metallization. (B) Cross section of a photoconductor detector with a thin film of the photoconductor material deposited on a substrate. The interdigitated conductor pattern is obtained by photolithographic techniques or by depositing through a metal mask.

crease in conductivity is due to intrinsic and/or extrinsic excitation of carriers in the photoconductor. The generation of these carriers depends on the wavelength λ of the light, the forbidden energy gap E_G of the intrinsic semiconductor, and the energy levels of the impurities, E_D (donors) and E_A (acceptors), in the extrinsic semiconductor.

For low-level optical detection, which is necessary in optical communications, the photoconductors are not as suitable as p-n junction photodiodes in the desired wavelength range of 0.8 to 1.4 μm because the thermal noise predominates at low-level detection, which makes the signal-to-noise ratio of the photodiodes much higher than that of the photoconductors. However, at high-level detection the thermal noise can be neglected, and the S/N of photoconductors and photodiodes are comparable. The photoconductor is a very-broad-bandwidth device; i.e., its detection efficiency beyond the long-wavelength limit for a given semiconductor material remains fairly constant at high-level detection of light. In the wavelength region above 3 μm the photoconductors are used extensively. No satisfactory alternatives have been found yet for this long-wavelength range, although the recent work with small Schottky barrier heights of silicon appears promising. In summary, photoconductor detectors are not suitable for optical communications in the wavelength range 0.8 to 1.4 μm. However, for longer wavelengths, photoconductors are used extensively.

Depletion Layer Photodiodes

The depletion-layer photodiodes, as broadly defined by Sze (5), are reverse-biased semiconductor diodes whose reverse currents are altered due to the absorption of light in or near the depletion layer. In such diodes, the applied reverse voltage is less than the breakdown voltage. Thus, the electron-hole pairs created by photon absorption do not multiply in number by impact ionization; they are simply swept by the electric field.

The absorption of light in a photodiode, which produces the photocurrent, depends on the absorption coefficient α_c of the light in the semiconductor used to fabricate the photodiode; α_c varies with the wavelength of light. Figure 4.3 shows a plot of α_c versus wavelengths for Si, Ge, and GaAs. Also shown in Figure 4.3 are the room-temperature wavelength ranges for several III-V ternary alloys, gas, and YAG–LED lasers. The type of devices fabricated with Si and Ge, GaAs, GaInAs, and GaAsSb, which are useful in different wavelength ranges, are also indicated in Figure 4.3. As in the case of photoconductors, the long-wavelength limits λ_c for p-n junction type photodiodes in Si, Ge, and GaAs are determined by the respective energy gaps. For Si, $\lambda_c = 1.1$ μm; for Ge, $\lambda_c = 1.7$ μm; and for GaAs, $\lambda_c = 0.88$ μm. For the Schottky barrier type of detectors, λ_c depends on the barrier height, and it is determined by

$$\lambda_c(V) = 1.24/\phi_{MS}(V) \tag{4.5}$$

where $\phi_{MS}(V)$ is the Schottky barrier height in eV, which varies with the reverse bias due to image forces, and λ_c is given in μm. As shown in Figure 4.3, the absorption coefficient α_c is very large for short wavelengths. Thus, at short wavelengths the absorption of light will occur close to the surface, and before the carriers can be swept out they recombine. This effect causes the quantum efficiency of the p-n photodiodes to drop significantly at short wavelengths.

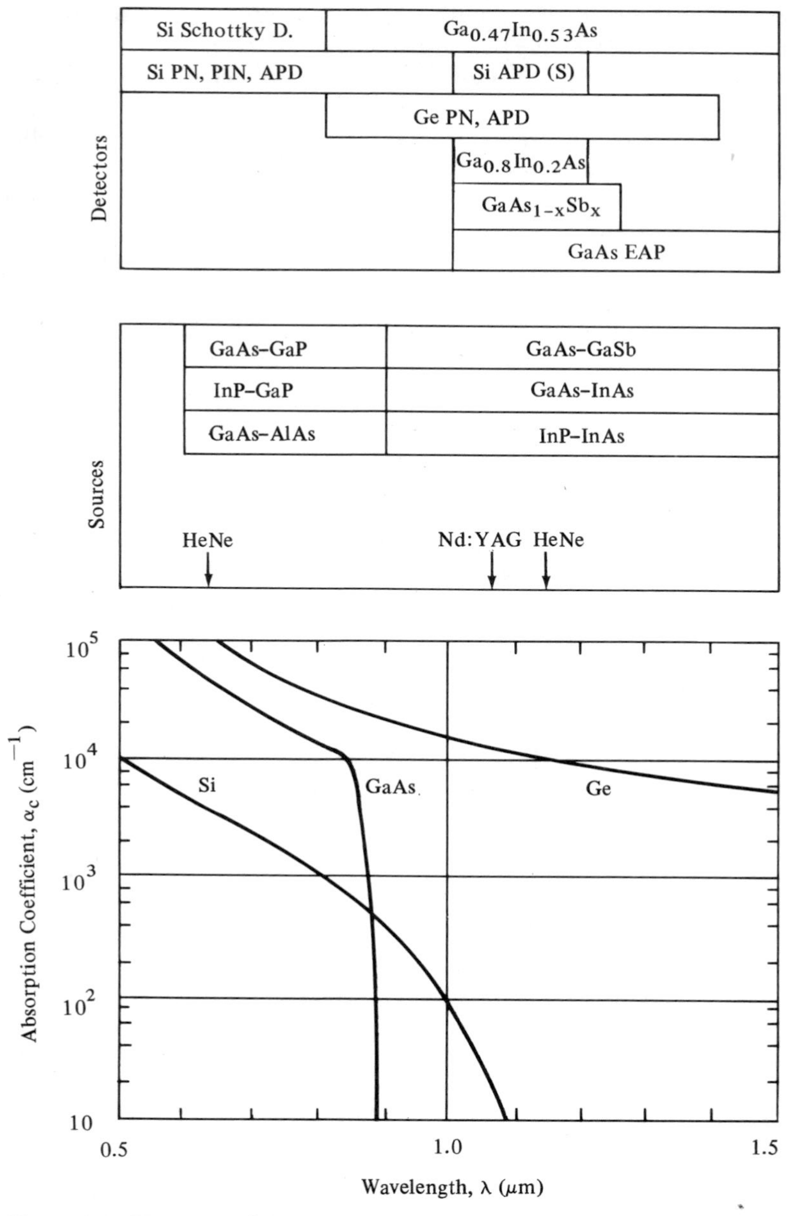

Figure 4.3 Variation of the absorption coefficient as a function of wavelength for Si, Ge, and GaAs. Indicated at the top of the illustration are the various group III–V alloys, gas, and Nd:YAG lasers used in different wavelength ranges. The types of photodetectors used at different wavelengths are also shown at the top.

However, Schottky barrier photodiodes tend to recover the quantum efficiency at short wavelengths because they are formed at the surface of the semiconductor. Figure 4.4 shows the quantum efficiency η of the p-n photodiodes on Si and Ge and its improvement on Schottky barrier diodes at short wavelengths. Also shown in Figure 4.4 are η values for $Ga_{0.8}In_{0.2}As$, $Ga_{0.47}In_{0.53}As$, and $GaAs_{1-x}Sb_x$ detectors.

PIN PHOTODIODE

The PIN photodiode is shown in Figure 4.5. The intrinsic or high-resistivity region (labeled i) essentially forms the depletion region. Varying the thickness and the resistivity of the i region allows optimization of the sensitivity and the wavelength response of the PIN diode. To peak its response at a par-

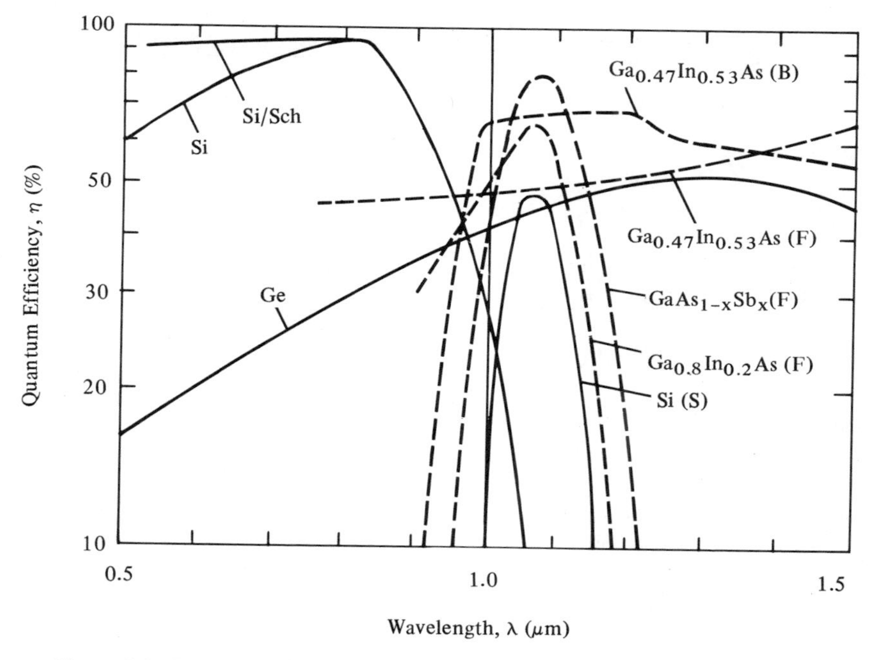

Figure 4.4 Variation of the quantum efficiency with wavelength for various types of p-n photodetectors made from Si, Ge, $Ga_{0.8}In_{0.2}As$, $Ga_{0.47}In_{0.53}As$, and $GaAs_{1-x}Sb_x$. Also shown are the improvement in η of a silicon photodiode at short wavelengths by using a Schottky barrier instead of a p-n junction. At long wavelengths, the efficiency of a Si p-n photodiode is increased by side illumination of the p-n photodiode, which is due to the increased path length of light in the photodiode. Legend: (——) elemental semiconductors; (– – –) alloy semiconductors; (F) front illumination; (B) back illumination; (S) side illumination; (Sch) with Schottky barrier.

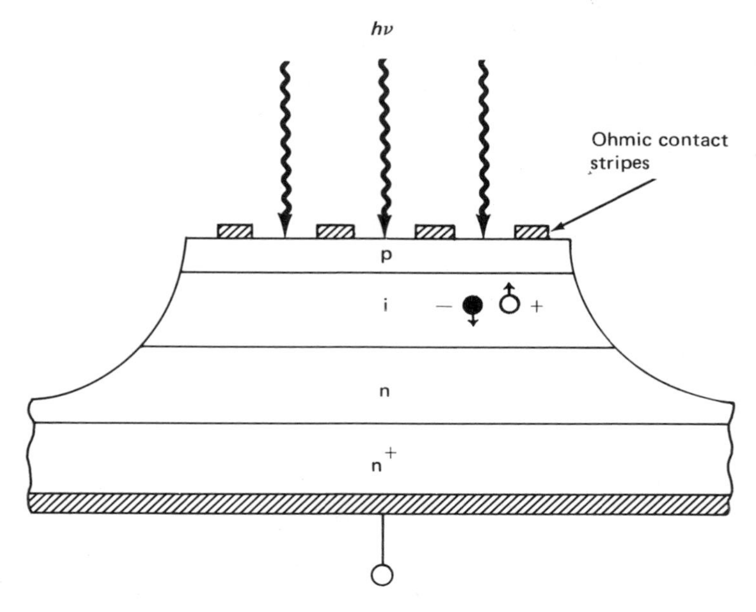

Figure 4.5 Cross section of a PIN photodiode.

ticular wavelength, the thickness of the i region is made equal to about $1/\alpha_c$.

For PIN detectors, the TNEP is calculated as

$$\mathrm{TNEP_{pin}} = (4h\nu B/q\eta)\,(2kTC)^{1/2} \tag{4.6}$$

where ν is the transmission optical frequency, B the bandwidth, T the diode-amplifier temperature, and C the diode capacitance.

PIN photodiodes are suitable for applications involving relatively high backgrounds and low bandwidths. Some internal gain mechanism in the photodiode is needed to obtain lower NEP values over wide bandwidths. Such gain will reduce the effect of the thermal noise of the circuit in which the photodiode is used. The mechanism that provides the gain is avalanche multiplication, as in avalanche photodiodes.

P-N PHOTODIODE

This is a p-n junction diode (PN detector) in which the absorption of the light occurs either in the n or the p region instead of in the depletion region. Figure 4.6 shows such a p-n junction structure, a mesa structure in which the edge of the p-n junction is exposed to the environment or the material in the package. Such mesa diodes are prone to surface leakages, and they are less stable than planar diodes, in which the edges of the junctions are protected by

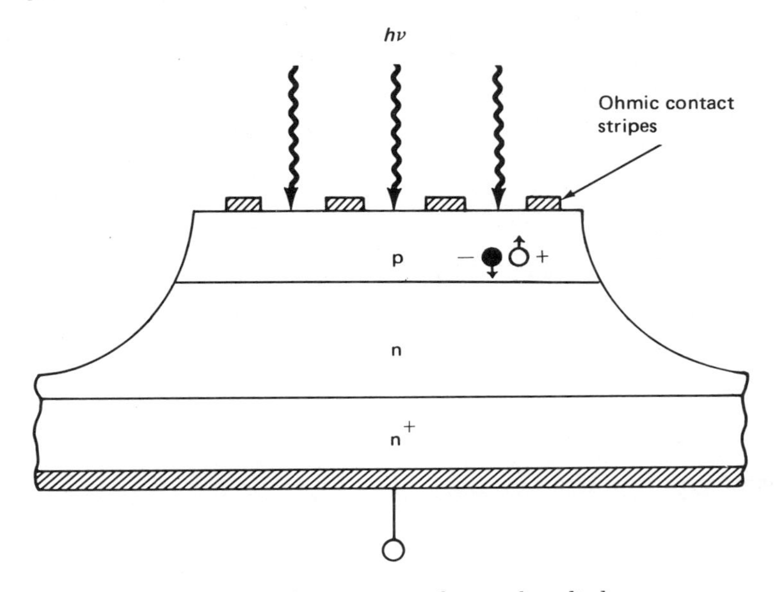

Figure 4.6 Cross section of a p-n photodiode.

the silicon dioxide layers. Guard rings are used to minimize the high-field edge effects which reduce the breakdown voltage and increase the dark current in the photodiode. Figure 4.7 shows a cross section of such a planar guard ring photodiode. For bulk resistivity greater than approximately 100 Ω cm and diffusion junction depths of less than about 2 μm, the performance of such p-n junction photodiodes is similar to that of PIN diodes. The choice of Si PIN, Ge PIN, or Ge p-n diodes depends on the wavelength to be detected and the modulation frequency (see ref. 5 on unpublished work by W.T. Lynch).

SCHOTTKY BARRIER PHOTODIODE

The Schottky barrier is a rectifying metal–semiconductor contact (7) which has been used in a variety of applications (8), e.g., clamps in transistor-transistor-logic (TTL) ICs, Schottky gate field-effect transistors (FETs), rectifiers, mixers, load resistors, photodiodes, phototransistors, etc. The conditions under which the rectifying contact is made are explained essentially by two models, i.e., work-function difference and surface-state models. These two models, and the anomalies in the current–voltage behavior of the Schottky barriers, have been reported earlier (9). Figure 4.8 shows a typical construction of a Schottky barrier photodiode with a p-n junction guard ring, using planar technology. As discussed, the long-wavelength cutoff limit λ_c of a Schottky photodiode can be varied with variation of the Schottky barrier height ϕ_{MS} by appropriate choice of metal and semiconductor. One of the

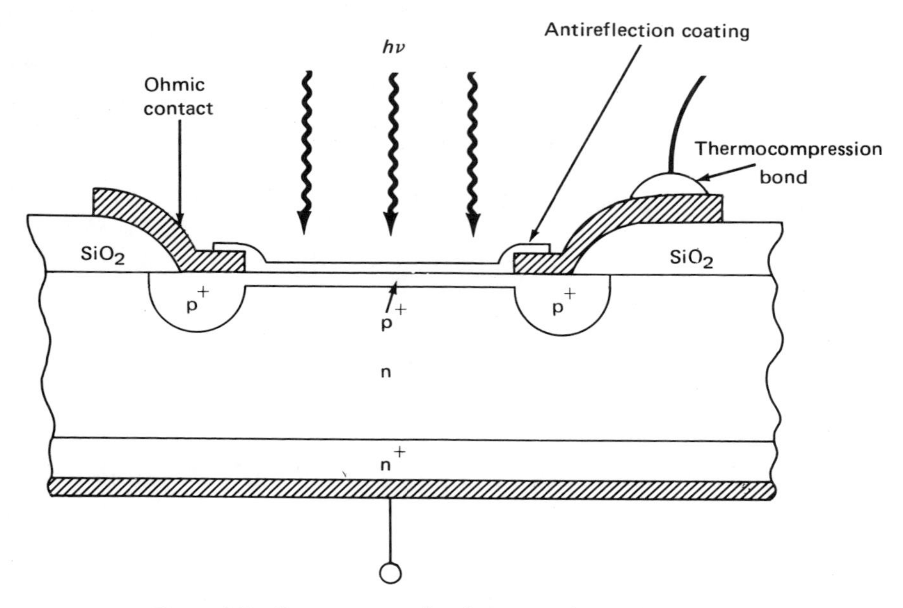

Figure 4.7 Cross section of a planar guard ring photodiode.

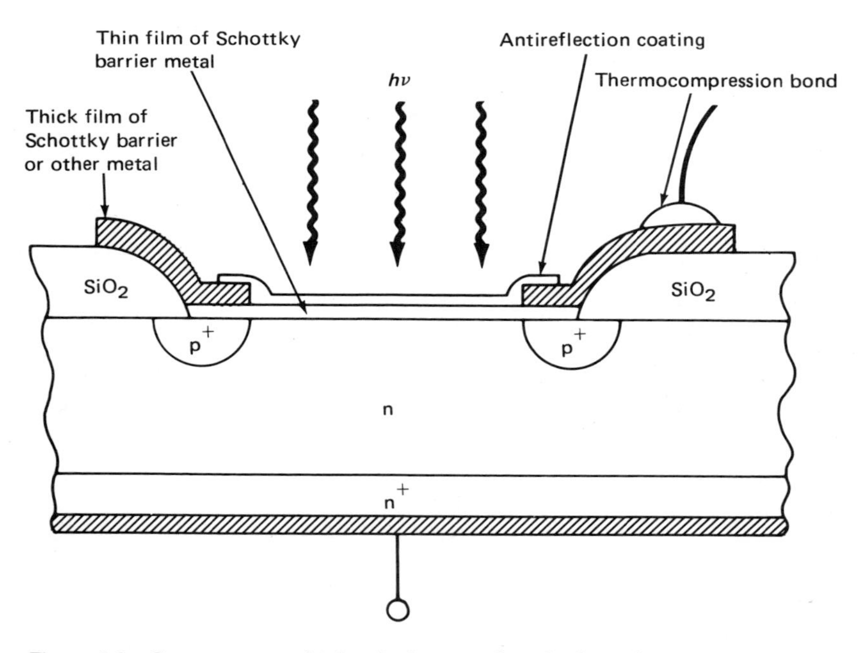

Figure 4.8 Cross section of Schottky barrier photodiode with a p-n junction guard ring using planar technology.

key advantages of the Schottky barrier technology is that no diffusion is necessary. In diffused junction technology, diffusion and the electrical properties of the dopant in the host semiconductor must be known to be able to fabricate good p-n junctions. In a well-understood semiconductor like silicon, the barrier height for a given metal can be varied by using ion implantation near the surface (8, 10). Schottky barrier photodiodes are expected to find many applications in optical communications.

p-n Junction Avalanche Photodiode

The basic principle of an avalanche photodiode (APD) is that the electrical signal on absorption of the photons is multiplied (avalanche multiplication) in the diode. The carriers generated by photon absorption are accelerated by the electric field, and they gain sufficient energy to generate additional, secondary free carriers by impact ionization of valence electrons into the conduction band, leaving free holes in the valence band. These secondary carriers in turn ionize other valence electrons by impact, thereby generating more electron–hole pairs. Such multiplication goes on to produce more and more electron–hole pairs, until the energy of the carriers is small enough so that they can no longer ionize. The ionization coefficients of electrons and holes are in general not equal, and they also depend on the electric field. These ionization coefficients, $\alpha_n(E)$ and $\alpha_p(E)$, respectively, are plotted in Figure 4.9 versus the reciprocal of the electric field for various semiconductors. The values of $\alpha_n(E)$ and $\alpha_p(E)$ depend not only on the material but also on the geometry of the photodiode because of the electric field dependence. It is beyond the scope of this review to discuss the detailed mechanisms for α_n and α_p and their effects on the performance of avalanche photodiodes. However, two special cases will be mentioned.

1. *Negligible ionization rate.* If the condition $\alpha_n = 0$ or $\alpha_p = 0$, equivalent to either $\alpha_n \gg \alpha_p$ or $\alpha_p \gg \alpha_n$, can be achieved in a semiconductor material because of its intrinsic properties and the geometry used, the possibility of obtaining the highest performance in an avalanche diode is enhanced.

2. *Equal ionization rates.* The condition $\alpha_n = \alpha_p$ is the worst case for performance in an avalanche diode. In this case, even though the gain or multiplication can be large, the statistical variations in the impact ionization process can produce large fluctuations in the gain and consequently considerable excess noise.

In construction, an avalanche photodiode is similar to a p-n diode having a mesa structure or a guard ring. Figure 4.10 shows cross sections of various types of structures of avalanche photodiodes (3). Also shown are the electric fields and the multiplication factors for each structure. The operating voltages of avalanche diodes are higher than those for p-n or PIN diodes, so that the breakdown fields are reached. The carriers multiply by impact ionization, and both electrons and holes undergo multiplication. The multiplication factor M defined in Equation (4.3) can be rewritten

A. Electrons

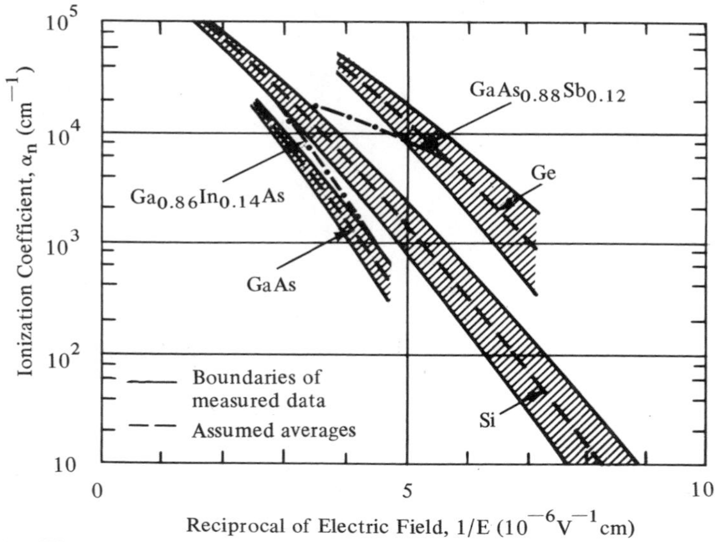

B. Holes

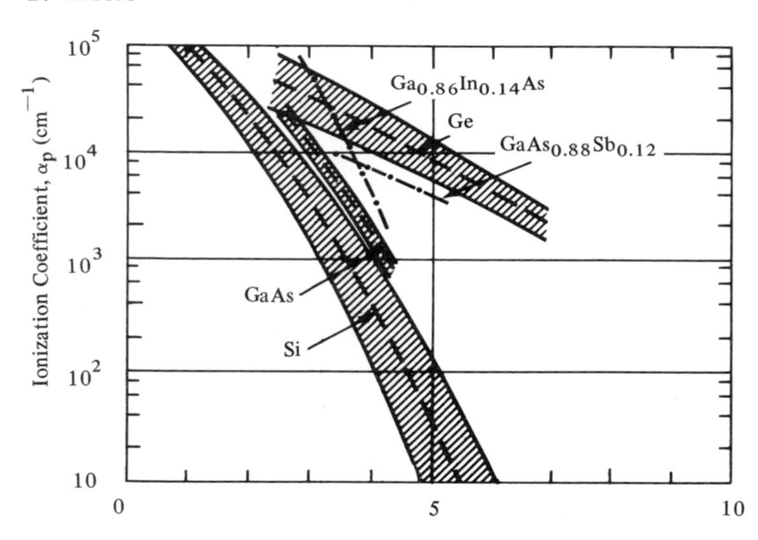

Figure 4.9 Variation of ionization coefficients for electrons, α_n (top), and for holes, α_p (bottom), as a function of the reciprocal of the electric field, $1/E$, for Si, Ge, GaAs, $Ga_{0.86}In_{0.14}As$, and $GaAs_{0.88}Sb_{0.12}$. The range of variation of α_n or α_p for Si, Ge, and GaAs is due to a large amount of experimental data for these materials obtained by various techniques; hence, they are shown as a broad band covering the entire range. The data of α_n and α_p for $Ga_{0.86}In_{0.14}As$ and $GaAs_{0.88}Sb_{0.12}$ are shown as single lines because not enough data are available to indicate the range of their variation.

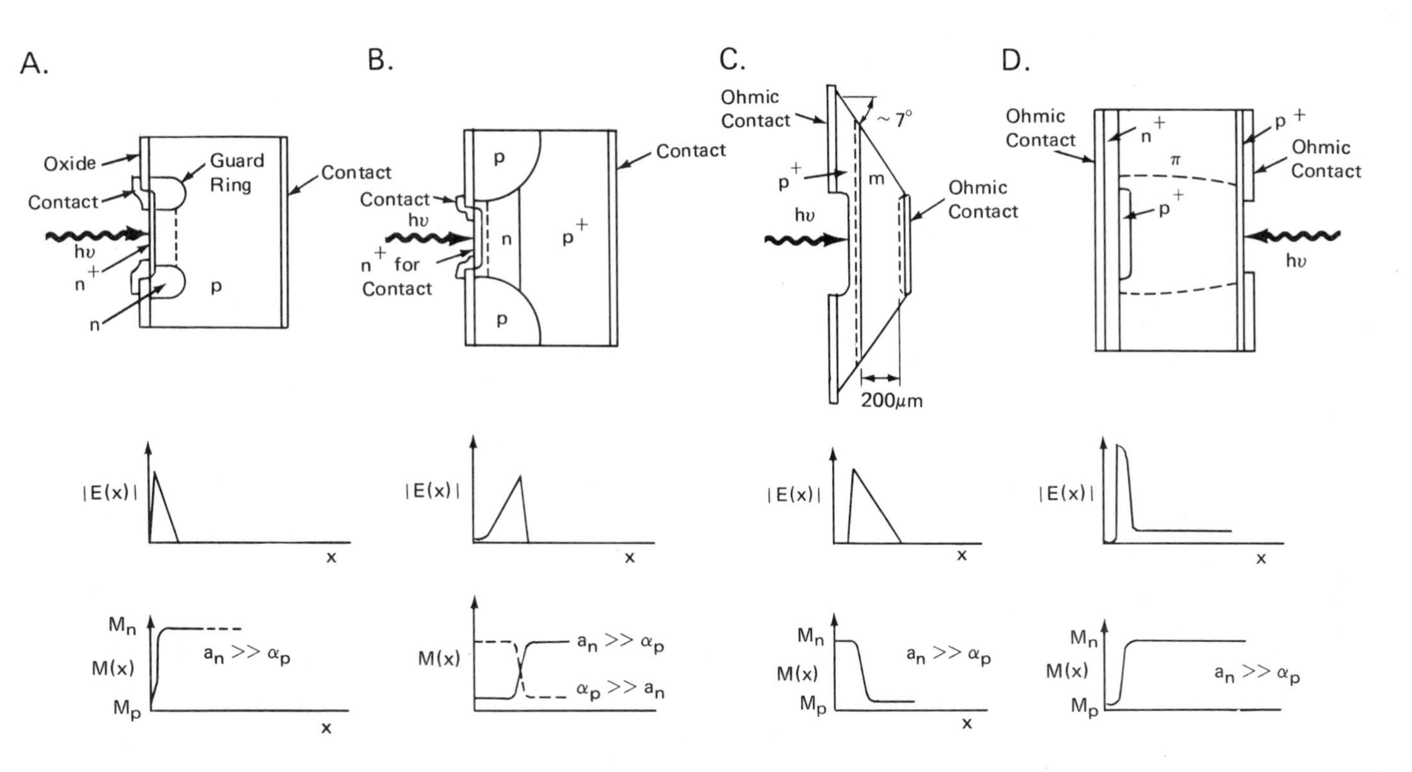

Figure 4.10 Various avalanche photodiode devices. (A) n^+-p guard ring structure. (B) Inverted, no guard ring, n^+-p structure. (C) Beveled-mesa or contoured-surface structure. (D) n^+-p-π-p reach-through structure (3).

$$M = \frac{1}{1 - [(V - IR)/V_B]^n} \tag{4.7}$$

where V is the applied voltage, V_B the breakdown voltage (depends on the geometry of the diffused junction, the resistivity, and the thickness of the active semiconductor layer), I the total output current, R the total effective series resistance (depends on area, thickness, and resistivity), and n a constant (depends on the semiconductor used, the doping profile of the diffused junction, and the wavelength of light being detected).

The majority of commercial avalanche photodiodes are fabricated from Si (3, 4, 11) because the technology of Si is known. Since both electrons and holes ionize and multiply in the avalanche process, the avalanche noise depends on the ionization rates of electrons and holes. The effective ratio k_{eff}, of these ionization rates,

$$k_{eff} = \alpha_p/\alpha_n \tag{4.8}$$

affects the excess noise factor (12, 13) F_N,

$$F_N \approx 2 + k_{eff}M \tag{4.9}$$

for large M and small k_{eff}. For Si, $k_{eff} \approx 0.02$–0.08. The avalanche noise decreases with k_{eff} (14).

In the long-wavelength region (1.0–1.4 μm), the quantum efficiencies of Si photodiodes drop rapidly, as shown in Figure 4.4, because of the rapid decrease in the absorption coefficient of light in Si in this wavelength region. To increase the quantum efficiency η of Si APDs in this λ range, side illumination of the diodes and trapping of the light in the diode by near total internal reflection are used. Figure 4.11 illustrates these two approaches (2). Figure 4.11A shows that the incident light enters the edge in a plane parallel to the junction and perpendicular to the electric field. By making the junction longer, the path length of the light to be detected is increased. This enhances the probability of absorption within the photodiode and hence increases the quantum efficiency. The same objective is also achieved by trapping the light in a thin-film structure by near-total internal reflection, shown in Figure 4.11B. Light enters through the junction in a conventional manner, but with appropriate back and front reflecting contacts the light is multiply-reflected within the depletion layer. Thus, the path length is increased, giving improved quantum efficiency.

APDs have been made in Ge (15) for use in the range $\lambda = 1.2$ to 1.4 μm. The energy band-gap of Ge is small, i.e., 0.8 eV at room temperature, and the effective ratio for Ge is ≈ 0.5 (16).

For an APD, the TNEP is calculated as

$$\text{TNEP}_{\text{APD}} = (h\nu B/q\eta G)\,(2kTC)^{1/2} \tag{4.10}$$

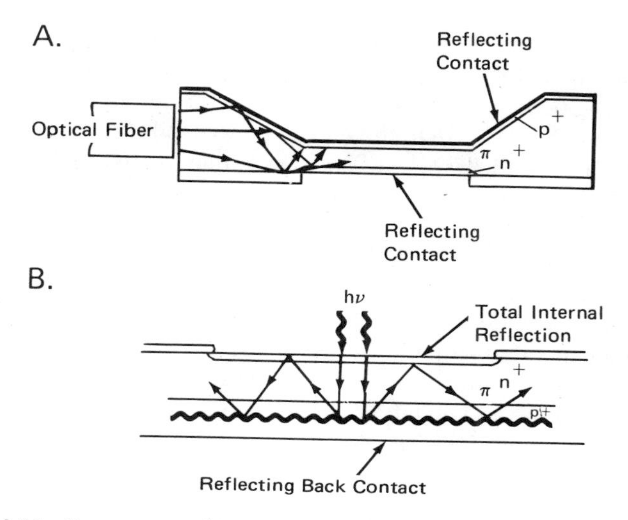

Figure 4.11 Two proposed structures for increasing the long-wavelength quantum efficiency of high-speed Si photodiodes. (A) Edge-illuminated structure. (B) Multiple-reflection structure (2).

where G is the detector gain at the operating voltage where the shot noise and the thermal noise at temperature T are equal.

Schottky Barrier APD

A Schottky barrier avalanche photodiode operates on the same principle as the p-n junction APD discussed above. A Schottky barrier replaces the p-n junction through which light enters. For the purpose of fabricating low–dark current Schottky barriers, PtSi on n-type silicon (17) and Hf on p-type silicon (18) can be used. To eliminate high-field edge effects, either mesa or p-n junction guard rings are employed.

The side illumination and total internal reflection techniques discussed above can be used to increase the quantum efficiencies of Schottky barrier photodiodes. These types of photodiodes are not yet available commercially. However, because of their improved performance and, relatively speaking, simple technology, they are expected to be available within a few years.

Electroabsorption Avalanche Photodiode

Electroabsorption avalanche photodiodes (EAPD) are avalanche photodiodes made from a direct-gap semiconductor material like GaAs having low carrier concentration ($<10^{15}$ cm^{-3}). In these diodes, the responsivity and quantum efficiency are high at wavelengths close to and beyond the absorption edge in GaAs, and the avalanche gain at these long wavelengths is much higher than the gain at shorter wavelengths. This improved performance of EAPDs is due to the Franz-Keldysh effect, i.e., shift of the absorption edge due to electric field. Qualitatively, the highly asymmetric values of electron (α_n) and hole (α_p) ionization coefficients for GaAs ($\alpha_p/\alpha_n \gg 1$) with low carrier concentrations ($<10^{15}$ cm^{-3}), gives rise to the enhanced spectral response at long wavelengths. For an excellent review and more quantitative explanations, the reader is referred to the comprehensive paper by Stillman and Wolfe (3).

GaAs EAPDs are still under development; however, they are expected to find important applications in optical communications and integrated optical circuits, where high quantum efficiencies and fast response are needed at long wavelengths. Materials other than GaAs which have asymmetric α_n and α_p and which exhibit strong Franz-Keldysh effects can also be used for EAPDs. However, next to Si and Ge, GaAs is the most studied semiconductor. Its electrical, optical, and metallurgical properties and fabrication technology are fairly well established now. Therefore, GaAs is the material which will most likely be used for EAPDs.

Other Photodiodes

New materials are needed to fabricate avalanche photodiodes for the wavelength region $\lambda = 1.0$ to $1.4\ \mu$m, where the transmission losses due to attenuation and dispersion in optical fibers are a minimum. An ideal detector material would have only one type of carrier undergoing ionizing collisions (1). However, this is almost impossible to achieve. The next best approach will be to find a material in which α_n dominates α_p or vice versa ($\alpha_n/\alpha_p \gg 1$ or $\alpha_p/\alpha_n \gg 1$). This poses a very challenging problem for workers in various disciplines. Some comment on these follows.

Solid-state physicists and chemists must elucidate why α_n and α_p differ in various materials. They must also elucidate scattering mechanisms and electric field and temperature dependences, as well as predict energy band gaps for various compositions.

Metallurgists and materials scientists need to develop the following: methods to grow single crystals and epitaxial layers of suitable compounds having as few defects as possible; dopants to fabricate p-n junctions; metals or

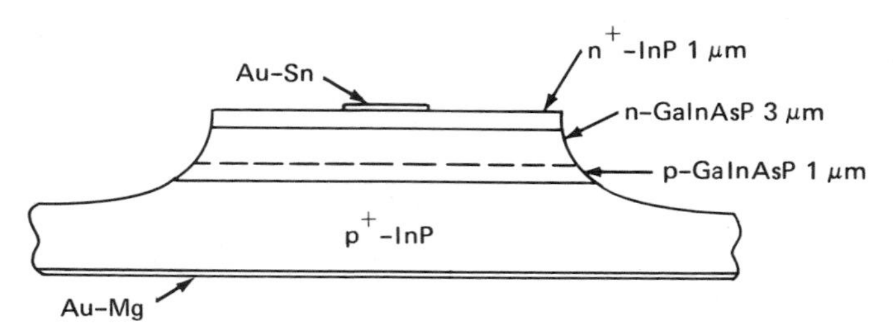

Figure 4.12 Schematic diagram of an inverted-mesa GaInAsP–InP avalanche photodiode.

metallic compounds to fabricate Schottky barriers; insulators and dielectric films to passivate surfaces of detectors; and ohmic contacts which can withstand high temperatures.

Solid-state technologists need to investigate the following: practical techniques to fabricate reliable detectors reproducibly at low cost; crystal slicing and polishing; etching; thermal diffusion; ion implantation; annealing; photolithography; dielectric and metal depositions; packaging; and testing.

Some of the new materials studied for photodetector applications are as follows: (1) $Ga_{0.21}In_{0.79}As_{0.52}P_{0.48}$–InP; (2) $Ga_{0.47}In_{0.53}As$; (3) $Ga_{0.86}In_{0.14}As$; (4) $GaAs_{1-x}Sb_x$; and (5) $Hg_{1-x}Cd_xTe$.

The quaternary compositions of $Ga_xIn_{1-x}As_yP_{1-y}$ give band gaps varying from 0.78 eV (1.6 μm) to 1.35 eV (0.92 μm); however, their lattice constants throughout this range match favorably those of InP. Thus, good epitaxial films of $Ga_xIn_{1-x}As_yP_{1-y}$ can be grown on InP as the substrate material. Hurwitz and Hsieh (19) have reported encouraging results on avalanche photodiodes fabricated from such epitaxial structures. Figures 4.12 and 4.13 show a cross section and the wavelength response, respectively, of these diodes. Pearsall et al. (20) have also reported good results on avalanche photodiodes fabricated from $Ga_{0.47}In_{0.53}As$. A comparison of the relative properties of photodetectors fabricated from this material versus those made from Si, Ge, and InAs is given (20) in Table 4.1.

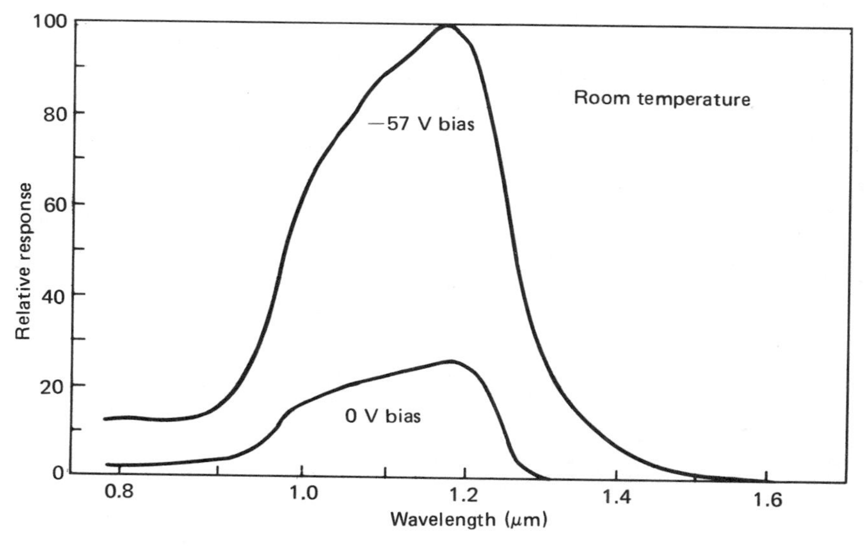

Figure 4.13 Wavelength response of a GaInAsP–InP photodiode with no bias and with sufficient bias to achieve an avalanche gain of about 4.5 (19).

TABLE 4.1. Comparison of Photodetector Properties of Several Materials

Material	λ_{max} response (μm)	η at 1.3 μm (%)	τ_R at 1.3 μm (s/cm^2)	Dark current density (A/cm^2)
$Ga_{0.47}In_{0.53}As$	1.6	75	2×10^{-7}	4×10^{-6}
Si	0.9	<0.01	—	2.5×10^{-6}
Ge	1.5	55	10^{-5}	10^{-3}
InAs	3.4	50	5×10^{-5}	10^{-1}

A large number of manufacturers in different countries produce photodetectors for a variety of applications, including fiber optics communications. From a reliability point of view, the most desirable and stable structures of the photodetectors are made by using the planar technology of Si. In this technology, the diffused p-n junctions are embedded under a grown oxide layer (silicon dioxide), which is an excellent insulator. Such junctions are almost unaffected by the ambient changes. The most popular types of photodetectors are PIN and APD silicon devices.

TABLE 4.2. Typical Commercial PIN Photodetectors

Manufacturer	Device	Responsivity (A/W) at λ_0 (nm)		Response speed (ns)	Bias voltage (V)
		PIN Diodes			
Bell-Northern	BNRD-5-1	0.55	(840)	1	100
EG&G	FND-100	0.62	(900)	1	90
EG&G	FOD-100	0.62	(900)	<1	90
Galileo	5105-031	0.50	(905)	5	50
Harshaw	538	0.62	(900)	7	100
Hewlett-Packard	5082-4205	0.45	(900)	1	20
Infrared Industries	7258	0.51	(1060)	8	50
RCA	C30808	0.65	(900)	5	45
Spectronics	SD5426	0.64	(905)	2	50
TI	TIXL80	0.55	(950)	15	100
UDT	PIN-3D	0.40	(850)	15	50

Manufacturer	Module	Responsivity (mA/μW)		3 dB bandwidth
		Preamp hybrid modules		
Bell & Howell	529	40	(800)	0.1
Centronic	OSI5LHSB	15	(900)	1
EG&G	MHZ-018Y	40	(900)	40
Meret	MDA425	15	(900)	10
RCA	C30818	300	(850)	40
TI	TIXL151	230	(900)	50

Source: Manufacturers' product sales literature, 1977 (for further and latest data, contact manufacturers). Reprinted from Elion and Elion, *Fiber Optics in Communications Systems*, p. 121, by courtesy of Marcel Dekker, Inc.

The characteristics of typical PIN diodes of a few manufacturers are listed in Table 4.2 (4). The characteristics of typical APD devices available commercially are listed in Table 4.3. (4). It can be observed that the responsivities of APDs are about one to two orders of magnitude greater than those of the PIN diodes. This is due to the avalanche multiplication gain in the former. A comparison of PIN and APD photodetectors is given in Table 4.4. (4).

PERFORMANCE

The performance characteristics of photodetectors can be described by noise, sensitivity, bandwidth, and dark current. Often, other performance parameters are used to evaluate the detector behavior, but usually they are of secondary importance. Detectors generally present fewer problems with respect to system performance than most other system components.

Most optical detectors offer reasonably good performance at 0.85 μm, but at larger wavelengths, such as at 1.06 and 1.27 μm, their performance is significantly degraded. Tables 4.5 and 4.6 summarize the characteristics of typical devices.

One of the most critical criteria in determining the usefulness of a detector is its noise characteristic. The presence of noise is one of the major limitations of system performance. For example, the maximum length of fibers dictated by signal pulse degradation is affected in part by detector noise. Consequently, it is generally desirable to avoid system operation in the noise-limited mode. The major detector types display a noise equivalent power in the range 10^{-14} to 10^{-13} W/Hz$^{1/2}$. The guard ring and reach-through devices offer better performance (21). High-speed quantum efficiencies typically range from 0.25 to 0.95 (including tail). The differences are due mainly to variations in depletion region width and/or to a diffusion tail affecting equivalent quantum efficiency at high speed. Bandwidth typically exceeds 100 to 200 MHz and is highest for avalanche devices.

TABLE 4.3. Typical Commercial APD Devices

Manufacturer	Device	Responsivity (mA/μW) at λ_0 (nm)		Bandwidth (MHz)	Bias voltage (V)
AEG	BPW28	0.20	(850)	200	170
EMI	S30500	0.04	(880)	200	165
GE	50CHS	0.10	(900)	100	200
RCA	30884	0.63	(900)	250	275
RCA	30817	0.16	(1060)	200	375
TI	TIXL451	0.20	(900)	200	100
TI	TIXL452*	75	(900)	40	230

[a]Source: Manufacturers' product sales literature, 1977. (For further and latest data, contact manufacturers.) Reprinted from Elion and Elion, *Fiber Optics in Communications Systems*, p. 123, by courtesy of Marcel Dekker, Inc.

*Fiber optics module amplifier unit.

TABLE 4.4. Comparison of Detector Types

Basic detector type	Sensing area (mm^2)	Sensitivity at 1 MHz (dBm)	Responsivity at peak λ_0 (A/W)	Bias voltage (V)
PIN	0.3–3	−58	0.4–0.7	10–100
APD	0.8–8	−70	10–70	250–350

The light-sensitive areas of optical detectors are comparable to typical fiber bundle cross-sectional areas, but they are significantly larger than those of single fiber core areas. Whereas the maximum cross-sectional area of a fiber core of 100 μm diameter is about 3×10^{-2} mm^2 and thus represents the maximum core area normally encountered for a single fiber, the sensitive area of most detectors is typically two to three orders of magnitude larger. As a result, the detection of information from fiber bundles requires the physical separation of the fibers during detection so that each fiber can transmit its data to its own detector. This presents a considerable inconvenience. Also, the fiber–detector area mismatch indicates a poor utilization of the optical power. Substantial interface losses are usually encountered with both PIN and APD detectors because of the fiber–detector separation caused by the conventional packaging methods and the divergence of the emission from the fiber ends. A very close fiber–detector spacing is required in most cases to reduce these

TABLE 4.5. Characteristics of State-of-the-Art Optical Detectors Suitable for Optical Communications

	Photoconductive detectors			Avalanche detectors	
Characteristic	PN diode	PIN diode (no GR*)	PIN diode (with GR)	APD (with GR)	Reach-through
Sensitive area (mm^2)	10–100	10–50	10–50	1–10	1–10
Reverse bias (V)	20–150	10–50	50–150	100–200	200–500
Capacitance (pF)	1–5	5–10	1–5	5–10	1–5
Noise, NEP (10^{-14} $W/Hz^{1/2}$)	10–15	10–20	5–10	1–2	0.5–1.0
Dark current† (nA)	10–30	20–50	5–20	1–5 (s) 0.05–1 (b)	50–200 (s) 0.05–0.2 (b)
Quantum efficiency	0.80–0.90	0.85–0.95	0.85–0.95	0.30–0.50	0.80–0.90
Bandwidth (MHz)	200–300	50–75	75–100	500–1000	100–200

*GR, guard ring.

†(s), surface; (b), bulk.

Dynamic range (dB)	Maximum data rate (Mb/s)	Lifetime range (hr)	Rise time (ns)	Peak λ_0 response (nm)
60	10	$10^4 - 5 \times 10^5$	1–5	870
20	90–150	$10^4 - 3 \times 10^5$	2–5	880

Source: Reprinted from Elion and Elion, *Fiber Optics in Communications Systems*, p. 124, by courtesy of Mareel Dekker, Inc.

TABLE 4.6. Advantages and Disadvantages of Principal Classes of Optical Detectors Useful for Fiber Optics Communications

Photodetector	Advantages	Disadvantages
p-n diode	Simple structure High quantum efficiency Moderately high bandwidth	High noise equivalent power Relatively high dark current Quantum efficiency and bandwidth affected by slow tail
PIN diode without guard ring	Low bias voltage High quantum efficiency	High noise equivalent power High dark current Low bandwidth High capacitance Large sensitive area
PIN diode with guard ring	Reduced noise equivalent power Low dark current High quantum efficiency	More complex structure Large sensitive area Low bandwidth High capacitance
Avalanche photodiode with guard ring	Low dark current Low noise equivalent power High bandwidth Small sensitive area	Low quantum efficiency High bias voltage High capacitance Quantum efficiency and bandwidth affected by slow tail Temperature sensitivity requires temperature compensating network
Reach-through avalanche photodiode	Low dark current Very low noise equivalent power Medium bandwidth High quantum efficiency Small sensitive area	Very high bias voltage Temperature sensitivity requires temperature–compensating network

coupling losses to a minimum. Considerations similar to those made for the source–fiber connections apply. Typically, coupling losses at the fiber–detector interface range from 0.5 to 1.5 dB per fiber.

Most optical detectors suffer from two major difficulties: the slow-tail frequency response of nonavalanching (PIN) devices and the temperature dependence of the avalanche process of avalanching (APD) devices (22). The speed of optical detectors is largely a function of the detector capacitance, which is composed of the device and package contributions. The device capacitance depends upon the width of the depletion layer and is therefore a function of the impurity concentration and the applied voltage. Since a large operating voltage effects a wide depletion layer and hence a smaller capacitance compared to an otherwise identical device operated at low voltage, APDs usually have a higher frequency response than PINs. The slow-tail response is encountered in nondepleted devices. It is a temporal response component which lags the main component and is due to photon absorption in the nondepleted bulk. In a typical nondepleted device, up to 50% of the pulse response may be due to the slow tail. It is responsible for intersymbol interference, low quantum efficiency, and other performance degradation.

Typically, the avalanche process has a temperature coefficient of 0.2 V/°C for the guard ring structure and of 1.8 V/°C for the reach-through structure. Therefore, the avalanche voltage is a sensitive function of temperature, so that avalanche photodetectors require the addition of temperature-compensating circuitry.

The quantum efficiency of a PIN detector can be maximized if the width of the active region is several times the light penetration depth within the semiconductor. On the other hand, the frequency response of the device decreases with the width of the active region. Hence, bandwidth and efficiency normally involve a trade-off. In order to obtain a detector with both high sensitivity and fast response, the paths for penetrating photons and generated carriers must be different. One way to achieve this is to illuminate the intrinsic layer parallel to the junction plane rather than from the top. Higher efficiency is achieved if the thick substrate of the device is removed and only a very thin contact layer is left, whereby the metallized back contact serves as a reflector; in this case, the light path is twice that of the generated carriers. When there is only a small opening in the top contact through which the light can penetrate under a large angle, the optical power is reflected up and down between two mirrors and completely absorbed within the semiconductor material. In practice, such a structure has a relatively high quantum efficiency (in excess of about 0.5 if only back–side reflection is used) and a fast response time (about 0.1 ns, corresponding to 10 GHz). Approximately 40% of the response time is caused by delay in the rather thick contact layers (23). Enhanced performance can be expected by the use of improved antireflection coatings and of back contact gratings to allow multireflection of the light within the detector (24).

Silicon avalanche photodiodes have a serious efficiency–frequency trade-off. This is a consequence of the indirect energy gap of silicon and its relatively low value, which results in opacity of the material. The efficiency–frequency product ηf of silicon (which has an optical absorption coefficient of 10 cm^{-1} at 1.06 μm) is about 40 MHz; it is assumed that the frequency is taken at the 3 dB point. This relatively small ηf for silicon can be contrasted to that of direct-gap compound semiconductors, which generally have a higher efficiency–frequency product. They are frequently transparent and hence have a small absorption coefficient, and they are characterized by high carrier mobility.

Generally, semiconductors whose energy gap E_G is larger than the photon energy E_{ph} are transparent, whereas materials whose energy gap is smaller than the light energy are opaque. Note that E_{ph} is wavelength dependent. This situation has implications for the usefulness of semiconductors as efficient optical detectors, since it is desirable to use opaque and hence light-absorbing materials in the active device region and transparent materials in the surrounding regions in order to achieve deep photon penetration down to the active layer. The relationship between photon energy E_{ph} and photon wavelength (wavelength of light) is given by:

$$E_{ph} = h\nu = hc/\lambda \tag{4.11}$$

where $h \cong 4.14 \times 10^{-15}$ eV s and $c \cong 3 \times 10^{10}$ cm/s. This is illustrated in Figure 4.14. Thus, photon energy is inversely proportional to wavelength and hence at larger wavelength, such as $\lambda = 1.27$ μm, the photons possess about 50% less energy than at $\lambda = 0.85$ μm to penetrate the semiconductor material. For reference, the energy gaps of a few applicable semiconductors are as follows: InSb, 0.2 eV; Ge, 0.7 eV; GaSb, 0.7 eV; Si, 1.1 eV; GaAs, 1.4 eV; CdS, 2.6 eV; and ZnS, 3.6 eV.

An example of a photodetector that is able to overcome the sensitivity problem associated with silicon by incorporating other materials is schematically shown in Figure 4.15. This device allows operation at 1.06 μm (corresponding to a photon energy of 1.17 eV). It is composed of GaAs–GaAsSb layers and contains both homo- and heterojunctions (25). Because the materials in this structure are chosen such that their energy gaps are larger than the energy of the incident light, they are transparent, except for the active n^-–$GaAs_{1-x}Sb_x$ layer, which is opaque and hence light-absorbing; the different layers have the characteristics shown in Table 4.7. The detector is not suitable for operation at 0.85 μm because at this wavelength the photon energy is 1.46 eV and consequently all layers absorb light and therefore reduce the quantum efficiency substantially. In this device the n^+ doping is low enough so as not to cause significant free carrier absorption. The $GaAs_{1-y}Sb_y$ layer acts as a buffer and defines the short-wavelength cutoff of the device. It also minimizes lattice mismatch and hence stress at the heterojunction. Typi-

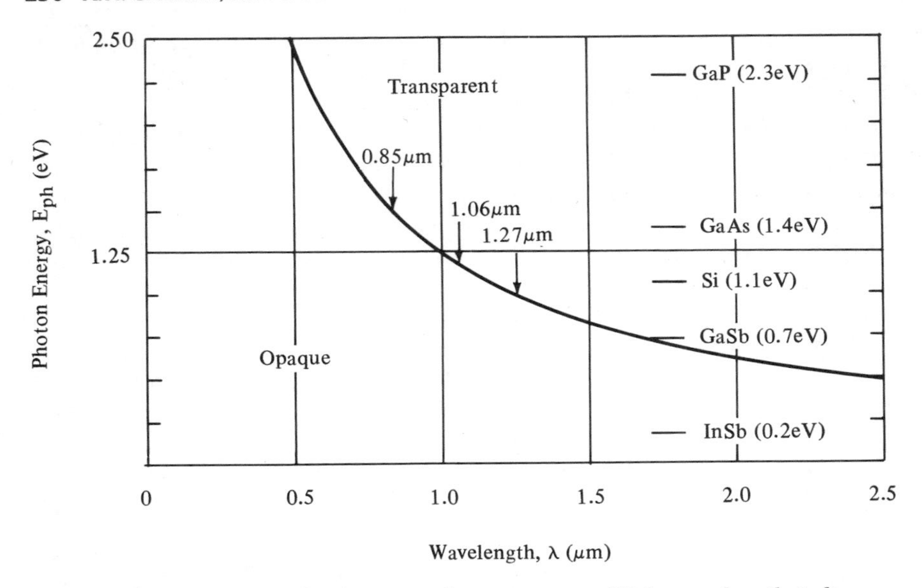

Figure 4.14 Relationship between photon energy and light wavelength, indicating the transition from opacity to transparency as the wavelength or the material energy gap increases.

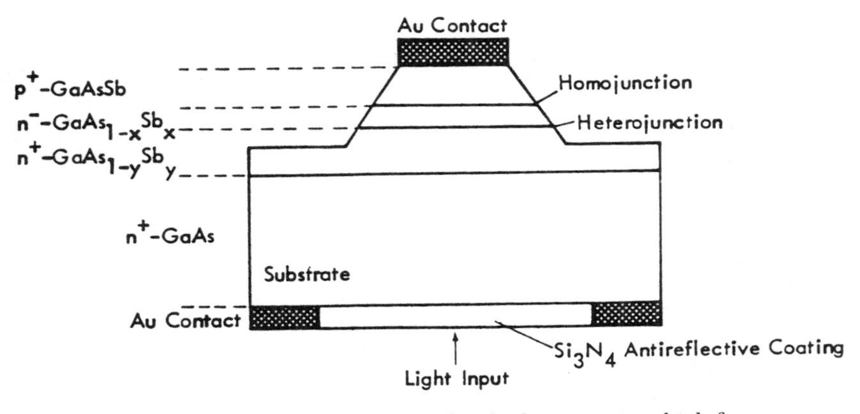

Figure 4.15 Schematic illustration of a high-sensitivity, high-frequency detector.

cal mole fractions are $x = 0.17$ and $y = 0.12$. The detector is capable of achieving a quantum efficiency of 0.95 at 1.06 μm and a data rate up to 1.5 Gb/s because of its small capacity.

Because optical detectors useful for operation at 1.06 μm and above are made of nonsilicon materials, they are not compatible with conventional integrated circuit techniques that would allow the eventual integration of the

TABLE 4.7. Properties of High-Sensitivity Photodetector

Material	Energy gap (eV)	Optical transmission	
		at 0.85 μm	at 1.06 μm
n^+–GaAs	1.4	Opaque	Transparent
n^+–$GaAs_{1-y}Sb_y$	>1.2	Opaque	Transparent
n^-–$GaAs_{1-x}Sb_x$	<1.1	Opaque	Opaque
p^+–$GaAs_{1-z}Sb_z$	>1.2	Opaque	Transparent

complete receiver together with the detector in the same chip. Hence, GaAs and GaAsSb detectors have the same economic drawbacks as nonsilicon light sources used at the transmitting end. However, such a disadvantage should be of only a temporary nature.

The electronic circuitry that follows the optic–electric conversion after the optical detector and the electrical preamplifier is similar for both optical and conventional coaxial communication systems (26). Therefore, of greater concern are questions relating to digital or analog signal handling, signal dynamic range, and bit detection. Typical digital and analog receiver interface configurations, illustrated in Figure 4.16, are discussed below.

The digital receiver shown is suitable for optimum detection of bits with small S/N and a small signal level dynamic range without automatic gain control (AGC). The detector, preamplifier, and voltage amplifier stages are used to increase the signal level, whereas the other circuits form a bit synchronizer function to achieve synchronized bit timing derivation and to effect bit decisions.

The analog receiver shown is similar to other low-level broad-band signal systems and consists of a preamplifier followed by one or several additional gain stages to increase the signal level, an AGC stage to avoid saturation of intermediate gain stages in case of a wide dynamic signal range, and an output buffer driver. Generally, the analog receiver is not designed for dc response because of offset compensation difficulties in detector and preamplifier circuits.

The preamplifier usually contains a temperature-compensating function if the detector is an APD that is expected to operate over a wide temperature range. The preamplifier is the only electrical circuit in both digital and analog receivers whose design depends upon the optical nature of the system. All other receiver circuits are similar to those used in coaxial systems and are therefore not discussed here. The design of the preamplifier must take into account the characteristics of the detector, such as its output shunt resistance, dark current level, capacitance, and noise. The design has a critical influence on receiver sensitivity and hence on the maximum amount of optical loss that can be tolerated in the optical cable. A series of preamplifier circuits exist that

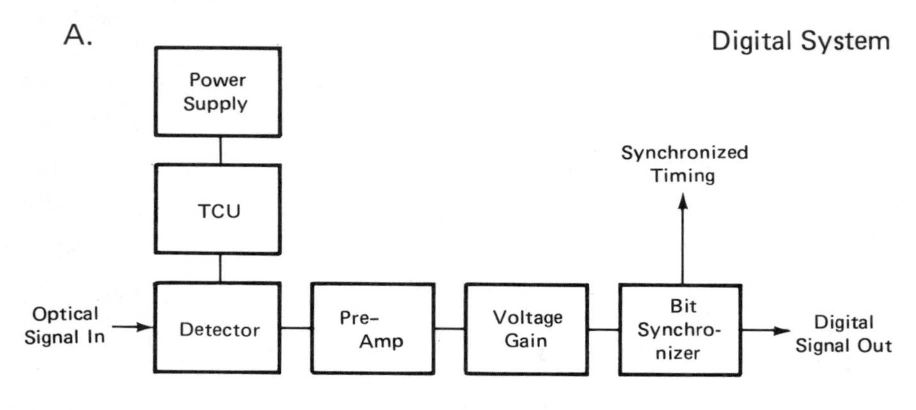

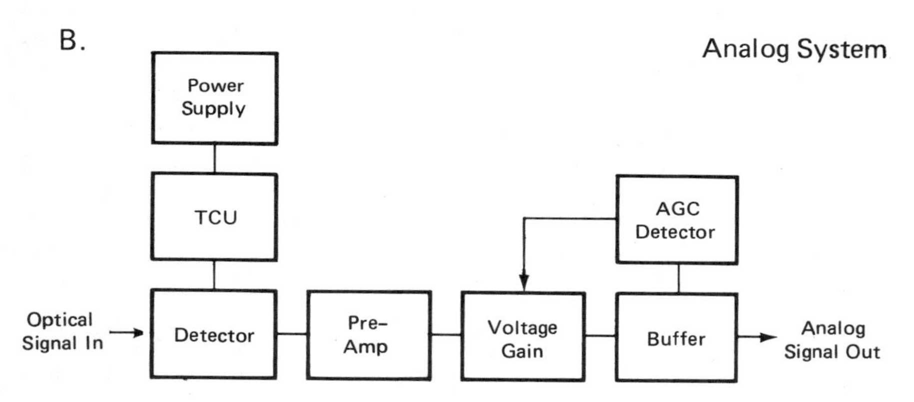

Figure 4.16 Block diagrams of (A) digital and (B) analog repeater configurations used in optical communications.

are able to achieve reasonable performance. Examples are the standard shunt feedback (SSF) and improved shunt feedback (ISF) preamplifiers, the input capacitance centralized preamplifier–detector (ICN), and the equalized high input impedance (EHI) preamplifier. The highest noise equivalent power (NEP) is associated with the SSF circuit, whereas the lowest NEP is achieved by the EHI circuit. In quantitative terms (expressed in 10^{-12} W/Hz$^{1/2}$) the NEPs of these preamplifiers are as follows: SSF, 2.0; ISF, 1.3; ICN, 2.0; and EHI 0.3. This assumes that these circuits are designed for a bandwidth of 10 MHz.

NOISE LIMITATIONS

One of the most stringent requirements on an optical detector is the reduction of the noise level. This is particularly important because the detector itself is normally the principal noise contributor in a fiber optics communications system. Noise presents one of the most severe systems limitations; it affects the maximum fiber length or the maximum allowable distance between repeaters. Therefore, it is of major interest in economic considerations. A more detailed analysis of the effect of noise on systems performance and cost will be given in Chapter 9, where systems aspects are discussed.

The treatment of detector noise and its effect on the operational limits of an optical system centers around an analysis of the ratio of signal power to noise power (S/N ratio) (26). Efforts to improve system performance thus are directed toward an increase of the S/N ratio as long as this is economically justifiable. The signal-to-noise ratio is a function of the allowed error probability or bit error rate and involves a complementary error function (erfc). The bit error rate or probability of error (ϵ) for a binary (on–off), i.e., digital, scheme is related to the signal-to-noise ratio by

$$\epsilon = \tfrac{1}{2}\ \mathrm{erfc}\left[(\mathrm{S/N})^{1/2}/\sqrt{2}\right]$$

$$\approx 0.4\ (\mathrm{S/N})^{-1/2}\ e^{-(\mathrm{S/N})/2} \text{ for S/N} > 10 \tag{4.12}$$

This is shown graphically in Figure 4.17. In typical systems a bit error rate of 10^{-9}, corresponding to S/N of about 36 dB, is desirable. The need for high S/N

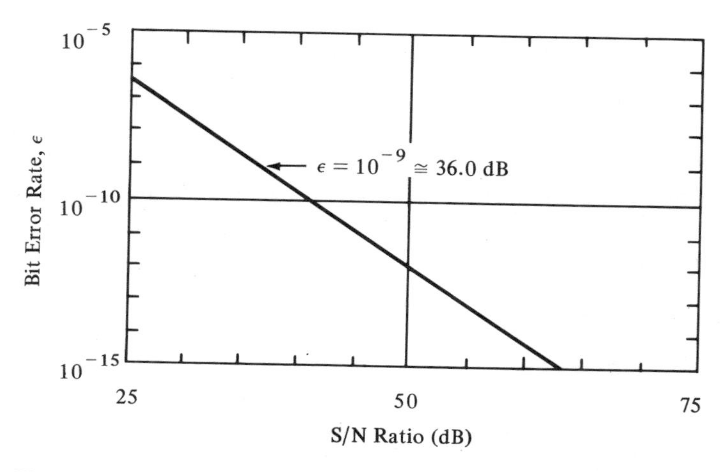

Figure 4.17 Relationship between error rate and signal-to-noise ratio.

in order to reduce the error probability is evident because of the strong relationship. For example, a reduction in the bit error rate from 10^{-6} to 10^{-9} requires an increase in S/N from about 17 dB to 36 dB.

The signal-to-noise ratio is a function of several noise-generating mechanisms. The signal power depends upon the source power, the fiber input and output coupling efficiency, and the fiber attenuation. The noise power depends upon the contributions from the various noise sources and the design of the detector circuits. From the general expression for the signal-to-noise ratio, we can distinguish two major cases, as follows:

Absence of background light. Depending upon whether the detector is a unity-gain (PIN) or a high-gain (APD) device, there are two different situations: In a unity-gain detector, thermal noise dominates and the signal-to-noise ratio is given by

$$\mathrm{S/N} = (\alpha P_o)^2 R_L / 4kTB \tag{4.13}$$

In a high-gain detector, shot noise dominates and the signal-to-noise ratio is given by

$$\mathrm{S/N} = \alpha P_o / qB \tag{4.14}$$

Presence of background light. In this case, the detector gain and type are arbitrary; background noise dominates and the signal-to-noise ratio is given by

$$\mathrm{S/N} = \frac{\alpha P_o R_L}{2qB(1 + P_b/P_o)} \tag{4.15}$$

In these expressions, α is the fiber attenuation, η is the detector quantum efficiency, m is the modulation index, P_o is the average optical power at the detector, P_b is the average optical power of the background, and B is the bandwidth of the detection system. In most applications where the fibers are coated by an opaque material the case for background light is negligible. In the case of dominant background noise, it is assumed to be much larger than the optical power, $P_b \gg P_o$.

The system operation thus may be limited by either of three types of noise. These are discussed below.

Quantum Noise. In this case, noise is the result of fluctuations in the average signal current according to

$$\mathrm{S/N} = \eta m^2 P_o / 4BFE_{\mathrm{ph}} \tag{4.16}$$

In this equation F is the noise increase due to detector current gain (applicable to APDs only), and E_{ph} is the optical energy received by the detector (on the order of 1 eV, depending upon the wavelength). Quantum noise represents the inherent limit in a system. In practice, this lower noise limit is usually not achievable, although it may be approached. Theoretically, S/N can be maximized by an increase in signal power and wavelength and by a decrease in detector bandwidth and amplifier gain. In practice, however, not all of these objectives are attainable or necessarily desirable because of potential trade-offs. An overall optimization normally leads to a system condition in which quantum noise is not minimized. In spite of this, quantum noise is usually secondary to the other two noise contributions. The resultant maximum pulse rate applicable to the quantum noise–limited case is approximately

$$R_{max} \cong A_q e^{-\alpha L} \tag{4.17}$$

where A_q is a proportionality factor.

Thermal Noise. In the absence of internal detector gain (such as in the case of PIN diodes), this noise contribution is mainly due to thermal fluctuations or shot noise:

$$\text{S/N} = \frac{(\eta m P_o)^2}{8\,(E_{ph}/q)^2 k T_e B / R_e} \tag{4.18}$$

The effective noise temperature T_e accounts for thermal and amplifier noise, and the equivalent load resistance R_e depends upon the detector geometry and the amplifier impedance. The terms q and k are electron charge and Boltzmann's constant, respectively. In the case of thermal noise, S/N is proportional to $(\eta P_o)^2$, whereas in the case of quantum noise, S/N is proportional to ηP_o. Both T_e and R_e depend on device and amplifier characteristics. If the amplifier is noiseless and has zero input admittance, then T_e and R_e are the physical temperature and the resistance of the load, respectively; if the amplifier contributes noise, then T_e is approximately the product of the amplifier noise figure and the ambient temperature. Based on the thermal noise model, the maximum data rate that can theoretically be expected is given by

$$R_{max} \cong A_t e^{-2\alpha L} \tag{4.19}$$

where A_t is a proportionality factor. Note the factor 2 in the exponent, in contrast to the factor 1 for the quantum noise model.

Shot Noise. In the presence of internal detector gain, given by the term G and found in avalanche photodiodes, shot noise contributes to the detector noise. It is minimized when the shot noise is amplified to the level of the thermal noise. In this case, S/N is equal to that in the thermal noise case multiplied by the factor $G_{opt}/\sqrt{2}$. If the detector gain exceeds the optimum, the detector sensitivity diminishes because of excess noise generated in the carrier multiplication process. Similar to the data rate limitation caused by thermal noise, the maximum data rate possible in the case of optimum detector internal gain is

$$R_{max} = A_s e^{-2\alpha L} \tag{4.20}$$

where the proportionality factor $A_s = A_t G_{opt}/\sqrt{2}$. Because the equivalent resistance is composed of the internal resistance of the detector and the load resistance, it is usually inversely proportional to the product of capacitance and maximum bandwidth of the detector circuit. If thermal noise limitation exists, then noise increases with the square of bandwidth. In the case of optimum gain the noise increase is proportional to bandwidth.

Factors Influencing Noise

It is desirable to separate the influences on the noise characteristics caused by the optical power from influences that are independent of the optical power. This means that S/N can be expressed

$$S/N = uP_o^2/(P_o + P_f) \tag{4.21}$$

where u is a factor that includes η, λ, B, and F. The denominator is the sum of a P_o-dependent and a P_o-independent term (P_f); P_f refers to fixed noise sources and excludes all contributions by P_o. For a given average optical power P_o, advantage can be gained by decreasing the pulse width at the expense of bandwidth to ensure that the detector is always limited by optimum gain; hence, bandwidth capability can be exchanged for optical power.

In the discussion above we distinguished between two cases. If $P_o \gg P_f$, then the S/N ratio varies exponentially with fiber length, according to S/N $\sim e^{-\alpha L}$; if $P_o \ll P_f$, then S/N $\sim e^{-2\alpha L}$. Thus, at a critical fiber length L_c the slope of a ln S/N-versus-L plot exhibits a change, meaning that at long fiber lengths the signal-to-noise ratio decreases faster with distance than at short fiber lengths. In other words, for long cables the magnitude of the slope is twice its value for short cables. This is illustrated schematically in Figure 4.18A.

This analysis indicates a P_o dependence of S/N in the case of quantum noise and a P_o^2 dependence in the case of thermal noise and optimum gain. This difference is of economic importance because it is related to the maximum distance between repeaters and the required power output of the source.

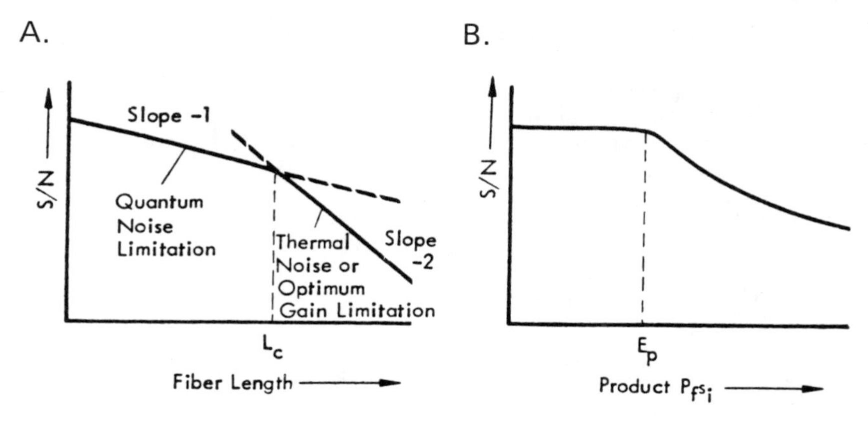

Figure 4.18 Schematic illustration of the variation of the signal-to-noise ratio with fiber length and product of pulse width and fixed-noise source energy.

The signal-to-noise ratio can be described in terms of the energy received by the light pulse, $E_p = P_o s_i$, under the assumption that the pulse width at the source, s_λ, has deteriorated to a value at the detector, s_i, whereby the bandwidth is defined as the inverse of s_i (so that $B = 1/s_i$):

$$\mathrm{S/N} = uE_p^2/(E_p + P_f s_i) \tag{4.22}$$

It is evident that for a given pulse energy E_p the signal-to-noise ratio can be increased by reducing the pulse width s_i. Thus, a laser source of small spectral width is more advantageous to use for high S/N ratio than an LED source. This assumes that signal noise has been limited by adjusting s_i such that $P_f s_i \ll E_p$. In this case, the pulse width can be decreased while the pulse energy remains constant. As Figure 4.18B indicates, decreasing the $P_f s_i$ product much below E_p does not result in a significant S/N enhancement. There is a seeming conflict with the requirement of a minimized bandwidth to reduce noise. It can be resolved by considering the fact that a smaller pulse width results in both a reduced receiver integration time and a smaller number of photon-generated electrons due to fixed noise sources, resulting in higher sensitivity.

REFERENCES

1. H. Melchior. *Phys. Today* **30**:32, 1977.
2. J. Conradi et al. *IEEE Trans. Electron. Devices* **ED-25**:180, 1978.
3. G.E. Stillman & C.M. Wolfe. Avalanche Photodiodes, in *Semiconductors and Semimetals*, Vol. 12, R.K. Willardson and A.C. Beer (Eds.) New York, Academic Press, 1977.
4. G.R. Elion & H.A. Elion. *Fiber Optics in Communications Systems*. New York, Marcel Dekker, Inc., Ch. 4, 1978.

5. S. Sze. *Physics of Semiconductor Devices*. New York, John Wiley & Sons, Ch. 12, 1969.

6. R.H. Bube. *AIME Trans* **239**:291, 1967.

7. H.F. Wolf. *Semiconductors*. New York, Wiley-Interscience, 1971.

8. A.N. Saxena. Conference on Physics of Compound Semiconductors, UCLA, California, February 1975.

9. A.N. Saxena. *Surf. Sci.* **12**:259, 1968.

10. J.M. Shannon. *Appl. Phys. Lett.*, **24**,369 (1974).

11. T. Misugi & H. Takanashi, IOOC 77, p. 33, Tokyo, July 1977.

12. P.P. Webb et al. *RCA Rev.* **35**:234, 1974.

13. R.J. McIntyre. International Electron Device Meeting, p. 213, Washington, D.C., 1973.

14. T. Kaneda et al. *J. Appl. Phys.* **47**:1605, 1976.

15. H. Melchior & W.T. Lynch. *IEEE Trans. Electron. Devices* **ED-13**:829, 1966.

16. R.A. Logan & S.M. Sze. *J. Phys. Soc. Jpn. Suppl.* **21**:434, 1966.

17. H. Melchior et al. IEEE Solid-State Device Research Conference, Boulder, Colo., June 1968.

18. A.N. Saxena. *Appl. Phys. Lett.* **19**:71, 1971.

19. C.E. Hurwitz & J.J. Hsieh. Topical Meeting on Integrated and Guided Wave Optics of the Optical Society of America, p. MC1-1, 1978.

20. T.P. Pearsall et al. Topical Meeting on Integrated and Guided Wave Optics of the Optical Society of America, p. MC2-1, 1978.

21. O. Krumpholz. Signal to Noise Ratio of Avalanche Photodiodes. *AEG-Telefunken Prog.* **44**:80, 1971.

22. O. Krumpholz & S. Maslowski. Avalanche Mesa-Photodiodes With Transverse Incidence of Light. *AEG-Telefunken Prog.* **44**:73, 1971.

23. S. Metz. The Avalanche Phototransistor—A Novel Type of Radiation Detector and its Potential Use in Optical Communication Systems. *IEEE Trans. Electron. Devices* **ED-22**:617, 1975.

24. K. Berchtold et al. Avalanche Photodiodes With a Gain-Bandwidth Product of More Than 200 MHz. *Appl. Phys. Lett.* **26**:585, 1975.

25. R.C. Eden. Heterojunction III-V Alloy Photodetectors for High-Sensitivity 1.06 μm Optical Receivers. *Proc. IEEE* **63**:32, 1975.

26. S.D. Personick. Receiver Design for Digital Fiber Optics Communications Systems. *Bell Sys. Tech. J.* **52**:875, 1973.

Bibliography on Photodetectors

1. L.K. Anderson & B.J. McMurtry. *Proc. IEEE* **54**:1335, 1966.

2. R.J. McIntyre. *IEEE Trans. Electron. Devices* **ED-13**:164, 1966.

3. H. Melchior & W.T. Lynch. *IEEE Trans. Electron. Devices* **ED-13**:829, 1966.

4. L.K. Anderson et al. *Appl. Phys. Lett.* **6**:62, 1966.

5. D.P. Kennedy & R.R. O'Brien. *IBM J. Res. Dev.* **10**:213, 1966.

6. R.J. Locker & G.C. Huth. *Appl. Phys. Lett.* **9**:227, 1966.

7. H.W. Ruegg. *IEEE Trans. Electron. Devices* **ED-14**:239, 1967.

8. H. Melchior. *Verh. Dtsch. Phys. Ges.*, **6-2**:57, 1967.

9. J.J. Chang. *IEEE Trans. Electron. Devices* **ED-14**:139, 1967.

10. R.B. Emmons. *J. Appl. Phys.* **38**:3705, 1967.

11. H. Kressel. *RCA Rev.* **28**:175, 1967.

12. M.P. Mikhailova et al. *Fiz. Tekh. Poluprovodn.* **1**:123, 1967 [English Transl. *Sov. Phys. Semicond.* **1**:94, 1967].

13. O. Krumpholz & S. Maslowski. Z. *Angew. Phys.* **25**:156, 1968.

14. H.C. Spriging & R.J. McIntyre, IEEE International Electron Devices Meeting, Washington, D.C., October 1968.
15. W.T. Lynch. *IEEE Trans. Electron. Devices* **ED-15**:735, 1968.
16. W.T. Lindley et al. *Appl. Phys. Lett.* **14**:197, 1969.
17. D.E. Persyk. *Laser J.* **1**:21, 1969.
18. W.K. Pratt in *Laser Communication Systems,* p. 209. New York, John Wiley & Sons, 1969.
19. D.P. Mathur et al. *Appl. Opt.* **9**:1842, 1970.
20. L.K. Anderson et al. in *Advances in Microwaves*, Vol. 5, L. Young (Ed.) New York, Academic Press, 1970.
21. H. Melchior et al. *Proc. IEEE* **58**:1466, 1970.
22. R.L. Bell & W.E. Spicer. *Proc. IEEE* **58**:1738, 1970.
23. R. Kuvas & C.A. Lee. *J. Appl. Phys.* **41**:1743, 1970.
24. R. van Overstraeten & H. DeMan. *Solid-State Electron.* **13**:583, 1970.
25. J.A. Lewis & E. Wasserstrom. *Bell Sys. Tech. J.* **49**:1183, 1970.
26. S.D. Personick. *Bell Sys. Tech. J.* **50**:167, 1971.
27. S.D. Personick. *Bell Sys. Tech. J.* **50**:3075, 1971.
28. P. Webb & R.J. McIntyre. *Proceedings of the Electro-Optical Systems Design Conference (East),* p. 51, 1971.
29. O. Krumpholz & S. Maslowski. *Wiss. Ber. AEG-Telefunken* **44**:73, 1971.
30. B.T. Desai & C.Y. Chang. *J. Appl. Phys.* **42**:5198, 1971.
31. W.E. Sayle & P.O. Lauritzen. *IEEE Trans. Electron. Devices* **ED-18**:58, 1971.
32. O. Krumpholz. *Wiss. Ber. AEG-Telefunken* **44**:80, 1971.
33. D.J. Coleman Jr. et al. *Proc. IEEE* **59**:1121, 1971.
34. R.J. McIntyre. *IEEE Trans. Electron. Devices* **ED-19**:703, 1972.
35. J. Conradi. *IEEE Trans. Electron. Devices* **ED-19**:713, 1972.
36. S. Maslowski. OSA Topical Meeting on Integrated Optics, Las Vegas, February 1972.
37. G.E. Stillman et al. *Solid State Res.* (MIT Lincoln Labs), 1972.
38. D.A. Soderman & W.H. Pinkston. *Appl. Optics* **11**:2162, 1972.
39. H. Melchior in *Laser Handbook,* Vol. 4, F.T. Arecchi & E.O. Schutz-Dubois (Eds.). Amsterdam, North-Holland, p. 725, 1972.
40. C.L. Anderson & C.R. Crowell. *Phys. Rev.* **B5**:2267, 1972.
41. D.L. Camphausen & C.J. Hearn. *Phys. Status Solidi (b)* **50**:K139, 1972.
42. H. Melchior. *J. Lumin.* **7**:390, 1973.
43. H. Melchior. Conference on Laser Engineering and Applications, Washington, D.C., 1973.
44. D. Gloge. *Opto-Electronics* **5**:345, 1973.
45. I.M. Naqvi. *Solid-State Electron.* **16**:19, 1973.
46. R.J. McIntyre. IEEE International Electron Devices Meeting, Washington, D.C., 1973.
47. R.A. Balliager et al. *J. Phys. C: Solid State Phys.* **6**:2573, 1973.
48. M.H. Woods et al. *Solid-State Electron.* **16**:381, 1973.
49. W.N. Grant. *Solid-State Electron.* **16**:1189, 1973.
50. G.E. Stillman et al. *Appl. Phys. Lett.* **24**:8, 1974.
51. H. Takanashi et al. *Fujitsu Sci. Tech. J.* **10**:119, 1974.
52. P.P. Webb et al. *RCA Rev.* **35**:234, 1974.
53. G.E. Stillman et al. *Appl. Phys. Lett.* **24**:471, 1974.
54. T. Shibata et al. *E.C.L. Tech. J. NTT* **22**:1069, 1974.
55. R.L. Moon et al. *J. Electron. Mater.* **3**:635, 1974.
56. G.E. Stillman et al. *Appl. Phys. Lett.* **24**:671, 1974.
57. J. Urgell & J.R. Leguerre. *Solid State Electron.* **17**:239, 1974.

58. J. Conradi. *Solid State Electron.* **17**:99, 1974.
59. A.D. Lucas. *Opto-Electronics* **6**:153, 1974.
60. T.P. Pearsall et al. *Appl. Phys. Lett.* **27**:330, 1975.
61. T. Koehler & A.M. Chiang. Proceedings of the Society of Photo-Optical Instrumentation Engineers, p. 26, San Diego, August 1975.
62. R.C. Eden. *Proc. IEEE* **63**:32, 1975.
63. T.P. Lee et al. *IEEE Trans. Electron. Devices* **22**:1062, 1975.
64. J. Conradi & P.P. Webb. *IEEE Trans. Electron. Devices* **22**:1062, 1975.
65. D.R. Decker & C.N. Dunn. *J. Electron. Mater.* **4**:527, 1975.
66. H. Melchior & A.R. Hartman. *IEDM Technical Digest*, p. 412, (December) 1976.
67. T. Yamaoka et al. *Fujitsu Sci. Tech. J.* **12**:87, 1976.
68. T. Kaneda et al. *J. Appl. Phys.* **47**:1605, 1976.
69. T. Kaneda et al. *J. Appl. Phys.* **47**:3135, 1976.
70. H. Kanbe et al. *J. Appl. Phys.* **47**:3749, 1976.
71. T.P. Pearsall et al. *Appl. Phys. Lett.* **28**:403, 1976.
72. M.P. Mikhailova et al. *Sov. Phys. Semicond.* **10**:509, 1976.
73. J.S. Escher et al. *Appl. Phys. Lett.* **29**:153, 1976.
74. H. Kanbe et al. *IEEE Trans. Electron. Devices* **25**:1337, 1976.
75. S.D. Personick in *Fundamentals of Optical Fiber Communications*, M.K. Barnoski, (Ed.). New York, Academic Press, (1976).
76. J.J. Hsieh et al. *Appl. Phys. Lett.* **28**:709, 1976.
77. M.P. Mikhailova et al. *Sov. Phys. Semicond.* **10**:578, 1976.
78. H. Melchior. *Phys. Today,* **30**:32, 1977.
79. T. Misugi & H. Takanashi. *IOOC 77*, p. 33, Tokyo, July 1977.
80. H. Fukuda et al. Meeting of the IECE of Japan, pp. 2–48, March 1977.
81. I. Jacobs & S.E. Miller. *IEEE Spectrum* **14**:32, 1977.
82. K. Takahashi et al. Meeting of the IECE of Japan, pp. 4–29, March 1977.
83. G.E. Stillman & C.M. Wolfe in *Semiconductors and Semimetals*, Vol. 12, R.K. Willardson & A.C. Beer (Eds.). New York, Academic Press, 1977.
84. J.J. Hsieh et al. in *Gallium Arsenide and Related Compounds* (St. Louis), L.F. Eastman (Ed.). Bristol, Institute of Physics, 1977.
85. C.C. Shen et al. *Appl. Phys. Lett.* **30**:353, 1977.
86. H.H. Wieder et al. *Appl. Phys. Lett.* **31**:468, 1977.
87. T.P. Pearsall & R.W. Hopson. *J. Appl. Phys.* **48**: 1977.
88. J. Conradi et al. *IEEE Trans. Electron. Devices* **ED-25**:180, 1978.
89. C.E. Hurwitz & J.J. Hsieh. Proceedings of the Topical Meeting on Integrated and Guided Wave Optics, p. MC1-1, 1978.
90. T.P. Pearsall et al. Proceedings of the Topical Meeting on Integrated and Guided Wave Optics, p. MC2-1, 1978.
91. T.P. Pearsall & R.W. Hopson Jr. *J. Electron. Mater.* **7**:133, 1978.
92. T.P. Pearsall & M. Papuchon. *Appl. Phys. Lett.* **33**:640, 1978.
93. J.R. Tucker & M.F. Millea. *Appl. Phys. Lett* **33**:611, 1978.
94. R.C. Miller et al. *Appl. Phys. Lett.* **33**:721, 1978.

OPTICAL CONNECTORS, COUPLERS, AND SWITCHES

John Joseph Esposito

Although connectors, couplers, and switches are often the last components considered in system design, these connecting elements are nevertheless essential. This fact is true for guided wave transmission at microwave frequencies and is equally true for data transmission systems utilizing optical waveguides.

If efficient methods of interconnection did not exist, the transmission of data for appreciable distances would become a more costly and complex problem. It would then become necessary to either manufacture unbroken cables kilometers in length or to detect and reamplify the signal at every break in the cable. Since reamplification is a costly method of attaining long-distance data transmission, efficient interconnecting devices are an economic necessity. Since connectors and couplers are usually passive devices, while repeaters use active components, the reduction of the number of repeaters in a system increases the reliability and decreases the energy requirement of the system. Thus, efficient interconnection components are cost effective.

Despite this demonstrated need for effective connectors, couplers, and switches for optical waveguides, their commercialization has lagged behind that of the other components. The reasons for this lag depend on the nature of the various elements making up a lightwave communications system. These elements are the transmission medium, the source, the detector, and the connecting components, i.e., connectors, couplers, and switches. Of these elements the connectors and couplers represent the newest technology. To illustrate this point, it is necessary only to review the history of the various devices.

Optical waveguides have existed in research laboratories since the 1950s and have been in commercial use since the 1960s. Thus, the attainment in the early 1970s of extremely-low-loss fibers represented an evolutionary step rather than a totally new development. Similarly, solid-state optical sources and detectors have existed for more than a decade and needed only to have certain of their parameters altered to become useful elements of an optical communications system. To a certain extent, optical switching has also predated the advent of optical communications systems because optical switching techniques developed in parallel with the development of the laser. On the other hand, optical couplers and connectors have been developed only after the feasibility of optical data transmission was demonstrated, which was relatively recent. Although it may be argued that optical couplers and connectors borrow some of their principles of operation from analogous microwave components, the technologies are nevertheless quite different.

But if the development of the coupling elements has lagged behind the development of optical sources, fibers, and detectors in the past, the gap is rapidly being closed. At the time of this writing, research that has been done during the past five years or so is quickly being applied to product development, and a number of companies have begun producing and marketing the connecting components. The objective of this chapter thus is to describe the principles of operation of the various components and to show how these principles have been or are likely to be applied to the production of hardware. In order to achieve these goals most efficiently, the chapter is divided into three major sections: connectors, then couplers, and, finally, switches. This sequence will lead to a logical progression where ideas initially covered will help the reader appreciate the material that follows.

CONNECTORS

If one traced a fiber optics communications system from beginning to end, three distinct types of connectors would be encountered. At the transmitting end of the communications link, source-to-fiber connectors would be used. Further down the line, where the link passed through a wall or bulkhead, a fiber-to-fiber connector would be encountered. Finally, a fiber-to-detector connector would be found at the receiving end of the link. Each of these three connector types requires different design criteria, and each will therefore be discussed separately.

Source-to-Fiber Connection

Figure 5.1 shows that optical fibers are not ideally compatible with Lambertian sources, sources whose off-axis intensity varies as $I = I_0 \cos\Phi$. Unfortunately, many light-emitting diodes emit as Lambertian sources. A number of methods of increasing coupling efficiency have therefore evolved.

When coupling light into optical fibers, it is usually desirable to maximize either of two quantities, the total power coupled into the fiber or the coupling efficiency. It may happen that both quantities can be maximized simultaneously, but that is not necessarily the general case (1). The total power coupled into a fiber may be increased by substituting a more powerful

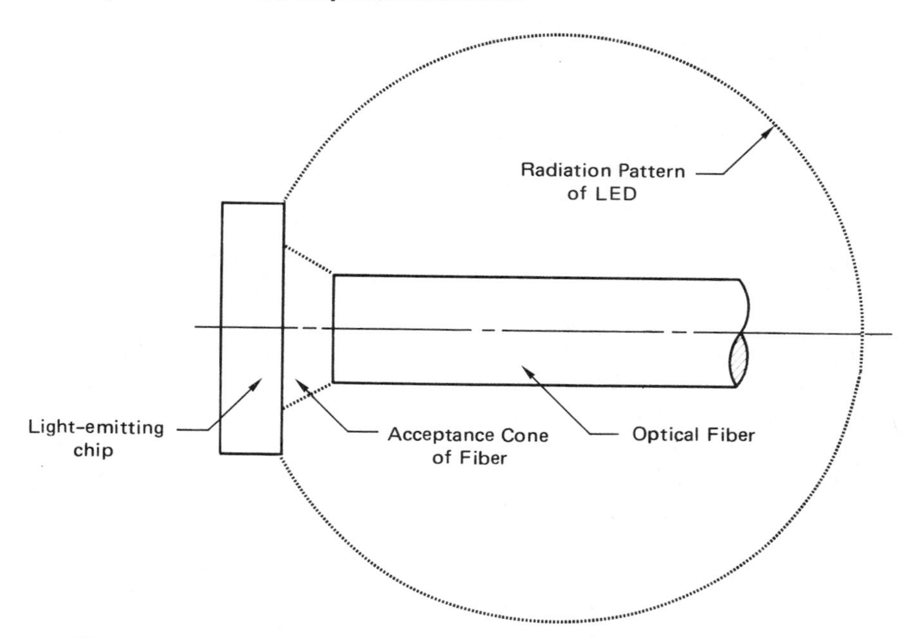

Figure 5.1 Mismatch between emission pattern of a Lambertian source and the optical fiber acceptance angle.

or a more directional source and leaving the coupling efficiency unchanged or even decreasing it. This seemingly contradictory statement will become clear as we proceed.

The simplest case with which to begin our discussion is that of a Lambertian source coupled to a step-index fiber with numerical aperture NA and core area A_{core}. With such a combination of components, the simplest method of transferring light energy is to butt-couple them, placing the fiber within a core diameter of the emitting surface of the diode. When the diode and fiber are properly butt coupled, the power injected into the fiber is given by

$$P_i = \pi T B_S A (\text{NA})^2 \tag{5.1}$$

where B_S is the brightness, or radiance, of the source, T is the transmissivity, which includes Fresnel reflections, and A is the area that LED and fiber core have in common. If the core area of the fiber is larger than the emitting area of the LED, $A = A_{\text{LED}}$; if the area of the source is larger than that of the fiber, then $A = A_{\text{core}}$. The optical coupling efficiency is then given by (2)

$$\eta = P_i/P_{\text{LED}} = T(A/A_{\text{LED}})\,(\text{NA})^2 \tag{5.2}$$

from which it is easy to see than when the fiber core area is larger than the area of the diode, the coupling efficiency is

$$\eta = T(\mathrm{NA})^2 \tag{5.3}$$

Examination of Equations (5.1) and (5.2) will reveal how the power injected into a fiber can be increased while the coupling efficiency decreases. Equation (5.1) shows that the power coupled into a fiber depends on the brightness of the source, the overlap area of the fiber core and emitting surface, and the numerical aperture of the fiber. The transmissivity has a constant value ranging from about 0.95 to 1.0, depending on the refractive index of the medium separating the fiber and the LED. Equation (5.2) shows that the coupling efficiency depends only on the area ratio and the numerical aperture of the fiber. Thus, the substitution of a large, bright source for a smaller, less radiant source would lead to an increase in injected power but a decrease in coupling efficiency.

The above discussion gives a fairly accurate analysis of the coupling of power from a Lambertian source into an optical fiber, but a more complete analysis must include the effect of skew rays. Skew rays are those rays that follow helical paths as they propagate down the fiber, and they usually have a higher numerical aperture than do meridional rays. To account for the effect of skew rays on the power coupled into a fiber, Equation (5.1) may be modified as follows (3):

$$P_i = TB_S A(\mathrm{NA}_s)^2 \tag{5.4}$$

where NA_S is numerical aperture of the fiber when skew rays are included. The calculated coupling efficiency is therefore increased when the effects of skew rays are included; this increase can be as great as 3 dB (1). However, not all of this power is useful, since skew rays leak from the fiber more readily than do meridional rays. In fact, skew rays are often referred to as leaky rays.

Many solid-state light sources have emission patterns that are non-Lambertian. The radiation patterns of such sources, however, can often be approximated by a power law variation of intensity of the form $I = I_0 \cos^{\sigma}\phi$, where I_0 is the axial intensity and σ is a numerical value characteristic of the source. Figure 5.2 is a comparison of the radiation pattern of a Lambertian emitter, $\sigma = 1$, with the pattern of a source with $\sigma = 20$. It should be fairly obvious that the less divergent, non-Lambertian emitter is more suitable for use with low-NA fibers.

For a source obeying a $\cos^{\sigma}\phi$ variation of intensity, the total power emitted by the source is (4)

$$P_{\mathrm{source}} = 2\pi A_{\mathrm{source}} B_S/(\sigma + 1) \tag{5.5}$$

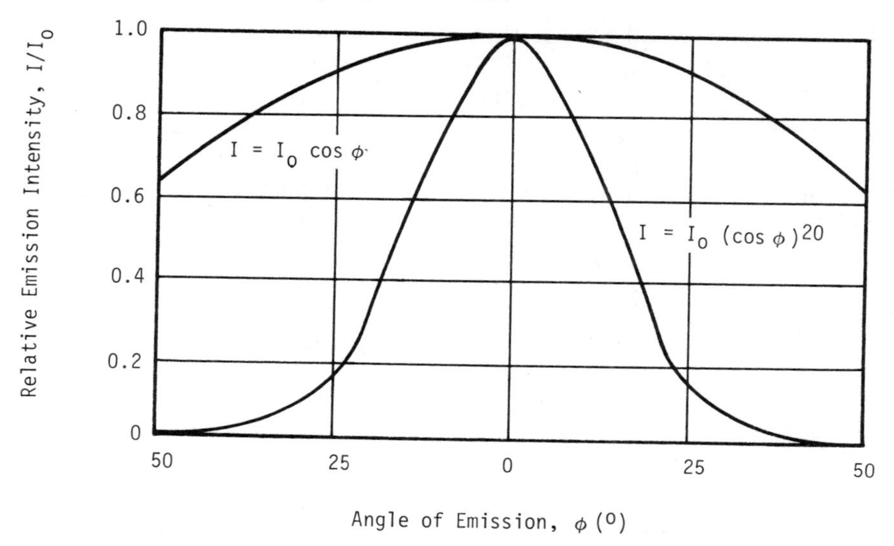

Figure 5.2 Radiation patterns of Lambertian and non-Lambertian sources.

For a Lambertian source, $\sigma = 1$, and Equation (5.5) reduces to

$$P_{\text{source}} = A_{\text{source}} B_S \tag{5.6}$$

The total power injected into a fiber by such a source is (4)

$$P_i = P_{\text{source}} f_p (1 - \cos^{\sigma+1} \phi_{\text{NA}}) \tag{5.7}$$

where f_p is the packing fraction. For small NA, Equation (5.7) becomes

$$P_i = \tfrac{1}{2} P_{\text{source}} f_p (\text{NA})^2 (\sigma + 1) \tag{5.8}$$

In both Equations (5.7) and (5.8), we have assumed that fiber bundle and source have the same areas. If this were not the case, both equations would have to be multiplied by the ratio of areas as was done in Equation (5.1).

The packing fraction f_p appearing in Equations (5.7) and (5.8) is a measure of the effective area of a fiber bundle. It is the ratio of the core areas of all the fibers in a bundle to the total area occupied by the bundle, and as such it tells what percentage of the light falling on the end of the bundle has a chance of being propagated down the bundle. For a single fiber, the packing fraction naturally is unity. Thus, the relationships given above are applicable generally and can be used for calculating the power coupled into either a single fiber or into a multifiber bundle by either a Lambertian or a non-Lambertian source. Examples of such non-Lambertian emitters are strip geometry light-emitting diodes and laser diodes. The subject of strip geometry sources is well covered in reference 1.

The above ray optics treatment of the problem of coupling light from luminescent sources into optical fibers is fairly adequate for step-index fibers, but a wave optics treatment is necessary to analyze the problem of coupling light-emitting diodes to graded-index fibers. Marcuse (5) has performed such an analysis. His approximate theory is valid for fibers with a square-law variation of refractive index of the form

$$n = n_0[1 - (r/a_1)^2]\Delta \tag{5.9}$$

with small Δ, where Δ is defined by Equation (2.2); r is the radial distance from the core and a_1 the core radius. Marcuse determined that the amount of power coupled into a graded-index fiber by an LED with emitting area larger than the fiber core area is

$$P_i = \pi(\pi a_1^2)B_S\Delta = \pi A_{\text{core}} B_S \Delta \tag{5.10}$$

We can see that this is very similar to Equation (5.1), which gives the amount of power that can be coupled into a step-index fiber. The only differences are that Δ has replaced $(\text{NA})^2$ and an index-matching substance has been assumed to fill the gap between fiber and source, thus causing the transmissivity to be unity.

We have thus far been speaking about butt-coupled components that are perfectly aligned, which is not the most general case. It is far more likely that

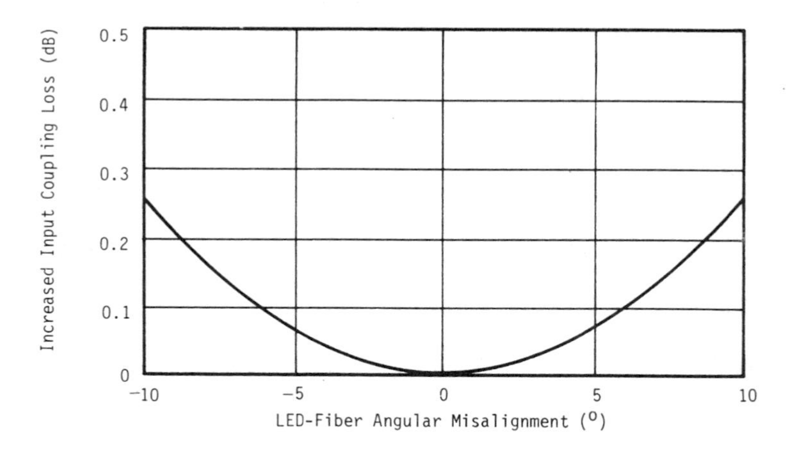

Figure 5.3 Increase in input coupling loss due to angular misalignment between fiber and LED (for experimental details, see ref. 4).

some degree of angular, longitudinal, or lateral misalignment will exist between the fiber and the LED. Each of these misalignments has a different effect on the coupling efficiency. These effects can best be demonstrated graphically.

Taken from Barnoski (4), Figures 5.3, 5.4, and 5.5 plot the coupling efficiency as a function of angular, longitudinal, and lateral misalignments for a

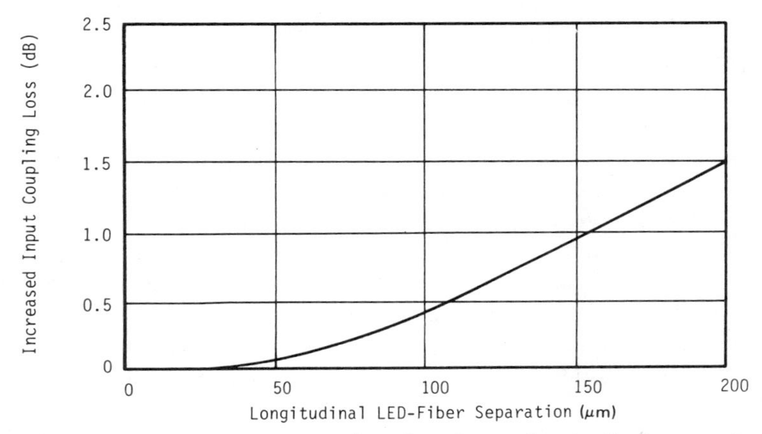

Figure 5.4 Increase in input coupling loss due to longitudinal separation of fiber and LED (for experimental details, see ref. 4).

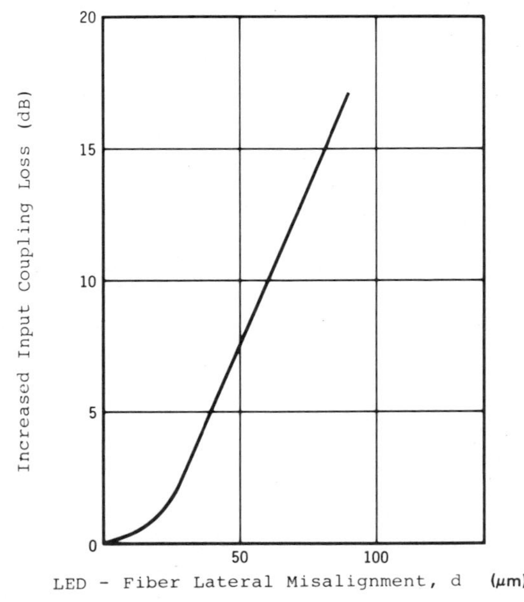

Figure 5.5 Increase in input coupling loss due to lateral misalignment between fiber and LED (for experimental details, see ref. 4).

50 μm LED coupled to a fiber of approximately the same size. Examination of this experimental data reveals that the LED-to-fiber coupling efficiency is surprisingly tolerant of angular misalignment and longitudinal separation of fiber and diode. Angles of 10° and separations of several core diameters can be tolerated without a drastic increase in input coupling loss. Lateral misalignments, on the other hand, can be extremely detrimental, as indicated by Figure 5.5.

Marcuse (5) has also dealt with the subject of lateral and longitudinal misalignments between source and fiber. His treatment was theoretical rather than experimental, however, and his results are quite complicated. These numerical results are therefore presented graphically.

Figures 5.6 and 5.7, which are taken from Marcuse's paper, show the calculated variations in injected power as a graded-index fiber is longitudinally and laterally displaced with respect to a surface-emitting source with an area equal to that of the fiber. We can again see that longitudinal dimensions are not as critical as lateral dimensions. The curves also indicate that lateral misalignment becomes less critical as longitudinal displacement is increased. This point should seem reasonable when one considers how the source's radiation pattern and the fiber's acceptance cone expand with increasing longitudinal separation.

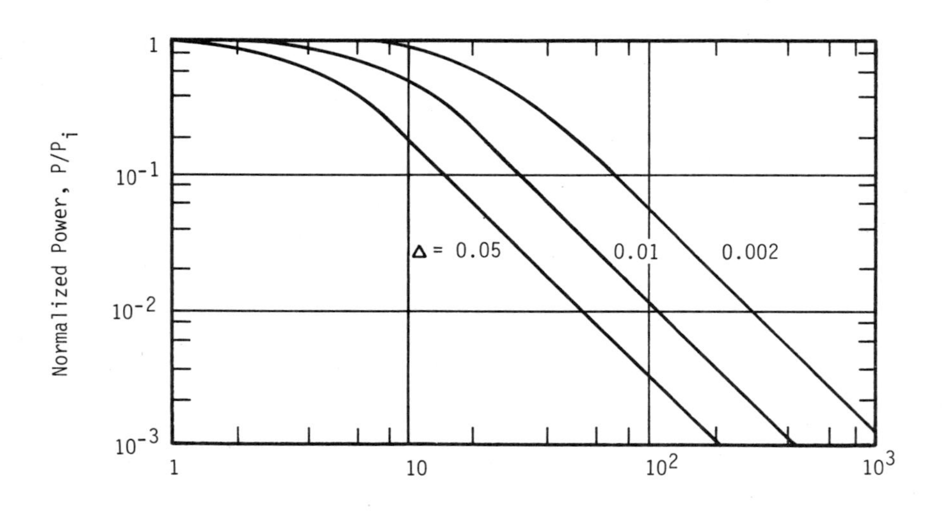

Figure 5.6 Normalized power injected into a graded-index fiber as a function of normalized separation distance between LED and fiber (5); a, core radius; z, separation distance; source and fiber core diameters are equal.

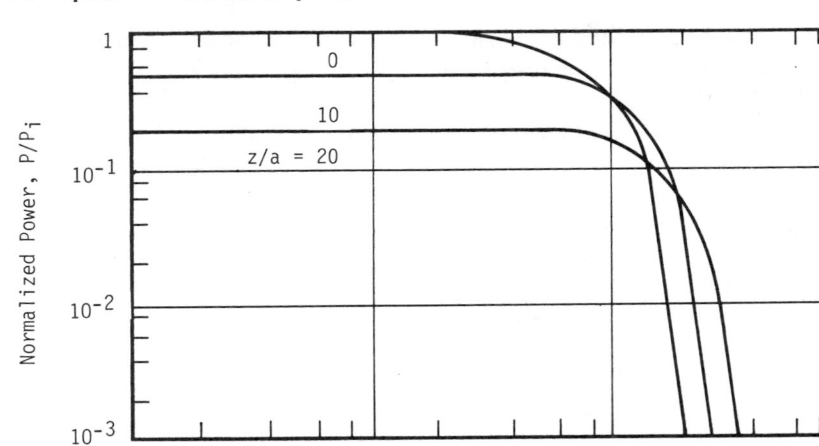

Figure 5.7 Normalized power injected into a graded-index fiber as a function of transverse displacement between LED and fiber ($\Delta = 0.01$) (5); a, core radius; d, displacement; source and fiber diameters are equal.

All of the references thus far noted contain some discussion of improving coupling efficiency with lenses, and the interested reader should refer especially to references 1, 4, and 5. Here we will present just some general comments about the use of lenses.

At first glance, it might appear that one need only image a large source onto a small fiber to attain a coupling efficiency of nearly 100%. Unfortunately, this is not the case. Figure 5.8 will help explain why such increased efficiency is not attainable. In Figure 5.8 we assume that a diode with emitting diameter h is placed at point B. The conservation of radiance dictates that if h' is smaller than h, then ϕ' is larger than ϕ. Thus, we can readily see that if

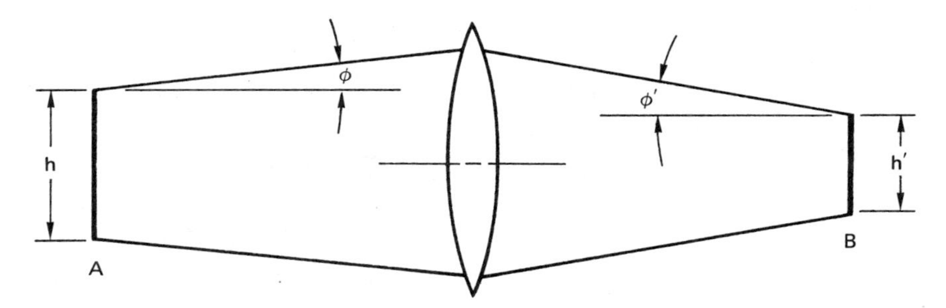

Figure 5.8 Schematic illustration of increasing the angle of incidence by image demagnification.

the source is demagnified, the maximum angle of incidence becomes greater than the acceptance angle of the fiber, thereby negating any gain in efficiency.

The laws of optics do not work totally against us, however. The reciprocity principle states that in most media, light must be able to travel in either direction along a given ray path. Thus, if we consider a small source placed at point *B* and a large fiber placed at point *A*, the magnification of the source would lead to a decrease in the angle of incidence. This fact has been used to reduce coupling loss into a fiber bundle to the packing fraction loss (4), and to obtain an input coupling loss of only 1 dB from a laser diode into a fiber (6). The single problem with lenses is that they complicate the inherently simple butt-coupling geometry.

Figures 5.9 and 5.10 show two possible designs for source-to-fiber connectors. By applying all of the information contained above, we will be able to determine which of the connector dimensions are critical.

As we have seen, longitudinal separation, angular misalignment, and lateral misalignment can affect coupling efficiency. Thus, the length of the sleeve into which are placed the fiber termination and source is a critical dimension, as is its inner diameter. Similarly, the length and outer diameter of the fiber termination are also critical dimensions. By precisely controlling these four quantities, longitudinal separations of about 25 μm and angular misalignments of less than 1° are attainable with current connector technology if the exact placement and angular orientation of the diode chip are known. The last requirement may be a stumbling block, however, because many existing diodes are not designed specifically for use in optical communications systems, and chip placement has consequently not been of primary importance. If improper chip placement has a detrimental effect on the controlling of longitudinal separation and angular misalignment, its effect on the

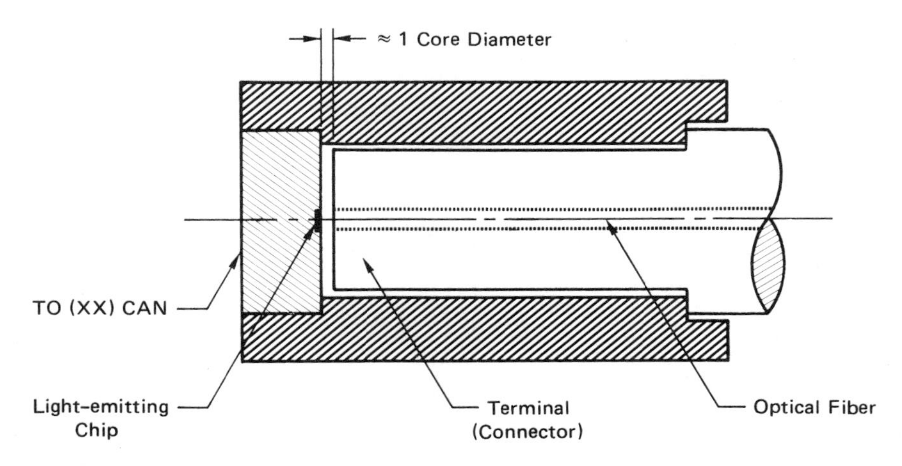

Figure 5.9 Illustration of a source-to-fiber connector allowing butt coupling.

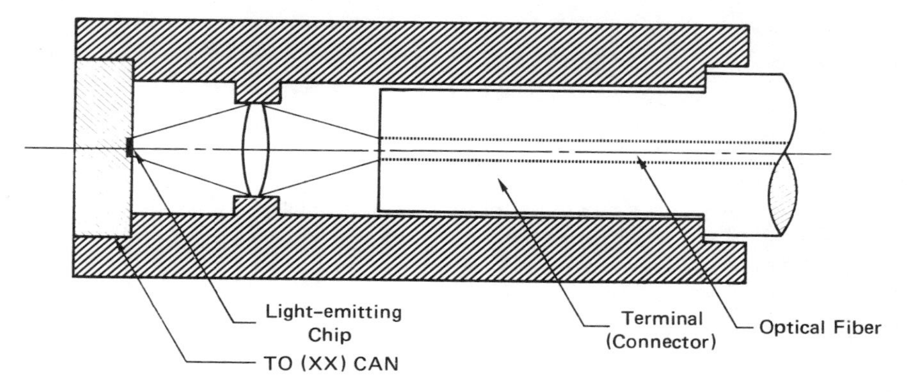

Figure 5.10 Illustration of a source-to-fiber connector allowing lens coupling.

controlling of lateral misalignment is disastrous. For many diode applications, there is no great need to have the emitting chip accurately centered on the base of the diode, and as a result many diodes have their emitting chips up to 500 μm off center. When one considers that many multimode single fibers have core diameters of only 50 μm, it is apparent that the attainment of efficient coupling is a hit or miss proposition. This situation may be alleviated in the future, however, as more sources are designed specifically for fiber optics communications. It remains to be seen whether or not the dimensions of such sources will be more precisely controlled or if either the user of the connector manufacturer will have to align source and fiber for maximum coupling.

The manufacturers of luminescent sources are taking steps that may eventually lead to the obsolescence of connectors of the type shown in Figures 5.9 and 5.10. Rather than accept all of the alignment problems associated with the use of a third component to align source and fiber, they are eliminating the independent connector completely. "Pigtailed" devices of the type shown in Figure 5.11 are becoming readily available, and there are signs that the whole industry may follow this route.

As Figure 5.11 shows, a pigtailed device is simply a source–fiber combination that has been aligned and permanently locked into place. Pigtails can be added to any type of LED or laser diode and are not confined to use with Burrus diodes, as illustrated in Figure 5.11. Manufacturers of such pigtailed devices specify the amount of power emitted by the fiber and may even supply the pigtailed device with a connector on the fiber. A fiber-to-fiber connection is then used to connect the device to the system. Since the coupling efficiencies of fiber-to-fiber connections are fairly well known, the approximate amount of optical power injected into the system is known.

Pigtailing has sufficient advantages so that it may ultimately become the accepted method of coupling light into a fiber. The emitting areas of sources

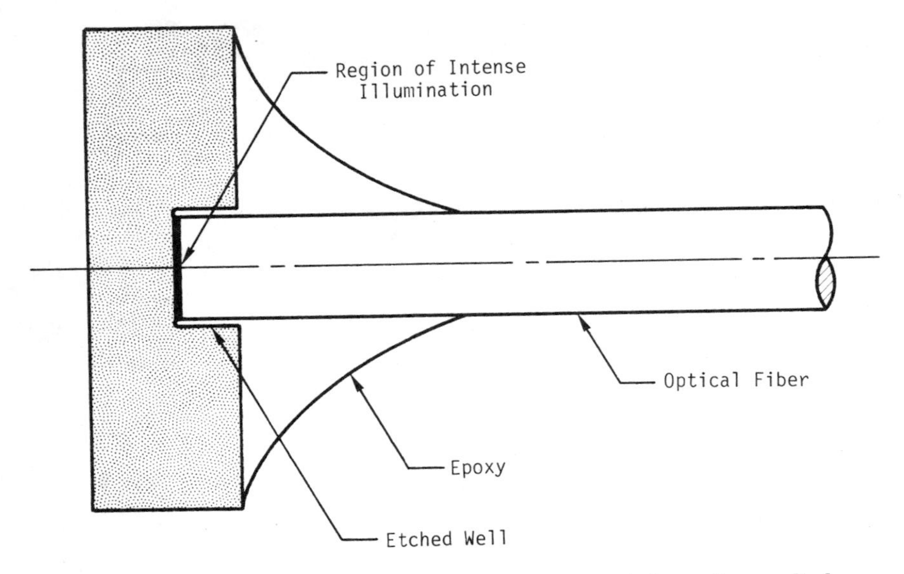

Figure 5.11 Illustration of a single optical fiber pigtailed to a Burrus diode chip.

may change, lenses may be formed on the ends of fibers, and other methods may be used to increase coupling efficiency, but the only outwardly apparent change in the pigtailed source may be the power coupled out of the fiber.

Fiber-to-Fiber Connection

Besides offering potential savings in size, weight, and cost, optical fibers are attractive as a communications medium because of their low loss and high bandwidth. Optical fiber connectors should therefore preserve both of these qualities. However, improperly designed optical fiber connectors can cause appreciable losses. Paradoxically, the fact that connectors introduce losses may indicate that they actually lessen pulse distortion. This is because the high-order modes suffer the bulk of the loss but are also responsible for a major portion of the dispersion. The dispersive effects of optical connectors have not received nearly as much attention as have the loss-inducing effects of the connectors. Thus, the following paragraphs will deal strictly with the mechanisms that cause loss at a fiber-to-fiber connection.

The losses that occur at a fiber-to-fiber interconnection can be grouped into several broad categories. These losses can be caused by variations in connector parameters, variations in fiber parameters, Fresnel reflections, and connection placement in a system. For the remainder of this discussion, these four categories will be referred to as connector-related losses, cable-related losses, reflection losses, and system-related losses, respectively.

The category of connector-related losses can be further divided into a number of subcategories, of which the most important are lateral misalignment losses, longitudinal misalignment, or end separation, losses, angular misalignment losses, fiber distortion, or microbending, losses, and end finish losses. Each of these subcategories will be separately discussed in the following paragraphs.

Connectors for multiple fiber bundles are susceptible to all of the loss-inducing mechanisms listed above plus an additional loss due to having a packing fraction less than unity. This packing fraction loss is due to the fact that a hexagonal array, as is shown in Figure 5.12, is the closest packing arrangement that can be achieved with fibers of circular cross section. Such an array would yield a maximum packing fraction of about 0.9 if the fibers were unclad. Bundles of clad fibers obviously have packing fractions of less than 0.9, and the range of 0.4 to 0.7 is more typical.

Figure 5.12 shows a possible configuration for a connector and adapter for a seven-fiber bundle. It can be seen that alignment of each fiber with an opposing fiber in a similar connector is possible only if some method of keying the connector exists. Such keying is not practical, however, because only bundles with a small number of fibers are likely to arrange themselves into a hexagonal array. A 1.15 mm diameter fiber bundle, for example, typically has 200 or more fibers, which are highly unlikely to arrange themselves into a hexagonal array. When two such bundles are terminated and butted together, the fibers align themselves in a random manner. Some fibers therefore inject their power into the cladding of opposing fibers or into the spaces between fibers. In either case, such light is wasted and constitutes the packing fraction loss.

Multifiber bundles typically suffer a loss from all sources that ranges from 2 to 3 dB. The largest portion of this loss may be due to the packing fraction,

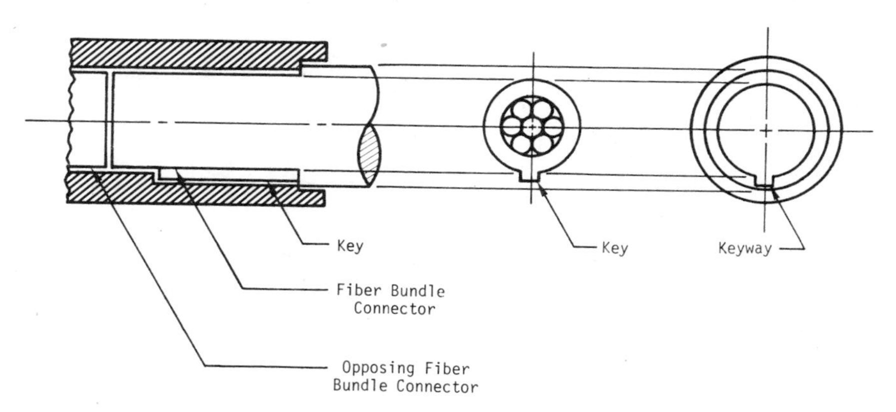

Figure 5.12 Possible keyed connector and adapter for small fiber bundles.

but it is difficult to separate the effects of the various loss mechanisms experimentally. The packing fraction loss tends to obscure the effects of the other sources of loss. Multifiber bundle interconnections are therefore far less sensitive than single-fiber interconnections to the various types of misalignment, although they do suffer higher average losses. A lateral misalignment of a core diameter is just as likely to align some fibers as it is to misalign others.

Several systems have been developed to decrease the packing fraction loss. Most do so by heating the fibers until they deform and fill the spaces that would have otherwise existed between them. Such termination procedures do offer some lessening of insertion loss, but they do so at the expense of added complexity. It is conceivable that improved multimode single-fiber technology will make bundle technology obsolete before such terminating systems become popular.

Lateral misalignment losses arise when the fibers being connected are not coaxial, the result of which is that the overlap region of the two fiber cores is reduced, as depicted in Figure 5.13. For uniformly excited fibers, it is the size of this overlap region that defines the lateral misalignment losses for step-index fibers. The coupling efficiency η_{lat}, and consequently the coupling loss, can be determined by simply calculating the ratio of the overlap region to the area of a fiber core, πa_1^2. This loss is (7, 8)

$$\text{Loss}_{lat} = -10 \log \eta_{lat} \tag{5.11a}$$

where the coupling efficiency is

$$\eta_{lat} = \frac{2}{\pi}\left\{\cos^{-1}\frac{x}{2a_1} - \frac{x}{2a_1}\left[1 - \left(\frac{x}{2a_1}\right)^2\right]^{1/2}\right\} \tag{5.11b}$$

and x is the amount of lateral misalignment.

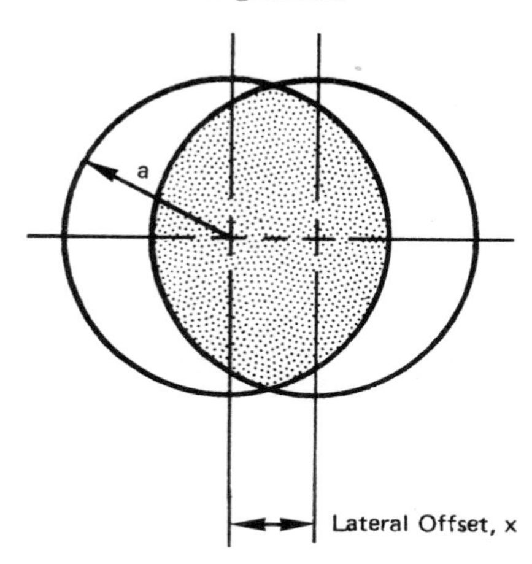

Figure 5.13 Reduced core overlap region due to lateral offset.

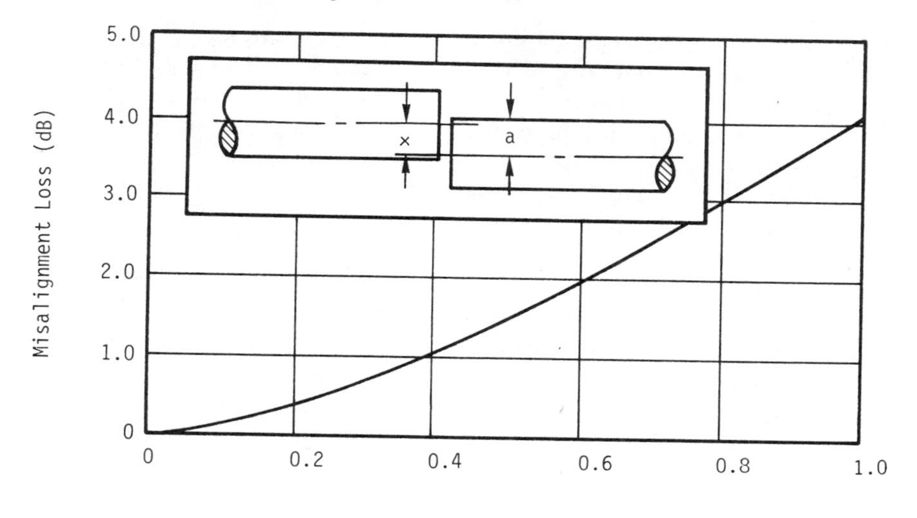

Figure 5.14 Lateral (axial) misalignment loss for step-index fibers (for a discussion of the theory, see ref. 7).

The lateral misalignment loss as calculated according to Equation (5.11) is plotted in Figure 5.14. This curve is in fair agreement with experiments (8), but still remains only approximately correct. The curve in Figure 5.14 will yield generally a pessimistic prediction of the amount of loss to be expected for a specified amount of lateral offset because it is derived under the assumption that all modes of the fibers are uniformly excited. Such an assumption is approximately correct only in certain regions of the transmission line. Connections made in these regions are therefore more likely to have losses that agree with the theoretical curve. This dependence on a placement of the connection in the line is an example of a system-related loss. Although it must be considered in the loss estimation, it is not used to choose the placement of a splice, which is more often determined by other physical factors, such as the presence of walls or bulkheads.

As stated above, Equation (5.11) is the result of a calculation based upon the assumption that the output of the transmitting fiber is uniformly excited. This assumption implies that all modes of the fiber are excited and attenuated equally and that no mode coupling occurs. Under such an assumption, the mode distribution appearing at the connection would be an attenuated replica of the distribution at the input of the transmitting fiber. Because this is not the general case, there have been discrepancies between theory and experiment. Several researchers (9, 10) have therefore examined this subject more closely.

Miller (9) also used the uniform excitation assumption to derive loss-versus-offset curves for both step- and graded-index fibers. His calculated curve for step-index fibers with equal core diameters agrees with the curve shown in Figure 5.14, but he used a different method to calculate the losses to be expected when parabolic index fibers are offset with respect to one another. For parabolic-index fibers, the evaluation of the overlap region is not sufficient to determine the lateral misalignment loss because the local numerical apertures of the fibers being connected are different when the fibers fail to overlap exactly. This problem arises naturally because of the position dependence of the local NA in any graded-index fiber. Thus, if the local NA of the transmitting fiber is greater than the local NA of the receiving fiber, the receiving fiber will not be able to accept the light that falls outside of its local NA. This causes a loss above and beyond the loss observed for step-index fibers, whose NA is constant across the cross sections of both fibers. Miller's calculated and experimental results are reproduced in Figure 5.15.

Because of the discrepancies between calculations and measurements in Miller's and earlier works, Gloge (10) reexamined the problem of lateral misalignment loss. He also began by assuming a uniform distribution of power among the various modes within the fiber, but also assumed that the offset x was only a small fraction of a fiber radius. For any fibers having a power law index of refraction variation of the form

$$n(r) = n_0(1 - 2\Delta r^i)^{1/2} \tag{5.12}$$

Gloge derived the following loss expression:

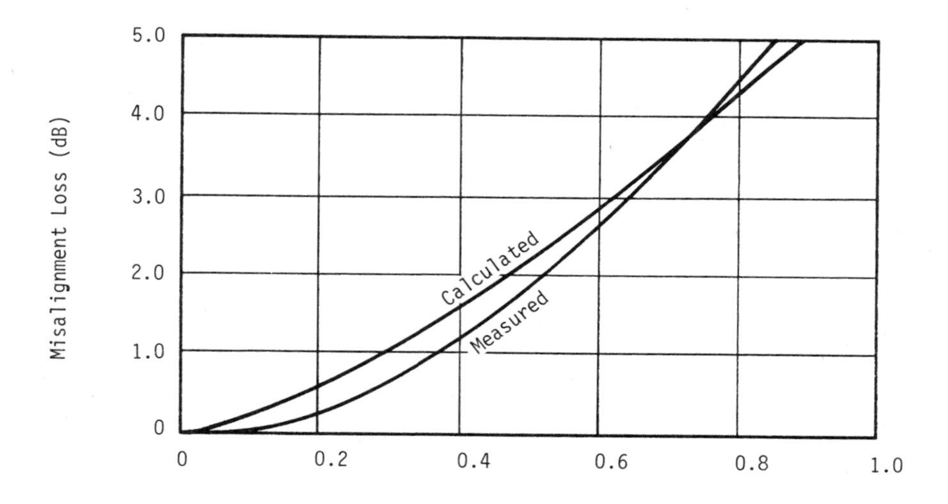

Figure 5.15 Lateral (axial) misalignment loss for graded-index fibers (9).

$$\text{Loss}_{\text{lat}} = \frac{(2x/\pi)\,(i + 2)}{i + 1} \tag{5.13}$$

For a parabolic variation of index of refraction $i = 2$, the loss is

$$\text{Loss}_{\text{lat}} = 8x/3\pi = 0.85x \tag{5.14}$$

If Equation (5.14) is plotted and compared with the curve in Figure 5.14, it is seen that Gloge's expression yields a fair approximation to the other curve in the region $0 \leq x \leq 0.2a_1$. But we have already stated that Figure 5.14 yields pessimistic results in many instances. Gloge therefore sought another method of explaining the discrepancy.

He next examined the situation where the incident power is distributed uniformly over both the trapped and the leaky modes of the fibers. The results of this analysis were uncomplicated and indicated that for parabolic index fibers the misalignment loss equals $0.85x$ if only trapped modes are considered and $0.75x$ if both trapped and leaky modes are considered. The difference is not adequate to bring theory and experiment into agreement. For step-index fibers, the results were that the misalignment loss equals $0.64x$ if only trapped modes are considered and $0.50x$ if both trapped and leaky modes are considered. Besides appreciating that the inclusion of leaky modes in the uniform excitation assumption does not fully explain the differences between calculated and observed lateral misalignment losses, we should also note that for the same amount of lateral offset, parabolic-index fibers are expected to yield greater losses than step-index fibers. This result agrees with Miller's work.

Because of mode coupling due to imperfections in any "real" fiber, the modal power distribution that exists at a connection that is formed even a fairly short distance from a uniformly excited input may be nonuniform. Thus, nonuniform power distributions are likely to be more common than uniform distributions. Consequently, upon learning that the exclusion of leaky modes is not the primary source of error in the earlier calculations of lateral misalignment loss, Gloge then examined the effects of nonuniform power distributions among the various fiber modes. He did so by assuming a uniform input distribution modified by a diffusion-type process. This is reasonable because coupling is most likely to occur between adjacent modes. Thus mode coupling would occur as a step-by-step process much as diffusion occurs.

The results of Gloge's calculations are expressed in terms of a parameter D, where $D^2 = x^2 + \sin^2\Theta/2\Delta$ and Θ is the angular misalignment. The results are therefore applicable to fibers that are either laterally or angularly misaligned or that suffer a combination of lateral and angular misalignment:

$$\text{Loss} = -10\,\log_{10}\eta_{\text{lat, ang}} \tag{5.15a}$$

where the lateral and angular coupling combination efficiency is

$$\frac{2D}{\pi}(1 + 2\gamma)\left[\left(\frac{3}{4 - 1.3\gamma/D} + \frac{8\gamma}{\pi D}\right)^{-1}\right] \tag{5.15b}$$

In this equation

$$\gamma = 0.17(L/L_c)^{1/2} \ll 1$$

L is the length between the input and the connection, and L_c is the coupling length. Gloge's results are shown reproduced in Figure 5.16.

To understand why Gloge predicts less loss for a given offset than did earlier researchers, one must examine the transformation of the fiber's modal power distribution as waves propagate down the fiber. The modes may be uniformly excited at the input, but mode coupling can occur at any imperfection. The energy in the mode at cutoff (i.e., at the critical angle) can scatter into the modes both above and below it. If the energy scatters to a lesser angle, it will be propagated at the angle; if the energy scatters to a greater angle, it couples to a radiation mode of the fiber and is lost. If one imagines this process occurring for every mode of the fiber, it is apparent that the net result of mode coupling is a slow loss of energy from the highest-order modes.

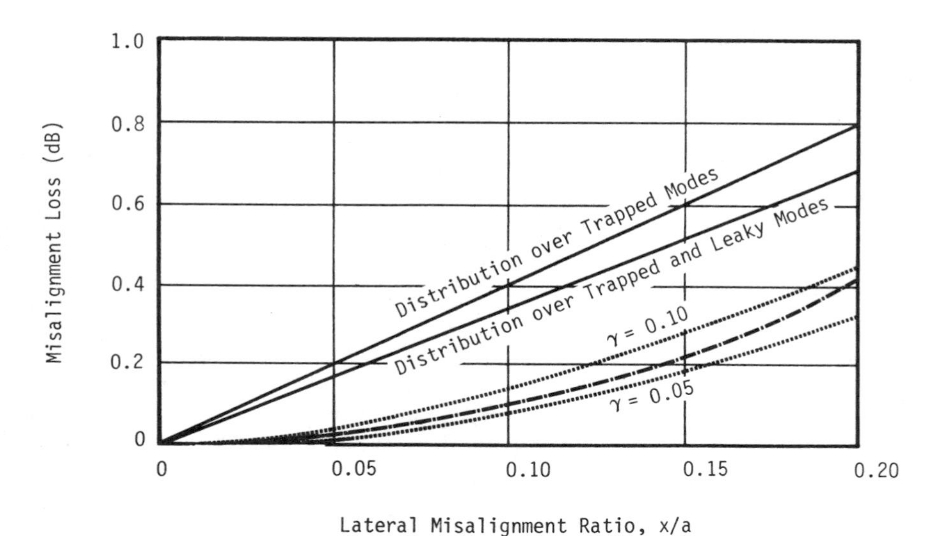

Figure 5.16 Loss due to lateral misalignment of graded-index fibers ($i = 2$) with nonuniform power distribution (10). Legend: (——) uniform distribution; (- - - - -) nonuniform distribution, Equation (5.15); (– - – - –) steady-state distribution, Equation (5.16). Assumption: $\Theta = 0$; hence, $D = x$.

But these are also the modes that suffer the greatest loss at a misaligned connection. Thus, a theory that assumes a nonuniform power distribution will predict lower misalignment losses and will be more accurate when the connection occurs an appreciable distance from an input.

As L approaches L_c, the modal energy distribution more and more closely approximates the steady state. Therefore, for a splice formed at a distance equal to or greater than the steady state length from the input,

$$\text{Loss}_{\text{lat, ang}} = u^4 D^2 / 8(u^2 - 4) \tag{5.16}$$

where $u = 2.405$ is the first root of the zero-order Bessel function that describes the steady-state distribution. Equation (5.16) predicts losses that fall between the losses calculated from Equation (5.15) with $\gamma = 0.05$ and $\gamma = 0.01$. The losses are lower than those calculated from the uniform power distribution models. It is a matter not of one model being right and the other models being wrong but of choosing the correct model for a particular application. Since the mode distribution occurring at an interconnection is in general not known, the uniform power distribution model and the steady-state power distribution model can be used to define the range into which the lateral misalignment losses can be expected to fall.

While the steady-state and the uniform power distribution models can yield a fair estimate of the range of splice losses to be incurred at a laterally misaligned interconnection, they are not the only models that have been suggested. Mettler and Miller (11, 11a) have assumed that the power distribution across a fiber's output radiation cone is Gaussian in form. They have also defined a point transmission function that depends on the numerical aperture of transmitting and receiving fibers. The product of the energy distribution and point transmission functions so defined is then integrated numerically over the area of core overlap, and it yields the splice loss due to both intrinsic parameter mismatch and lateral misalignment.

Another interesting point mentioned by Mettler and Miller (11, 11a) is that the loss suffered at a splice between parabolic index fibers depends not only on the length of the fiber between the source and the splice but also on the length of the fiber between the splice and the detector. If a long length of fiber lies between splice and detector, greater losses will be incurred. This is because some of the power that is initially coupled from the transmitting fiber to the receiving fiber will be lost in reestablishing steady-state conditions. Mettler and Miller therefore generalized the point transmission function to make it applicable to cases where either long or short fibers follow the splice. In both cases, fairly good agreement with experiment was observed.

Marcuse (12) has presented results for the calculation of lateral misalignment losses in single-mode optical waveguides. His analysis is based upon proof that the energy in both step- and parabolic-index single-mode fibers propagates as Gaussian or near-Gaussian beams. Then the transmission factor for offset Gaussian beams can be derived from the works of earlier researchers (13, 14). The loss associated with laterally offset single-mode fibers with the same parameters is therefore

$$\text{Loss}_{\text{lat}} = -10 \log e^{-x^2/w^2} \tag{5.17}$$

where w is the beam width parameter. Equation (5.17) is plotted in Figure 5.17.

By allowing for different beam widths in the two fibers being connected, it is possible to predict losses for misaligned fibers differing in core diameter or index profile. That generalization is not presented here, since it falls into the class of cable-related coupling losses. However, there is one final point to be mentioned here. Marcuse (12) derived an expression, much like the uncertainty principle of quantum mechanics, that shows that single-mode fibers can be made more tolerant of lateral offset only at the expense of increased tolerance to angular misalignment, or vice versa. This point had already been demonstrated in an experiment dealing with the splicing of single-mode fibers (15).

To summarize, lateral alignment of the fibers has been found to be an extremely important parameter. Under even the most favorable conditions, offsets of more than about 30% of a core radius will cause about 0.5 dB loss. Under the least favorable conditions, offsets of only about one-half this magnitude will yield the same loss. When one considers that low-loss multimode fibers typically have core radii in the range 30 to 60 μm, and typical single-

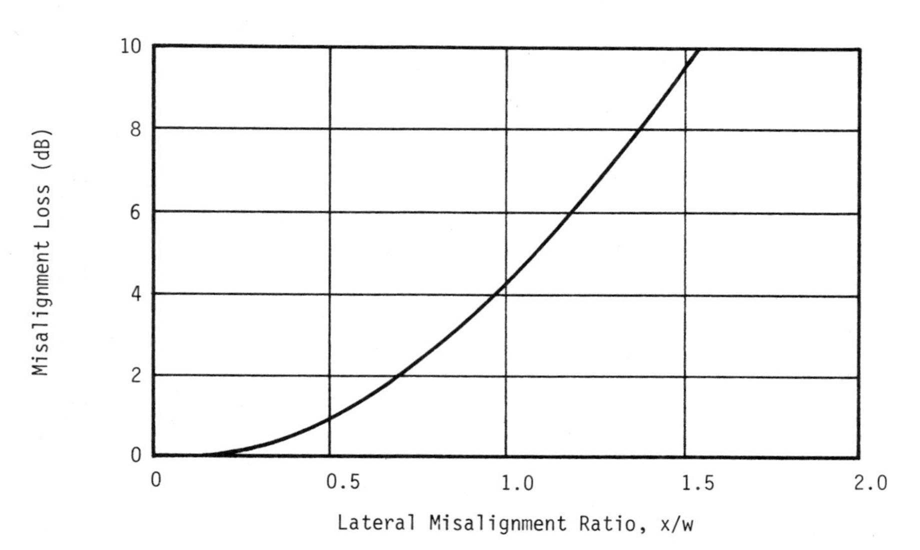

Figure 5.17 Loss due to lateral misalignment of monomode step-index fibers (12), where w = beam width.

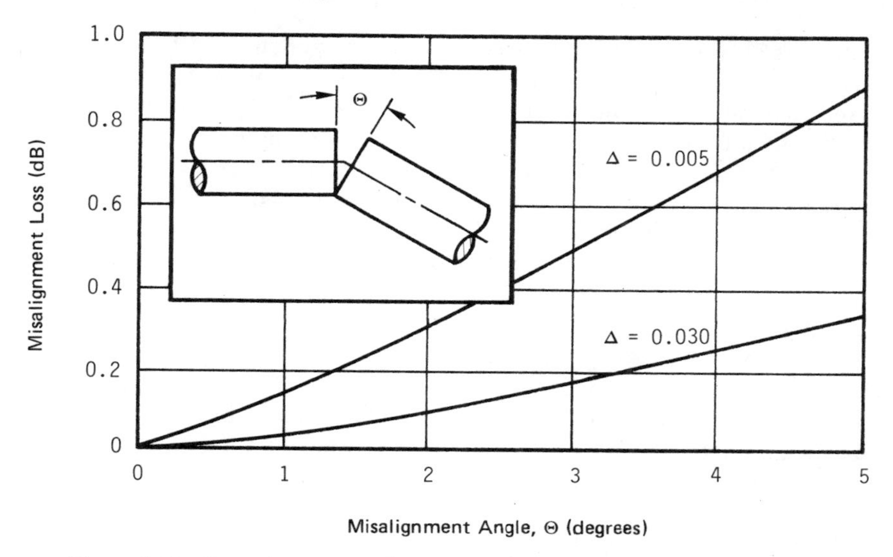

Figure 5.18 Loss due to angular misalignment as a function of misalignment angle (8).

severe restrictions are placed upon the connectors for fibers, particularly for monomode fibers.

If an angle exists between two fibers that are butt-coupled, the coupling efficiency is decreased. It has already been shown that one can use Equation (5.15) to calculate this loss when graded-index fibers are coupled. Separate treatments of tilt-induced loss have also been presented for multimode step-index fibers and single-mode fibers.

Tsuchiya et al. (8), have examined the phenomenon of angular misalignment–induced loss both experimentally and theoretically. Their theoretical analysis is approximately correct for multimode step-index fibers. The approximations used include the assumption that the modal power distributions is uniform and the assumption that only meridional rays propagate. We have already seen that theories that assume a uniform modal power distribution yield results that are pessimistic in some cases, but Tsuchiya et al. can claim good agreement between their experimental and theoretical analyses. For an angular misalignment of Θ, the predicted loss is

$$\text{Loss}_{\text{ang}} = -10 \log\left[1 - n\Theta/\pi n_1(2\Delta)^{1/2}\right] \tag{5.18}$$

where n_1 is the refractive index of the fiber core, n is the refractive index of the medium filling the gap separating the fibers, and Θ is expressed in radians. The relationship expressed in Equation (5.18) is plotted in Figure 5.18 for fibers having a core refractive index of 1.50 and Δ of 0.005 and

0.030. It is observed that for a specific angular misalignment, the losses are greater for fibers with small values of Δ. This is intuitively reasonable because small values of Δ imply low NA; for low-NA fibers, the portion of the output cone affected by a particular angular misalignment is proportionately greater than it would be for a fiber with a larger NA.

For single-mode fibers that are tilted with respect to each other, the result for angularly misaligned Gaussian beams can be applied (12). For the coupling of fibers with identical parameters, this result reduces to

$$\text{Loss}_{\text{ang}} = -10 \log\{\exp[-(\pi n_2 w \Theta/\lambda)^2]\} \tag{5.19}$$

where Θ is the tilt angle, λ is the wavelength of the propagating radiation, n_2 is the refractive index of the cladding, and w is the beam width.

The approximate Gaussian beam width as a function of a fiber's v value has been calculated numerically by Marcuse (12), whose results are reproduced in Figure 5.19. Since single-mode fibers must have v values below 2.405, the interesting region of Figure 5.19 lies below that value. Choosing a v value of 2.400, we use Figure 5.19 and Equation (5.19) to calculate the tilt loss for both graded- and step-index single-mode fibers. These results are presented in Figure 5.20.

The discussions above indicate that for both parabolic- and step-index multimode fibers the angle between the endfaces of the fibers should not be more than about 1°. The same is true of single-mode fibers. Such angular tolerances can be maintained with current connector technology.

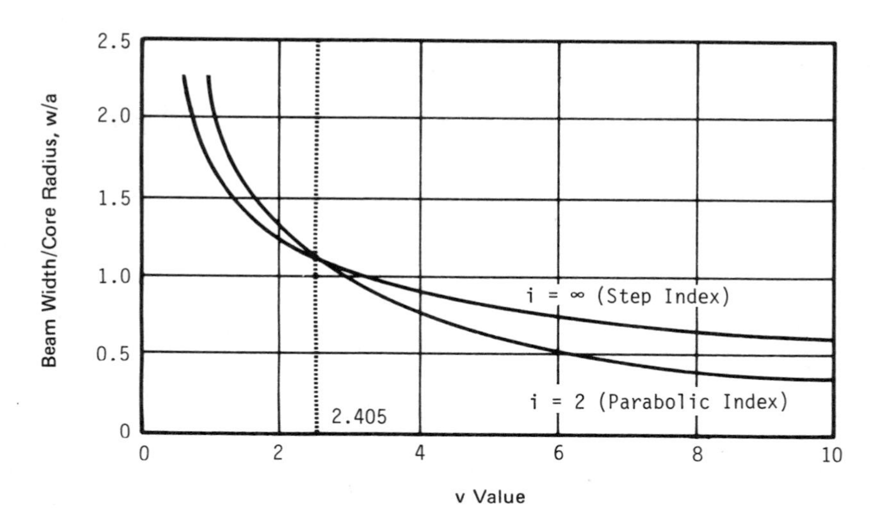

Figure 5.19 Ratio of beam width and core radius for step-index and graded-index fibers as a function of v value (12).

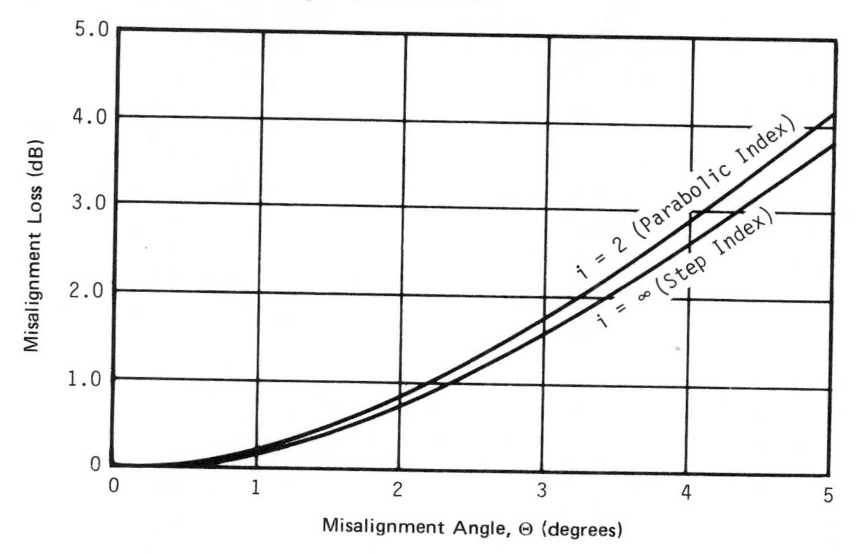

Figure 5.20 Angular misalignment loss for step-index and graded-index monomode fibers (results calculated; see ref. 12). Assumptions: $n_1 = 1.50$, $n_2 = 1.49$, $\lambda = 0.90\ \mu$m, $a = 1.998\ \mu$m, $v = (n_1^2 - n_2^2)2\pi a/\lambda = 2.4$; and, from Figure 5.19, $w = 2.067\ \mu$m for $i = \infty$ and $w = 2.147\ \mu$m for $i = 2$.

Because of the divergence of the light beam emitted by an optical fiber, the length of the gap separating two butt-coupled fibers affects the insertion loss experienced at that joint. As the receiving fiber is moved further away from the transmitting fiber, it intercepts proportionately less of the cone of light projected by the transmitting fiber, thus giving rise to end separation loss. It is certain that this end separation, or longitudinal misalignment, loss has less effect than either lateral or angular misalignment losses; the longitudinal coupling efficiency can be expressed by

$$\text{Loss}_{\text{long}} = 10 \log \eta_{\text{long}} \tag{5.20}$$

However, there is some disagreement about the actual magnitude of the effect and hence about the composition of the term η_{long}. One group of authors (7) claims that for multimode step-index fibers separated by less than one core radius, the longitudinal coupling efficiency is given by

$$\eta_{\text{long}} = (a_1^2 - a_1 z \tan\beta_0 + \tfrac{1}{6} z^2 \tan^2\beta_0)/(a_1^2 + \tfrac{1}{3} z^2 \tan^2\beta_0) \tag{5.21a}$$

where β_0 is defined by the NA of the fiber and z is the separation between fibers. Another group (8) claims that for the same type of fibers, the longitudinal coupling efficiency is given by

$$\eta_{\text{long}} = 1 - (z/4a_1)\,(n_1/n)\Delta^{1/2}(2)^{1/2} \qquad \text{or} \qquad (2\Delta)^{1\ 2} \qquad (5.21b)$$

where n_1 is the index of refraction of the fiber cores and n is the refractive index of the medium filling the gap between fibers.

Both groups of researchers assumed uniform cross-sectional and angular power distributions, and both groups predict higher losses for high-NA fibers. This prediction is reasonable because the larger divergence of high-NA fibers indicates that they project larger light cones for a given separation. Although the two groups agree on certain assumptions and predictions, they disagree on the magnitude of the expected losses. Equation (5.21a) predicts losses appreciably higher than those predicted by Equation (5.21b).

The only reason this potentially confusing point is mentioned is that curves based upon Equation (5.21a) have received wide publication. The fact that both groups have claimed good agreement with experiment further confuses the issue but is perhaps understandable because of the number of variables involved in making this type of measurement. As we have seen, the length of the cables used in the experiment might have an effect on the measurements, as would small amounts of lateral or angular misalignment in the test fixtures. We have therefore plotted both equations in Figure 5.21 and add

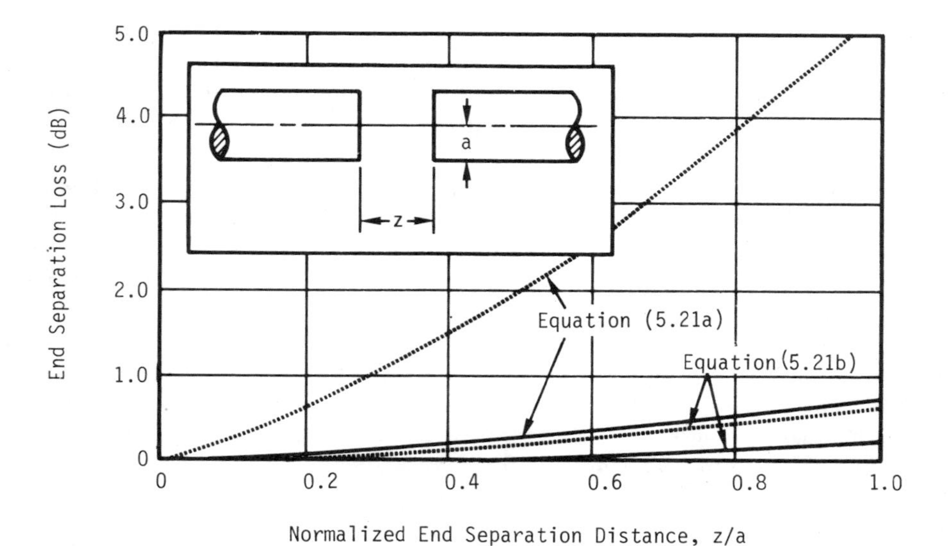

Figure 5.21 End separation loss as a function of normalized separation distance for step-index fibers (7, 8). Legend: (——) NA = 0.17; (- - -) NA = 0.61.

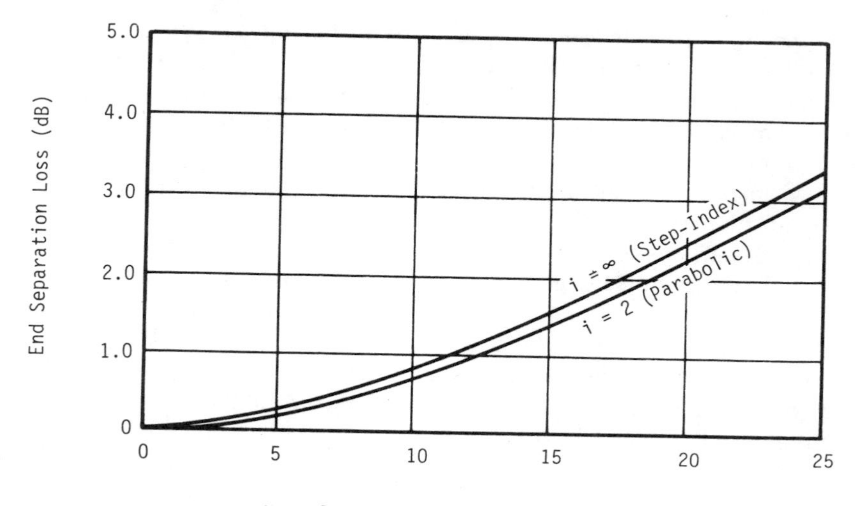

Figure 5.22 End separation loss as a function of normalized separation distance for monomode fibers (12). Assumptions: $a = 1.988\ \mu m$, $w = 2.067\ \mu m$ for $i = \infty$, $w = 2.147\ \mu m$ for $i = 2$.

the final comment that a third group of investigators (16) have presented data that agrees more closely with those of Tsuchiya et al. (8).

Measurements (17) have shown that an end separation of one core radius increases the loss suffered at a splice between parabolic-index fibers by about one-eighth of a decibel. This loss is less than that predicted by the curves in Figure 5.21. One must remember, however, that the curves in Figure 5.21 are applicable to step-index fibers and are based on theory, while the results presented in ref. 17 were empirically derived.

For single-mode fibers, the result for longitudinally misaligned Gaussian beams can be applied (12). For identical fibers, this result is

$$\text{Loss}_{\text{long}} = -10 \log \frac{4\,(4z^2 + 1)}{(4z^2 + 2)^2/4 + z^2} \tag{5.22}$$

where z is the length of the gap between the fibers. Equation (5.22) is plotted in Figure 5.22.

No matter whether we accept Equation (5.21a) or Equation (5.21b) as correct for multimode fibers, we can see that insertion loss is relatively less dependent upon end separation than on the other two types of misalignment. We also note that both theories predict that end separation becomes more important as NA increases. Thus, if the length of cable used causes the exit NA to be smaller than the calculated NA, both theories would yield pessimistic results. In addition, the lengths of manufactured parts are more easily controlled than are concentricities. Thus, current connector technology can control end separation losses in multimode fiber connectors so that they have a minimal impact on total insertion loss. Furthermore, single-mode fibers, although an order of magnitude smaller in size, are only slightly more susceptible to longitudinal misalignment losses.

For maximum transmission across a butt-coupled joint, both fiber endfaces must be flat, smooth, and perpendicular to the fiber axes. There are a number of imperfections associated with the various termination procedures that might prevent the attainment of such surfaces. The endfaces might be cut at a slight angle or polished to a slightly convex profile. The endfaces might also be rough compared to the wavelength of light, and distortion, or microbending, losses might be suffered in the connectors. Termination procedures that require the fiber to be epoxied to the connector and then polished are susceptible to all of the above imperfections, while termination procedures based on scribing, breaking, and crimping the fibers are susceptible to losses due to fiber endface tilt, endface roughness, and fiber distortion.

Although microbending losses can be described both mathematically and experimentally, distortion losses due to fiber termination procedures occur at random because uneven pressures may be exerted on a fiber by improper crimping or epoxy curing techniques. The occurrence of such processes is difficult to predict mathematically. Fortunately, such a description is not necessary, since a statistical analysis of the excess insertion loss due to distortion is equally acceptable. We are not aware of any such results in the literature, but we can suggest what such an experiment might reveal. It is expected that a plot of loss due to microbending versus the percentage of terminations falling into each loss category would show that with proper termination procedure the average microbending loss occurring in the connectors could be kept below several tenths of a decibel. It is also expected that such a curve would have an exponential tail, revealing that a few percent of such terminations would suffer excess microbending losses of several decibels. This is true even though buffered fibers suffer greatly reduced microbending losses, since the buffering is usually removed in the vicinity of a connector.

Approximate mathematical results have been presented to describe the effects of fiber endfaces that are polished to a convex profile or that are polished or broken at an angle (8). The endface tilt loss can be described by

$$\mathrm{Loss}_{\text{tilted endface}} = -10 \log \eta_{\text{tilted endface}} \tag{5.23a}$$

For multimode step-index fibers that are polished at an angle, the additional loss is

$$\mathrm{Loss}_{\text{tilted endface}} = -10 \log \frac{1 - n \mid n_1/n - 1 \mid (\Theta_1 + \Theta_2)}{\pi n_1 (2\Delta)^{1/2}} \tag{5.23b}$$

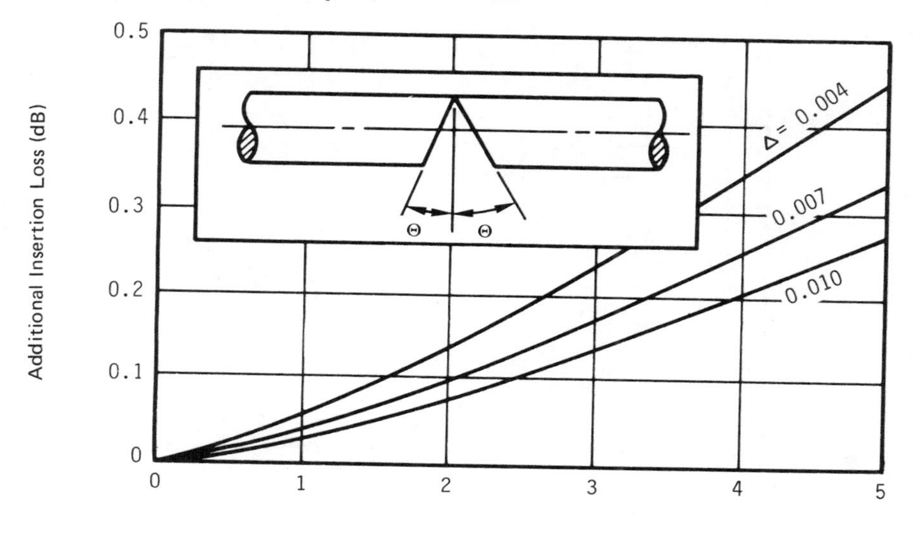

Figure 5.23 Additional insertion loss caused by polished fiber endfaces with a tilt (8); $n = 1.0$, $n_1 = 1.46$; $\Delta = 0.004$, $\Delta = 0.007$, and $\Delta = 0.010$.

where Θ_1 and Θ_2 are the angular deviations of the two fiber endfaces from the line that is perpendicular to their axes. This relationship is plotted in Figure 5.23, which shows that the effect is more detrimental to low-NA fibers, as expected. For multimode step-index fibers polished to a convex profile, the derived result is

$$\text{Loss}_{\text{convex endface}} = -10 \log \left(1 - \frac{n \mid n_1/n - 1 \mid [d_1 + d_2]}{2n_1 a_1 (2\Delta)^{1/2}}\right) \quad (5.24)$$

The radii of curvature, R_1 and R_2, are approximately $R_1 = a_1^2/2d$ and $R_2 = a_1^2/2d_2$; d_1 and d_2 are defined in Figure 5.24.

The same authors (8) also investigated experimentally the dependence of insertion loss on the roughness of the fiber endfaces. Their results, which concur with our own unpublished data, indicate that improvements in insertion loss are obtained as the size of the polishing grit is reduced to the 1 μm range. Polishing with grits smaller than 1 μm yields no significant improvement in insertion loss. It is interesting to note that 1 μm is approximately the wavelength of light used in typical fiber optic systems.

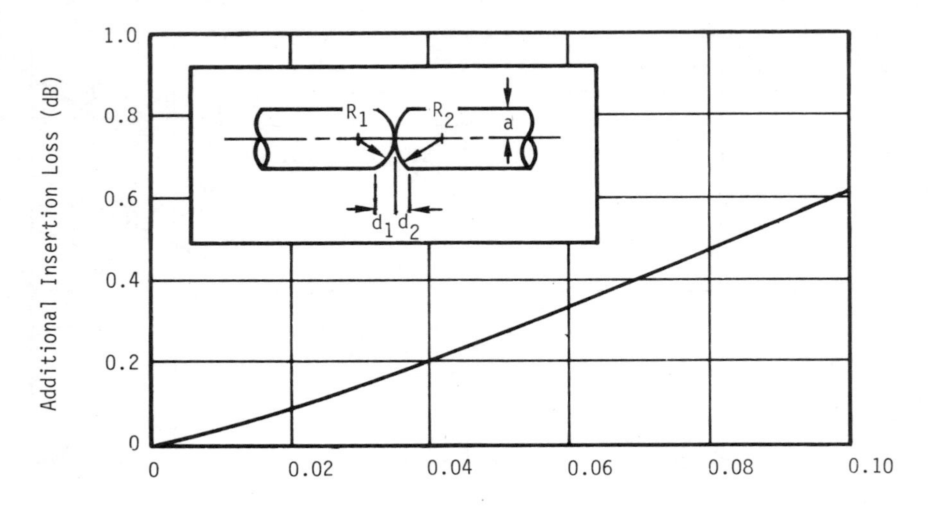

Figure 5.24 Additional insertion loss caused by polishing fiber endfaces to convex surfaces (8); $n = 1.0$, $n_1 = 1.46$, $\Delta = 0.007$.

A fluid with a refractive index that matches that of the cores of the fibers being coupled can be used to eliminate the losses due to imperfect endfinish. Such a fluid can also potentially eliminate Fresnel reflection losses that typically add 0.3 to 0.4 dB to a junction that does not have an index-matching substance in the gap between the fibers. The actual magnitude of the Fresnel reflection, which is caused by the reflection of electromagnetic radiation from the boundary between dielectrics with dissimilar refractive indexes, is given by

$$\text{Loss}_{\text{Fresnel}} = -10 \log\{1 - 2[(n_1 - n)/(n_1 + n)]^2\} \tag{5.25}$$

where n_1 is the refractive index of the fiber core and n is the refractive index of the substance filling the gap between fibers. It is obvious that when $n = n_1$ there is no reflection. Any imperfections in the fiber endfaces are also masked by the presence of the fluid. The effects of the various misalignments, however, cannot be totally eliminated; in fact, the effect of angular misalignment is worsened by this technique.

Losses suffered at a fiber-to-fiber junction depend not only on the alignment and fiber endfinish requirements mentioned above but also on the waveguiding characteristics of the fibers themselves. The fiber into which power is injected must be capable of propagating as many modes as is the fiber from which power is emitted. If this is not the case, losses will be suffered even at a joint at which perfect alignment is attained and at which an index matching fluid is used. The magnitude of this loss is (18)

$$\text{Loss}_{\text{mode mismatch}} = -10 \log (N_{m_r}/N_{m_t}) \tag{5.26}$$

where N_{m_r} and N_{m_t} are the number of modes propagated by receiving and transmitting fibers, respectively. The number of modes any particular fiber can support is (19)

$$N_m = [i/(i + 2)]\, a_1^2 k_1^2 \Delta \tag{5.27}$$

where $k_1 = 2\pi n_1/\lambda$ and i is the index profile parameter. The normalized refractive index difference Δ is related to the square of the numerical aperture of the fiber. Thus, we see that the total number of modes capable of being propagated by a fiber will be altered if the core diameter, index profile, or peak numerical aperture are allowed to vary. In practice, there are statistical variations of these parameters which, according to Thiel and Hawk (18), typically yield an average cable-related loss of less than 0.5 dB, but a small number of splices have cable-related losses of up to several decibels even if otherwise perfect. The average performance of fibers can, of course, be improved by increasing the controls on the fiber production process, but the tail of potentially high-loss junctions will most likely remain, although it may be shifted to lower loss values.

Equation (5.26) indicates that cable-related losses are experienced only when the mode volume of the receiving fiber is smaller than the mode volume of the transmitting fiber. Using Equations (5.26) and (5.27), the losses due to mismatches in peak numerical aperture, NA(0), index profile, and core diameter can be determined (18). Assuming that two of the parameters are held constant while the third is varied, these losses are

$$\text{Loss}_A = -10 \log (a_r/a_t)^2 \quad \text{for } a_t > a_r \tag{5.28}$$

$$\text{Loss}_N = -10 \log [\text{NA}_r(0)/\text{NA}_t(0)]^2 \quad \text{for } \text{NA}_t(0) > \text{NA}_r(0) \tag{5.29}$$

$$\text{Loss}_I = -10 \log \frac{(i_r/i_t)\,(i_t + 2)}{(i_r + 2)} \quad \text{for } i_t > i_r \tag{5.30}$$

Because Equations (5.28) and (5.29) are similar, both are represented in Figure 5.25; Equation (5.30) is plotted separately in Figure 5.26.

As we have mentioned, optical fibers are designed to have certain nominal parameters, but there are statistical variations about these nominal values. If, for example, the design of a given fiber allows ±2% standard deviation about its nominal diameter, this implies that some samples of fiber could deviate from that nominal diameter by amounts greater than 2%. A very small

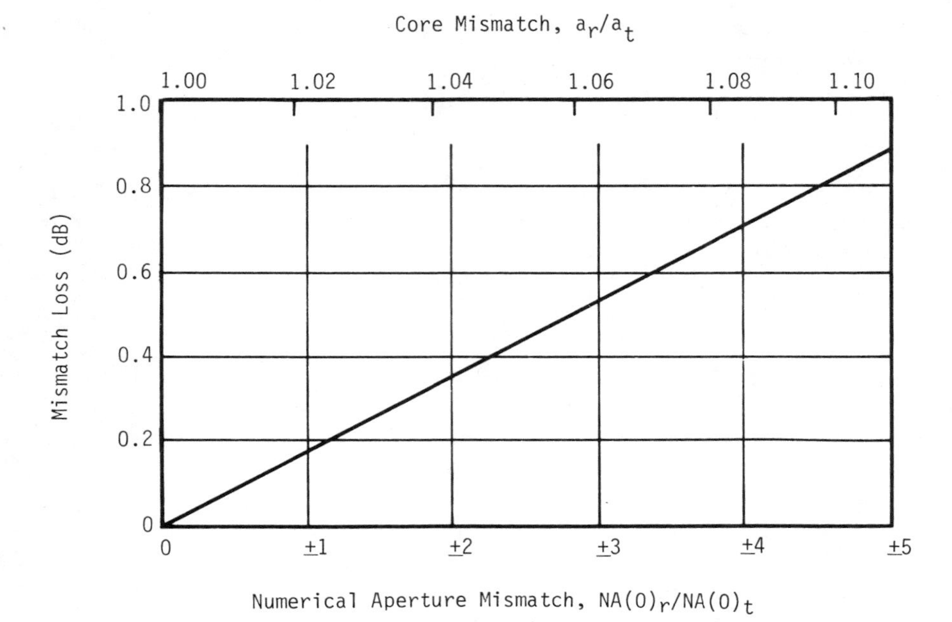

Figure 5.25 Loss due to mismatch in either core diameter or peak numerical apertures (18).

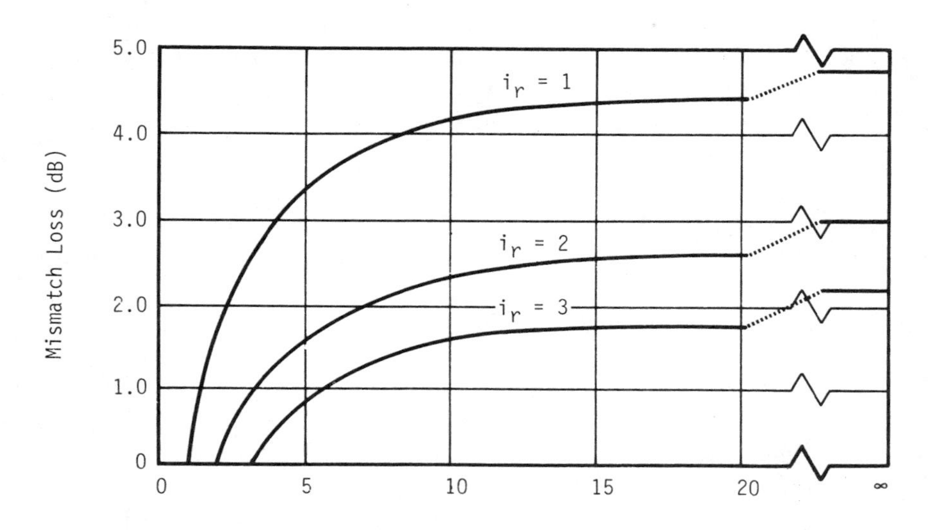

Figure 5.26 Loss due to mismatch in fiber index profiles (18).

number of fibers could have parameters quite far from the nominal values. These variations would cause high losses when such fibers were joined to fibers with nominal parameters. When joined in worst-case fashion, which is luckily a statistically unlikely occurrence, the splice losses would be even higher. Of course, fiber production processes are constantly becoming more sophisticated, and control of the fiber parameters is likely to continue improving in the future.

The above discussion has shown that a number of variables can contribute to the insertion loss suffered at a fiber-to-fiber connection. We have also demonstrated that the largest contributors to this insertion loss in a low-loss connector are likely to be lateral misalignment losses, losses due to coupling fibers with different parameters, and Fresnel reflection losses. The reflection losses can easily be eliminated by using an index-matching fluid, but these fluids can trap dust, which causes increased scattering losses. For this reason, index-matching fluids are often ignored and the 0.3–0.4 dB additional loss accepted. Cable-related losses are also ignored; although they can be considerable, they are beyond the control of the connector designer.

Consequently, given that butt coupling is the most efficient method of joining fibers, lateral alignment of the fiber cores within a fraction of a core diameter is the most difficult task. A number of methods of achieving lateral alignment have been tried, and we reference a few examples of each type for the interested reader. Lateral alignment has generally been achieved through the use of any one of the following devices: precision sleeves (7, 20–26), v-shaped grooves (7, 27, 28), overlapping rods (16, 18, 29), or elastic pressure (16, 18, 30). Other designs have attempted to skirt the need for tight lateral alignment tolerances by altering the optical path (31, 32), or by adjusting the connector for minimum loss after assembly is completed (8, 33). Multichannel single-fiber connectors have also been developed (18, 28).

Fiber-to-Detector Connection

The coupling of an optical fiber to an optical power detector is a task that is more easily accomplished than either coupling power into the fiber initially or coupling power from fiber to fiber. This is because in the former case the geometry is much more favorable than in either of the latter cases. In the case of coupling light sources to fibers, the efficiency increases as the size of the source decreases to approximately the size of the fiber. Thus, an extremely small fiber must be aligned to a source that is equally small. Similarly, for efficient fiber-to-fiber coupling, accurate alignment must be achieved between two equally small fibers.

On the other hand, high coupling efficiency between fiber and detector can be achieved even if the active area of the detector is many times larger than the cross-sectional area of the fiber. Such a detector is capable of intercepting nearly all the light emitted by the fiber; the only losses are caused by reflections from the various surfaces encountered by the emitted light and by the angular response of the detector. Alignment is not critical because even very small detectors have active areas that are 0.25 mm square, whereas typical multimode fibers have cores that have diameters of 0.060 to 0.125 mm. In addition, detectors with active areas of up to 1 cm^2 are routinely available. Because large-area detectors also have higher capacitances and slower response times, the chosen detector must not only couple easily to the fiber but must also be compatible with the bandwidth requirements of the entire system.

Therefore, to couple a fiber to a detector, it is only necessary to build a fixture that allows the fiber to butt against the detector's active area, as shown in Figure 5.27. If the geometry of the detector does not allow efficient coupling in this manner, then lenses may be used to focus the light onto the detector, as shown in Figure 5.28. This scheme has the added advantage that the angle of the light can be made narrower if the detector area is larger than the fiber area. Angular response loss is thereby decreased.

The pigtailing of fibers to detectors has also been tried, and this technique may be used more frequently in the future. Unless increases in transmitted data rates dictate that detectors smaller than those currently available must be developed, it is not certain that the increased complexity of pigtailed devices can be justified economically.

COUPLERS

Optical couplers are components, usually passive, that are used to allow input to or output from a main transmission line without totally interrupting the signal carried by that line. The signal thus tapped may simply be used to monitor

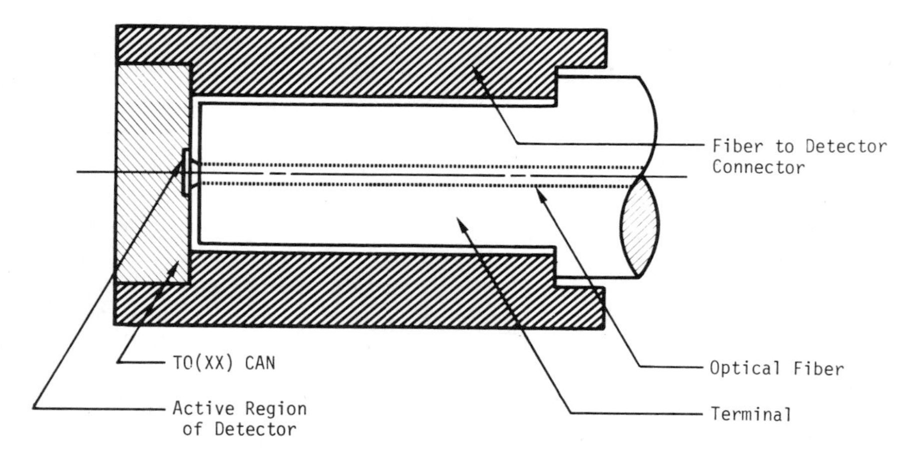

Figure 5.27 Example of an arrangement for butt coupling an optical fiber to an optical power detector.

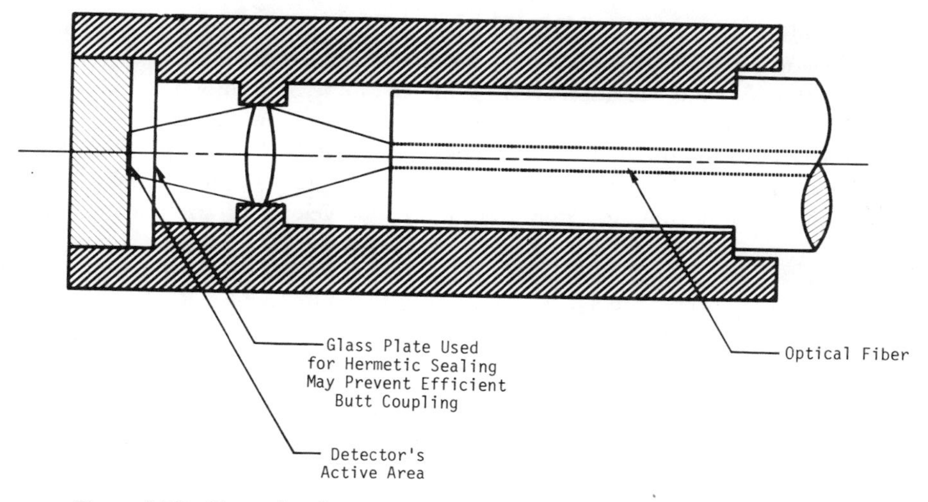

Figure 5.28 Example of an arrangement for lens coupling an optical fiber to an optical detector.

the system status or to distribute data to a remote location. A number of such couplers may be used in the construction of a data bus, which is an essential element in modern communications and control systems.

Although there are many variations of each, there are two basic types of optical couplers: the T coupler and the star coupler. Both types of couplers can be used to distribute data to locations remote from a central processing unit. Both types are normally passive and therefore allow data to be distributed without the use of repeaters, which are subject to electrical failure. Both types have particular advantages that we will explore.

T Coupler

Data buses usually allow communications between each terminal and every other terminal and a central processor. If an attempt were made to accomplish this by dedicating a transmission line to communications between every pair of terminals, the number of such dedicated lines would soon be overwhelming because the number of lengths of cable needed to connect n terminals with a dedicated line for each terminal pair is $\frac{1}{2}n(n - 1)$. Data distribution systems that use T couplers rather than dedicated lines therefore enjoy a large savings in both cable cost and cable bulk because T couplers allow the n terminals to be coupled to one main transmission, or bus, line.

A number of T couplers have been demonstrated, but they are nearly all based upon variations of three basic coupling concepts. Most T couplers so far demonstrated use bifurcation and/or mixing rods; beam splitters and/or focusing elements; or evanescent wave coupling. Space will not allow discussion of all possible variations of each type of coupler, but a representative, although not exhaustive, list of references is offered the interested reader.

Although many variations of a T coupler can be built, they must all share certain properties. We will therefore first discuss a generalized T coupler. We will then describe the three basic types of T couplers.

Consider the T coupler shown in Figure 5.29. It is simply a "black box;" we know nothing about its internal structure, but we can still use it to describe the basic properties of a T coupler.

Our black box has four ports to which we can connect various types of optical cables. At port 1, we attach an optical cable out of which is emitted light with a power P_{o1}. We assume that only a fraction of this power is actually coupled to the device, and label this power P_{i1}. The coupling loss from cable to black box is therefore

$$\mathrm{CL}_1 = -10 \log(P_{i1}/P_{o1}) \tag{5.31}$$

where CL_1 may include Fresnel reflection losses, packing fraction losses, misalignment losses, or losses due to any other mechanism that will lessen the amount of power transferred from cable to black box.

Out of port 2 is emitted light with a power P_{o2}. Therefore, the loss due to transit through the coupler is

$$\mathrm{TL} = -10 \log(P_{o2}/P_{i1}) \tag{5.32}$$

The light emitted by port 2 is then coupled to an optical cable, and we assume that the amount of power accepted by the cable is P_{i2}. The loss incurred in coupling our T coupler to the second piece of cable is therefore

$$\mathrm{CL}_2 = -10 \log(P_{i2}/P_{o2}) \tag{5.33}$$

where CL_2 will usually be equal in magnitude to CL_1 owing to the symmetry of most coupler designs.

We define the throughput loss as the total loss incurred by coupling cable 1 to cable 2 through the T coupler:

$$\text{Throughput loss} = \mathrm{CL}_1 + \mathrm{TL} + \mathrm{CL}_2 \approx 2\mathrm{CL}_1 + \mathrm{TL} \tag{5.34}$$

We assume that our tapped signal is extracted at port 4, where a power P_{o4} is emitted. The ratio of this power to the power actually injected to the T coupler we will call the tap, or coupling, ratio. Expressed in decibels, this ratio is

$$\text{Tap ratio} = -10 \log(P_{o4}/P_{i1}) \tag{5.35}$$

The power emitted by port 4 can be either injected into another piece of cable or allowed to fall on an optical detector. In either case, there is likely to

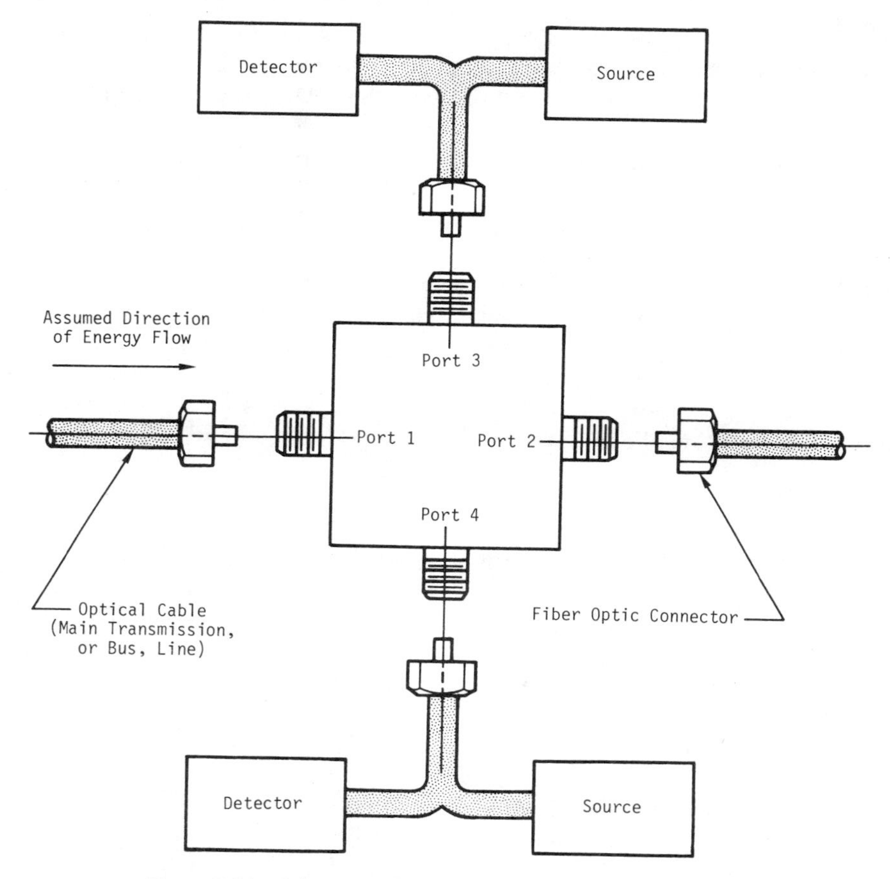

Figure 5.29 Schematic diagram of a generalized T coupler.

be some coupling loss, but the magnitude of this loss will depend on the actual system design. If the power is injected into another piece of cable, the loss is likely to be about equal to CL_1, but if the power falls on a detector, the magnitude of the loss is likely to differ from CL_1. In either case, the loss is still given by

$$CL_4 = -10 \log(P_{o4}/P_{i4}) \tag{5.36}$$

where P_{i4} is either the power accepted by the optical cable or by the detector.

Port 3 will usually be an input port for signals traveling from port 1 to port 2. Because of the bidirectionality of the coupler, port 3 is the output port for signals traveling from port 2 to port 1. Thus, for the case that we have been describing, transit from port 1 to port 2, no power should be emitted at port 3. However, some coupler designs allow greatly attenuated signals to be emitted by port 3; for such couplers, the ratio of the power emitted by port 3 to P_{i1} is termed the directionality of the coupler. Light emitted at port 3 when the direction of the transmitted signal indicates that port 3 should be acting as an input port is lost.

When port 3 is used as an input to the main transmission line, the coupling loss is

$$\mathrm{CL}_3 = -10 \log(P_{i3}/P_{o3}) \tag{5.37}$$

where P_{o3} is the power emitted by either a light source or illuminated cable situated at port 3 and P_{i3} is the power accepted by the T coupler. Again, it is obvious that CL_3 will be approximately equal to CL_1 if a cable is connected to port 3 and very different from CL_1 if an LED or laser is situated at that port.

Conservation of energy demands that

$$P_{o2} + P_{o4} = P_{i1} \tag{5.38}$$

if the coupler is ideal, which will not necessarily be the case. If the relationship shown in Equation (5.38) is not satisfied, it is violated because of a loss mechanism not associated with the division of the input power. This additional loss is often termed excess loss and includes power emitted by port 3 or radiated or absorbed in the interior of the coupler:

$$\text{Excess loss} = -10 \log\left[(P_{i1} - P_{o2} - P_{o4})/P_{i1}\right] \tag{5.39}$$

We can now see which parameters should be compared when evaluating couplers of different designs. An ideal coupler would have zero excess loss and a throughput loss that depends only on the tap ratio. The tap ratio should also be a parameter that can easily be changed if system requirements so demand. Even if such an ideal coupler were produced, it would still impose certain limitations on system designers. To understand why this is so, consider the following example:

Assume that 1 mW of power is injected into a data bus and that the detector at the opposite end of the bus line can detect signals up to 60 dBm while maintaining the required bit error rate. Assume also that the T couplers used are ideal; the only power loss is due to the tapping of the input signal. Assume that at each coupler 10% of the incident power is tapped while 90% is transmitted. This implies that the throughput loss is 0.458 dB. How many of these idealized couplers can be placed between the source and the detector if coupling losses and attenuation losses in the cables are ignored? Solution: The system can stand a 60 dB loss before it ceases operating correctly, and a loss of 0.458 dB is suffered at each coupler. Division of these two numbers indicates that a maximum of 131 couplers can be placed in series.

Therefore, energy conservation places an upper limit of 131 couplers on the idealized system described in the example. If we use a more realistic throughput loss value of 3 dB (i.e., 1 dB each for input and output connector losses and 1 dB for combined excess and tapping losses), the number of couplers that can be used is reduced to 20. When cable attenuation losses are included, the number of couplers that can be used drops even further. We thus see that although T couplers are inherently modular, there is a definite limit to how many can be used in a particular system.

"Furcation" comes from the Latin word *furcatus* and means branched, like the prongs of a fork. Bifurcation therefore means split in two, and a bifurcated fiber bundle is therefore a bundle that has been split into two branches. Using this concept of bundle splitting, a device such as that shown in Figure 5.30 can be constructed. Let us examine its properties (34–39).

For convenience, let us assume that we start with a fiber bundle containing 300 fibers. At both ends of the bundle, we split the bundle into two branches containing 200 and 100 fibers, respectively, and put connectors on the four branches so created. If the device is properly constructed, 100 fibers connect port 1 to port 2, another 100 fibers will connect ports 1 and 4, and the last 100 fibers will connect ports 2 and 3.

If we assume that the main transmission line is connected to ports 1 and 2, we can see that only one-half of the light entering port 1 actually exits at port 2. In addition, the light entering port 1 from the main line has suffered a 1–3 dB coupling loss due to reflections and packing fraction. As the light reenters the main line at port 2, it again suffers a coupling loss. This indicates that a throughput loss of 5.0 to 9.0 dB has been suffered during transit through the coupler; as we shall see, even greater losses will be incurred with such a coupler.

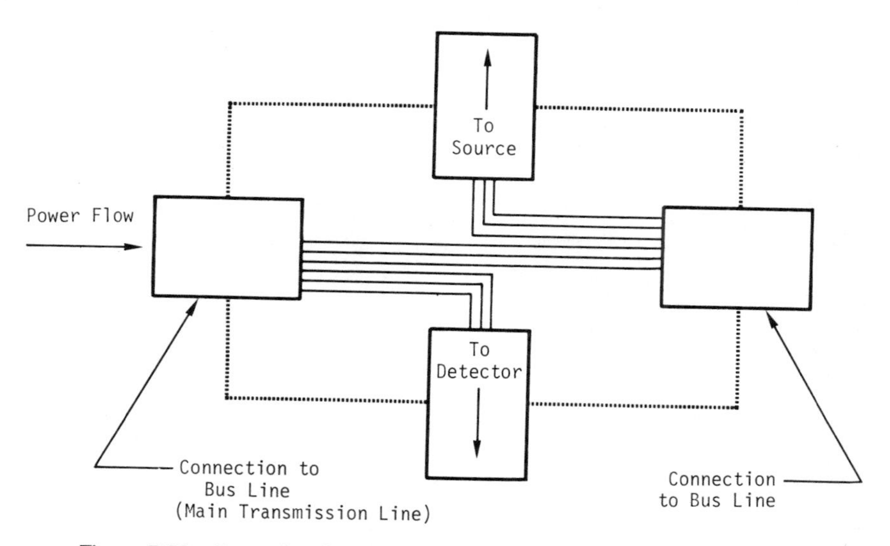

Figure 5.30 Example of a unidirectional coupler using furcation of a fiber bundle.

If we assume that the main transmission line has 300 fibers, as does the bundle used to construct the coupler, then at port 1 a bundle with 300 fibers will be joined to a bundle with 200 fibers. Such a situation leads to a loss above the normal 1–3 dB connecting loss assumed earlier. We can, of course, use a main transmission line with 200 fibers to improve the coupling at port 1. But then at port 3, 100 fibers will be trying to illuminate 200 fibers. Unless some means is used to spread the radiation from port 2 of the T coupler, many fibers in the main line will not be illuminated. This will obviously lead to a disastrous loss at the junction with the next T coupler. A coupler such as that shown in Figure 5.30 is therefore impractical.

Although it is admittedly true that the coupler and the main transmission line need not use the same type of fibers and that 100 larger fibers may be used to illuminate 200 smaller fibers fairly uniformly, this would not be a useful approach because it would limit a particular coupler to use with a limited variety of cables. It would be more reasonable to design a coupler that can be used with any manufacturer's cable. However, to do so requires a means of taking the power emitted by a small number of fibers and using it to illuminate a larger number of fibers uniformly. This is exactly what a mixing or scrambling rod does.

A scrambling rod is a dielectric rod, usually clad, that has the property that partial illumination of its input face causes full illumination of its output face. This is accomplished by designing the mixer in such a way that nearly every input ray suffers at least one internal reflection before it reaches the output face of the rod. With such mixing rods, we can overcome some of the problems associated with the T coupler shown in Figure 5.30. We can use two of the rods to construct a device such as that shown in Figure 5.31. Now 100 illuminated fibers can be used to illuminate uniformly the 200 fibers of the

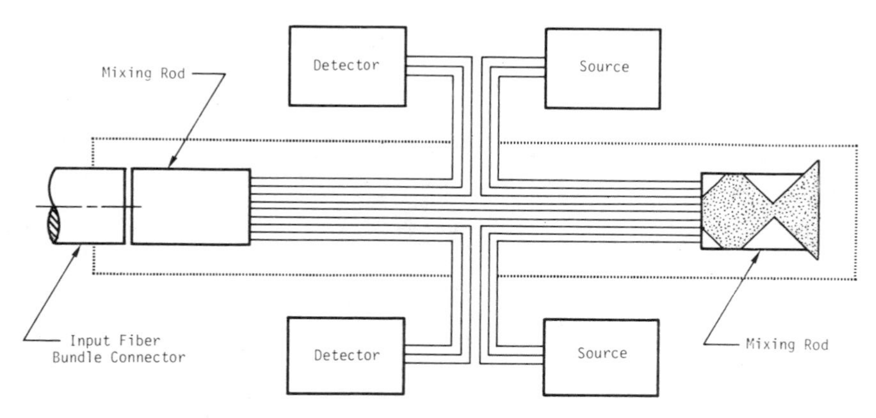

Figure 5.31 Example of a bidirectional coupler using bifurcation and mixing rods.

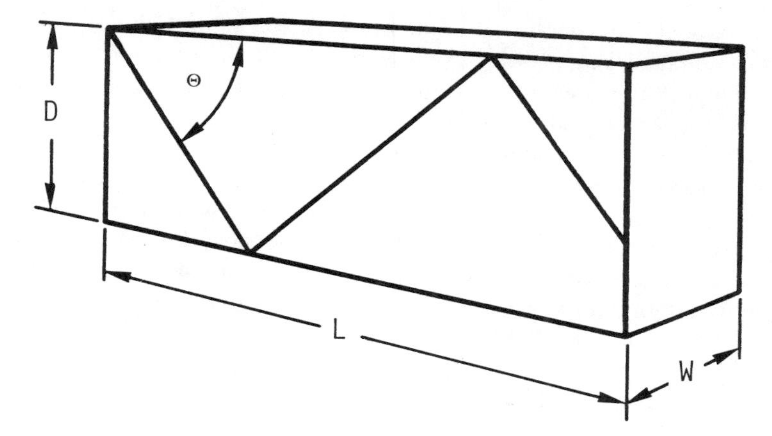

Figure 5.32 Schematic diagram of a mixing rod (analyzed by Milton and Lee, ref. 39); $D \gg W$.

main transmission line. Packing fraction and other coupling-related losses are still encountered, and additional losses can be caused by transit of light through the mixing rods, but the use of the mixer rods allows a number of couplers to be placed in series.

The configuration shown in Figure 5.31 is neither unique nor optimum, and many possible configurations of rods and bundles can be conceived. In addition, rods of various geometries can be used. They may be rectangular, cylindrical, or tapered. No matter what their shape, however, mixing rods must share certain properties. They must cause fairly low throughput loss, and they must distribute the input signal in such a way that it exits the mixer's output face distributed uniformly both spatially and angularly. The NA of the rod should match that of the fibers being used; if the NA of the mixer is much larger than that of the fibers, the fibers will not accept light that falls outside of their own numerical aperture, and extra loss will be incurred. The reverse argument is also true. Spatial uniformity is needed to ensure that all fibers at the rod output are uniformly excited.

For a mixer rod of particular geometry, the optimum dimensions must be determined by ray trace analysis or experiment. Milton and Lee (39) have performed such an analysis for a specific geometry. They used a rectangular rod like the one illustrated in Figure 5.32 and assumed that the width was much larger than the thickness. They were therefore analyzing mixing in one dimension. Their results indicate that for this particular geometry, the minimum length of the rod should be

$$\text{Rod length} \geq D \tan \Theta_m \tag{5.40}$$

where Θ_m is the maximum angle that the input light takes with respect to the axis of the rod. This angle can be determined by dividing the arcsine of the NA of the fibers being used by the index of refraction of the mixer rod's core. Experiments (39) showed that longer rods provided better spatial uniformity at the cost of increased throughput loss.

One last comment should be made about T couplers using bifurcation. The bending of an optical fiber causes radiation loss and changes the output radiation pattern of the fiber. Therefore, regardless of whether the bifurcation is achieved by actually bending a fiber bundle (34,35) or by forming bent arms on mixing rods (36–38), such problems can be encountered. At least one design (38) has therefore used a prism in conjunction with mixing rods to achieve beam splitting. This design could easily be grouped in the next category of T couplers.

Lenses, prisms, and partially reflective mirrors have been used to collimate and divide optical beams for years. Such components can be used in fiber optic systems in ways similar to their use in larger optical systems (40–46). Of course, the small size of the fibers indicates that misalignments, aberrations, and component fabrication may impose limitations on the performance and economy of couplers using such components.

Advantages of couplers using beam splitters and/or lenses are as follows:

1. Both single-fiber and multifiber bundles can be used with such a coupler.
2. The strength of the tapped signal can be varied by varying the transmission–reflection ratio of the beam splitter.

The major disadvantage is, as mentioned, that fabrication is normally difficult.

A simply constructed variation of this type of coupler is described by Kuwahara et al. (45) and illustrated in Figure 5.33. In this particular coupler, the partially reflecting surface is formed by actually polishing the fiber ends at an angle and injecting a fluid into the gap. The tap ratio can be varied by replacing the fluid in the gap with a fluid that has a different refractive index. Careful inspection of this particular coupler shows that although it is simply constructed it requires very careful alignment.

The coupler illustrated in Figure 5.33 is obviously only one of a large variety of such couplers that can be imagined. For other examples of this class of T couplers, the interested reader is directed to the other references cited.

T couplers using evanescent wave coupling recall the fact that optical fibers are waveguiding structures. In fact, parallel, single-mode fibers that are placed within one or two wavelengths of each other behave in much the same manner as do parallel microwave waveguides. Slab or cylindrical single-mode fibers that are coupled together and excited by coherent radiation can exhibit the phenomenon of beating.

When beating is exhibited, there can be a complete transfer of energy from the waveguide initially excited to the waveguide initially unexcited. This transfer takes place in a distance $\frac{1}{2}L_b$, where L_b is defined as the beat

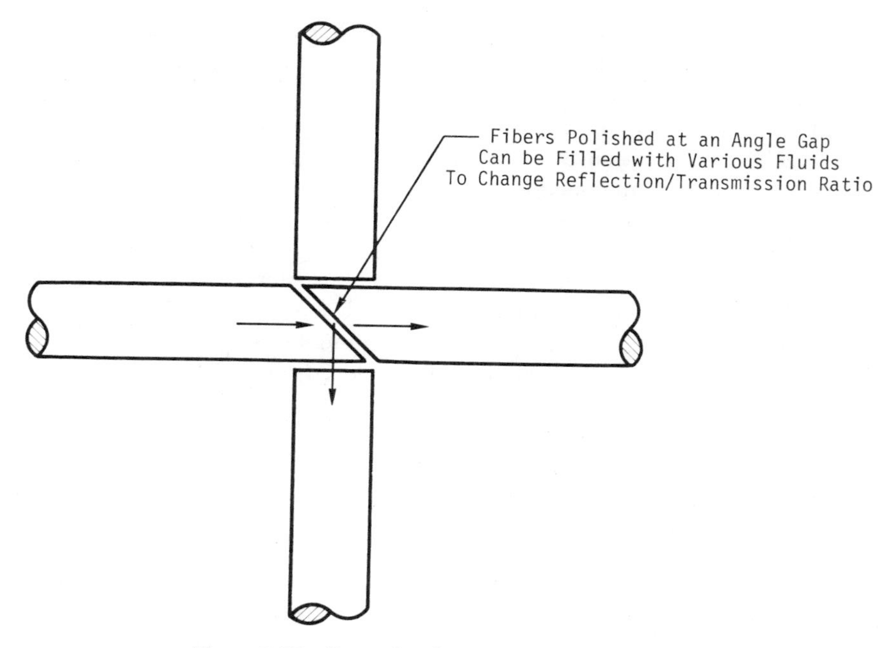

Figure 5.33 Example of a practical T coupler (45).

length. The power then transfers back to the initial waveguide after a further distance, $\frac{1}{2}L_b$, has been traversed. This beat length is defined by the diameter of the fibers, the center-to-center spacing of the fibers, the refractive indexes of the cores, the refractive index of the medium separating the cores, and the wavelength of the exciting radiation. The beat length is given by (47):

$$L_b = \pi/\Delta\beta \tag{5.41}$$

where $\Delta\beta$ is the difference between the propagation constants of the normal modes of the coupled waveguide system and the normal modes of isolated waveguides. The dependence of the beat length on all of the parameters mentioned earlier is, of course, buried in the value of $\Delta\beta$.

Therefore, it can be seen that for single-mode slab or cylindrical waveguides, highly efficient couplers with various coupling ratios can be produced in integrated form. It is necessary only to couple the guides for a distance less than $\frac{1}{2}L_b$ to achieve power transfer of less than unity. It should also be obvious that if it can be controlled electrically, such a coupler can be used as an optical switch. Before integrated optics and single-mode fibers are routinely used, there is likely to be an interim period during which multimode single fibers will be the transmission media of choice. It is therefore desirable to have multimode fiber T couplers that use evanescent wave interaction as their coupling mechanism. A number of evanescent wave couplers of both the single-mode and multimode varieties have been described (47–57).

It is possible to transfer energy from an excited optical fiber to an adjacent, parallel fiber because not all of the energy propagated by a fiber is confined to the core region. An appreciable portion of the energy is propagated in the cladding by the evanescent wave. This evanescent wave decays exponentially with increased distance from the core and is usually strongest within one or two wavelengths of the core–cladding interface. Most coupling schemes therefore try to position the cores as close together as possible.

The mathematical description of the coupling between two single-mode waveguides is highly complicated, and the description of the coupling between two multimode fibers is almost unwieldy. Such descriptions often use a large number of approximations and still must be computed numerically. For readers interested in such analyses, references 57 through 59 are offered. In place of such an analysis, we provide a qualitative description of how evanescent wave coupling occurs and a report on the best results thus far achieved.

There are at least two ways of interpreting the coupling that occurs between parallel optical fibers. One may imagine that the entire mode volume of one fiber is excited while the second fiber remains completely unexcited. Energy propagates only in the first fiber until the interaction region is encountered. In that region, each propagating mode sees a composite fiber that has twice the mode volume of either fiber alone. Modes with the same propagation constants and sufficient overlap of their evanescent waves can couple from excited fiber to unexcited fiber.

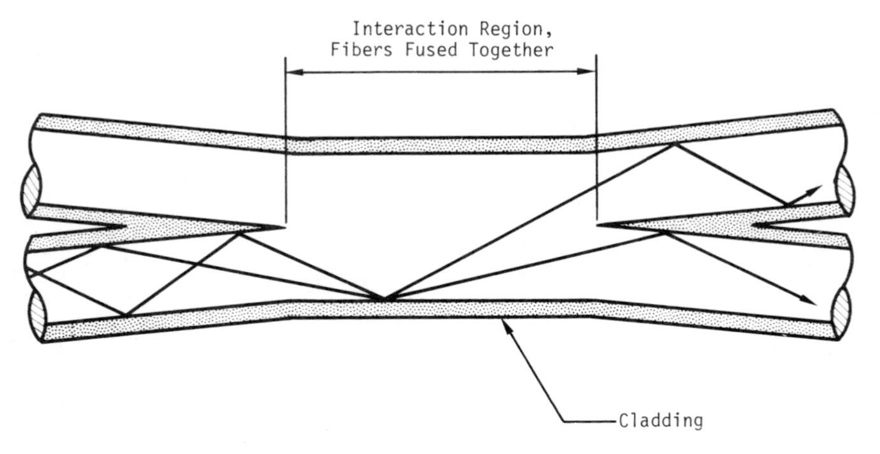

Figure 5.34 Ray optics model of evanescent wave coupling. Higher-order modes suffer a greater amount of reflection and therefore have a greater probability of coupling across the interaction region.

An alternate interpretation of what is occurring can be obtained through use of the ray optics picture. Figure 5.34 shows two identical fibers joined core to core for a certain interaction length. Lower-order modes propagate at shallower angles and are therefore more likely to be unaffected by the presence of the interaction region. Higher-order modes, on the other hand, propagate at steeper angles and are thus more likely to couple into the unexcited fiber.

Calculations indicate that maximum coupling is achieved when the fiber cores are actually touching, and Owaga (57) has estimated that as much as 72% (−1.5 dB) of the incident power may be transferred between multimode fibers. Coupling efficiencies as high as predicted have not yet been observed, but Kawasaki and Hill (50) have reported coupling efficiencies of 24% (−6.2 dB) for fibers actually fused together over their interaction length. The same authors have also reported excess losses as low as −0.11 dB. It should be obvious that such low values of insertion loss indicate that in a properly designed system many such couplers can be used in series.

Star Coupler

Although T couplers have been used at microwave frequencies for years, the unique nature of light can be employed effectively in a different type of data distribution system, i.e., the star coupler (60–65).

Star couplers can also be used in the construction of data buses. Such data buses require far less cable than systems using dedicated lines. Star couplers use one mixing rod to distribute light evenly (if all cable lengths are approximately equal) to a large number of terminals. Star couplers can be applied effectively with either multifiber bundles or single fibers. Star couplers can generally be grouped into two basic categories: the reflective star coupler (Figure 3.35A) and the transmissive star coupler (Figure 5.35B). Star couplers can be designed to allow communication between every terminal of the system or to allow every terminal to communicate with a central processing unit.

It is simple to understand how a star coupler operates. The coupler's mixing rod is designed so that a fiber placed at any position on the input end of the rod will fully illuminate the output end after the light emitted by the fiber completes either one transit (i.e., transmissive star) or two transits (i.e., reflective star) through the rod. To design a rod able to accomplish this task, it is really necessary to resort to a ray trace analysis of the particular rod–fiber combination, but Hudson and Thiel (60) report that a length-to-diameter ratio of 7:1 works well for cylindrical mixing rods used to build reflective star couplers. This information would seem to indicate that cylindrical rods designed to be used in transmissive star couplers should have length-to-diameter ratios of greater than 7:1. Of course, rods that are longer than the minimum length should also provide efficient scrambling action, but they may cause increased absorption losses.

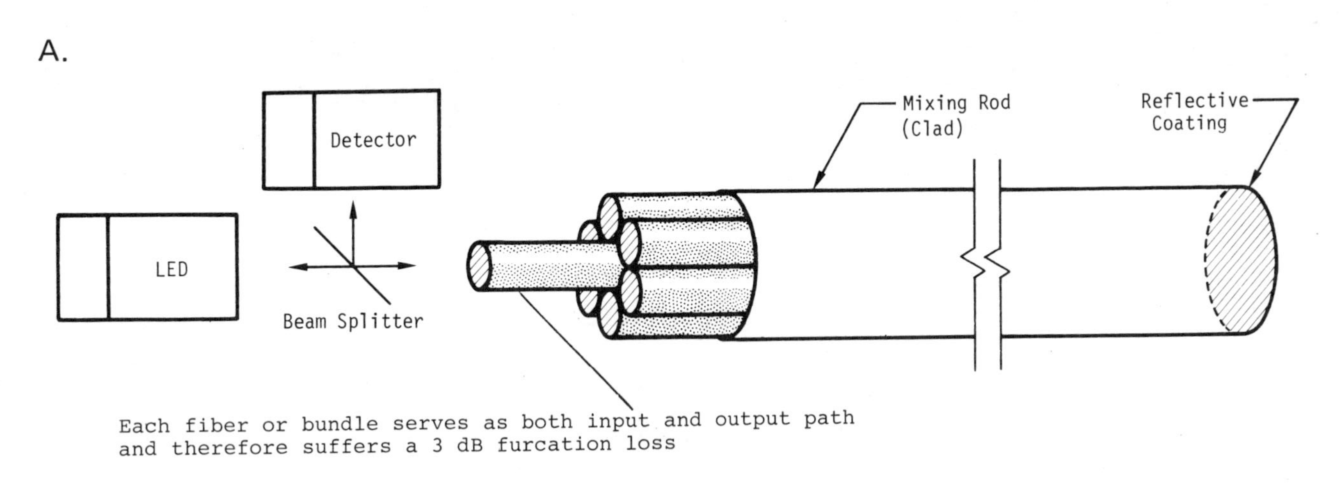

Figure 5.35 Examples of star couplers: (A) reflective and (B) transmissive.

As we did with our analysis of T couplers, let us first examine the limitations that physics imposes on a generalized star coupler.

Our generalized star coupler is a transmissive device because its workings are perhaps easier to visualize, although the analysis of this coupler is applicable also to reflective star couplers. The generalized star coupler is illustrated in Figure 5.36. It is a black box, as was our generalized T coupler, but instead of having four ports, the star coupler has $2N$ ports (i.e., a reflective star would have N ports because each piece of cable used in a reflective star carries both input and output signals). We will initially assume that connection and transmission losses are zero, and we will examine only losses associated with the fractionalization of the input energy.

We assume that the input ports are located on the left-hand side of the coupler illustrated in Figure 5.36, but the symmetry of the coupler indicates that the position of input and output transducers can be interchanged; star couplers are inherently bidirectional. We also assume that at one of the input ports is placed either a fiber bundle or a single fiber carrying light energy. If the coupler has been properly designed, this energy will exit the output face evenly distributed. At the output face, N bundles or fibers are butt-coupled to the star coupler. If the ray angles of the input light have not been transformed in such a way that a portion of the energy falls outside the numerical aperture of the output fibers and if the packing fraction is unity, then each output fiber or bundle receives a fraction $1/N$ of the input energy. If the number of output ports is doubled, then each output fiber receives a fraction $1/2N$ of the input energy. Therefore, by doubling the number of ports while keeping all other parameters constant, the loss between any two terminals is increased by only 3 dB.

For the idealized star coupler which we have been discussing, the losses between any two terminals are determined only by the number of terminals served by the coupler. In reality, however, packing fraction losses, absorption losses, connection losses, cable losses, and losses due to less than optimum

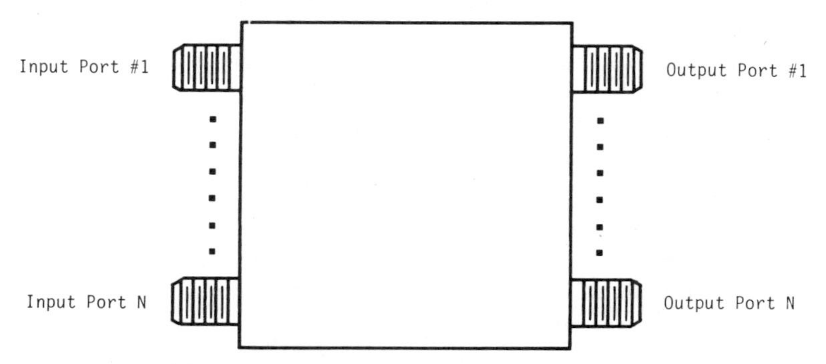

Figure 5.36 Generalized star coupler.

spatial and angular distribution of incident energy will contribute to the losses incurred in the coupling of any two terminals through a transmissive star coupler. Light transmitted through a reflective star coupler suffers all of the above losses plus an additional 3 dB due to the bifurcation needed to allow input and output transmission over the same paths. Losses such as these can be likened to the throughput losses of T couplers, while the $1/N$ loss corresponds to the tap ratio loss of T couplers. The advantage of star couplers is that losses are suffered only once as energy is transmitted between any two terminals, whereas with the T configuration losses are incurred at every coupler linked in series.

Comparison of T and Star Couplers

The above discussion indicates that there are a number of T and star configurations, with much variation even within the various categories of T and star couplers. T couplers that use bifurcation and mixing rods can be used with fiber bundles but have thus far exhibited high insertion losses. Beam-splitter and lens-type T couplers can be used with single fibers or bundles but usually require strict alignment tolerances. T couplers that use evanescent wave coupling show great promise for use with single fibers, but it must be demonstrated that the relationship between coupler parameters and interaction length is understood well enough that couplers with specific properties can be repeatably manufactured. Star couplers, although usable with both single fibers and fiber bundles, require low connection and packing fraction losses to achieve optimum performance.

We have seen that data buses that employ T couplers need far less cable than systems using dedicated lines. Such data buses, however, pay for this decreased usage of cable through an increased vulnerability to breakage. A break at any point on a series data bus will cut communications between terminals on opposite sides of the break. A data bus that uses a star coupler, on the other hand, will be little affected by the breakage of any one line. The only effect would be cessation of communications between that terminal and the others, but communications between all other terminals would not be affected. However, such a system does use more cable. It should also be obvious that damage to the coupler itself would render the bus inoperative.

The breakage differences between T- and star-coupled systems listed above may be important for military applications, but for the majority of applications the losses incurred by light passing through the system are probably used as the criterion for choosing between the systems. Let us therefore examine the differences in the losses exhibited by each system.

Data buses using either T or star couplers exhibit some amount of loss due to the use of the couplers. The magnitude of these losses depends on the types of connectors used, the type of cable used, whether the T coupler uses bifurcation and fiber bundles or single fibers, etc. Connection techniques and mixing rod design will continue to improve, and the losses associated with these variables will continue to decrease. The best way to compare couplers is therefore to ignore all of these variables by comparing an idealized single-fiber T coupler with an idealized single-fiber star coupler and by using conservation of energy to analyze the physical limitations of the two systems.

For the T coupler, we assume (a) no excess loss (i.e., that transmission through the coupler is determined only by the tap ratio), and (b) no connector and cable losses.

The first assumption is justified by the fact that excess losses as low as −0.1 dB have already been achieved (50). The second assumption is justified by the fact that we can imagine an experimental setup where all connections between cable and couplers are made with translation stages and index-matching fluid and only short lengths of extremely low-loss cable join the couplers.

For the star coupler, we assume (a) a transmissive star is used so there is no bifurcation loss at input and output, (b) connector and cable losses are zero, (c) transmission and angle-dependent losses due to the mixing rod are zero, and (d) input light is distributed uniformly at the coupler's output.

The use of the first assumption does not affect the reasoning, since specification of a reflective star would only add a constant to the total value of loss. The second assumption can be justified if we assume that all fibers are mechanically aligned to the mixing rod's input and output faces, index-matching fluids are used, and short lengths of low-loss cable join the coupler to input and output transducers. The third and fourth assumptions are harder to justify, but it is expected that as the technology matures, rods with the specified properties will be routinely manufactured.

Using the above-mentioned assumptions, we assume additionally that we transmit a portion t of the signal entering each T coupler. This means that after transmission through the first coupler, a power $P_i(t)$ can enter the second coupler, where P_i is the amount of power that entered the first T coupler. After transmission through the second coupler, the power transmitted is $P_i(t^2)$. Thus, after passage through N couplers, the power propagated is t^N, where P_i has been normalized to unity. The throughput loss due to transmission through N couplers is therefore

$$\text{Throughput loss} = -10 \ \log t^N = -10N \ \log t \tag{5.42}$$

Loss as a function of the number of terminals served is plotted in Figure 5.37; curves for both $t = 0.8$ and $t = 0.9$ are displayed.

We can include packing fraction, connector, and excess losses in the above formulation by assuming that each of these quantities decreases the transmission through the coupler. As we can see by comparing the curves for

$t = 0.8$ and $t = 0.9$, the only significant change is the slope of the curve, whereas the linear dependence of the loss on the number of terminals would remain unchanged.

For a transmissive star system, the amount of power coupled from one input to any output terminal is determined by the ratio of core area of the output fiber to the area of the output endface of the mixing rod. This ratio depends in turn on both the number of output terminals served by the coupler and the packing fraction. To prove this, let us examine the following example:

We are using seven low-loss fibers with core diameters of 85 μm and fiber diameters of 125 μm. With these fibers, we are using a transmissive star coupler with a cylindrical mixing rod with a core diameter of 375 μm. Therefore, the loss due to the ratio of fiber core area to mixing rod core is $-10 \log(85^2/375^2) = -12.89$ dB. However, the loss due to splitting power into seven equal parts is $-10 \log \frac{1}{7} = -8.45$ dB. Furthermore, the loss due to packing fraction is $-10 \log(7 \times 85^2/375^2) = -4.44$ dB. For our example $(-4.44) + (-8.45) = -12.89$, which implies that for our idealized star coupler

$$\text{Total loss of star coupler} = -10\left[\log f_p + \log(1/N)\right] \tag{5.43}$$

In Figure 5.37, we have also plotted the losses associated with a star coupler with a packing fraction loss of zero; this is admittedly unrealistic, but it is illustrative for comparison with the curve for a star coupler with a packing

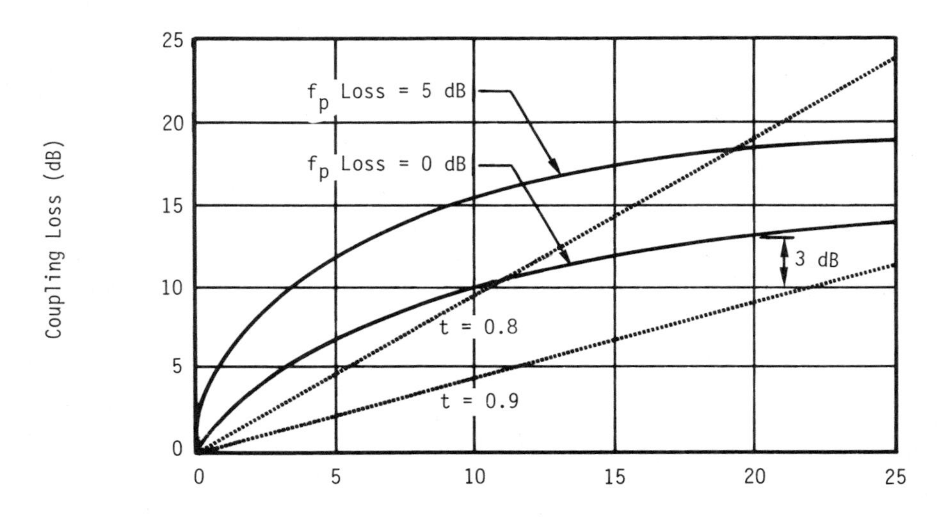

Figure 5.37 T-coupler (- - - - -) and star-coupler (——) losses as a function of number of terminals served by the couplers.

fraction loss of 5 dB. We see that the shapes of the curves are the same; the former curve is simply shifted upward by the amount of the packing fraction loss. Thus, if we include connector losses, packing fraction losses, and excess losses, the curve shifts upward by an amount that depends on these variables. For a T coupler, on the other hand, in addition to this upward shift, the slope of the curve becomes steeper as these additional losses cut down the transmission through each coupler. Simply stated, we are comparing in a linear dependence on the number of terminals with a logarithmic dependence on the number of terminals. At some point, the coupler with the logarithmic dependence must provide lower losses.

The actual crossover point at which a star coupler will begin to outperform a number of T couplers will be determined by the magnitudes of the losses that we have ignored in our idealized couplers. Given a knowledge of the magnitude of the losses that we have ignored, an analysis similar to that performed above would allow us to predict the crossover point.

SWITCHES

In the introduction to this chapter, we stated that optical switching techniques were developed in parallel with the development of the laser. We also stated that the bulk optical switches already developed needed only to be scaled down to become useful elements of integrated optical communications systems. We did not mean to imply, however, that this was an easy task. Quite to the contrary, the task has proven formidable. A number of research laboratories have been pursuing research into integrated optical switches for a number of years, yet it still does not seem likely that such switches will be commercially available for some time.

What we therefore hope to do here is to provide some insight into how an optical waveguide switch should work. We will thereby provide the reader with some criteria by which optical switches can be judged once they do become available. We will also discuss some of the switching schemes that are likely candidates for commercial development.

Definition

Before proceeding with the remainder of our discussion of optical switching, the concept of the optical waveguide switch has to be defined. An optical waveguide switch could be a device that turns the power in a waveguide on and off, but this function is most simply performed electrically at the light source itself. The more general definition of an optical waveguide switch is a device for diverting the flow of power from one optical waveguide to any one of a number of adjacent optical waveguides.

Although there are variations in each category, there are three basic ways that optical signals may be switched from one optical channel to another.

Mechanical Switching. The first method is the simplest and perhaps the least useful. A beam-diverting mirror, prism, or other mechanical device may be physically placed in the path of the beam to be deflected. The insertion and extraction of the beam diverter may be controlled electromechanically, but the presence of moving parts implies that switching speeds are slow. Although there are some special applications for such low-speed switches, most conceivable communications and data processing applications will require faster switching speeds.

Electronic Switching. To obtain faster switching speeds, an already developed technology can be used. An optical signal can be converted into an electrical signal by a photodetector. This electrical signal can then be switched by transistor to another communications channel, where it would be reconverted into an optical signal. There seems to be no reason why such switching methods cannot be made to switch signals nearly as rapidly as the light source can be "chopped" at the transmitting end of the system. If electronic circuits can be made to switch the light source on and off very rapidly, then similar types of circuits should be capable of switching the converted optical signal rapidly. This switching option may be viable in the short run but will probably not remain competitive if efficient methods of switching light itself are developed, all known electrical-to-optical conversion techniques have limits on their conversion efficiencies. Thus, excessive losses are incurred with the use of this type of switch.

Optical Switching. Switches that utilize the deflection of the optical power itself therefore represent the third basic type of optical waveguide switch, and they are the topic of interest in the remainder of this chapter. In systems using such switches, conversion of optical power to electrical power or vice versa is accomplished only at the system inputs and outputs. There are a number of conceivable ways of constructing such switches; before we review them, let us describe what properties an optical waveguide switch should have.

Optical Waveguide Switches

Before preceeding with our discussion, we should first define several terms. Refer to the two-channel switch illustrated in Figure 5.38. As with the T coupler discussed earlier, it is a four-port device. Since we are at this point not yet concerned with its internal workings, it is displayed as a voltage-controlled black box.

To illustrate how an ideal switch would work, let us assume that an optical signal is injected at port 1. If no voltage is applied to the switch, then all of the injected power is emitted at port 3. If, on the other hand, the proper voltage is applied to the switch, the entire injected power is emitted at port 4. Whereas this describes the operation of an ideal two-channel switch, optical switches behave differently in practice. We must therefore consider the behavior of a practical two-channel switch. It would normally be designed to operate in approximately the same manner as the ideal switch, but its performance would fall short of the ideal in several respects. The two most important differences concern power loss and crosstalk.

Power Loss. The first point of difference between practical and ideal switches is that with no voltage applied to the practical switch, not all of the power injected at port 1 is emitted at port 3. This is because a "real" switch attenuates the signal transmitted through it. This attenuation is caused by the absorption and scattering losses incurred in the optical material used in the construction of the switch, and it should be less than about 1 dB/cm for the switch to be useful (66).

Crosstalk. The second point of difference between practical and ideal switches is that if the power dissipation in the practical switch were added to the power emitted at port 3 of that switch, the total power would not necessar-

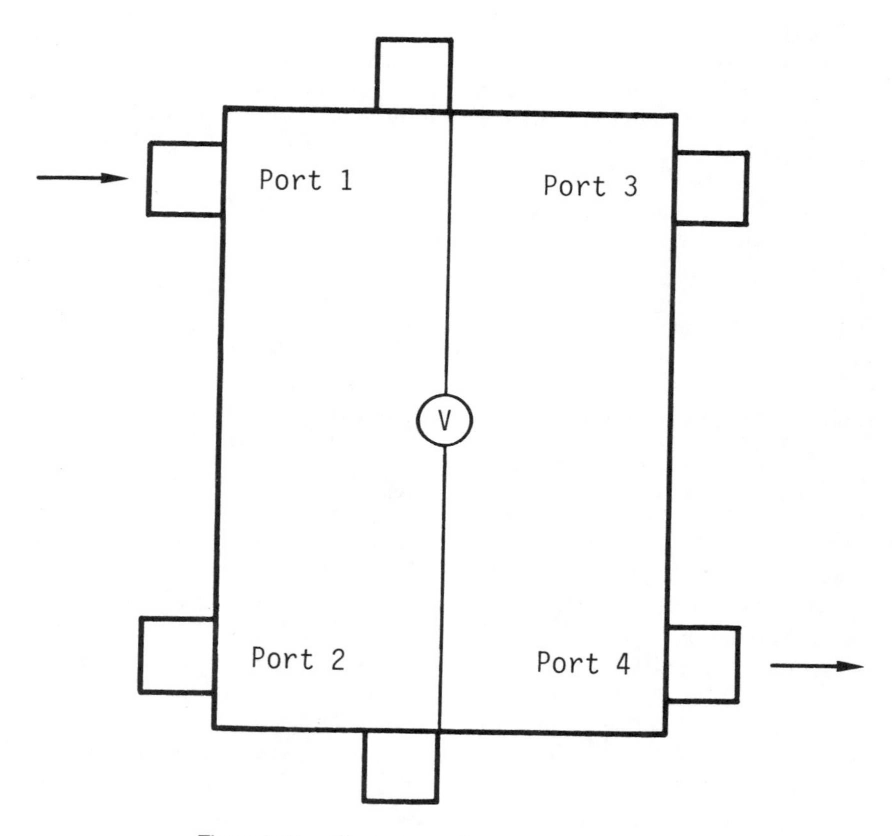

Figure 5.38 Illustration of a two-channel switch.

ily equal the power injected at port 1. The power that is neither emitted at port 3 nor absorbed in the switch can be emitted at port 4. Such power is termed off-state crosstalk, and it is clearly undesirable. Similarly, if power is emitted at port 3 while voltage is applied to the switch and power is injected at port 1, the power emitted by port 3 is called on-state crosstalk. If crosstalk cannot be totally eliminated, a switch can be designed to at least keep crosstalk below some maximum allowable level. This level can be set arbitrarily low if power consumption and expense are of no concern, but the level chosen will often represent a compromise between the need to minimize crosstalk and other considerations.

Since the idealized two-channel switch is an unnecessarily restrictive example, let us consider what properties a more generalized optical waveguide switch might possess:

Low value of crosstalk in both on and off states.

Combined absorption and scattering losses of less than 1 dB per cm of switch length.

Compatibility with cylindrical optical waveguides of both single- and multimode designs.

Capability of switching to a fairly large number of channels.

Capability of switching rapidly, which in turn imposes a limit on the value of switch capacitance.

Reasonable voltage and power requirements.

Ease of fabrication.

Bidirectionality.

Of course, all of these properties need not be possessed by every switch. If, for example, a switch is designed for use in a single-mode system, it would not be necessary for the switch to be compatible with multimode fibers. However, certain of the above-mentioned properties should be possessed by all optical waveguide switches. We will examine some of these characteristics in more detail.

Optical waveguide switches are usually constructed as films of high refractive index that are either deposited upon a substrate or sandwiched between a substrate and a superstrate of lower refractive indexes. Such waveguiding layers may confine the light in one transverse dimension (i.e., planar waveguides) or in both transverse dimensions (i.e., channel waveguides). We have already mentioned how a switch could be constructed from parallel single-mode channel waveguides (67). Let us consider the operation of such a switch in greater detail.

Consider the waveguide switch illustrated in Figure 5.39. It consists of two single-mode channel waveguides that are formed parallel to each other on some optical substrate. We have already seen that if the waveguides are placed parallel to each other for the proper interaction length L, then power will couple completely from the intially excited channel to the initially unexcited channel. Suppose, however, that instead of being formed in a passive optical material, the waveguides are formed in an optically active material such as $LiNbO_3$ or $LiTaO_3$. If waveguides of the proper length were formed in such material, power would transfer completely from the excited to the unexcited channel when no voltage is applied. If voltage were applied, the refractive index of the cladding material would change (68), thereby destroying the coupling between the channel waveguides. Thus, we would have a two-channel switch that would either allow a signal to be transmitted undisturbed or would switch it to the other channel, depending on whether or not voltage was applied to the switch.

There are several problems associated with the construction of such a switch. First, the coupling length for complete energy transfer between channels has a precise value. Deviations from the value cause the switch action to be less than 100% efficient; i.e., crosstalk occurs. This need for a precisely determined coupling region implies that fabrication of a zero-crosstalk switch would be difficult. This is not the only problem encountered in switch design; polarization also causes crosstalk (69–71). Let us examine why.

Even in single-mode fibers, there are two orthogonal polarizations that may propagate: the TE and TM modes. The problem that arises in switch construction is that in optically active materials the propagation constants for TE and TM modes are generally different. Therefore, in calculating the

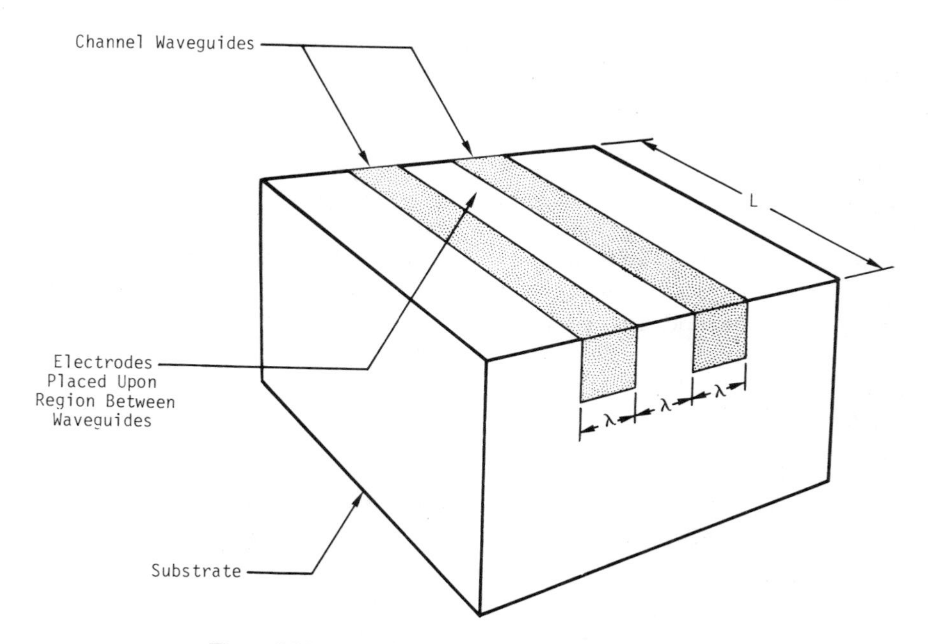

Figure 5.39 Possible two-channel switch design.

proper interaction length for 100% coupling efficiency, one must optimize for either TE or TM propagation. However, this implies that one of the two polarizations will not switch with 100% efficiency, thereby introducing crosstalk in both on and off states. This problem is serious enough for one group of researchers to state that "integrated optical switches demonstrated to date are incompatible with glass fiber transmission lines" (70).

One would believe that polarization effects should not be a problem in single-mode fibers, since such fibers theoretically have the ability to transmit linearly polarized light. However, core ellipticity and stress in "real" fibers cause linearly polarized light that is injected at the fiber input to be transformed into elliptically polarized light by the time it is emitted by the fiber (72). Of course, one can inject linearly polarized light into the switch by placing a polarizer between the fiber and the switch, but this would introduce an average loss of 3 dB above and beyond the coupling and throughput losses normally incurred. Several researchers have therefore tried to minimize these polarization-induced problems through other means.

Figures 5.40 and 5.41 will help illustrate how polarization effects in switches may be overcome without sacrificing 50% of the input light. In Figure 5.40 the effective refractive indexes of TE and TM modes in $LiNbO_3$ are plotted as a function of waveguide thickness. For a single-mode switch, only the TE_0 and TM_0 modes can propagate; the difference in their propagation constants is clearly illustrated. If we compare Figures 5.40A and 5.40B, we see a marked difference. In Figure 5.40B there are certain values of waveguide thickness for which the TE and TM propagation constants are the same: they are phase matched. This difference between the curves in Figures 5.40A and 5.40B is due to the fact that they represent propagation at 0° and 4° with respect to the optic axis. Figure 5.41 shows that each mode number requires a different propagation angle if a difference in propagation constants of zero is to be attained. Several researchers (69,70,73) have suggested that these facts be used to provide polarization insensitivity and ease of fabrication simultaneously. By cutting the crystal substrate at the proper angle, the tolerances on guide fabrication are loosened and polarization-induced crosstalk is minimized. Another suggestion (71) is to use a multiple-electrode design that provides alternating regions of opposing electric fields. Fine tuning of the switch is thereby facilitated and crosstalk reduced.

We have already spoken directly of the need for low values of crosstalk in optical waveguide switches and indirectly of the problems of switch fabrication and compatibility with glass fiber transmission lines. However, other characteristics that an optical waveguide switch should have must also be considered.

The growing of crystals large enough to be used as substrates presents a fabrication problem. Furthermore, the materials with the best electrooptic

A.

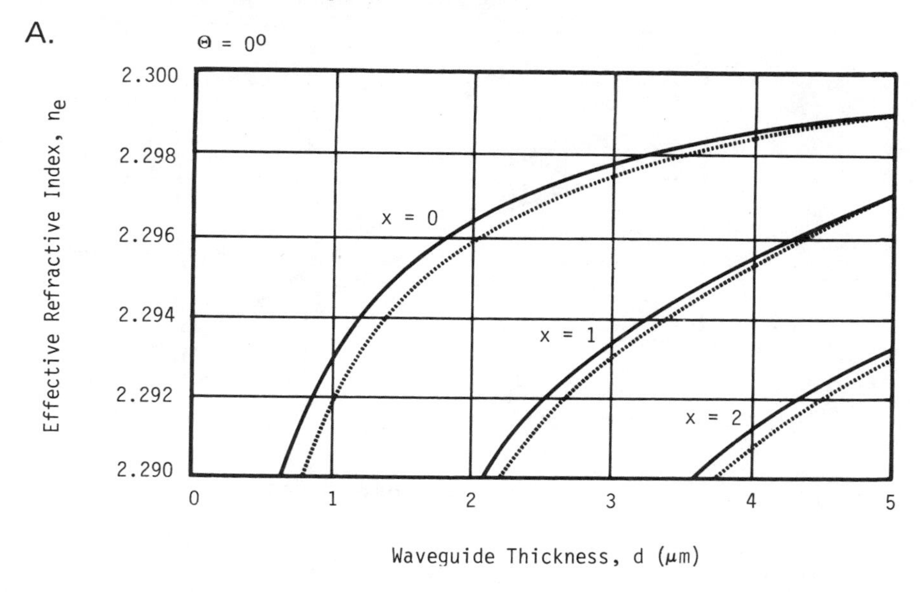

B.

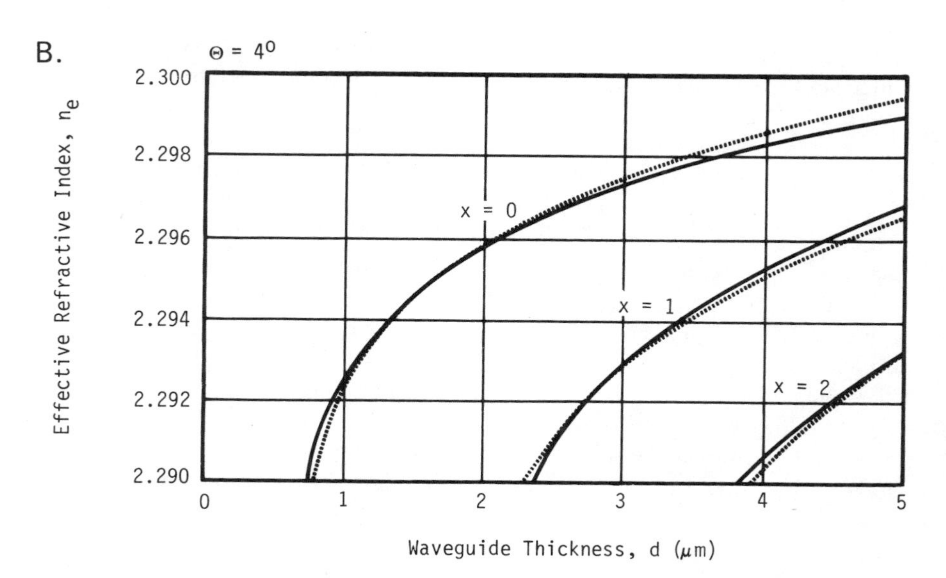

Figure 5.40 Effective refractive index as a function of waveguide thickness for light propagating at angles of 0° or 4° off the optical axis ($\lambda = 0.633\ \mu m$) (77). Solid line, TE_x mode; $n_e = \beta/k_0$. Dotted line, TM_x mode.

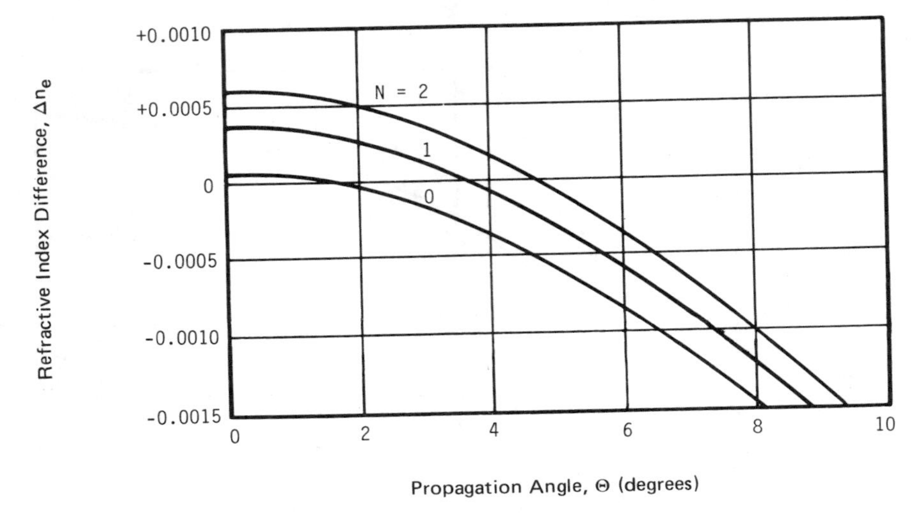

Figure 5.41 Variation of effective refractive index caused by propagation off the optical axis (77); $\Delta h_e = (\beta_{\mathrm{TEN}} - \beta_{\mathrm{TMN}})/K_0$.

properties and absorption spectra may not be used if materials that are less expensive to use and easier to handle prove to be fair substitutes. GaAs may prove a useful material from which to fabricate switches (75) because of the large body of knowledge that exists about this material. As for switch capacitance and voltage and power requirements, computer studies and experiments (74–78) have indicated that capacitances in the 10 to 100 pF range, voltages in the 10 to 100 V range, and powers in the 1 to 10 mW/MHz range can be expected. In terms of these three parameters, thin-film switches enjoy large advantages over their bulk predecessors. Therefore, there is reason to expect that optical waveguide switches may eventually become common elements of integrated circuits.

The final switch characteristic to be discussed is the ability to switch a signal to N locations, where N is greater than 2. Figure 5.42 illustrates how a number of two-channel switches may be ganged to provide switching to 2^n locations, where n is the number of switch elements that a particular signal traverses in its transit through the composite switch. Such a switching network will perform as described, but it has a number of inherent disadvantages, such as high power consumption and low speed due to transit through a number of switching stages. If the concept is taken to extremes, the switch may actually become impractical. Therefore, it is desirable to have a single switch that is capable of diverting an input beam to any one of a number of locations. We will discuss how this can be accomplished, at least in theory, but let us first imagine that it can be accomplished. How many locations can be served by such a switch?

We know that due to the small size of waveguides light diffracts as it exits a waveguide switch. If the maximum diffraction half-angle is Θ_m and the maximum deflection angle caused by the switch is Θ_D, then the maximum number of spots that can be resolved (or the maximum number of locations to which the beam can be switched) is (77)

$$N = \Theta_D/\Theta_m \tag{5.44}$$

Such a switch would allow the input beam to be deflected through discrete angles between Θ and Θ_D. The deflection angle would be determined by the electrical input.

There are a number of ways in which a deflection-type switch could be constructed. We will examine several of them.

REFRACTIVE SWITCH

A refractive-type switch can be constructed from electrooptic material if a special electrode design (78) is used. The basic idea behind the design of this switch is that an optical beam can be refracted by an electrooptic crystal if a

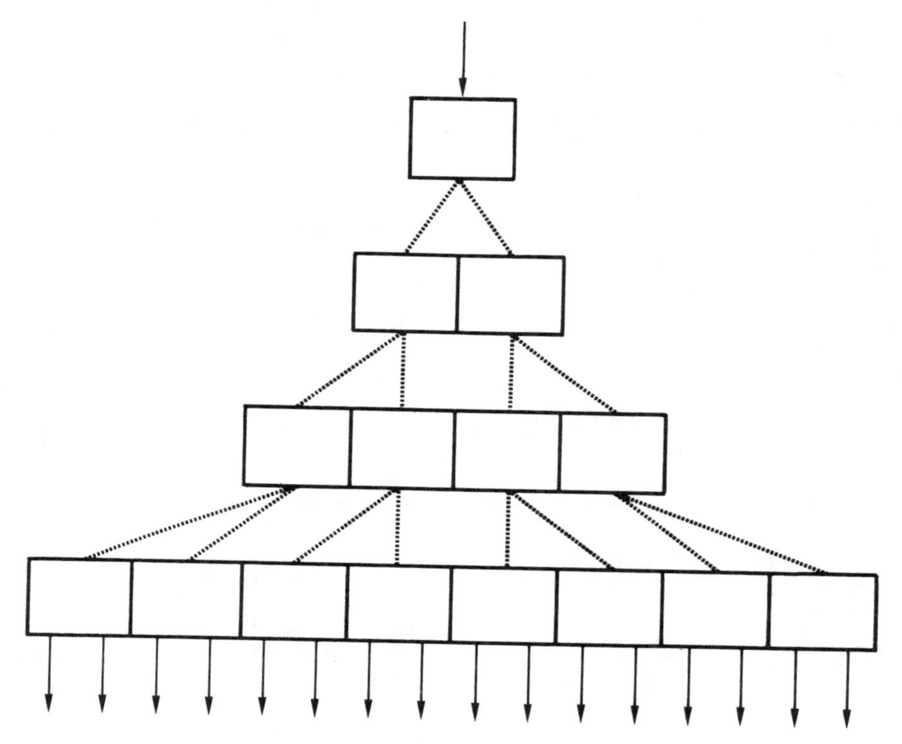

Figure 5.42 Illustration of the switching of a signal to N locations.

prismlike region can be induced in the crystal. The prismlike region is illustrated in Figure 5.43. The beam deflection angle is controlled by the applied voltage. If the design of the electrodes and their polarities are properly chosen, regions of refractive index varying by $\pm \Delta n$ around the refractive index's zero voltage value n can be created. The total deflection angle can be increased by inducing a number of prismlike regions in series.

Because of the problems encountered in coupling single-mode fibers, Soref et al. (77) have examined how a refractive design might perform as a multimode switch. They demonstrated that a multimode input beam could be resolved into two spots by the switch if the input fiber had a small numerical aperture and a small core diameter (i.e., propagated few modes). However, because of the multimode (i.e., thick film) design of the switch, the voltage needed for switching was $\pm$ 650 V, which is too high to be of practical utility. Switching rates of 10^7 to 10^8 s^{-1} were anticipated.

DIFFRACTIVE SWITCH

Diffractive-type switches (77,79–81) are among the most promising types of potential waveguide switches. A Bragg diffraction switch is shown in Figure 5.44. This type switch uses either the acoustooptic (68,73) or the electrooptic effect to diffract light through the required angle. If the electrooptic effect is used, the light is diffracted by regions of discontinuity in the refractive index of the crystal. This gratinglike effect is caused by the electric fields applied to the crystal by specially designed electrodes. Such electrodes are described by Hammer (81). When the applied field is zero, the beam propagates undeflected; when the proper voltage is applied, this input beam is diffracted at the Bragg angle Θ_B. The angle is determined by the periodicity Λ and the length l of the electrode-induced discontinuity. In the small-angle approximation, the Bragg angle is given by (77)

$$\Theta_B = \lambda/2n\Lambda \tag{5.45}$$

where λ is the wavelength of the input light.

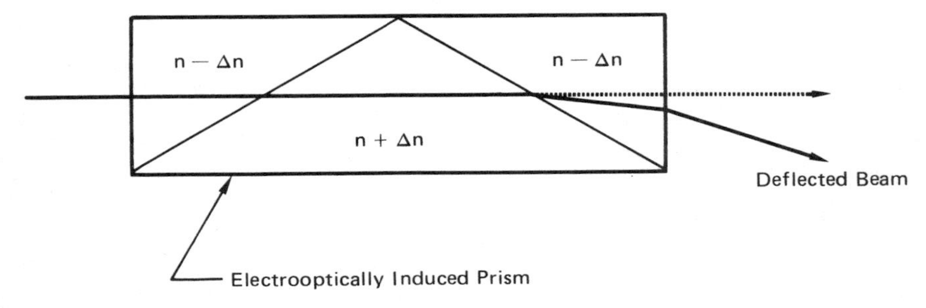

Figure 5.43 Illustration of a switch design using prismlike regions.

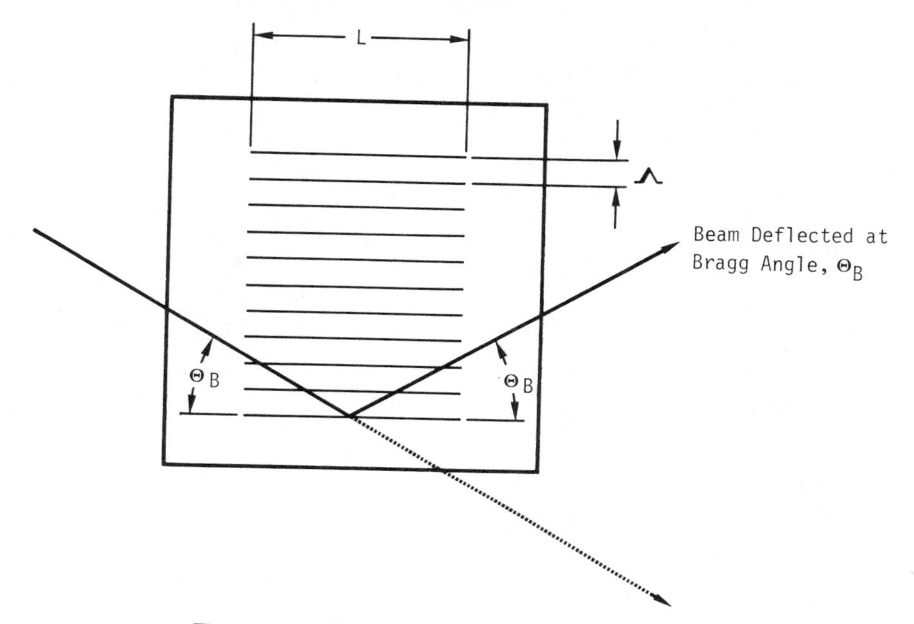

Figure 5.44 Illustration of a Bragg diffraction switch.

Bragg diffraction switches can also be constructed from photoelastic materials such as Ta_2O_5. In such a case, the light interacts with a traveling sound wave of wavelength Λ. This indicates that a variety of deflection angles are attainable by altering the frequency of the sonic transducer. A multiple-input, multiple-output switch using this concept has already been patented (80).

The Future

We have discussed the basic concepts needed to understand optical waveguide switching, but we have avoided speaking in too much detail about any particular switching technique. As far as is known, there are presently no commercially available optical waveguide switches. While many of the experiments performed thus far indicate that the construction of such switches in integrated format is feasible, substantial work remains to be done.

In the future, monomode switches made from $LiNbO_3$ or $LiTaO_3$ may be readily available. Switches made from more exotic materials such as $Sr_{1-x}Ba_xNb_2O_6$, which has the largest refractive index changes per unit applied voltage, may become available if problems associated with the fabrication of such crystals can be overcome. If multimode graded-index fiber systems develop in parallel with single-mode systems instead of being made obsolete by them, multimode switches (76,77) made from such materials as

CdTe, GaAs, and GaP may also be developed. At this point, it is too early to predict which type of switch will dominate; certainly, concepts not yet conceived are distinct possibilities.

REFERENCES

1. K.H. Yang & J.D. Kingsley. Calculation of Coupling Losses Between Light Emitting Diodes and Low Loss Optical Fibers. *Appl. Opt.* **14**(2):288, 1975.

2. M.L. Dakss. Coupling Light Sources to Fibers. *Laser Focus* 31, Dec. 1975.

3. R.L. Gallawa. *A User's Manual for Optical Waveguide Communications*, Office of Telecommunications, OT Report 76 (PB 252 901) Washington, D.C., 1976.

4. M.K. Barnoski. *Fundamentals of Optical Fiber Communications.* New York, Academic Press, 1976, pp. 92–93.

5. D. Marcuse. Excitation of Parabolic-Index Fibers With Incoherent Sources. *Bell Sys. Tech. J.*, **54**(9):1507, 1975.

6. C.C. Timmermann. Highly Efficient Light Coupling From GaAlAs Lasers Into Optical Fibers. *Appl. Opt.* **15**(10):2432, 1976.

7. K. Miyazaki et al. Theoretical and Experimental Considerations of Optical Fiber Connector. OSA Topical Meeting on Optical Fiber Transmission, p. WA4–1 Williamsburg, Va., January 1975.

8. H. Tsuchiya et al. Double Eccentric Connectors for Optical Fibers. *Appl. Opt.* **16**(5):1323, 1977.

9. C.M. Miller. Transmission Vs. Transverse Offset for Parabolic-Profile Fiber Splices With Unequal Core Diameters. *Bell Sys. Tech. J.* **55**(7):917, 1976.

10. D. Gloge. Offset and Tilt Loss in Optical Fiber Splices. *Bell Sys. Tech. J.* **55**(7):905, 1976.

11. S.C. Mettler & C.M. Miller. A Gaussian Power Model for Optical Fiber Splices. Conference on Laser and Electro-Optical Systems, San Diego, February 1978.

11a. C.M. Miller & S.C. Mettler. A Loss Model for Parabolic-Profile Fiber Splices. *Bell Sys. Tech. J.* **57**(9):3167, 1978.

12. D. Marcuse. Loss Analysis of Single-Mode Fiber Splices. *Bell Sys. Tech. J.* **56**(5):703, 1977.

13. H. Kogelnik. Coupling and Conversion Coefficients for Optical Models in Quasi-Optics. *Microwave Research Institute Symposia Series, 14.* Polytechnic Press, New York, 1964, pp. 333–347.

14. J.A. Arnaud. *Beam and Fiber Optics.* New York, Academic Press, 1976.

15. J.S. Cook et al. Effect of Misalignment on Coupling Efficiency of Single-Mode Fiber Butt Joints. *Bell Sys. Tech. J.* **52**(8):1439, 1973.

16. K.J. Fenton & R.L. McCartney. Connecting the Thread of Light. Electronic Connector Study Group Symposium, 9th Annual Symposium Proceedings, p. 63, Cherry Hill, N.J., October 1976.

17. T.C. Chu & A.R. McCormick. Measurements of Loss Due to Offset, End Separation, and Angular Misalignment in Graded-Index Fibers Excited by an Incoherent Source. *Bell Sys. Tech. J.*, **57**(3):595, 1978.

18. F.L. Thiel & R.M. Hawk. Optical Waveguide Cable Connection. *Appl. Opt.* **15**(11):2785, 1976.

19. D. Gloge & E.A.J. Marcatili. Multimode Theory of Graded-Core Fibers. *Bell Sys. Tech. J.* **52**(9):1563, 1973.

20. H. Hasegawa et al. Separable Optical Fiber Connector and Method of Manufacturing the Same. U.S. Patent 3,861,781, Jan. 21, 1975.

21. A. Bridger et al. Optical Fiber Connector. U.S. Patent 3,871,744, March 18, 1975.

22. D. Schicketanz. Connecting Plug for Optical Glass Fibers. U.S. Patent 3,870,395, March 11, 1975.

23. J.F. Dalgleish et al. Connectors for Optical Fibers. U.S. Patent 3,972,585, Aug. 3, 1976.

24. O. Nakayama. Connector for Light-transmitting Cables. U.S. Patent 3,982,815, Sept. 28, 1976.

25. J.F. Dalgleish et al. Method of Connecting Two Optical Fibers in End to End Relationship. U.S. Patent 4,005,522, Feb. 1, 1977.

26. J.F. Dalgleish et al. Optical Fibre Connectors. U.S. Patent 4,008,948, Feb. 22, 1977.

27. J.F. Dalgleish et al. Optical Fibre Connectors. U.S. Patent 3,885,859, May 27, 1975.

28. C.M. Miller. A Fiber Optic Cable Connector. *Bell Sys. Tech. J.* **54**(9):1547, 1975.

29. J.J. Esposito. A New, Simple Connector for Single Fiber Optical Cables. Electronic Connector Study Group Symposium, 10th Annual Symposium Proceedings, p. 152, Cherry Hill, N.J., October 1977.

30. W.L. Schumacher. Fiber Optic Connector Design to Eliminate Tolerance Effects. Electronic Connector Study Group Symposium, 10th Annual Symposium Proceedings, p. 163, Cherry Hill, N.J. October 1977.

31. R.L. McCartney. Single Optical Fiber Connector. U.S. Patent 3,990,779, Nov. 9, 1976.

32. R.L. McCartney. Optical Coupler. U.S. Patent 3,995,935, Dec. 7, 1976.

33. S. Zemon et al. Eccentric Coupler for Optical Fibers: A Simplified Version. *Appl. Opt.* **14**(4):815, 1975.

34. R.E. Love & F.L. Thiel. Optical Communications System. U.S. Patent 3,883,217, May 13, 1975.

35. A.F. Milton. Multiple Optical Connector. U.S. Patent 3,936,141, Feb. 3, 1976.

36. A.F. Milton. Optical Connector With Single Scrambling Volume. U.S. Patent 3,933,410, Jan. 20, 1976.

37. L.W. Brown. Optical Access Coupler for Fiber Bundles. U.S. Patent 3,902,786, Sept. 2, 1975.

38. A.F. Milton. Mirrored Optical Connector. U.S. Patent 3,901,582, Aug. 26, 1975.

39. A.F. Milton & A.B. Lee. Optical Access Couplers and a Comparison of Multiterminal Fiber Communications Systems. *Appl. Opt.* **15**(1):244, 1976.

40. J.G. Racki & F.L. Thiel. Optical Coupler. U.S. Patent 3,870,396, March 11, 1975.

41. F.L. Thiel. Variable Ratio Light Coupler. U.S. Patent 3,874,779, April 1, 1975.

42. R.E. Love. Passive Coupler for Optical Communications System. U.S. Patent 3,870,398, March 11, 1975.

43. A.F. Milton. Single Fiber Access Coupler. U.S. Patent 3,937,560, Feb. 10, 1976.

44. T. Hawkes & J.C. Reymond. Coupler for Optical Communications System. U.S. Patent 4,011,005, March 8, 1977.

45. H. Kuwahara et al. A Semi-transparent Mirror-type Directional Coupler for Optical Fiber Applications. *IEEE Trans. Microwave Theory Tech.* **23**(1):179, 1975.

46. Y. Suzuki & H. Kashiwagi. Concentrated-type Directional Coupler for Optical Fibers. *Appl. Opt.* **15**(9):2032, 1976.

47. N.S. Kapany. *Fiber Optics—Principles and Applications.* New York, Academic Press, 1967, p. 73.

48. M.K. Barnoski & H.R. Friedrich. Fabrication of an Access Coupler with Single-Strand Multimode Fiber Waveguides. *Appl. Opt.* **15**(11):2679, 1976.

49. T. Ozeki & B.S. Kawasaki. *Appl. Phys. Lett.* **28**:528, 1976.

50. B.S. Kawasaki & K.O. Hill. Low-Loss Access Coupler for Multimode Optical Fiber Distribution Networks. *Appl. Opt.* **16**(7):1794, 1977.

51. H. Kuwahara et al. Power Transfer of a Parallel Optical Fiber Directional Coupler. *IEEE Trans. Microwave Theory Tech.* **23**(1):178, 1975.

52. H.F. Taylor. Passive Frequency-selective Optical Coupler. U.S. Patent 3,957,341, May 18, 1976.

53. S.E. Miller. Directional Optical Waveguide Coupler. U.S. Patent 4,019,057, April 19, 1977.

54. M.M. Ramsay. Method of Making Optical Fiber Optical Power Divider. U.S. Patent 4,008,061, February 15, 1977.

55. H.P. Hsu & A.F. Milton. Single Mode Optical Fiber Pickoff Coupler. *Appl. Opt.* **15**(10):2310, 1976.

56. H. Fujita et al. Optical Fiber Wave Splitting Coupler. *Appl. Opt.* **15**(9):2031, 1976.

57. K. Owaga. Simplified Theory of the Multimode Fiber Coupler. *Bell Sys. Tech. J.* **56**(5):729, 1977.

58. A.W. Snyder. Coupled-Mode Theory of Optical Fibers. *J. Opt. Soc. Am.* **62**:1267, 1972.

59. A.W. Snyder & P. McIntyre. Crosstalk Between Light Pipes. *J. Opt. Soc. Am.* **66**:877, 1976.

60. M.C. Hudson & F.L. Thiel. The Star Coupler—A Unique Interconnection Component for Multimode Optical Waveguide Communications Systems. *Appl. Opt.* **13**(11):2540, 1974.

61. R.E. Love. Coupler for Optical Communications System. U.S. Patent 3,874,780, April 1, 1975.

62. M.C. Hudson. Coupler for Optical Communications System. U.S. Patent 3,883,223, May 13, 1975.

63. F.L. Thiel. Coupler for Optical Communications System. U.S. Patent 3,874,781, April 1, 1975.

64. F.L. Thiel. Tapered Coupler for Optical Communications System. U.S. Patent 3,901,581, Aug. 26, 1975.

65. A.F. Milton. Star Coupler for Single Mode Fiber Communications Systems. U.S. Patent 3,937,557, Feb. 10, 1976.

66. I.P. Kaminow. Optical Waveguide Modulators. *IEEE Trans. Microwave Theory Tech.* **23**(1):57, 1975.

67. E.A.J. Marcatili. Dielectric Rectangular Waveguide and Directional Coupler for Integrated Optics. *Bell Sys. Tech. J.* **48**(7):2071, 1969.

68. A. Yariv. *Quantum Electronics.* New York, John Wiley & Sons, 1975, pp. 327–366.

69. R.A. Steinberg & T.G. Giallorenzi. Performance Limitations Imposed on Optical Waveguide Switches and Modulators by Polarization. *Appl. Opt.* **15**(10):2440, 1976.

70. R.A. Steinberg & T.G. Giallorenzi. Design of Integrated Optical Switches for Use in Fiber Data Transmission Systems. *IEEE J. Quantum Electron.* **13**(4):122, 1977.

71. R.A. Steinberg et al. Polarization-insensitive Integrated-optical Switches: A New Electrode Design. *Appl. Opt.* **16**(8):2166, 1977.

72. W.A. Gambling et al. Birefringence and Optical Activity in Single Mode Fibers. OSA Topical Meeting on Optical Fiber Transmission, Williamsburg, Va., February 1977.

73. P.L. Adams et al. Acousto-optic Interactions. Stanford University G.L. Report 2523 (July 1976).

74. V. Ramaswamy & R.D. Standley. A Phased, Optical, Coupler-Pair Switch. *Bell Sys. Tech. J.* **55**(6):767, 1976.

75. K.W. Loh et al. Bragg Coupling Efficiency for Guided Acoustooptic Interaction in GaAs. *Appl. Opt.* **15**(1):156, 1976.

76. A.R. Nelson & R.A. Soref. Electrooptic Polarization Switch for Multimode Fibers. *Appl. Opt.* **16**(1):119, 1977.

77. R.A. Soref et al. Integrated Optical Switches. Sperry Research Center Report SCRC-CR-75-12 (July 1975).

78. I.P. Kaminow & L.W. Stulz. A Planar Electrooptic-Prism Switch. *IEEE J. Quantum Electron.* **2E-11**(8):633, 1975.

79. L. Kuhn et al. Deflection of an Optical Guided Wave by a Surface Acoustic Wave. *Appl. Phys. Lett.* **17**:265, 1970.

80. M.L. Dakss. Optical Switch. U.S. Patent 3,990,780, Nov. 9, 1976.

81. J.M. Hammer. Digital Electro-optic Grating Deflector and Modulator. *Appl. Phys. Lett.* **18**:147, 1971.

6

ECONOMIC ASPECTS

Roger A. Greenwell

In the evaluation of any new system or technology, the two main criteria used to determine its usefulness are performance and cost. Although both cost and performance are usually of prime concern, the required cost for subsystem or system implementation dominates. In most cases, depending upon a particular application, designers strive to optimize the product of cost and performance, whereby the definition of both factors may vary from case to case. This chapter, which deals with the composition, estimation, and reduction of fiber optics systems cost, is intended to provide the reader with an overview of the appropriate cost-evaluation methods. The general discussion will be complemented by two timely examples, taken from the commercial and the military sectors.

INTRODUCTION

Cost is the primary factor which influences a manager's decision on future technological design alternatives. For a program manager in fiber optics, the principal aim of a cost analysis is to estimate the cost of various electrooptical equipment design alternatives that can then be employed to make cost-effective system trade-offs. Since design selections can affect more than the simple cost of purchasing components, it is required that the manager choose a cost analysis methodology which allows for comparison of the alternatives on a broader cost basis.

The selected methodology is also a function of the type of analysis to be performed: absolute analysis, a comparative analysis, or some combination of the two. In an absolute analysis, for example, the important factor is the generation of extremely credible cost estimates used in budgetary reports. In a comparative analysis, the emphasis is on the relative costs of the selected designs, and the need for accuracy is of less concern as long as the cost elements are consistent. In most economic analysis studies, both types of analysis are required.

For fiber optics technology, some of the major economic techniques are cost-effectiveness and cost-benefit analysis methods. These economic methods (generally termed economic analysis) are intended to provide management with priorities and alternatives which substantiate their decisions. Cost elements and measures of effectiveness are acquired from various methods in light of economic criteria; it must be realized that one of the most difficult factors in an economic analysis is the correct choice of economic criteria. The scope of economic analysis is constrained in two ways:

1. Its role is not to effect a comparison of various solutions whose classification is independent of cost (e.g., technical, psychological choices, etc.).

2. Its usefulness ceases when it begins to compare economic with extra-economic variables.

Economic Analysis

The objective of an economic analysis for fiber optics systems is to supply the manager with a justification for allocating resources to meet future systems requirements. The general process involves the identification of costs and benefits of each fiber optics design alternative. After defining objectives, making assumptions, and identifying alternatives, one may select a preferred economic method. The economic choices are the maximization of benefits for a given cost or the achievement of a given performance level at minimum costs.

Figure 6.1 illustrates in a schematic way the economic analysis process. The various major steps required to arrive at a decision as to whether or not a system meets the required economic criteria are indicated. Of these steps, the first one is usually the most important and also normally the least understood: the definition of the objective. Most simply stated, an objective is some fixed standard of accomplishment. In every instance, whether the objective is to maintain logistic support, produce an effective system, or identify an organization capable of functioning in terms of quality, time, or degree, management is required to choose the best course to accomplish the task. The analysis process can be categorized according to its impact on the objective of an organization.

Once the objective of a fiber optic program has been defined, the next step is to determine all feasible alternatives which meet the objective and given criteria. It is the economist's task to study all feasible alternatives and to present to the decision maker those alternatives which are most beneficial. This requires a systematic approach on the part of the analyst and a certain amount of interaction with the decision maker so that the correct a priori judgments can be made. Often, one who performs an economic analysis is directed to select alternatives in keeping with certain constraints, e.g., manpower, facilities, or funding limitations. This in itself often eliminates many alternatives. Despite the a priori rejection of some alternatives, it is only through the reiteration of the analysis for many alternatives that the analyst may be able to formulate and justify the final recommendations.

Assumptions are statements made to justify and reasonably constrain the level of the effort. Because an assumption is a "given" as opposed to a "fact"

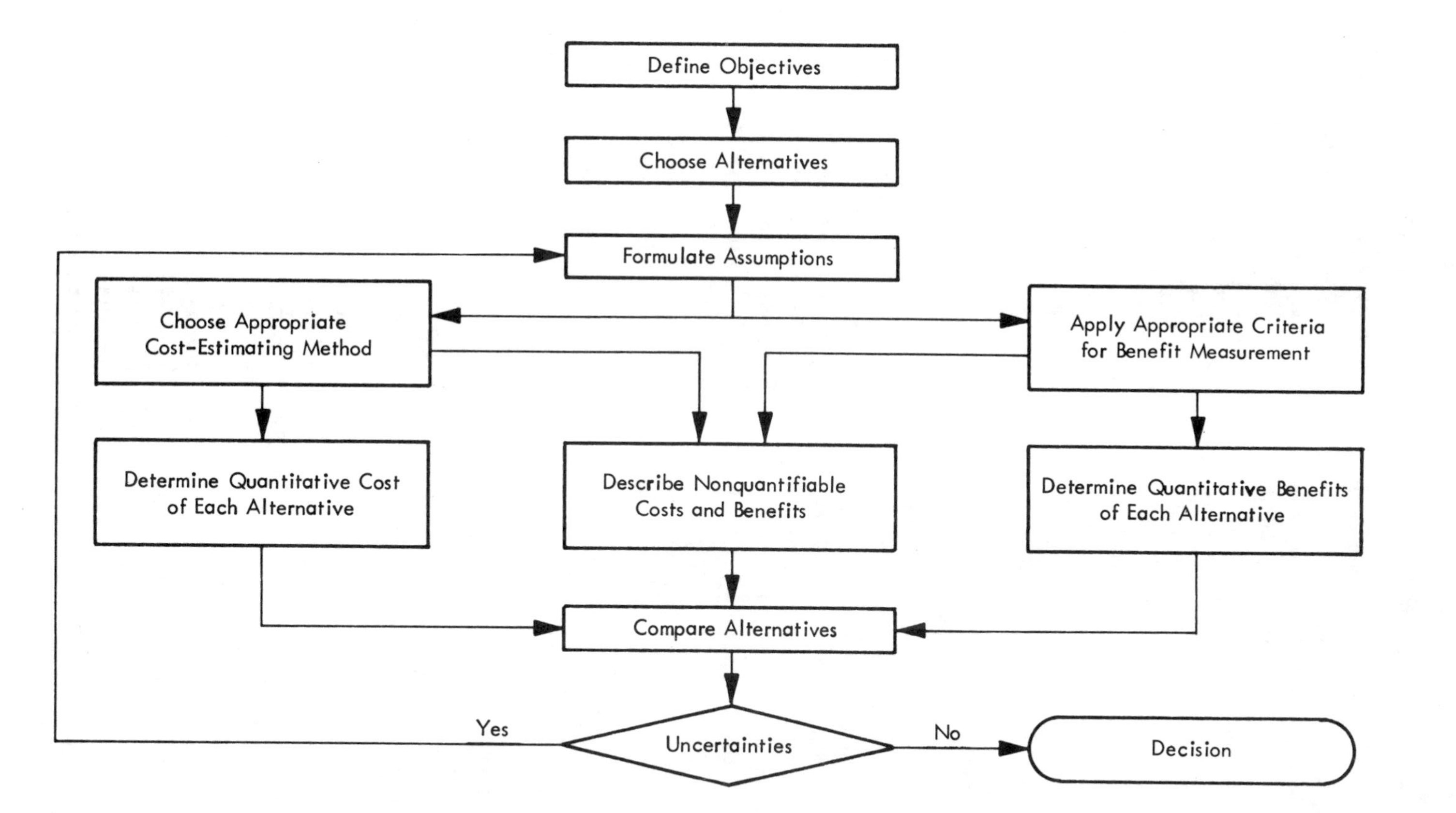

Figure 6.1 Schematic illustration of an economic analysis process.

and projects a future occurrence, it involves a degree of uncertainty. It is therefore necessary to record all assumptions in order to add to the credibility of the analysis. Additionally, it is important that assumptions not be utilized when factual data can be documented.

There are several economic techniques which can be utilized in a cost analysis. Two techniques are supported by a defined mathematical process. Another is largely dependent upon the judgment of fiber optics experts. The two mathematical processes are the parametric method and the analogy method. The appropriate estimating method must be selected and systematically developed by the analyst to arrive at the estimated cost of each alternative. Mathematical costing techniques generally require the skills of a costing specialist. However, adequate cost estimates may be made without the use of a cost analyst. The value of the cost results must be evaluated by the manager within the context of the problem.

Specifying economic benefits implies that they be measured in quantifiable terms, which is relatively difficult to accomplish most of the time. Some of the economic criteria usually considered for fiber optics systems benefits are operating efficiency (throughput time), accuracy, and reliability, as well as several others. However, nonquantifiable benefits must frequently be identified, particularly when different alternatives yield equal benefits in quantifiable terms. An economic analysis is most effective when applied to situations in which benefits can be defined in terms of physical yield. However, it may be applied with less precision where the benefits are nonquantifiable and must be defined and measured in terms of relative benefits. Effectiveness or benefits may be derived through personnel, equipment, or systems performance.

Costs and benefits are measured for each alternative through a given method of comparison. If one alternative is strictly dominant over all time periods and for all levels of effectiveness and cost, then there is no need for comparison. However, this is not realistic, and it thus becomes necessary to constrain the problem in a way that makes one alternative preferable to the others. When time is involved, the analysis requires the use of the technique of discounting, which allows one to take into consideration the time value of money associated with investment decisions. The time value of money is considered by calculating present value costs through an applied discount rate to the time-phased expenditure amounts. Assuming everything else equal, the alternative with the lowest present value cost is the most preferable in terms of the most efficient allocation of resources.

The type of economic comparisons made are (a) least cost for a given level of performance, (b) greatest level of performance for a fixed cost, and (c) variations in both cost and performance, where neither can be fixed. The alternative with the longest economic life may determine the discount comparison method. However, the manager or economist may opt for a shorter period consistent with the objectives and assumptions of the study. Whether the longest or shortest economic life is selected, adjustment of unequal life is required. If the shortest life is selected, the residual values of the alternatives with longer lives must be recognized in the cost computation for those alternatives. Should the longest life be used to establish the time period of the analysis, the cost of extending the benefit-producing years of those alternatives with a shorter life must be recognized.

The uncertainties associated with an economic analysis may be attributed to a number of factors. When addressing the future, in addition to economic uncertainties, there are always political and technological uncertainties. However, one can generally trace the source of these uncertainties to the assumptions that are defined during the analysis, and these can sometimes be clarified through the process of iteration. When uncertainties cannot be resolved explicitly, analytical techniques such as contingency analysis, a fortiori analysis, and sensitivity analysis are applied.

Contingency Analysis. This is the investigation of how the setting of priorities ("prioritization") of alternatives remains consistent when a major change in criteria for evaluating the alternatives is postulated or a relevant change in the general environment is assumed.

A Fortiori Analysis. This analysis is applicable to fiber optics systems in which generally accepted intuitive judgment strongly favors one alternative, but when, based on preliminary analysis, it appears to the analyst that this alternative might be a poor choice and another most advantageous.

Sensitivity Analysis. Sensitivity analysis can be applied to an optical fiber project where there are a few key parameters about which the analyst is very uncertain. Instead of using expected values for these parameters, the analyst may use several values (say, high, medium, and low) in an attempt to see how sensitive the prioritization of the alternatives is to variations in those few parameters.

Finally, it should be emphasized that an economic analysis does not produce a decision. It merely provides information for the manager upon which the decision can be substantiated from data based on accepted economic theory.

Fiber Optics Economic Analysis Sequence

From the above discussion, a logical sequence can be developed for an economic approach for solving fiber optics problems. This sequence involves the following considerations:

Establish systems criteria such as electrooptical component cost data and performance parameters.

Evaluate existing fiber optic data as to type, system level and function, purpose rigidity, and applicability.

Choose between economic models to evaluate the alternatives developed during conceptual design. (a) If cost-benefit analysis is most applicable, utilize the techniques of life-cycle costing, present value analysis, cost-estimating relationships, cost forecasting, and exponential smoothing, as well as other useful cost–benefit tools to evaluate fiber optic systems. (b) If cost-effectiveness models are preferable, there are many mathematical techniques that are useful. Some of these techniques are linear programming, regression analysis, sensitivity analysis, queueing theory, and other statistical–probability models which can be utilized to evaluate fiber optic systems.

Conduct the study utilizing the most applicable techniques and tools and generate a systems model or tabular–matrix presentation of the answers. The interrelationship of these tools and techniques is illustrated in Figure 6.2.

Report the results and analyze the merits of each alternative fiber optics system design.

The study results should emphasize the purpose, objectives, model, recommendations, and conclusions.

TRENDS IN FIBER OPTICS TECHNOLOGY

Projecting trends in any future technology is an area laden with uncertainty. It has been reported from the early stages of technological development in the field of fiber optics that these first predictions encountered many uncertainties. Attempting to predict the market potential of fiber optics technology is one effort that many experts have attempted. In fiber optics several levels of technology are emerging, and the developments are occurring so rapidly that cost and demand predictions must be considered with substantial skepticism. Listed below are several of the changing factors that will influence future market potentials and fiber optic applications.

Levels of Technological Sophistication Choices. Single-fiber or multifiber technology; low-, medium-, or high-loss cables; single-mode or multimode fibers with rugged strength or small size; lightweight cables; data rate capabilities of kilo-, mega-, or gigabits per second; discrete, hybrid, monolithic, or integrated optical circuits; LED or laser injection diode transmitters; power splitters or multiport couplers which are either passive or active; single- or multipin connectors for single or bundle fiber cables; bandwidth.

Systems Design Requirements and Systems Applications Choices. Short- or long-haul systems; point-to-point or distributed data systems; benign or extreme environmental changes to include temperature, shock, humidity, EMI/EMP (electromagnetic interference/pulses), and nuclear attacks effects; redundant or single-path systems with very low system reliability or highly sophisticated reliability and survivability; military or commercial applications.

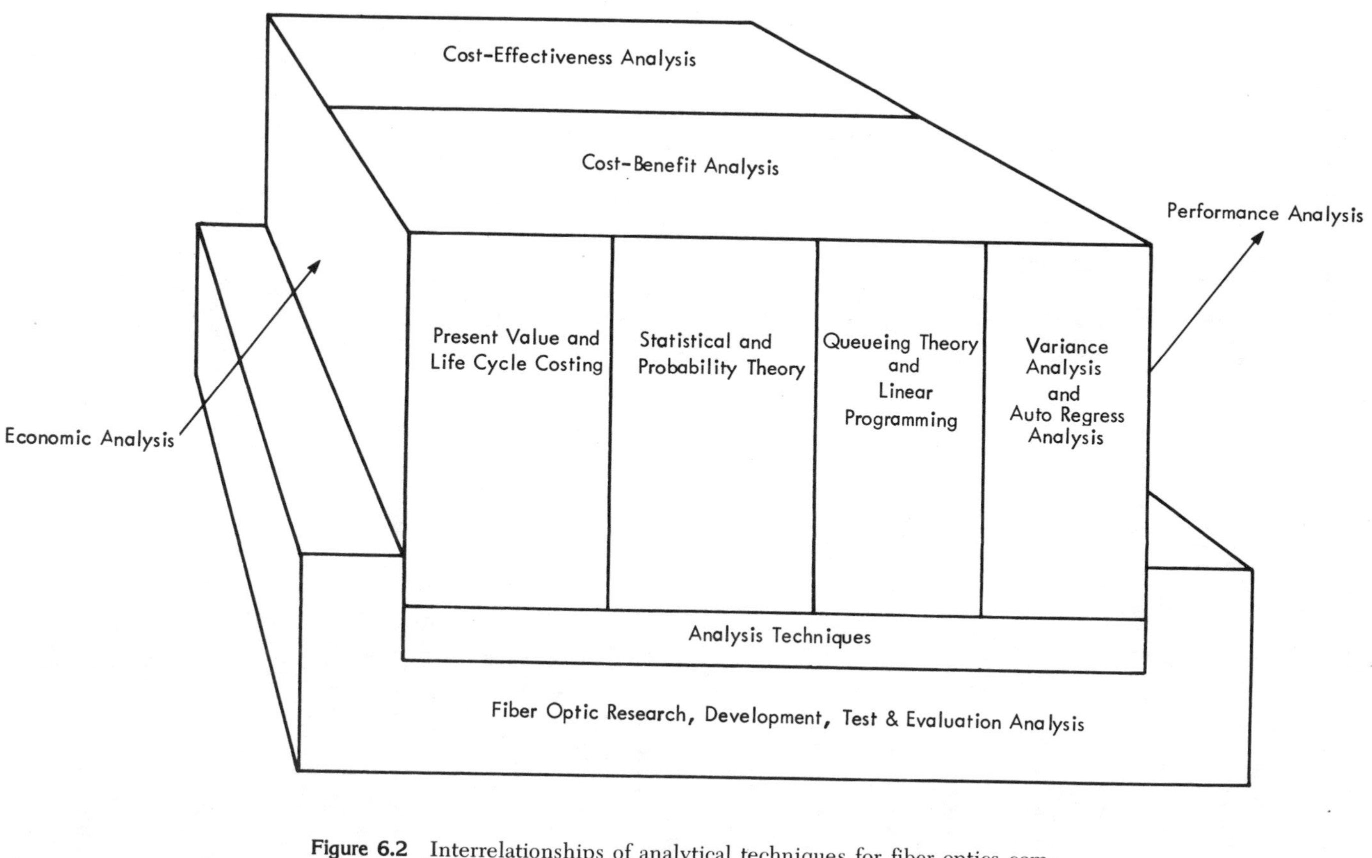

Figure 6.2 Interrelationships of analytical techniques for fiber optics communications studies.

Timing of Applicability. When will each of the technological developments listed above be available off the shelf? When will technological advancement sufficiently level off to preclude an existing generation of fiber optics components from approaching either real or perceived obsolescence, as happened in the case of first-, second-, and third-generation computers? When will there be sufficient market potential to induce fiber optics producers to mass produce components? Does even the strongest possibility of a military "go-ahead" in this area offer enough incentive for industry to establish a production base for mass production? What market potential (measured in millions of dollars and/or millions of feet of cable) will be sufficient inducement to industry? When will military design requirements force military decision makers to utilize fiber optics in order to meet EMP/EMI immunity requirements? When (and how much) will government-sponsored R&D funds be made available to industry and/or the military for continuing research and development? In addition, there are many other and similar questions to be addressed.

Trends in Fiber Optics Utilization

The advances made in the second half of the 1970s in fiber optics communications have delighted even the most optimistic. New systems have done better than expected in rigorous field trials. Improved components (most efficient light sources, less noisy photodetectors, and more rugged cables) have become available off the shelf. Single-fiber cables, plus matching connectors, are starting to surface, while production techniques are attracting the kind of attention due only to mature technologies.

Commercial applications, in fact, have leapt into prominence both in the United States and overseas, and the military remains as dedicated as ever to the pursuit of the interference-free, lightweight attractions of fiber optics systems. For the time being, to be sure, cost is uncertain. Other difficulties include a lack of standards (although several groups are already at work in this area) and the unfamiliarity of the wider engineering community with such a new technology. However, there seems to be no doubt that by the 1980s optical fibers will often be a practical alternative to copper wire. [Based on data collected in November 1977.]

As will be discussed in more detail in Chapter 12, this multifaceted technology will reach billion-dollar heights before the 1980s are over (1). By that time, however, the major user of fiber optics will be the telecommunications industry. Figure 6.3 attempts to predict the fiber optics potential growth by market type and also attempts to project the distribution of dollars by time periods as they relate to research and development (R&D) and production. As indicated in this illustration, the majority of the R&D efforts are conducted in military establishments, but the full production potential will be in the commercial communities.

No country currently appears to dominate fiber optics technology, although Japan and the U.S. seem to be expending more research and development funds in fiber optics than other countries. The United States has considerable support from the Federal Government as well as from private enterprises. Major companies are pursuing independent research and development efforts, and many new fiber optics companies have recently been established. Large R&D efforts are underway in Japan in all phases of civilian fiber optics telecommunications. The Japanese Government actively supports the commercial interests of the telecommunications industry by establishing captive markets, with trade policies, and through technology transfer from Government laboratories to commercial industry. Japanese industry is still second to that of the U.S. in overall fiber optics technology, although it is gaining ground quickly and is pushing ahead with several experimental systems.

The United Kingdom presently supports domestic civilian and military efforts in fiber optics technological development. Its military applications are nearly equal to those of the United States, and R&D in the commercial companies is continually being expanded. In France there is a strong Government emphasis on systems implementation. The French Government is second only to the Japanese in supporting its domestic industry in the international market. The French fiber optics industry has no apparent domestic source of optical fibers; this is potentially a serious handicap on the international market. The West German R&D efforts are concentrated on long-term civilian

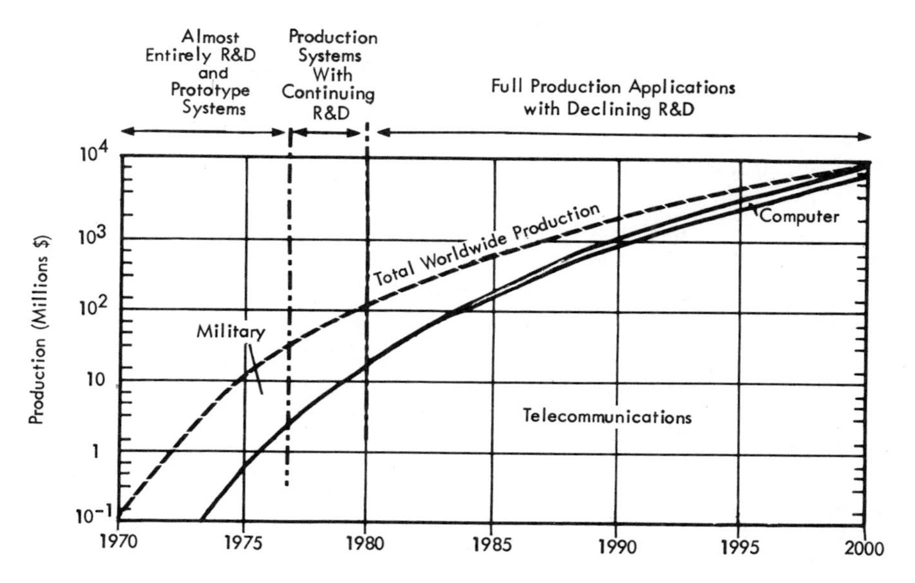

Figure 6.3 Estimated worldwide fiber optics system production by major market categories.

applications, aimed primarily at the post-1980 markets and involving large-scale, high-capacity networks. As with the other European countries, the German Government actively protects internal markets for domestic industry; it is less active regarding international markets than the Japanese and French.

Trends in Component Costs

Perhaps the single most important barrier to widespread use of fiber optics systems in the future is the present high cost of components. The first country to acquire large-volume sales of optical fibers or systems will be able to mass produce them, resulting in a substantial cost reduction. The reduced cost, along with reputation gained, agreements made, and the tendency of customers to continue using components compatible with the first fiber optics systems they install, will give the dominant competitor of the next five years a sales advantage in later years.

Forecasting the cost of individual fiber optic components must not only include the technical progress and economic benefits of fiber optics, but the long-term sociologic impact must be considered as well. Current efforts are being made to improve the manufacturing process of existing components while the development of new and improved components is still pursued. Based on a survey of current off-the-shelf equipment and R&D efforts, an attempt will be made in the subsequent discussion to project the cost of future fiber optics components.

The cost trends of the major types of fiber optics components are shown in Figure 6.4. These components are classified generically by type rather than by specialized classes. Many tables would be required if one were to attempt such a task, and the validity of the cost projections would be somewhat suspect for the later years.

Any form of cost comparison between existing and future technologies is a most difficult task. Indeed, even in the comparison of different metallic cable systems, compromises in choice have to establish typical cost values for diverse items of plant and for different installation methods and conditions. However, economic comparisons must be made to verify that the new technique will indeed lead to cost benefits. Without this assurance, fiber optics systems are unlikely to proceed far. Fortunately, enough work has been done to provide reasonable confidence in such comparisons, and the advantages of fiber optics are sufficiently great to permit a realistic approach.

First, in a study performed by Gnostic Concepts, Inc. (2), attempts were made to project the number of channels that could be available with a given technology and based on projected demands. The comparisons were made between twisted shielded wire pairs, coaxial cable, and fiber optics. Based on projected demands and channel capacities, relative systems costs were estimated. Inflation was also included in these relative costs. Figure 6.5 presents

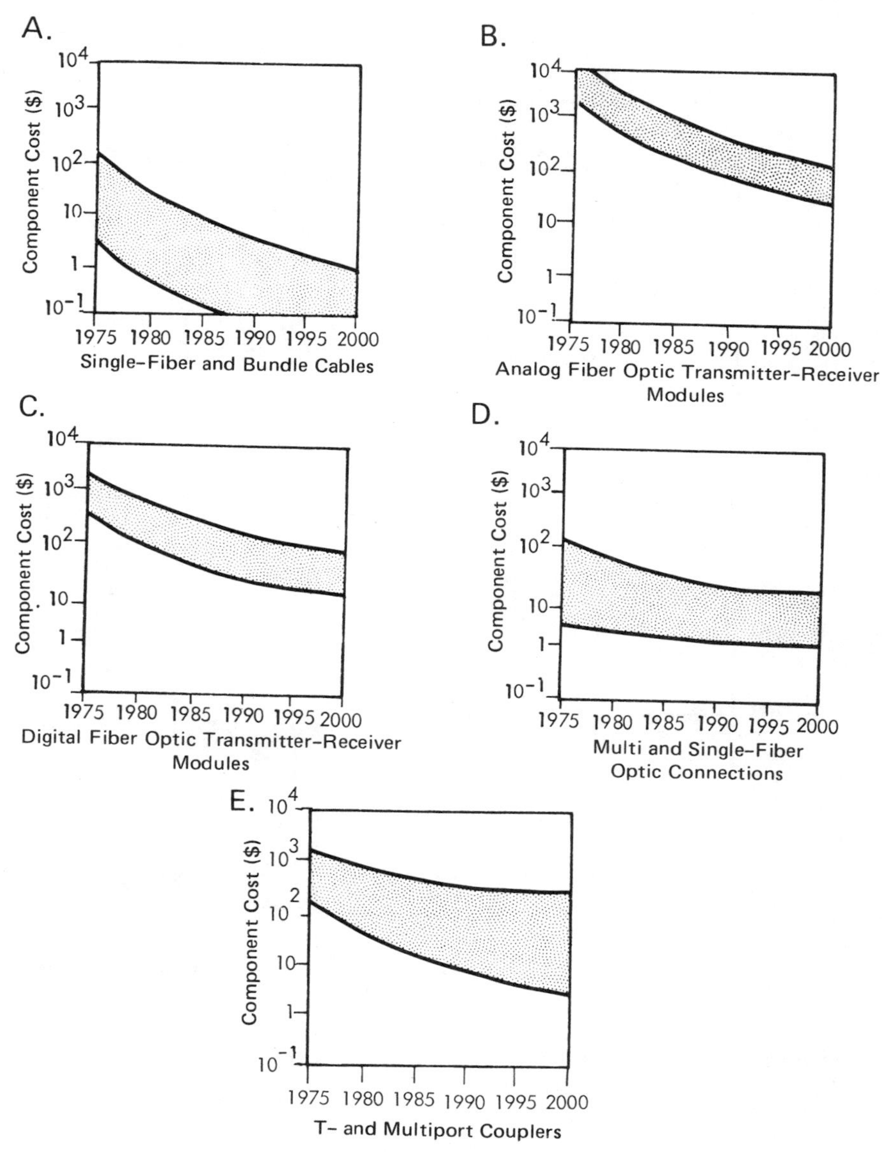

Figure 6.4 Trends in component costs for fiber optics systems.

these data. Based on these values, a hypothetical system life cycle cost estimation was calculated as a function of data rate. This is shown in Figure 6.6. Fiber optics appears to be more beneficial as data rate increases above 10 Mb/s.

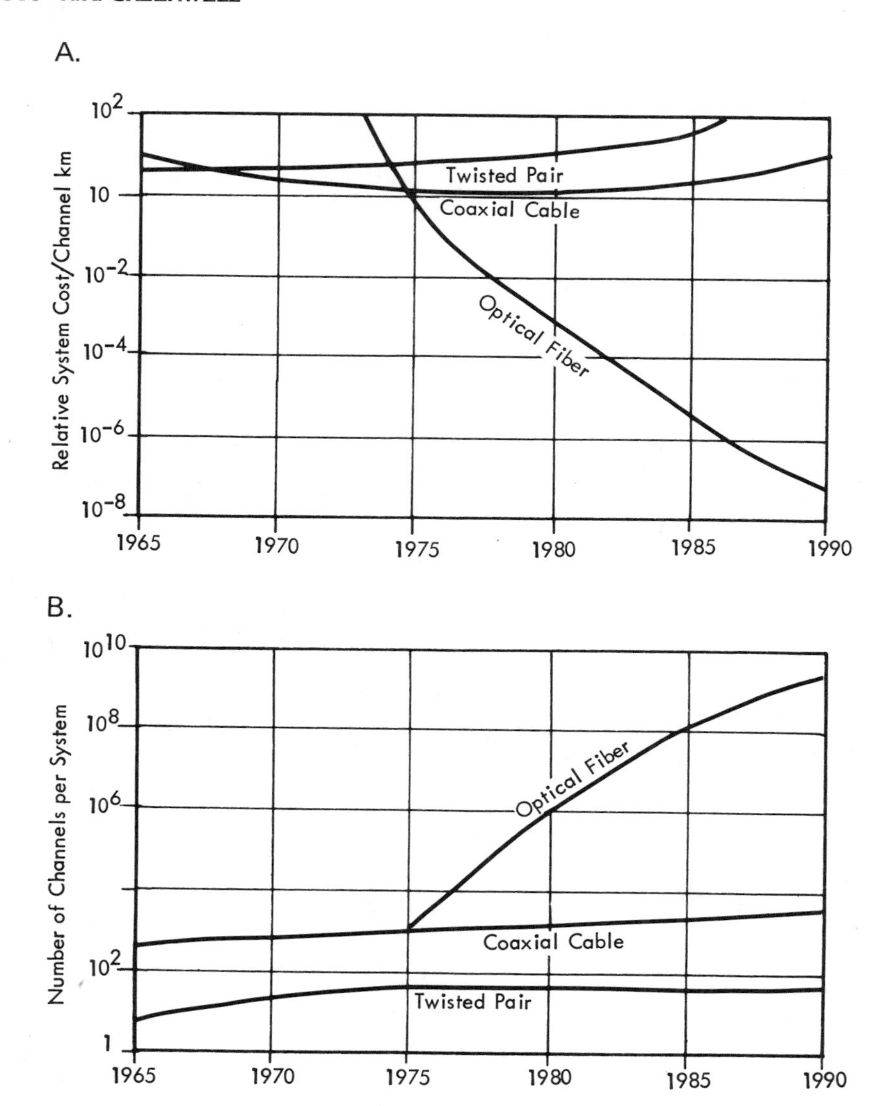

Figure 6.5 Comparison of trends (A) of system cost per channel kilometer and (B) of channel capacity for major cable types.

SYSTEMS COST ESTIMATION METHODS

Various types of evaluation criteria are used by industrial and Government organizations to measure the actual performance or to estimate the value of projected fiber optics costs. The two most common evaluation methods used by industrial companies are return on investment (ROI) and profit margin,

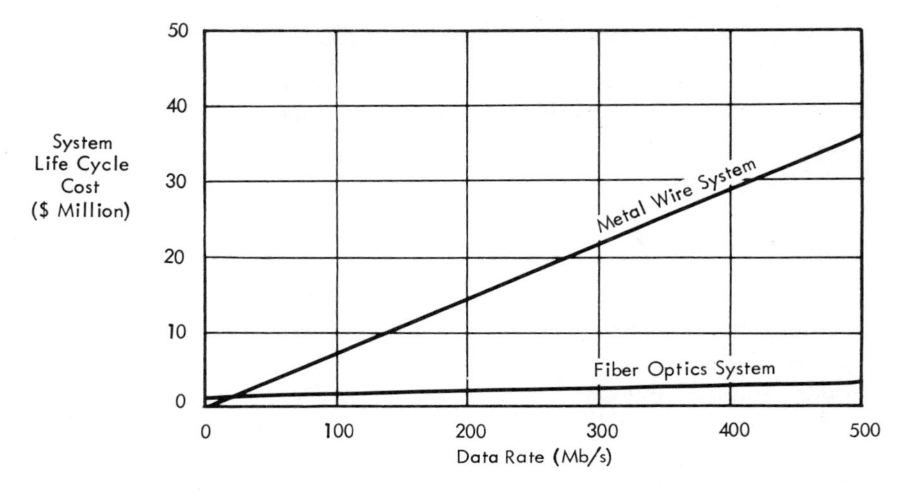

Figure 6.6 Hypothetical comparison of system bandwidth and cost for fiber optics and metal wire systems.

which measure earnings as reflected in accounting records in relation to investment and sales, respectively. Although the Federal Government does not profess to be a profit-making organization, so that it would be difficult to measure profit margin, the Government does invest its capital and is definitely interested in its ROI.

In industry, ROI may represent total earnings divided by gross investment (total investment before deducting accumulated depreciation and working capital). Return on investment applicable to Air Force and Army organizations has been presented as the total savings or benefits achieved with fiber optics technology divided by the investment costs in a fiber optics technology program. Benefit–cost situation and savings–investment ratio are other terms which basically define or describe the same situation as return on investment. However, these ratios do not provide for comparison or a rank ordering of the programs. In all of these definitions, a ratio greater than unity indicates that more benefits have been achieved than funds expended.

Supplemental techniques will permit a more sophisticated assessment of a future investment than that provided by ROI. These techniques generally relate to the timing and amounts of projected cash investments and benefits and include present worth, annual cost/benefit, and internal rate of return. These methods are particularly useful in investment decisions that involve noncapitalized expenditures, such as research and development, tooling and setup costs, and other nonrecurring investments, as well as capitalized disbursements. These methods are explained below, and their more important advantages and limitations are described.

Return on Investment (Benefit–Cost Ratio)

Return on investment is similar to the benefit–cost ratio or the savings–investment ratio (SIR). It is the total quantity of benefits (or savings) divided by the total costs and can be expressed as

$$\mathrm{ROI} = \mathrm{SIR} = B/C = \sum_{j=1}^{n} \frac{B_j}{C_j} \tag{6.1}$$

where B_j is the sum of all the benefit contributions of the jth period, C_j is the sum of all the cost contributions of the jth period, and n is the total number of periods. These benefits and costs are expressed in terms of actual dollars which are nondiscounted; also, working capital and nonaccumulated depreciation are not deducted.

An evaluation of ROI provides a direct and simple method of evaluating a program because accounting data are readily available. When gross investment (as contrasted to depreciated investment) is used, the resulting ratio tends to be conservative and can be viewed as a worst-case (a fortiori) analysis. However, in evaluating an investment program, ROI can be calculated for each year, as an average over the years of the program life, and/or as a cumulative total for the life of the investment, depending on the manager's decision. It is not sensitive to the timing of cash disbursements and benefits and therefore may not be valid for present cash-flow projections.

To calculate the benefit–cost ratio for a given time period, several equations must be considered, depending on the cash–flow disbursements. For a discrete end-of-year cash flow with a single cost–benefit at the end of the time period, the B/C ratio is

$$\mathrm{ROI} = \sum_{j=1}^{n} \frac{B_j}{C_j} (1 + i_u)^{-j} \tag{6.2}$$

where i_u is the uniform interest rate. For discrete, end-of-year uniform series cost–benefit, the B/C ratio is:

$$\mathrm{ROI} = \sum_{j=1}^{n} \left(\frac{B_j}{C_j}\right) \left(\frac{(1 + i_e)^{j-1}}{i_e(1 + i_e)^j}\right) \tag{6.3}$$

where i_e is the discrete end-of-year interest rate. The most complicated, yet most common, benefit–cost ratio is a uniform series, continuous compounding, and continuous cash flow, which is written

$$\mathrm{ROI} = \sum_{j=1}^{n} \left(\frac{B_j}{C_j}\right) \left(\frac{1}{\ln(1 + i_c)}\right) \left(\frac{(1 + i_c)^{j-1}}{(1 + i_c)^j}\right) \tag{6.4}$$

where i_c is the continuous compound interest rate.

The essential element of the return on investment technique is almost trivial and is misleading in its simplicity. An increment of investment is justified only if the benefits are in excess of the total costs. It is clear that if benefits must exceed costs, then the ratio of benefits to costs must exceed unity. That is, for C > O, if B > C, then B/C > 1.0; most Government programs, in fact, prefer a ratio in excess of 2:1. In this case, the program can be considered beneficial; otherwise, the program provides no significant benefits which would justify consideration (B/C < 1).

Present Worth (Present Value)

The essential features of the present worth method is that a preselected discount rate is established and that all cash flows expected to result from an investment decision within a given time frame are expressed in current dollars. That is, in order to satisfy the basic requirement that alternatives be compared only if money consequences are measured at a common point in time, the period in which the project life begins is arbitrarily selected as the base year. The net present value (NPV) is not determined by discounting all cash flows and benefits not already occurring at time zero and then summing these benefits and cost differences:

$$\mathrm{NPV} = \sum_{j=1}^{n} (B_j - C_j)(1 + i_e)^{-j} \tag{6.5}$$

where i_e is the discrete end-of-period interest rate, and B_j and C_j are the benefits and costs, respectively, occurring at the end of period j for $j = 1, 2, \ldots, n$. Net present value is therefore a cost–benefit difference method rather than a cost–benefit ratio method as is return on investment, as is evident from a comparison of Equations (6.2) and (6.5). As is obvious from Equation (6.5), the NPV must be greater than zero to consider the program beneficial.

The most common net present value method is called continuous series present worth, where investment costs and benefits are assumed to be flowing uniformly and continuously throughout each period from the base year through the final year. Costs and benefits are averaged over each period and then discontinued to the base year. Thus, the net present value can be expressed (3)

$$\mathrm{NPV} = \sum_{j=1}^{n} (B_j - C_j)\left(\frac{1}{\ln(1+i)}\right)\left(\frac{(1+i)^{j-1}}{(1+i)^j}\right) \tag{6.6}$$

which should be contrasted to Equation (6.4). Thus, the result is expressed in current dollars as the excess (or deficit) of discounted benefits less discounted costs. It provides an indication of the relative magnitudes of alternative investment proposals.

The discount rate to be used in present worth analysis by Government agencies has been established and estimated to be 10%; it can be found in DOD Instruction 7041.3 of October 18, 1972. As stated in this DOD Instruction, "The discounting technique is based on the premise that no public investment should be undertaken without explicitly considering the alternative use of the funds which it absorbs or displaces." One way for DOD (Dept. of Defense) to assure this is to select a discount rate which reflects the foregone private sector investment opportunities. The rate reflects the preference for current and future money sacrifices that the public exhibits in nongovernment transactions. A 10% discount rate is considered to be the most representative overall rate at the present time.

A discount rate of 7% is prescribed by the Office of Management and Budget for the lease or purchase of general-purpose real property. When a constant dollar price deflator of 2%–3% is applied, the effective discount rate is 9%–10%, which should be used for comparative cost studies. Thus, the 10% discount rate represents an estimate of the average rate of return on private investment before corporate taxes and after adjusting for inflation.

Equivalent Uniform Annual Cost or Benefit

Alternatives cannot be compared unless their respective costs and benefits are first translated to comparable points in time through discounting. This is being accomplished by a method that establishes the relationship between the equivalent uniform annual costs and the equivalent uniform annual benefits (EUAC/B). This annual cost method translates all cash flows and benefits to an equivalent uniform series. The relative economic desirability of each alternative is a direct function of the amount by which (equivalent) end-of-period benefits exceed (equivalent) end-of-period cash flows. Equivalent uniform annual cost–benefit should be applied when alternatives have different lives.

The calculation is made by dividing the total present value cost–benefit ratio by the sum of the discount factors for the years in which an alternative produces benefits. This calculation yields the average cost per year of production. The technology with the smallest average cost or greatest average benefit per year is considered to be the most efficient alternative.

Internal Rate of Return

This method calculates the effective interest rate, net of tax, earned by an investment. It determines the cash flow against discounted benefits. The internal rate of return (IRR) is calculated when the net present value is equal to zero. The rate of return takes into consideration the amount as well as the timing of expected costs and benefits. The end-of-period cash flow at end-of-period compounding is

$$\sum_{j=1}^{n} (B_j - C_j)(1 + i)^{-j} = 0 \tag{6.7}$$

where NPV = 0; solving for i allows the determination of IRR. In an equivalent uniform and continuous series cash flow during each period of continuous compounding of interest, again solving for i, the equation is

$$\sum_{j=1}^{n} \frac{(B_j - C_j)\, i(1 + i)^{-j}}{n(1 + \mathrm{i})} = 0 \tag{6.8}$$

The calculated internal rate of return for a given program must then be compared to some minimum attractive rate of return (MARR) which could be earned if the program's funds were invested elsewhere.

IRR is one of the best techniques for comparing one investment against another and provides a direct comparison against the net cost of borrowing. Hence, this method is most appropriate for estimating the cost of fiber optics systems.

Techniques Conclusion

As discussed above, the four major techniques used in cost analysis are return of investment (ROI) or benefit–cost ratio, present worth or net present value (NPV), equivalent uniform annual cost–benefit (EUAC/B), and internal rate of return (IRR). Each of these methods is subject to some constraint for initial acceptance. The benefit–cost ratio must be strictly greater than unity; the net present value must be greater than zero; the equivalent uniform annual cost must be less than zero; and the internal rate of return must be greater than the minimum attractive rate of return. NPV and IRR can be represented graphically as areas of acceptance and rejection, as illustrated in Figure 6.7.

COST– BENEFIT IDENTIFICATION

The tools and techniques discussed in the previous section are easy to apply to any fiber optics study after the costs and benefits of each alternative have

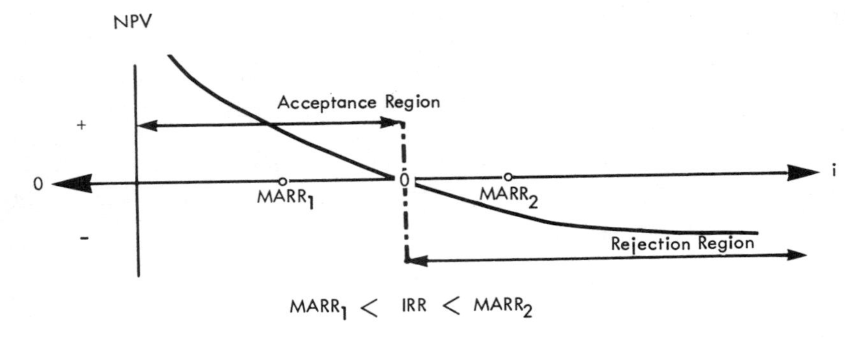

Figure 6.7 Comparison of initial acceptance and rejection methods, where i is the rate of interest.

been identified and the timing of the cash flows has been determined. The difficulty lies in collecting the cost data and defining as well as quantifying the benefits. Current cost data are relatively easy to obtain through basic accounting procedures; there are also techniques that allow the projection and estimation of future costs. It is also necessary to estimate future benefits, and in some cases they can be quantified or equated to current dollars. However, some benefits are nonquantifiable and can be stated only subjectively. Estimating techniques have been developed for those future costs and benefits which are quantifiable. The most common are the statistical and mathematical models, where data have been collected, and the Delphi method, where subjective evaluation is required. In the following analysis a detailed review of the cost and benefit identification procedure will be given.

Cost-Estimating Relationships

The most important part in any cost analysis is estimating the cost contributions. Although it is true that without the preceding steps in cost analysis no credible estimating could be performed of the cost, analysis probably would not be performed if cost estimating did not require it. A typical cost analysis procedure is to utilize cost-estimating relationships (CERs) on the major cost drivers for fiber optics systems procurements and to estimate costs in logistic support areas.

CER is a numerical expression of the connection between a physical characteristic, natural resource, demography, or technical parameter and a particular cost factor associated with it or of the link between different but related costs.

An example of the first type of CER would be a function relating the procurement cost of a fiber optics component to its weight, size, or failure rate. In estimating the yearly cost of maintaining this component, a CER of the second type could be used. This would estimate the maintenance cost as a fixed percentage of the original component procurement cost.

There are essentially three general approaches to developing CERs for use in estimating future costs. They are the engineering approach, the analog approach, and the statistical approach.

Engineering Approach. The engineering approach identifies essential development efforts in proposed future programs and then makes estimates of the funds required to solve these tasks. The accuracy of this procedure in estimating large-scale research and development projects has been very low in the past. The main reason for this is that the approach depends too much on the capability and experience of the estimator. Thus, on large-scale projects, where many estimators are involved, the consistency of the estimates varies widely, to the point where the results have an excessively large deviation. This type of estimating is therefore more suited to small projects, where one estimate will suffice and the consequences of inaccuracies are not significant.

Analog Approach. The analog approach makes use of known costs of similar systems, identifying their essential development efforts, their production efforts, and their documented costs. This approach is most applicable to systems for which there is no historical precedent (i.e., costs cannot be extended) but for which there are related systems costs available.

Statistical Approach. The statistical approach calculates cost-estimating relationships (CERs) based on historical costs, performance data, and physical data from predecessor equipment. When historical data are available, this approach should have precedence over the other approaches in the development of costs.

Because of the overriding importance of the statistical approach, the following discussion will give further details of this procedure. It will briefly make use of known statistical techniques. In the statistical approach, CERs may be calculated from simple linear statistical regressions. A regression equation is an equation that estimates a dependent variable (for example, Y_1) from independent variables ($X_1, X_2, \ldots$). In the functional notation it is sometimes written $Y = f(X_1, X_2, \ldots)$. Linear regression equations may be one dimensional (having a single independent variable) or multidimensional (having two or more independent variables) (4). These two cases may be expressed by the following equations:

Multidimensional case: $$Y = \alpha_0 + \sum \beta_i X_i \tag{6.9a}$$

One dimensional case: $$Y = \alpha_0 + \beta_1 X_1 \tag{6.9b}$$

where Y is the value of the cost measure to be estimated, α_0 is a constant, and β_i is the relative factor assigned to the ith input parameter (independent variable).

A regression equation represents a regression plane (hyperplane if the equation has three or more independent variables), which is the generalization of the regression line in the one-dimensional case shown in Equation (6.9b).

Statistical regression analysis applies the least-squares method by fitting a set of data points. This is achieved by minimizing the squared distance between the data points and the regression plane. When this distance is minimized, the optimal β_is ($i = 0, 1, 2, \ldots, n$) have been found.

Both Equations (6.9a) and (6.9b) are linear. As one would expect in the real world, not all cost-estimating relationships can be determined linearly. However, many nonlinear relationships can be transformed into linear relationships. Transformation examples which take a logarithmic relationship and a simple square of a variable for a new variable are shown below.

Functional form	Transformation	Linear function
$Y = \alpha_0 X_1 \beta_1$	$\log Y = \log \alpha_0 + \beta_1 \log X_1$	$Y = \alpha + \beta_1 X_1$
$Y = \alpha_0 + \beta_1 X_1 + \beta_2 X_2^2$	$Z_1 = X_2^2$	$Y = \alpha_0 + \beta_1 X_1 + \beta_2 Z_1$

(6.10)

Almost any cost relationship can be transformed into a linear relationship so that the linear regression theory may be applied. The regression line in the two-dimensional case or the plane in the multidimensional case is, in effect, an average relationship. An important statistical parameter which adds to the credibility of the CER is the closeness of the fit or the level of significance of the cost relationship to the historical data.

As a schematic example, Figure 6.8 shows two diagrams with the linear regression line drawn. The data points in the first diagram are much closer to the line than in the second. Hence, the degree of confidence in the cost prediction made from these CERs will be much higher in the first case than in the second.

The correlation coefficient (r; multiple correlation coefficient in the multidimensional case) is a signed measure of the closeness of fit or significance of the relationship for each particular input parameter. The square of the correlation coefficient (r^2), sometimes called the coefficient of determination (R), is an unsigned measure of the variance that measures the significance of the dependent variable to the relationship of all the independent variables. The first example in Figure 6.8 would have a higher r^2 than the second. In the multidimensional case the simple correlation coefficient must be replaced by its multiple equivalent.

A more advanced multiple-regression technique for developing CERs is known as the two-stage least-squares method. This method uses basic multiple-regression techniques but is capable of dealing with several equations

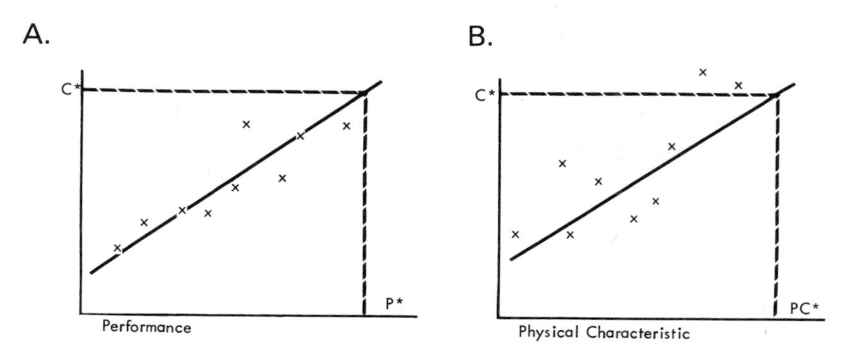

Figure 6.8 Linear regression diagrams; (A) performance; (B) physical characteristic. X, previous fiber optics components; P* (PC*), expected performance (physical characteristic) of fiber optics components to be estimated; C*, estimated component cost as derived from CER.

simultaneously, allowing the independent variable of one equation to be included among the dependent variables of another. Since this technique yields more reliable results, it is best suited for developing CERs for a life-cycle cost model (5).

The standard procedure for cost estimating with this statistical technique consists of three identical portions.

Relevant Variables Identification. Identifying the relevant variables consists of limiting the number of independent variables to be considered as predictors in the derivation of cost-estimating equations. Tests for significance aid in limiting the number of independent variables.

Historical Data Collection. Some variables will also be eliminated during the data-collection phase. Historical data may not be available for certain variables, whereas for others the data may be insufficient to yield acceptable results. Poor data and inconsistency from one source to another are further factors that can also eliminate some variables.

CER Calculation. After the variables and data are selected, the cost-estimating relationship can be calculated by applying one of the standard multiple-regression computer programs. The programs usually compute some or all of the following statistical terms: (a) weight regression coefficients (β_i); (b) means (μ); (c) standard deviations (σ); (d) coefficient of determination ($R = r^2$); (e) multiple-correlation coefficient (r); and (f) regression constants (α).

These results yield the CER by defining the numerical relationships, but the important analysis involves deciding how much significance may be assigned to the results and the estimated error attached to the CER.

Judgmental Techniques

In fiber optics, few historical data exist to establish firm technological, price, or demand trends. Sufficient data may not be available to apply regression, sampling, smoothing, or other mathematical analyses to develop cost estimates. Hence, cost estimates must be formulated from the opinions of experts.

In cases where there is little or no historical information on the specific alternative, or when the cost estimate is required so quickly that an extensive data search is precluded, cost estimates must be based entirely on expert judgment. Even in cases where historical information and time are available and a standard method of costing can be applied, judgment must be used to reach conclusions not directly supported by data. Expert judgment may be used to construct CERs or to check their behavior when they extend significantly beyond the data base and when the data base is too small to be statistically significant. There are basically three major judgmental techniques that have found widespread use in the past and that are applicable to the estimation of fiber optics systems costs: the analogy method, the Delphi method, and the learning method.

ANALOGY METHOD

The analogy method is a specialized technique of judgment that may be used to estimate costs by making direct comparisons with historical information on like or similar existing alternatives or their components. It is, in fact, the most widely used method of analysis to date, although it is certainly not the most accurate. Caution must be exercised in applying the analogy method because it is basically a judgment process and, as a consequence, requires a considerable amount of expertise if it is to be used successfully. Two types of the analogy method have been utilized in the past. One is based upon similar products and the other upon similar concepts. Similar products can be compared, such as using cost data on commercial aircraft. Second, when the cost of a new concept or a new system must be estimated, experience gained on a different product may be used. An example of this is estimating fiber optics production costs based on integrated circuit production experience.

The necessity of using experienced judgment to fill voids in data has long been recognized. In some cases the major portion, or even the entirety, of a cost estimate must be based on judgment. The complexity of the problem, the predisposition of the manager, the point of view of the analyst, the importance of the project (in terms of both mission and finances), and the availability of qualified statistical analysts determine the extent of analysis necessary. The keynote in using judgment is that it be reasonably tempered with a large dose of impartiality. Moreover, judgment must always be identified as what it is: a guess, albeit an educated guess.

DELPHI METHOD

The Delphi method is another judgmental technique which requires the polling of inputs in a particular field; for example, fiber optics. Delphi, as a technological forecasting method, has been credited to O. Helmer, T.J. Gordon, and N.C. Dalkey of the RAND Corporation; preliminary work was done by Helmer as early as 1959. Helmer's publication of "Report on a Long-range Forecasting Study" by the RAND Corporation in 1964 explained the Delphi technique in detail. It discusses the now well-known method of soliciting forecasts from a panel of experts in order to deal with specific questions, such as when a new technology will gain widespread acceptance or what new developments will take place in a given technological area. Instead of a panel of experts gathered together to discuss or debate the questions, each individual usually answers assigned questionnaires through written or other formal means, such as the questionnaire developed by the Naval Postgraduate School for the A-7 ALOFT Project (6).

There are, of course, several major advantages and disadvantages to any judgmental technique which employs questionnaires for objective evaluation. The Delphi method is considered advantageous for estimating fiber optics systems costs for the following reasons:

1. Fiber optics is an emerging technology which is laden with not only technological uncertainties but also demand uncertainty. In addition, fiber optics is comprised of many different technologies.

2. The experts in the fiber optics industry are few and can be easily identified.

3. Users and producers alike can benefit from the results of a Delphi study. It is to their mutual advantage to cooperate in efforts to realize the potential benefits of this emerging technology.

4. Improved forecasts or estimates of future demand quantities, industry growth rates, technological advances, and component prices are expected to decrease the uncertainty of the estimates for these results.

EXPERIENCE CURVE METHOD

Besides the Delphi technique and the analogy method, experience curve theory is also a qualified forecasting technique to estimate the future cost behavior of fiber optics components. Experience curve theory is related to, but should not be confused with, the well-known learning curve theory. Learning curve theory predicts cost reductions from two cost factors, labor and production inputs (material), whereas experience curve theory predicts cost reductions from all cost elements, including labor, development, overhead, capital, marketing, and administration. Experience curve theory is a much broader theory concept which incorporates learning curve theory. Because of the relative value of experience curve theory and the general familiarity with learning curve theory, the following discussion compares both theories, noting their similarities and differences.

Experience Curve

Both the experience curve and learning curve theories are expressed as cost–quantity relationships, stating that each time the total quantity of items produced doubles, the cost per item is reduced to a constant percentage of its previous cost.

LEARNING CURVE THEORY

The history of learning curve theory dates back to 1925, when in the aircraft industry learning patterns of aircraft production personnel were first observed by the Commander of Wright-Patterson Air Force Base. The Commander observed the constant reduction in direct labor hours required to build airplanes as the number of aircraft being built doubled. Subsequently, learning curve theory has been documented and used in many industries to predict cost reductions for direct labor and raw material, or production inputs (7). The semiconductor industry provides one of the best-known examples for the practical application of the learning curve theory. Typical learning curve slopes range from 75% to 90%. Some of the elements that account for direct labor and material cost reductions can be summarized as follows:

Job familiarization by workmen, resulting from the repetition of manufacturing operations.

General improvement in equipment experience, job shop assembly, and engineering liaison.

Development of more efficiently produced subassemblies.

Development of more efficient tools.

EXPERIENCE CURVE THEORY

Experience curve theory was established in 1965 (8). It considers the full range of costs, which includes development, capital, administration, marketing, and overhead, as well as labor costs. Raw materials costs, which were considered in learning curve theory, are not included in this list. The cost of raw materials usually depends upon factors such as the availability of supply. Strictly speaking, correct measurement of the experience effect therefore requires that expenditures be calculated net of the cost of raw materials, i.e., on value added to the product. In general, experience curves do not apply if major elements of cost, or price, are determined by patent monopolies, natural material supply, or Government regulation. The experience curves apply to products in industries with multiple producers who interact as rivals as well as to other products in purely and perfectly competitive industries. The factors that cause the experience curve effect include the following: (a) "learning curve effect"; (b) competition among producers in a given product market; (c) economies of scale and specialization, the "scale effect;" and (d) investment in capital to reduce cost and increase productivity.

The learning effect, i.e., people learning by doing, will be discussed in detail below. It is the major factor which causes reduction in labor costs. The second factor, competition and rivalry among producers, forces each manager to find techniques of lowering total average costs in relation to competitors. The successful producer will then be able to lower prices and induce a situation which causes a "shakeout" of those producers who have been unsuccessful in reducing costs. This will give the low-cost producer an increased market share. With increased market share, the third factor, economies of scale, can be realized. With scaled-up volume due to increased market share, it is possible to use more efficient equipment and amortize fabrication cost over a sufficient number of produced units so that both labor and overhead costs are reduced. Increased volume may also make it possible to consider alternative materials and alternative manufacturing methods which were uneconomic on a small scale. The final factor, investment in capital, is a further attempt to reduce cost by displacement of less efficient factors of production. This can be achieved by automating various steps in the production process, thus reducing labor costs.

Learning curve theory has primary application and adds important significance to major acquisition programs. The per-unit component costs, and therefore the program costs, should reflect the fact that unit cost decreases as the quantity produced increases. This is because production facilities develop improved methods, install more appropriate equipment, and learn how to do the job better and faster as more components are built. The larger the quantity of manufactured units, the more consideration is given to maximizing the operation for mass production.

The learning curve is based on the supposition that as the total quantity of units produced doubles, the cost required to produce the cost per unit of this doubled quantity will be reduced by a constant percentage. The complement of this percentage of reduction is commonly referred to as the slope.

Table 6.1 and Figure 6.9 show the learning curve for a hypothetical fiber

TABLE 6.1. Example of Component Cost Reduction With Cumulative Production Increase

Cumulative number of components ($\times 10^6$)	Cost per component	Percent reduction compared to previous size (percent)
1	10.0	10
2	9.0	10
4	8.1	10
8	7.3	10
16	6.6	10
32	5.9	10

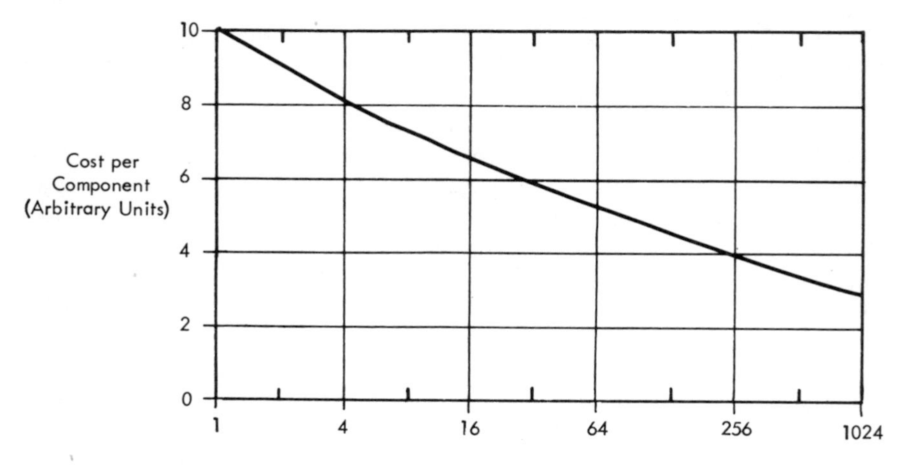

Figure 6.9 Example of a learning curve.

optics component. If one component were procured, the cost would be \$10; if another component were added, the per-unit price is reduced by 10%. If four components are procured, the per-unit cost is 10% lower than if two components were procured. If this same example with the 90% slope, as it is sometimes called, were plotted on log–log graph paper, the curve becomes a straight line. Thus the learning curve is a log-linear (semilog) curve. The general form of the equation for the learning curve is

$$y = ax^b \tag{6.11}$$

where y is the cost per unit, a the cost to produce the first unit, x the cumulative output, and b the slope of the learning curve.

Benefits

The tools and techniques for estimating fiber optics systems costs must not only include the various cost factors but must also be contrasted to the achievable benefits. This will be shown below. It is well known that present-day metallic wire and coaxial cables have operational limitations that cannot be cost-effectively overcome because of their inherent characteristics. Because optical fibers are fabricated totally of dielectric materials their characteristics are fundamentally different from those which require conductive materials. It has been demonstrated that optical fibers can improve on or entirely eliminate the deficiencies of conductive cables. As has been discussed in more detail in Chapter 1, some of the important benefits of optical fibers are the following (it must be kept in mind that some of these benefits are difficult or impossible to quantify and must hence be considered in a qualitative manner):

Electromagnetic Compatibility. With proper design, a jacketed optical cable will neither emit nor pick up electromagnetic radiation, ensuring secure communications and immunity from interference and crosstalk to a degree not possible with electrical cables. Suitably designed fiber optics interfaces have been proven to be immune to all forms of electromagnetic compatibility problems, i.e., to crosstalk, short-circuit loading, EMI, radio frequency interference (RFI), ground loops, ground isolation, ringing, and echoes. Fiber optics systems also offer increased security by eliminating signal leakage and increasing significantly the difficulty of surreptitious tapping of communications lines.

Weight and Size. The fiber optics transmission line is much smaller in size and lighter in weight than an electrical line of equivalent bandwidth. When electrical lines must meet nuclear attack hardening and EMP-immunity requirements, the weight and size advantages of fiber optics lines can be multiplied.

Bandwidth Capacity. The wide bandwidth capacity of fiber optics can far surpass the 1 MHz limit of twisted-shielded wire pairs and the 20 MHz limit of coaxial cable. Fiber optics offers excellent multiplexing bandwidth possibilities; depending on the fibers and the source–receiver combination, the bandwidth–link products of a fiber optic link at present range from 20 MHz/km to 40 GHz/km. The bandwidth capacity increase that can be achieved with optical fibers is related to the potential for improved multiplexing capability. Electrical-line multiplexing is limited by problems of crosstalk, EMI, and impedance matching, problems alien to fiber optics.

Reliability and Maintainability. Even though the communications process in fiber optics systems requires the added step of converting electrical signals into light and back again, the increased multiplexing capability of fiber optics will reduce the quantity and complexity of interfaces to enhance system reliability and maintainability correspondingly. The single optical junction of the fiber optics line with source–detector should be an extremely reliable interface that does not require electrical or mechanical contacts. Fewer, more reliable, and less complex components of the fiber optics system means troubleshooting and preventive/corrective maintenance, along with simplified maintenance support.

Improved Ability to Survive "Battle Damage". Fiber optics is inherently less susceptible to battle damage than electrical wires and their required interfaces. The electromagnetic fields created by a nuclear weapon burst or lightning can generate large electric pulses in wires and cables. These pulses can cause circuit burnout or temporary errors which can degrade mission performance. Electromagnetic pulses do not affect optical fibers or their required interfaces. Thus, fiber optics data links can provide electromagnetic pulse immunity and "survivability" improvement for all weapons and systems. The ability to survive battle damage can be enhanced further by routing a redundant fiber optics interface system, a practicable, expedient approach because of the reduced weight and bulk of fiber optics components. In addition, fiber optics is capable of withstanding higher temperatures, reduces the possibility of sparks and shorts occurring in hazardous areas, and reduces significantly electromagnetic interference problems.

Safety. The curving or breakage of fiber optics lines during transmission does not produce sparks or other hazards. This offers safety advantages for fiber optics systems located in areas which contain fuel, ammunition, oxygen, etc., as well as during catastrophic situations to which a military platform may be subjected.

Disadvantages

There are some potential disadvantages of optical fiber cables, but most have been or are expected to be overcome through ongoing developmental efforts.

Radiation Damage. Optical fibers are subject to damage from strong X rays, γ rays, and neutron radiation that are generated during a nuclear weapon explosion. Efforts are being made to find fiber materials which are less susceptible to radiation damage.

Connectors and Splices. The losses at connectors and splices in fiber optics systems are generally greater than those in conventional systems. Moreover, the connectors for single-fiber systems require more precision in fabrication and installation than those of conventional communication systems. The fiber optics cable design will be strongly influenced by the method of connections and splicing.

System Design. Optical fibers are not being considered for the transmission of power. Separate power supplies will be required. Generally, fibers provide better performance with digital modulation than with analog modulation. Thus, it is preferable to design systems with the intention of using fiber optics communications than to refit present systems with fiber optics.

New Technology. Fiber optics is a new technology. Designers, installers, and operators of fiber optics systems will require training in new procedures for handling, fault finding, and maintenance of fiber optics. Currently, there also remain unknowns, such as the aging of fiber, the long-term effect of moisture on fiber strength, and the effect of small amplitude vibrations on fiber bending loss and the resulting noise in the transmission channel.

Little doubt remains that fiber optics implementation in nearly every information transfer requirement will result in many significant systems improvements. The rate at which the evolution of fiber optics applications progresses is a function primarily of the aggressiveness at the management level and the timely solution of technical deficiencies.

RISK AND UNCERTAINTY

Any presentation of cost procedures would be inadequate without some discussion of risk and uncertainty. Uncertainty here refers to the accuracy of cost projections or errors in cost estimates over time. Uncertainty occurs when not all the relationships are known between a simple or compound event and its correlated outcome.

Cost Analysis Uncertainty

The two major sources of uncertainty in the cost analysis of future units are requirements uncertainty and cost-estimating uncertainty. Requirements uncertainty refers to variations in cost estimates stemming from modifications in the configuration of the equipment being estimated. Cost-estimating uncertainty refers to variations in cost estimates of equipment where the configuration of the equipment is essentially constant. These latter variations arise for such reasons as differences in cost models, errors in the basic data used, and errors in cost-estimating relationships.

Requirements uncertainty is the major source of uncertainty in cost analysis of most equipment. An initial estimate is often smaller than a subsequent one, and the difference may sometimes be quite significant. Requirements uncertainty exists because of changes in price level, quantity, or configuration. For example, the price level may increase during the time interval between two estimates, and the quantity (size or number) of the equipment assemblies may be larger in the second estimate. Price and quantity usually do play a significant role. However, the inequality still remains in many instances, even after adjusting for these two factors by putting the estimate and actual cost on a comparable basis through corrections of price level and quantity changes. The primary reason for this uncertainty is the change in equipment configuration, which causes major errors in the estimates.

Configuration modification refers to alteration in hardware characteristics and/or change in the operational concept. Many reasons may be given for changes in equipment configuration. The following are a few examples for requirements-estimating uncertainty:

Original design may fail to produce the desired performance characteristics, thus resulting in hardware configuration modification;

Errors and omissions may cause changes in specifications;

Attempts to obtain the equipment sooner than was originally intended by substituting resources for time;

Changes in personnel skills and availability;

Changes in the specified mission.

The most important sources of cost-estimating uncertainty are (a) schedule slippages; (b) competitive optimism; (c) timing of cost estimates; (d) technical complexity of the equipment; and (e) technological improvement in the components of the equipment.

Although all these sources of uncertainty present potential problems to some degree, those configurations that will be operative for many years in the future have troubles with the last three in particular. One study indicates that the average cost overruns (i.e., actual cost minus the estimated cost) tend to decrease as the estimates are made later in the project (8).

In most cases the technical complexity of the equipment will be an important factor in achieving accurate estimates. For example, estimating the cost of a technically complex fiber optics transmitter module is considerably more difficult than estimating the cost of a nonrugged fiber optics bundle cable. Technical complexity may consist of a large number of technical problems for new equipment, the interrelation among the technical problems, or the large number of assemblies in the equipment in light of reliability requirements.

The last potential problem, which is different from the others because it is the only one that will tend to make the estimates higher than the actual costs, is the improvements to existing technology that may be expected to occur within the next two decades. Here, the problem is forecasting rapid technological advances that may drastically alter reliability, performance, obsolescence, and, especially, cost (9).

Outstanding uncertainties occur when evaluating future international, political, and economic conditions, rate of progress of technical state-of-the-art developments, and a host of other factors bearing directly or indirectly on fiber optics component and systems requirements, capabilities, and productivity. However, it is less important to recognize how much uncertainty exists than that it exists at all and that it must be considered during advanced planning.

Sensitivity Analysis

Sensitivity analysis is a method of determining the degree of risk possible without jeopardizing the validity in the results. Cost sensitivity analysis is the iterative analytical process with different quantitative values for cost elements, operational assumptions, or other estimates that determine variations in the estimated fiber optics systems costs. It is a systematic examination of fiber optics systems cost changes and key input characteristics as these inputs are varied over a relevant range. If a small change in an assumption results in a proportionally greater change in the results, then the results are said to be sensitive to that assumption or parameter. For example, if the reliability of a particular fiber optics component in terms of mean time to failure (MTTF) has a significant effect on cost, it is important to determine what happens to the cost if the reliability specifications are not achieved. Other possible cost sensitivity analyses might evaluate the change in fiber optics systems costs by varying the weight of given components where investment cost has been determined as a function of weight, or by varying the slope of the experience curve over the number of units of components to be produced.

Cost sensitivity analysis is especially useful in answering the question of how inaccuracies in the estimate of a parameter affect total cost. Assume, for example, that the cost of a component is reduced by 40% and then increased by 200%, and this only has a 0.1% effect on total system cost. Then the analysis says that this cost element or parameter is not sensitive to inaccuracies, and it is not worth the additional effort necessary to obtain a better estimate. On the other hand, if a plus or minus move of 10% cost element translates into a 20% change in total system cost, then the cost element is extremely sensitive to inaccuracies and proportionally more effort should be concentrated on improving the uncertainties for this cost element.

The use of the single-valued estimate makes it difficult to deal effectively with problems of uncertainty. Most cost analysts prefer single-valued estimates because they are simple to grasp and easy to manipulate in calculations. However, a single-valued estimate by itself implies a high degree of confidence in the accuracy of the estimate, which is rarely justified. Simply stated, the single-valued estimates hide too much information about the uncertainties involved in assuming that it is correct. Many approaches have been used to compensate at least partially for this failure of the single-valued estimate. Reducing the error by a more accurate forecast is a worthy objective, but there is a limit to the accuracy attainable in forecasting the cost of procuring and operating new technological systems fifteen or more years from the time at which the estimate is made.

An empirical adjustment factor based on past experience in component cost estimating is sometimes used to yield better estimates. If actual fiber optics component costs have historically exceeded the original estimates by a given percentage, then consideration is given to the possibility of increasing cost estimates by an equivalent adjustment factor. Although this adjustment does not answer any questions concerning estimate variability, its single-valued estimate is often closer to the actual cost than the original estimate.

The three-level estimate (high, medium, and low) gives the broad range over which the total system cost may be expected to vary. Although the three-level estimate is a definite improvement, it still does not go far enough in evaluating uncertainty to permit valid comparisons between alternatives.

The Monte Carlo technique solves most of the problems inherent in other methods. Basically, the technique combines the projected variations inherent in all relevant cost elements. A simulation of the way these cost elements combine is the key to extracting the maximum information from available forecasts. Monte Carlo simulation consists of three basic steps, as follows:

1. Estimate or determine the probability distribution and range of each input parameter.

2. Select at random one value from each cost element distribution and combine the costs of all elements into a total cost.

3. Repeat the total cost calculation many times to define and evaluate the odds of the occurrence of each possible cost.

The results may be used to give high, medium, and low values for the total cost, the most likely value, a probability distribution for the total cost, or the likelihood that the total cost will be less than a certain value. Although this method does not improve the accuracy in a cost estimate, it reduces the uncertainty and provides a useful indication of its reliability. However, the application of this method requires a considerable expenditure of time and resources.

EXAMPLES

The previous discussion will be illustrated by two specific examples which are considered to be representative of the economic aspects of an operational fiber optics system in the commercial and the military sectors. These are early applications of optical communications systems and thus have achieved an adequate amount of later verification. First, the applicable cost model will be detailed; then, a comparison will be made of the cost estimates of a commercial fiber optics system and of a comparable coaxial cable system. Finally, an optical system installed in military aircraft will be contrasted to its metal cable

TABLE 6.2. Cost Elements Selected for Analysis

Symbol	Definition	Symbol	Definition
C_{AS}	Cost of assembly	H_H	Man-hours of special skills
C_{EH}	Cost of automated equipment for special handling and processing	H_I	Man-hours of inspection of all components in production run
C_{E_i}	Cost of ith peculiar support and test equipment	H_S	Man-hours of setup
C_{EQ}	Cost of equipment procured for testing and quality assurance	H_{TE}	Man-hours of testing
C_{FA}	Cost of fabrication and assembly	H_{TO}	Man-hours of tooling
C_{M_i}	Cost of ith material purchased	i	Number ranging from 1 to n
C_{MP}	Cost of purchased material	Q_{E_i}	Quantity of ith peculiar support and test equipment procured
C_{MW}	Cost of wasted material	Q_{M_i}	Quantity of ith material used
C_{PS}	Cost of production setup	Q_{UP}	Quantity of units produced
C_{QC}	Cost of quality control and inspection	R_A	Labor rate of assembly
C_{SE}	Cost of peculiar support and test equipment	R_E	Labor rate of engineering
C_{SH}	Cost of special handling and processing	R_F	Labor rate of fabrication
C_{SM}	Cost of set-up material	R_H	Labor rate of special skills
C_{TM}	Cost of tooling material	R_I	Labor rate of inspection
H_A	Man-hours of assembly	R_S	Labor rate of setup
H_E	Man-hours of engineering	R_{TE}	Labor rate of testing of all components in production run
H_F	Man-hours of fabrication of all components produced	R_{TO}	Labor rate of tooling
		U_E	Utilization rate of equipment
		U_{QC}	Utilization rate of test and quality control equipment

equivalent. In both cases it is estimated that the benefits as well as the cost savings that can be realized definitely tend to favor the fiber optics approach.

Manufacturing Technology Cost Model

In preparing an economic analysis report for a manufacturing technology program in fiber optics, certain essential cost elements must be considered. All requirements, procedures, constraints, and hardware must be made by both the fiber optics alternatives and the other alternatives under consideration. A full-scale production effort must be assumed with standard schedule and delivery dates. The cost elements in Table 6.2 have been selected for fiber optics manufacturing technology as defined below. Since this is a manufacturing process, all research and development costs are generally amortized in the manufacturing costs.

General equations for the manufacturing cost elements are given below, whereby a distinction is made between material, production, assembly, control, overhead, and other cost factors. In all equations given here, the terms defined in Table 6.2 have been used.

PRODUCTION MATERIAL

This includes the type, quantity, and cost of the materials required in the fabrication of the particular component. Special consideration should be given to the identification of any excess material costs expected because of minimum-buy requirements. The materials cost is defined as the sum of the products of all material quantities purchased and the cost per quantity unit per unit production:

$$C_{MP} = \sum_{i=1}^{n} \frac{Q_{M_i} C_{M_i}}{Q_{UP}} \tag{6.12}$$

This cost can be defined as the sum of quantity of material purchased multiplied by cost of quantity per unit production. In addition to the cost of the materials actually used, the cost of the materials wasted must be considered. The total cost of the material waste per unit produced is the difference between the quantity of the materials purchased and the quantity of the materials used times the cost per quantity per unit production, integrated over all n cost elements:

$$C_{MW} = \sum_{i=1}^{n} \frac{(Q_{M_i} - U_{M_i}) C_{M_i}}{Q_{UP}} \tag{6.13}$$

Thus, the difference between materials purchased and materials used determines the cost of material waste.

PRODUCTION SETUP

This includes the tooling, test fixtures, instrumentation, assembly line, and facilities necessary to produce a given production run of fiber optics components. The labor hours, skill levels, and personnel costs are identified for each function in a production setup. This cost element includes the engineering labor hours necessary to translate the technical data package into a production line startup. The production setup cost is a one-time expense and is composed of tooling and labor costs as well as setup times:

$$C_{PS} = (H_{TO}R_{TO} + C_{TM} + H_S R_S + H_E R_E)/Q_{UP} \qquad (6.14)$$

Therefore, the production setup cost can be described by

tooling man hours
× tooling labor rate
+ tooling materials cost
+ equipment setup and checkout labor hours
× setup labor rate
+ engineering labor hours
× engineering labor rate per unit production

SPECIAL HANDLING AND PROCESSING

This includes the amount of manual versus automated handling and processing required during the fabrication of the components. The special handling and processing cost is determined by the cost of special personnel as well as the cost of utilization of automated equipment:

$$C_{SH} = (H_H R_H + C_{EH} U_E)/Q_{UP} \qquad (6.15)$$

This cost element can therefore be described by

specialized personnel labor hours
× special labor rate
+ automated equipment monitor cost
× utilization rate per unit production

ASSEMBLY

This item contains the labor hours and labor rate required to assemble a given component. Times and labor rates must be separated by assembly function and personnel skill. Assembly cost is composed of assembly time and assembly labor rate of the system, integrated over all appropriate processes:

$$C_{AS} = \sum_{i=1}^{n} \frac{H_{A_i} R_{A_i}}{Q_{UP}} \tag{6.16}$$

It can therefore be explained by the summation of assembly labor hours multiplied by assembly labor rate per unit production.

QUALITY CONTROL

Quality control refers to the tests and methods required to ensure high yield and satisfactory performance. Costs for each test must be kept separate for all methods. Quality control is that function of management relative to all procedures, inspections, examinations, and tests required during procurement, production, and storage to provide the user with an item of required quality.

QUALITY CONTROL AND INSPECTION COST

The quality control and inspection cost element is a singular item and is a function of inspection labor time and rate as well as equipment cost and utilization. It can be expressed

$$C_{QC} = (H_I R_I + H_{TE} R_{TE} + C_{EQ} U_{QC})/Q_{UP} \tag{6.17}$$

and be explained by the following:

inspection labor hours
× inspection labor rate
+ test labor hours
× test labor rate
+ equipment utilization hours
× equipment utilization rate per unit production

OVERHEAD

Overhead includes expenses indirectly related to the production of the component, such as the cost of management and executive offices, staff services for the legal and accounting departments, public affairs expenses, and other overall business expenses. Overhead is normally a fixed percentage of direct costs. Neither fees nor profits are included in this cost element. It is the estimated burden rate for a given company and is a percentage of the direct labor production costs. The estimated burden rate includes supervision, clerical, and other indirect labor hours; insurance, vacation and holiday pay, and other fringe benefits; miscellaneous materials and manufacturing expenses; and certain fixed charges relating to property rentals and taxes.

MANUFACTURING

This includes the direct labor costs incurred during the fabrication, processing, subassembly, final assembly, reworking, modification, and installation of parts and equipment to an end item. It also includes the cost of any special test devices, checkout equipment, automatic machines, and assembly tools necessary in manufacturing of the components. The manufacturing cost is determined by the sum of fabrication and assembly worker-hours and the appropriate labor rates per unit of production:

$$C_{FA} = (H_F R_F + H_A R_A)/Q_{UP} \tag{6.18}$$

Hence, per unit production is given by
fabrication worker-hours
× fabrication labor rate
+ assembly worker-hours
× assembly labor rate

PECULIAR SUPPORT AND TEST EQUIPMENT

This cost element describes the equipment and tools required to maintain and care for the system or portions of the system while not directly engaged in the performance of its mission and which have application to a given material item. It includes, for example, vehicles, equipment, and tools used to service, transport, repair, overhaul, assemble, disassemble, test, inspect, or otherwise maintain the mission equipment and its production line. Peculiar support and test equipment cost is the sum of equipment expenses and labor and material cost:

$$C_{SE} = \sum_{i=1}^{n} \frac{Q_{E_i} C_{E_i}}{Q_{UP}} \tag{6.19}$$

It can be described by the sum of
quantity of required peculiar support equipment
× cost per equipment item procured or 0.25
× (manufacturing labor + materials cost + purchased equipment and parts for production) per unit production

Digital Long-Haul Telecommunications Link

Although fiber optics is a relatively new technology, trial systems have already demonstrated its technological feasibility, and wide use is expected in the near future. Cost is an important factor that must be considered in the implementation of any system, especially when a trade-off of alternative systems is required. The primary vehicle for determining cost is a cost-estimating relationship (CER), which translates key technical parameters into system life-cycle cost. Expected costs are derived from empirical cost and performance data; future costs are assumed to vary with technical parameters in the same manner as they have in the past.

The independent user has a number of choices in the selection of a telecommunications system. The user may lease communications lines or can construct any variety of telecommunications systems. Some of the major technical parameters influencing cost of applicable telecommunications systems are as follows:

1. Leased communications systems are influenced by access–trunk channel miles, terminations, bit rate, and multiplex hierarchy.

2. Microwave systems are influenced by length, hop distance, and number of hops.

3. Cable systems are affected by length, bandwidth, repeater spacing, and modulation method.

This study concentrates on a trade-off between coaxial cable and fiber optics transmission systems. The cost constituents of the transmission systems such as the transmitter, repeater, receiver, and cable are represented as functions of systems parameters. These parameters are the overall system length, repeater spacing, and bandwidth (10). Short-haul telecommunications systems which are considered to be less than 1 km in length and which generally do not require repeaters are excluded from this analysis.

It may be many years after long-haul fiber optics communication becomes feasible before the technique is fully implemented by the telephone companies because of existing plant investments. The Government may be in a position to push the introduction of fiber optics technology by first providing their own telecommunications cable systems over a few selected heavy-traffic routes. The telephone companies, on the other hand, are likely to counter such a move with tariff charges similar to those presently proposed which distinguish between high- and low-density routes. Such tariffs offset the competitive advantage of the independent common carriers on high-density routes. Thus, it is likely that in this specific area, future fiber optics long-haul systems may be more dependent on legal and political factors than on technical capabilities. It appears that fiber optics systems implementation will occur through the efforts of private telecommunications companies that can meet the needs for special private concerns.

The most meaningful method of comparing costs of different types of implementation is through life-cycle costs. Life-cycle cost factors are the economic and scheduling variables that offset the systems cost. For the purposes of this cost model, procurement cost includes the cost to purchase the equipment together with the cost of installation, integration and assembly, common and peculiar support equipment, testing and training, system–project management, and initial spare parts and documentation. Maintenance costs include the personnel required to maintain the system as well as management and engineering personnel. In addition, other maintenance costs such as replacement parts, utilities, support services, training, and supplies are considered.

All assumptions are based on 1978 technology and all dollar values are calculated in constant 1978 dollars. Improvements in the design and production of technologically new products which have historically resulted in reduced costs are assumed to continue to take place but are not factors in this particular model.

In any cost analysis there are certain considerations which must be stressed. Among these are the following:

Recognition of the uncertainties inherent in the specifications and costing of future systems.

Recognition that the central issue of the costing approach is the development of the specification of what is to be costed.

Emphasis in costs of relative rather than absolute accuracy.

Realization that valid cost-estimating relationships (CERs) are vital to good cost-estimation results but are often very difficult to derive. Many are based on empirical data and subject to frequent revisions.

As mentioned previously, the critical input parameters in estimating the total cost of a data transmission system are distance of complete system, repeater spacing, and bandwidth. For coaxial cable systems, historical data indicate that the following empirical cost-estimating relationship applies (11):

$$C_{\text{coax}} = 2350(14.3 - 0.838L_r)L(B/3.1)^{1/3} \tag{6.20}$$

where L_r is the repeater spacing (in km), L is the total system distance (in km), and B is the bandwidth (in MHz). This CER includes the acquisition costs of the cable terminals and repeaters, a 15% acquisition installation cost for equipment and cable, a 100% nonrecurring support cost, and a 2% per annum equipment maintenance cost over a ten-year operating period. On the basis of these empirical estimates and a ten-year life cycle, cost after acquisition for a 100 km system operating at a digital multiplexed bandwidth of 83.7 MHz per channel, with repeaters spaced 1 km apart along the entire line, results in a total system cost $C_{\text{coax}} \approx \$9.5$ million.

For the repeater spacing changes, the bandwidth will remain constant at 83.7 MHz. For ease of mathematical implementation and for bandwidth changes the repeater spacing will be assumed to be constant at 1 km. The system distance remains 100 km.

Establishment of an acceptable CER based on limited data is difficult and generally has a low level of confidence. The fiber optics data transmission system life-cycle cost is based on a limited number of cost–input values from recent civil–military systems. Kao and Collier (12), in a comparison of capital cost per voice-circuit kilometer, shows that, based on repeater spacing and channel capacities, the relative cost per circuit is greater for fiber optics if the system capacity is less than 5 Mb/s, and copper cable systems are more expensive for data rates greater than 5 Mb/s. These costs, however, are not total life-cycle costs and therefore provide only baseline information. Comparisons of repeater spacings and channel capacities were also presented by Gallawa (13). Fiber optics repeater spacings then become a function of system power budget loss contributions, receiver sensitivity, power output, and desired signal-to-noise ratio. For fiber optics digital systems, costs generally vary with a fractional power of the system bandwidth along with a direct cost relationship with repeater spacing and total system distance. The following cost-estimating relationship (CER) was developed for long-haul fiber optics data transmission systems:

$$C_{fo} = 425(20 - L_r)L(B/5)^{1/3} \tag{6.21}$$

Similar installation, nonrecurring support, and maintenance costs were also estimated empirically for this model. Maintenance costs for fiber optics equipment were assumed to be 4% per annum over a ten year operating period. Realistically, most long-haul data transmission systems are expected to operate over a twenty to thirty year time frame. However, given a ten year life-cycle cost after acquisition for a 100 km system operating at 20 Mb/s (40 MHz) per channel with repeaters spaced 5 km apart, the total system cost is $C_{fo} \approx$ \$1.28 million.

A few Army and Navy systems costs have provided inputs for this cost model (14). Figure 6.10 presents the life-cycle cost results for variations in repeater spacing and bandwidth, respectively. As variations are given for repeater spacing, the bandwidth remains constant at 40 MHz, and for variations in bandwidth the repeater spacing was assumed constant at 5 km. The system distance is held constant at 100 km.

In comparing this cost model to an actual data sheet for long-haul communications, the results appear to be very similar. The proposed system is a three-channel, base band frequency time division multiplexed system, operating over 100 km. The cable is a single graded-index fiber with repeater spacings of 5 km, and the repeater utilizes injection laser diodes and PIN detectors. The system is operating at 88 MHz per channel and is assumed to have a system MTTF greater than 10,000 hours and a total system life of ten years with replacement spares.

The estimated life-cycle cost for this long-haul fiber optics digital communications system is $C_{fo} \approx$ \$1.66 million, or an average cost of approximately \$17,000 per kilometer for a ten year operating period. This yields an approximate initial procurement plus installation cost of \$4480 per kilometer, as compared to the data sheet estimate of \$4380 per kilometer. This is better

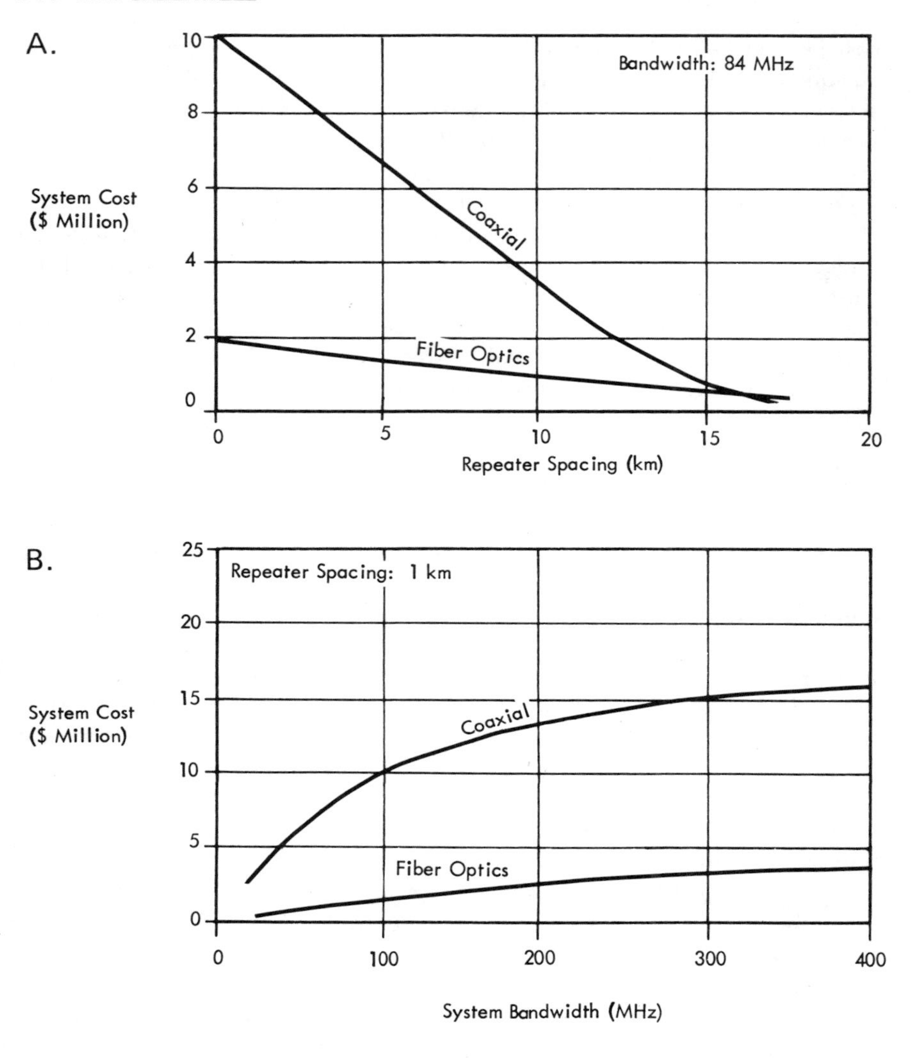

Figure 6.10 Relationship of cost and repeater spacing for coaxial and fiber optics data transmission (1978).

than a 95% confidence factor in these cost results. The model is therefore valid but also suspect due to the limited number of data samples for model sensitivity.

A similar coaxial system transmitting at the same bandwidth over the 100 km route would require repeater spacings of 1 km. The estimated life-cycle cost of this system is $C_{\text{coax}} \approx \$9.65$ million, corresponding to an average cost of over $96,000 per kilometer. This is a cost ratio of greater than 5:1 for the two systems, which implies a substantial benefit and cost offset for fiber optics digital long-haul communications applications.

ALOFT Cost Model

An economic analysis of fiber optics technology for the A-7 aircraft was undertaken in 1975. The project, Airborne Light Optical Fiber Technology (ALOFT), was assigned to the Naval Ocean Systems Center (NOSC), San Diego, to manage the flight test and evaluate the effectiveness of a fiber optics interface communications system in an operational aircraft (2). The economic analysis developed credible cost projections for three performance-equivalent cable alternatives: coaxial, twisted shielded pair, and fiber optic. NOSC coordinated the efforts of Naval Postgraduate School (NPS) personnel and McDonnell Aircraft Company (McAir) engineers.

The necessity of an economic analysis is presented in most Government program and planning documents. Economic analysis, as defined by the Defense Economic Analysis Council (DEAC) and as explained in DOD Instruction 7041.3 and DOD Directive 5000.28, as well as many others, is "the process which assists the decision maker in the allocation of resources through the determination of the costs and benefits of each future course of action." Prior to any production or procurement decisions, selected alternatives must be compared with total life-cycle costs and total benefits. All the risks and uncertainties of each alternative and their associated sensitivities must also be evaluated.

The ALOFT program, sponsored by the Naval Air Systems Command, Washington, D.C., provided a meaningful demonstration of fiber optics systems for use as internal aircraft signal data transmission links. Signal wiring in the navigation weapon delivery system (NWDS) of a Navy A-7 aircraft was replaced by electronic multiplexing circuits and fiber optics interface components. The system was first tested in the laboratory and then installed in the A-7 aircraft, in which nearly 150 flight test hours were conducted on the fiber optics interconnect configuration. Definitive comparisons have been made of the original A-7 wiring and the improved fiber optics interconnect configuration to show EMI immunity, reduction in system transient pulses produced by high magnetic induction fields, increased reliability, and total cost offsets.

The approach of the A-7 ALOFT Economic Analysis Program was to develop valid cost estimates for performance-equivalent digital data transfer systems utilizing conventional wire and fiber optics. The analysis developed credible cost projections for three performance-equivalent cable alternatives: coaxial, twisted shielded pair, and fiber optic. Cost estimates were generated by an approach which utilized two techniques. One technique computes research and development [R&D, investment, and operating and support (O&S) costs for the fiber optics and wire-interconnect data transmission subsystems ("bottom-up" model)]. The other technique computes the total weapons systems costs as a result of changes in weight of the respective subsystems ("top-down" model).

Therefore, the "bottom-up" model can be defined as one that places primary importance on the system cost, whereas the "top-down" model is used to estimate the system benefits derived from technologies to be compared. The "bottom-up" model outputs became inputs to the "top-down" model, which yielded the total life-cycle cost (LCC) data. Results from these two models were then consolidated, tested for sensitivity, and evaluated to provide the total LCC for each alternative. The results must be bounded, since they are constrained by the basic assumptions of the program. The major assumptions were as follow:

All costs were applicable to a new program, not a retrofit program, since all alternative configurations were new design subsystems for the A-7.

Cost elements would be developed only where differences between conventional wire and fiber optics costs occurred for the "bottom-up" approach.

Fiber optics presented no serious development, reliability, or production problems, and the fiber optics components were environmentally qualified with a life expectancy equal to that of comparable conventional wire components.

For the coaxial, twisted shielded pair, and fiber optics components, actual costs were used wherever possible. Specifically, the cost of materials was requested from several sources, although it was not always received. Where actual data were unavailable, engineering judgment was exercised.

Several other basic parameters had to be established before data could be applied to the "bottom-up" model. Production schedules and quantities had to be established for each alternative design configuration. Escalation, strategic commodity rate increases, and experience curve estimates had to be established for each alternative.

The base year for the economic analysis was established as beginning January 1, 1977, with a three year period assigned to perform the research, development, test, and evaluation (RDT&E) of a subsystem design. The next four years were assigned to the acquisition of the subsystem, and the final ten years were assigned as anticipated operational life without a service life extension program (SLEP). The basic A-7 Navigation/Weapon Delivery Subsystem (N/WDS) was the baseline design in a total production schedule of 812 A-7E aircraft. Of these 812 aircraft, 12 were test vehicles, the costs for which were included in RDT&E fabrication costs. The remaining 800 aircraft would meet the following delivery schedule: 80 in 1980, 240 in 1981, 240 in 1982, and 240 in 1983. It was also assumed that of the 800 aircraft, 675 would be operational vehicles. The utilization rate was assumed to be 35 hours per month for nine of the ten years of operation. The remaining year was considered a wartime operational environment and the operation rate was assumed to be 12 hours per day. A-7E aircraft attrition rates in Southeast Asia were also assumed for "survivability" analysis.

"Bottom-Up" Model (15). The Naval Postgraduate School began a directed research effort in 1975, under the management of NOSC, with a thesis on the history, technology, and costing of fiber optics (16). A followup effort emphasized life-cycle cost elements and the development of a differential cost model (17). The final report, completed in 1976, comprises data collection, data analysis, and calculation of cost elements.

Many difficulties arose in the life-cycle costing (LCC) model development. No data base on production-unit costs existed other than those for model-shop work and prototype development, and no cost models existed for component level development, such as cables and connectors, and particularly the development of fiber optics.

There was no demand for fiber optics and therefore no large-scale production, and there was no existing operational fiber optics system. These difficulties meant that standard analytical techniques could not be applied to the study and that many uncertainties had to be considered. Highlighting these difficulties in terms of future uncertainties was the recognition that current research would lead to a rapid change in the technology base. A decision made at the wrong time would freeze technological design and later production. Such a decision would also create uncertainty in overall demand, which, in turn, would affect cost. Because the design-freeze decision always provides a basis for performance uncertainty and directly affects schedule

TABLE 6.3. Equations Used for R&D, Test, and Evaluation Analysis

Cost Item	Equivalent
Engineering	Engineering worker-hours (design + development + integration) × engineering labor rate
Fabrication	Fabrication worker-hours × production labor rate × number of flight test aircraft + Assembly worker-hours × production labor rate × number of flight test aircraft + Material cost of fiber optics per foot × number of feet per aircraft × number of flight test aircraft + Sensor cost × number of sensors per aircraft × number of flight test aircraft + Tooling labor cost + Tooling material cost
Contractor development test	Engineering worker-hours test × engineering labor rate cost
Test support	Test-support worker-hours × engineering labor rate cost
Peculiar support and test equipment	[See equations used for nonrecurring investment]
Government tests	Government-test worker-hours × government engineering labor rate cost

TABLE 6.4. Equations Used for Nonrecurring Investment

Cost Item	Equivalent
Manufacturing support equipment	[Support equipment will be same as fabrication RDT&E]
Technical support	Product support for RDT&E + Cost factor × engineering worker-hours including avionics integration + Cost factor × sustaining engineering for flight test vehicles × product support labor rate
Initial spares and repair parts	Cost of spare × (number of demands per month) (repair cycle time in months) × number of operational vehicles
Maintenance training	Training cost for specialty × number of persons per aircraft × number of operational aircraft
Peculiar support	Cost factor × (cost of airframes + engines + avionics and test equipment + subsystems)
Training devices	Cost factor × (cost of airframes + engines + avionics and equipment + subsystems)
Maintenance training	Cost to train a worker × number of workers
Instructor training	Cost to train an instructor × number of instructors

uncertainty, the conceptual solution that alleviates some uncertainty involves considering only those costs that are relevant to the problem at hand. Hence, only the significant cost differences between coaxial (or wire interconnect) and fiber optics systems were considered.

These significant cost elements are presented as the NPS LCC "bottom-up" model. The cost elements are separated into four major categories: (a) research, development, test, and evaluation (RDT&E), (b) investment (non-

TABLE 6.5. Equations Used for Recurring Investment Analysis

Cost item	Equivalent
Manufacturing	[Same as fabrication RDT&E except for material and use number of production vehicles]
Purchased equipment and parts	Cost of equipment × number per aircraft × number of production aircraft
Subcontractor items	Subcontractor equipment items cost × number per aircraft × number of production aircraft
Sustaining engineering	Cost factor × production labor worker-hours × cost factor × engineering labor rate
Quality control and inspection	Labor worker-hours × quality control rate

recurring), investment (recurring), and (c) operating and support. Each cost element is defined by an estimating equation which may be an existing cost estimating relationship (CER), an engineering estimate, or a Delphi-structured estimate (17). The established cost equations are given in Tables 6.3–6.6.

TABLE 6.6. Equations Used for Operating and Support Analysis

Cost item	Equivalent
Organizational maintenance	$\left(\text{Preventive maintenance time} + \text{corrective maintenance time}\right) \times \text{cost of organizational maintenance personnel/hour} \times \text{quantity of operational equipment}$
Corrective maintenance time	$\text{Number of operating hours/year} \times \dfrac{\text{mean time to repair}}{\text{mean time between failures in flight hours}}$
Support equipment maintenance	$\text{Support equipment maintenance factor (may use 10\%)} \times \text{cost of common and peculiar support equipment}$
Spare parts and repair material	Average repair parts cost × number of demands per month × number of operational vehicles × 12 months per year × 10 years + condemnation rate × number of demands per month × number of operational vehicles × 12 months per year × 10 years × unit price per spare
Inventory management	$\text{Number of new FSN items} \times \dfrac{\text{FSN item first year cost} + \text{FSN item recurring cost} \times \text{number of years per life cycle}}{\text{number of years per life cycle}}$

Cost factors	Value	Units
Number of new FSN items		Items
FSN item first year cost	From table below	\$/item
FSN item recurring cost	From table below	\$/item/year
Number of years per life cycle		Years

Inventory Line Item Management Costs

FSN dollar value	Introduction costs	First year cost*	Annual recurring costs
\$25,000 + Over	\$680	\$1070	\$720
\$10,000–\$24,999	530	770	420
\$ 2,500–\$ 9,999	450	580	130
Under \$2,500	430	460	110
Weighted average	480	510	160

*Includes introduction cost.

Initial cost data were gathered with the use of Delphi questionnaires and experience curve theory for fiber optics life-cycle cost elements. Appropriate Delphi questionnaires were distributed both to aircraft and fiber optics manufacturers, complemented by telephone and personal interviews with representatives of manufacturers and other organizations, to finalize the data collection. From the data collection effort, cost factors were calculated for the fiber optics cost elements. These cost factors are summarized in Table 6.7 for the A-7 ALOFT project.

For example, in a fully multiplexed aircraft, twisted shielded pair and fiber optics data transfer configurations were compared. Appropriate components were selected and optimum circuits were designed to meet the data transfer requirements for a 100 V/m electromagnetic environment. The rationale for selecting the quantity of multiplexed twisted shielded pair and fiber optics signal data transfer cables was based on the most advanced electronic and electrooptic design considerations. Table 6.8 summarizes the total cost results. These costs are based on the above assumptions for the A-7 ALOFT "bottom-up" cost model. Fiber optics multiplexed systems contribute to an overall cost offset of nearly $80 million. This equates to a $100,000 cost savings per procurement aircraft over its life cycle.

"Top-Down Model" (3). The second major effort undertaken was the definition, quantification, and evaluation of system effectiveness—in other words, the determination of what benefits are received for the dollars expended. The comparison of total costs of each alternative subsystem meets

TABLE 6.7. Composition of Major Cost Contributions (Based on New Production of 800 Aircraft, Assuming 675 Operational)

Cost contribution	Cost elements	Cost factor
R&D, test, evaluation (RDT&E)	Design engineering cost	0.80
	Fabrication cost (test aircraft)	0.95 (labor) 1.05 (material)
	Development test costs	$100,000
	Test support costs	$100,000
	Test equipment costs	$100,000
Nonrecurring investment (NRI)	Initial spares and repair parts	0.83
	Maintenance training (contractor)	$4000
	Peculiar support test equipment	1.30
	Training devices costs	2.00
	Maintenance training (Government)	$8000
	Instructor training (Government	$8000
Recurring investment (RI)	Manufacturing costs	0.80
	Purchased equipment and parts	0.83
	Sustaining engineering	0.80
Operating and support (O&S)	Organizational maintenance	0.80
	Support equipment maintenance	0.80
	Spare parts and repair material	0.50
	Inventory management costs	1.60

TABLE 6.8. Example of Differential Cost Analysis of Metal and Optical Fiber Cables (1977)

	Cost (millions of dollars)			
	Metal cable*		Fiber optics cable†	
RDT&E				
Engineering	2.6		2.5	
Fabrication labor	1.5		2.2	
Fabrication material	1.4		0.7	
Technical support	0.2		0.2	
	5.7	5.7	5.6	5.6
Investment nonrecurring				
Initial spares	57.7		42.2	
Peculiar support equipment	34.3		25.1	
Training devices	0.4		0.3	
	92.4	92.4	67.6	67.6
Investment recurring				
Manufacturing labor	47.2		55.2	
Manufacturing material	3.6		3.8	
Purchased parts	86.6		41.4	
Sustaining engineering	6.0		7.1	
Quality control	7.8		9.1	
	151.2	151.2	116.6	116.6
Operating				
Organizational maintenance	1.1		1.1	
Support equipment maintenance	34.3		25.1	
Spares and repair parts material	15.8		7.5	
	51.2	51.2	33.7	33.7
Totals:		300.5		223.5

*Twisted shielded pair.

†Fiber optics bundle cable; multiplexed interconnection, 100 V/m electromagnetic environment.

only half the requirements for an economic analysis. Benefits such as improved mean time to failure (MTTF) and mean time to repair (MTTR) may result from one alternative or the other. Immunity to EMP, EMI, or RFI may also be achieved by one alternative. Signal bandwidth capacity may be increased, cable redundancy may be improved, weight and volume may be reduced, and many more benefits may be achieved. Each of these effectiveness parameters must be quantified, ranked, verified, and revised in terms of cost offsets and levels of attainment. An advanced concept cost model estimates costs and benefits as functions of designs and weight requirements. Two effectiveness areas were pursued for total cost impact from the "top-down" model: weight and MTTF. In order to relate these effectiveness factors into some quantifiable measures, a direct cost relationship approach was applied.

Normally, MTTF does not vary without some equipment modifications, improvements, replacements, etc. that may have resulted from a weight or requirement change. A direct cost relationship can be estimated from a change in MTTF. An increase in MTTF will reduce total maintenance worker-hours and replacement spares required, which is easily equated to a dollar value. Changing the MTTF of a particular avionics–electrical subsystem has a cascading effect over the entire aircraft. The subsystem failure rate impacts the system of which it is a part and the total aircraft MTTF. MTTF changes were estimated only for procurement and O&S cost changes, while RDT&E costs remained unchanged. The MTTF–cost relationship is nonlinear, but for purposes of general cost estimation a factor of \$500 ± \$100 per MTTF hour per aircraft may be used for major aircraft developments. Thus a ten hour increase in a \$2 million cost offset for the entire life cycle of the program that produces 400 aircraft, as is evident from Figure 6.11.

Changes in weight will result in two types of cost analysis evaluation (18). If the economic analysis is performed during the concept formulation of the project, a complete design-level cost analysis applies where a change in weight of the avionics–electrical subsystem has a cascading effect on the entire aircraft size. Retrofit or new design changes in the engineering development phase of a program will affect only those costs which will be directly altered as a result of reduced weight. Thus, operational design weight-saving cost deltas will be related only to such factors as fuel costs, tanker support costs, and other support elements. This cost estimation equals 10–15¢/kg per flight hour. Thus, a 7750 kg aircraft which operates on the

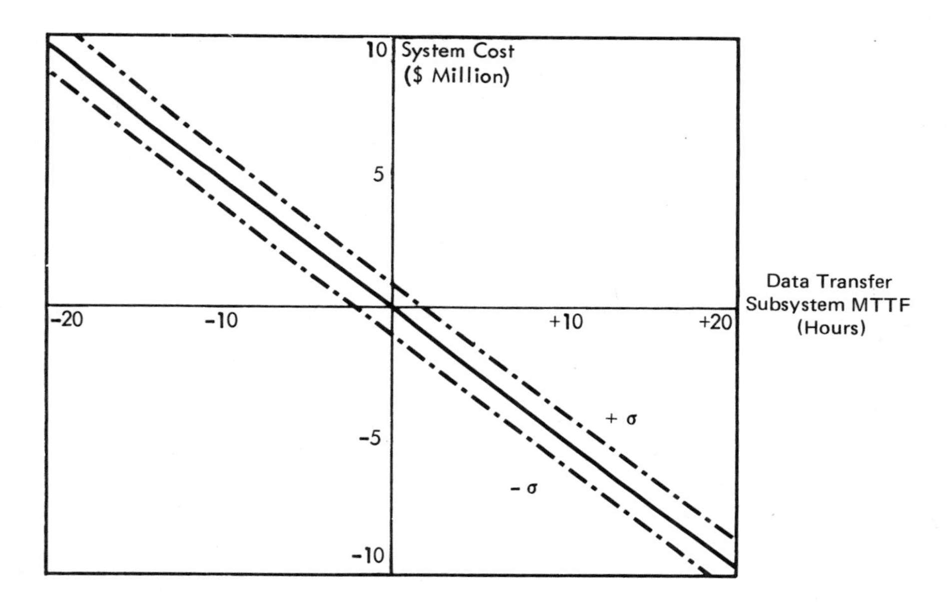

Figure 6.11 Procurement and operating cost delta values for major aircraft.

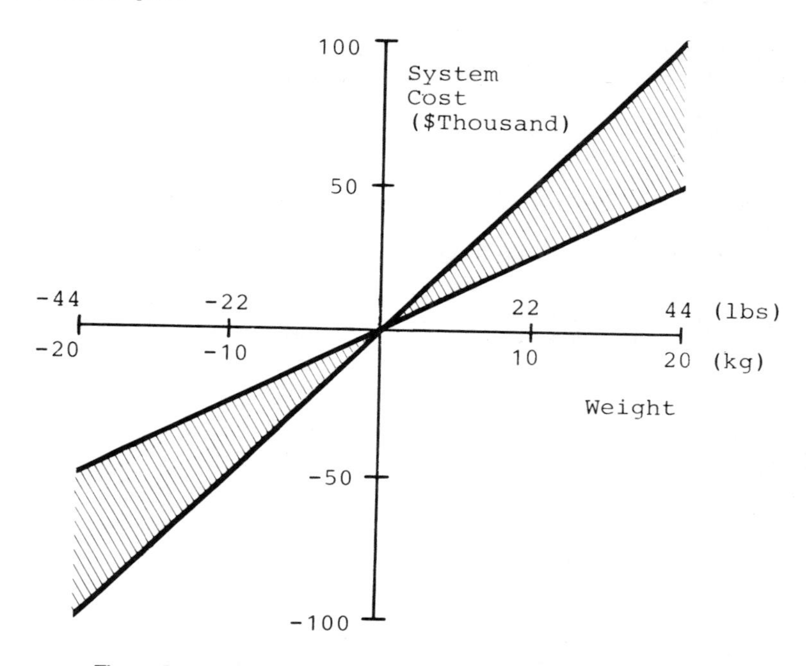

Figure 6.12 Variation of system cost with system weight.

average of 30 flight hours per month and 250 ground maintenance hours per month for a ten year operational period will expend approximately $2 million to $4 million in fuel, tanker support, and other support costs. That same aircraft at an advanced design level will have a cost estimation factor of $2000 to $4500 per kilogram, as seen from Figure 6.12, which equates to a procurement expenditure of $17 million to $34 million per aircraft. Thus, a cost offset of $120,000 to $200,000 can be realized in a production of 400 aircraft if 1 kg of weight is saved in the electrical subsystem during the procurement and installation cycle. A kilogram of weight saved during the conceptual development phase will yield a cost offset of $800,000 to $1,800,000 for the same production of 400 aircraft.

For example, in the A-7 ALOFT economic analysis, completely multiplexed fiber optics and twisted shielded pair interconnects were designed and their component weights were calculated. These weights were then compared to the baseline system, and the weight deltas were determined. This analysis is again based upon a systems design which will operate in an electromagnetic environment of 100 V/m. Table 6.9 presents the weight savings.

For this particular example, the weight analysis was estimated during the advanced design level of the overall program. Therefore, the weight has a cascading effect over the entire aircraft size and affects research and development costs as well as procurement and operating costs. Thus, Table 6.10

TABLE 6.9. Example of Weight Impact of Metal and Fiber Optics Cable for A-7 Aircraft Interconnect Designs in 100 V/m Electromagnetic Environment

Cable type	Weight (kg)	Weight saving compared to metal cable (kg)
Unmultiplexed baseline, copper wire configuration	286	0
Multiplexed twisted shielded pair configuration	150	−136
Multiplexed fiber optics configuration	137	−149
Weight savings of multiplexed fiber optics configuration compared to twisted shielded pair		−13

TABLE 6.10. Advanced Design Cost Results of Multiplexed A-7 Aircraft Interconnect Designs in 100 V/m Electromagnetic Environment (1977)

	Cost per multiplexed system (millions of dollars)	
Cost category	Twisted shielded pair	Fiber optics
Research and development	23.8	26.0
Procurement	129.8	142.0
Operating and support	97.8	105.4
Life-cycle cost (baseline)	251.4	273.4
Life-cycle cost (fiber optics)	—	−22.0

represents the entire cost offsets for the two multiplexed alternative interconnect configurations.

In comparing a retrofit twisted shielded pair system to a fiber optics system, the fiber optics system has the same weight advantage, but its cost is considerably less. Only fuel and tanker support costs are estimated, but reduced components and improved vulnerability could be projected as well with the proper extensive analysis. For a production of 800 aircraft, a fiber optics interconnect multiplexed configuration could save as much as $5 million in fuel and tanker support costs. In an advanced design cost analysis, a $22 million cost offset was estimated.

NOTES

1. Experience curve theory is primarily credited to Bruce Henderson, founder and president of Boston Consulting Group, Inc., a management consulting firm specializing in developing corporate strategy.

2. This example has been extracted from AGARD/NATO Paper 16–1, "A-7 ALOFT Economic Analysis and EMI-EMP Test Results," by R.A. Greenwell and G.M. Holma, May 1977.

3. This example has been extracted from NELC TR 2024, "A-7 Airborne Light Optical Fiber Technology (ALOFT) Demonstration Project," by R.D. Harder, R.A. Greenwell, and G.M. Holma, Feb. 3, 1977.

REFERENCES

1. J.D. Montgomery. The Demand Explosion of Fiber Optics. Proceedings of the Society of Photo-Optical Instrumentation Engineers, Vol. 77, Fibres and Integrated Optics, pp. 2–7, 1977.

2. H.F. Wolf & J.D. Montgomery. *Fiber Optics and Laser Communications Forecast 1975–1990*. Menlo Park, Gnostic Concepts, Inc., 1976.

3. G.A. Fleischer. Capital Allocation Theory: The Study of Investment Decision. New York, Meridith Corporation, 1969.

4. T. Yamane. *Statistics: An Introductory Analysis*. New York, Harper & Row, 1967, Chapters 14 and 22.

5. J. Kmenta. *Elements of Econometrics*. New York, The MacMillan Company. 1971.

6. C.R. Jones. Life Cycle Costing of an Emerging Technology: The Fiber Optic Case. Naval Postgraduate School, Monterey, Calif., NPS-55JS76031, 1976.

7. W.B. Hirschman. Profit From the Learning Curve. *Harvard Business Rev.* Jan.-Feb., p. 17, 1964.

8. R.D. Carter. *A Survey of Techniques for Improving Cost Estimating of Future Weapons Systems*. Analytical Services, Inc., March 1965.

9. L.H. Bean. *The Art of Forecasting*. New York, Random House, 1969.

10. T.A. Alper. Cost Model for an Optical Fiber Communications System. STC PP–134, Netherlands, SHAPE Technical Center, 1977.

11. Philco-Ford/IBM. *All Digital 1980 DCS Study, Appendix 10A: Cost Model*. April 1970.

12. K.C. Kao & M.E. Collier. Fiber Optic Systems in Future Telecommunications Networks. *Telecommunications* **2**(4):27, April 1977.

13. R.L. Gallawa. Telecommunications via Certain Guided Wave Structures. U.S. Department of Commerce, Department of Transportation Report 73–4, March 1973.

14. L. Dworkin. Progress Toward Practical Military Fiber Optic Communications Systems. SPIE Conference Presentation, p. 122–10, San Diego, June 1977.

15. R.A. Greenwell. A-7 ALOFT Life-Cycle Cost and Measures of Effectiveness Models. NELC TR 1982, March 1976.

16. J.M. McGrath & K.R. Michna. An Approach to the Estimation of Life-Cycle Costs of a Fiber Optic Application in Military Aircraft. NOSC Publication, July 1975.

17. E.W. Knoblich & R.J. Johnson. The A-7 ALOFT Cost Model: A Study of High Technology Cost Estimating. NOSC Publication, Dec. 1975.

18. R.A. Greenwell. Cost Analysis Methodology: The Fiber Optic Case. Presented at the International Telecommunications Conference, **2**:658, Atlanta, Oct. 1977.

7

COMPONENT APPLICATIONS

Jeff D. Montgomery

The basic fiber optics cable system requires only a few components: a transmitter, receiver, fiber optics cable, and connectors for attaching the transmitter and receiver to the cable. The transmitter converts the electronic signal to a light signal; the receiver converts the signal back from light to electronic and amplifies as needed. In some cable systems, additional components, such as splices, couplers, and repeaters, may be required.

As the fiber optics communications field has developed, the basic system component categories have evolved into a wide variety of types. These may be classified as either active or passive. The three active components are as follows:

Transmitters: LED (light-emitting diode), ILD (injection laser diode).

Receivers: PIN (PIN junction diode), APD (avalanche photodiode).

Repeaters.

Four types of passive components are as follows:

Fiber cables: graded index, step-index glass on glass (multimode; monomode), plastic coated silica, plastic fiber, bundled fiber, mixed fiber, and hybrid fiber–wire.

Connectors: single fiber, multiline, and hybrid.

Splices: single fiber, multiline.

Couplers: T, multiport (or star).

The fiber optics cable system is designed as a complete assembly which may be substituted directly for a conductive cable assembly. Thus, it accepts an electronic signal at the input end and provides a high-quality electronic signal at the output end. A selection of the compatible components given above may be used to make up the fiber optics cable system, as illustrated in Figure 7.1.

A wide range of parts and materials is required for fabrication of the fiber optics components. The transmitter, for example, requires an LED or ILD,

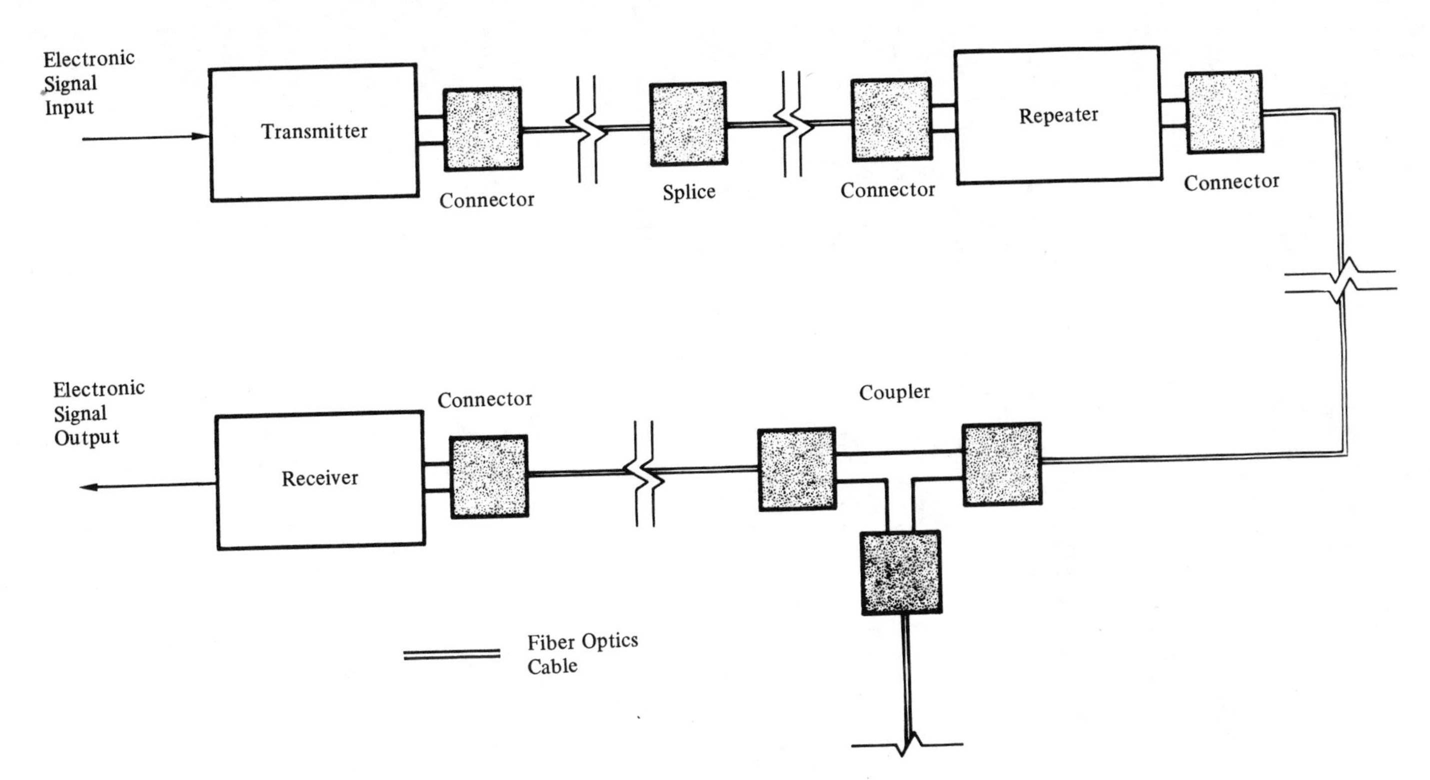

Figure 7.1 Schematic illustration of a typical fiber optics cable system.

silicon integrated circuits, connectors, and possibly passive components such as resistors, capacitors, and others. The fiber optics cable requires one or more optical fibers. It also includes typically some type of protective covering for each fiber and a protective sheath or jacket for the overall cable. Strength members, consisting of a wire or a plastic cord which runs the length of the cable, are often added. The coated fibers also may be enclosed in a metal tube or braided wire tube for mechanical protection. The cable may contain only fibers of a single type, a mixture of various types, or a mixture of fibers with conductive wire and coaxial cable.

The cable may contain a number of fibers, ranging from a single fiber to hundreds, although most cables contain between 2 and 20 fibers. Thus, an almost limitless variety of fiber optics cables may be fabricated. Many cables, for future applications, will be designed for the specific application, comparable to the variety of conductive cables found in telecommunication and similar applications. A wide range of standard cables also is being developed.

ACTIVE COMPONENTS

The active component category includes transmitters, receivers, and repeaters, which are receiver–transmitter combinations. The transmitter and receiver modules are sometimes called transducers, since they convert from one energy form to another. Repeaters are sometimes called transceivers. Each module includes one or more emitter or detector diodes, plus any required circuits for modulation or demodulation, amplification, timing, bias control, and other functions. Earlier design units incorporated a separately packaged emitter or detector mounted on a printed circuit board with numerous packaged integrated circuits, resistors, capacitors, and other components. These designs, however, are evolving into a hybrid configuration, with emitter or detector chips plus silicon IC chips mounted directly on a ceramic substrate. Other required circuit components, such as resistors or capacitors, are screened on the substrate or are chips soldered in place.

The hybrid packaging provides a smaller, lighter-weight module. The module is also more reliable because pressure contacts are replaced by soldered contacts, the number of component leads is minimized, and the lower mass make the unit less affected by shock and vibration. Hybrid circuits also are less expensive than discrete circuits when manufactured in quantities larger than a few hundred units. A disadvantage of hybrid circuits, however, is that design changes are more difficult and expensive to make. Ultimately, both photoelectric and electronic functions will be fabricated directly on a single integrated substrate, within integrated optics technology, but such monolithic integration of complete transmitter and receiver modules is at least a decade away from large-volume production feasibility.

Transmitters

Transmitter designs may use a conventional LED, a high-radiance LED, or an injection laser diode, with a wide range of supporting circuitry. The choice of emitter and related circuits depends on the level of performance required in the overall fiber optics system.

The simplest transmitter is a conventional LED with power supplied to operate at the cutoff level, using the input signal to drive the LED into emission. This is the same concept that is used in packaged optical isolators, which are now produced at rates of about a million per year. Light output coupled into the fiber is relatively low, the 0.8 to 0.9 μm emission wavelength is not an optimum match to the characteristics of graded-index fiber, and the feasible modulation rate is low, below 1 Mb/s. There is a very wide range of high-volume applications, however, for which this performance is satisfactory. These are applications that require mainly a low data rate and transmission over a maximum of a few meters. The conventional LED, together with a plastic or glass fiber and a conventional PIN detector, provides a very-low-cost link. The performance of such a link is comparable to the performance of an optical coupler, since the fiber introduces only a few dB of loss. Such links, by the mid-1980s, can be produced at a cost of less than $1.00 each in quantities of 100,000. Potential major applications for such low-performance transmitters include:

Automotive interconnection from sensors to microcomputers.
Machine controls.
Interconnection inside equipment, displacing optical isolators.
Telecommunications secure local subscriber hookup.
Instrumentation.

Transmitters incorporating high-radiance LEDs and relatively sophisticated circuitry will gain the greatest usage, on a value basis. Emission of these devices, typically at 0.82 μm, falls below the rejection band of graded-index fiber. The rise time of high-radiance LEDs has been reduced to a few nanoseconds, so that with appropriate modulation circuits they can transmit digital signals at rates up to 50 Mb/s with very low error. These devices couple an order of magnitude more light into the fiber, making them suitable for long-distance transmission. Potential major applications for which high-radiance LED transmitters are being used include

Telecommunications (interexchange trunks and local distribution).
Military communications.
Military data bus.
Computer systems interconnection.
Process controls.

Office electronic systems.
Cable television.

Transmitters incorporating injection laser diodes are reserved for applications involving extremes of transmission requirements, e.g., data rates above 50 Mb/s or distances of several kilometers between transmitter and receiver, or both. The ILD transmitter is much more expensive than the high-radiance LED transmitter because of the much higher price of the diode plus the requirement for more sophisticated circuits. Also, there is greater concern with the lifetime of the ILD compared to the high-radiance LED. Most ILD transmitters are used, in the following:

Telecommunications long-line transmission.
Military communications long-distance systems.
Very-high-data-rate computer systems interconnection.

Receivers

Receiver modules are based on PIN diode or avalanche photodiode detectors, plus very limited use of photomultiplier tubes in some experimental installations. Receivers almost always are paired with transmitters at a comparable performance level. The PIN receiver provides no signal gain; the APD receiver provides 20 dB or more gain. The APD module, however, is more sensitive to environmental and voltage changes, requires higher voltage, and is much more expensive.

A conventional PIN detector, without support circuitry, may be used with a conventional LED in applications of very short length and low data rate, such as automotive and machine control. A higher-performance PIN detector with support circuitry, sometimes including an amplifier, typically is used with the high-radiance LED transmitter in medium-performance links, up to a few hundred meters in length and data rates of a few Mb/s. The APD receiver is used with the ILD transmitter or a high-radiance LED transmitter in transmission systems of over 1 km and data rates of over 50 Mb/s.

Repeaters

Systems designers avoid the use of repeaters wherever possible. When designing a system having a combination of transmission line length of several kilometers and data rates above 50 Mb/s, designers first choose fiber with the lowest available loss and dispersion, coupled with a high-radiance LED or ILD transmitter and an APD receiver. Only when this combination of high-performance components fails to provide the desired safety margin will the designer resort to a repeater.

The use of repeaters is limited to a few of the longest interexchange trunks, plus commercial and military long-lines transmission and underseas cables. Economics do not favor the use of a low-performance repeater in a low-performance transmission line system, much less in a high-performance system. Repeaters therefore incorporate either a high-radiance LED or an ILD as the emitter and an APD as the detector. In addition to the usual receiver and transmitter functions, the repeater incorporates pulse shaping–regeneration circuitry and other functions.

PASSIVE COMPONENTS

The basic fiber optics transmission line is composed of passive components. The minimum assembly includes the fiber optics cable plus connectors at each end. For installations of several kilometers, fiber splices may be used, depending upon the nature of the installation. Data bus applications may also use couplers.

Fiber Cables

Although a few standard cables are emerging, most cable production is custom designed to fit a unique user application. The cable may include anywhere from one to dozens of independent fibers, or one or more bundles of fiber, or some combination of fibers plus wire and coaxial cable. The individual fibers may be any of several basic types, such as graded index, step-index glass, step-index plastic-coated glass, or all plastic. Any fiber type is available in several different sizes, depending upon supplier. These various combinations of fibers, and possibly wire, can then be made into cables incorporating a wide variety of strength members, individual fiber protection, and overall cable protection. The cable may be round, plait, or some other cross-sectional design. The cable may be divided into categories according to installation environment, as follows:

Flexible pull-in (duct, conduit or unprotected).
Aerial (pole or tower mounted, with or without messenger cable).
Buried.
Submarine (underseas).
Surface (military, temporary).

There are thousands of feasible and reasonable combinations of number of fibers, types of fibers and wire, type of interior cable construction, and type of external cable protection. A variety of standard cables will evolve, each suited to high-volume applications, with many needs continuing to be met by custom designs.

For applications requiring transmission lines longer than 1 km, economics favor the use of low-loss fiber. Depending upon the desired safety margin and on whether the length is small (1–2 km) or large (10–20 km), fibers with loss as high as 6 dB/km may be used, or loss below 2 dB/km may be specified. If high data rates are involved, low dispersion must also be specified. Graded-index fibers usually meet these requirements. As longer-wavelength emitters and detectors in the 1.1–1.5 μm range become more generally used in the mid-1980s, the loss of graded-index fiber can be pushed to the 1 dB/km region. For cross-country long-line applications, monomode fiber, with losses below 1 dB/km and bandwidths above 1 GHz, will ultimately be used.

Low-loss fiber cable sometimes is the optimum choice for shorter transmission links, in the range 100–1000 m. The use of low-loss graded-index fiber, in contrast to medium-loss step-index fiber, may permit the use of lower-cost transmitters and receivers; a cost trade-off analysis is needed for planning such systems.

Transmission line systems in the range of ten to several hundred meters typically use step-index fiber with loss in the range 10–40 dB/km. The main advantage of step-index fiber, compared to graded-index fiber, is its significantly lower cost, although the cost differential will shrink in future years. For short transmission lines and relatively low data rates, step-index fiber is adequate. Typical applications include the following:

Military data bus (aircraft, shipboard/submarine, surface weapons systems)

Computer systems interconnection

Process control

Telecommunications local distribution

Many designers who are working primarily with long transmission line systems, using graded-index fiber, also specify graded-index fiber for shorter applications to simplify logistics, style of cable, and connectors for all parts of the system. Conversely, for the same reasons, designers working with short to medium-length systems using step-index fiber also specify step-index fiber for very short parts of the system, although plastic fiber could meet the low-end needs.

For systems up to 100 m in length, plastic fiber with loss in the range 100–500 dB/km is often chosen. Compared to step-index glass fiber, plastic fiber has the following advantages:

Larger diameter; thus, easy to couple in more light from the LED.

Shorter bend radius without microcracking (which increases loss and reduces strength of glass fiber).

Lower cost.

Plastic fiber, however, has the disadvantage of softening at lower temperatures than glass, making its use debatable in very-high-temperature environments, such as in automotive engine compartments. Also, unprotected plastic fiber is more subject to chemical damage than glass.

The advantage of plastic fiber is substantial in a number of high-quantity, low-price applications. Nearly all of these applications are for lengths of less than 10 m, and many are for lengths of less than 2 m, so that a loss of even 500 dB/km is not significant. The leading applications for plastic fibers include automotive interconnection; machine controls; appliance controls; instrumentation; and interconnection inside equipment enclosures. Automotive and other vehicle applications, for example, could use tens of millions of plastic fiber optics links per year by 1990.

Telecommunications cables typically are directly buried in rural and suburban areas and pulled through ducts in metropolitan areas. Military communications cables typically are directly buried or installed on the surface. Cable television lines frequently are aerial. Computer systems interconnection cables are commonly flexible and are installed under false flooring or in overhead spaces without further protection. Many data bus fiber optics cables, especially for airborne use, will be made up as special harnesses, with various branches and with braid or additional molded protection at stress points.

Connectors

The variety of fiber optics connectors required will be almost as great as the cable variety. Standard multipin connectors will not be available until the multiline fiber optics cables become established as standards. Many multiline cables will be interconnected by use of single-pin connectors, one for each fiber. This makes it possible to open a single line without disconnecting the others. It also bypasses the need for design and fabrication of special connectors for special cables.

Bulkhead connectors, which mount on a panel and mate to a fiber optics connector on both sides, will be widely used. These will be predominantly single lines, although multipin devices will be developed for a few applications. Applications for the bulkhead feedthrough connector include backplane uses (bringing fiber connections into equipment), and bulkheads of aircraft avionic bays, feeding through fuel tanks, and similar uses.

Multipin connectors will be used widely in military electronics and computer systems interconnection. Single-pin connectors will be used commonly in telecommunications and industrial systems.

Splices are not widely used. Fiber drawing continues to improve, so that fibers and fiber cable will be available in lengths of several kilometers, meeting nearly all needs without splicing. Splices, however, will be required for (a) very long lines, above a few kilometers; (b) installation in ducts or conduits that include a large number of sharp bends; and (c) salvaging leftover short lengths of cable, to reduce scrap loss.

Making a low-loss permanent splice is considerably less difficult than making a demountable low-loss connector. Two main splice methods have been developed: (a) mating the fibers in a precision-machined or etched groove, using epoxy to hold the assembly together; (b) fusion, melting the two fiber ends together. Splices with less than 0.5 dB loss are easily made, at a cost of about 10% of the cost of a demountable connector pair.

Telecommunications will be the main user of fiber splices. Military fiber optics cable systems will use demountable connectors for field connection of multiple links, so that no field assembly to the fiber will be necessary. Computer systems interconnections, military data buses and most other connections are short enough that splices are not needed.

Couplers

Signal transmission between segments of a data system may be accomplished in parallel fashion by providing a separate transmission line from the processor to each remote point. Alternately, the data may be transmitted in serial fashion, multiplexing all data onto a single data bus, with tap-offs provided along the bus, and demultiplexing at each receiving point. The optimum economic choice depends on complexity, and thus cost, of the multiplex–demultiplex circuits compared to transmission line cost, which is a function of its length and other factors. For most applications, parallel transmission in fiber optics systems will be most economic. Some applications, however, can best be served by serial transmission, requiring couplers for the fiber optics line. For such applications, couplers may be of the T type, splitting off a certain percentage of the signal power into a single side branch, or multiport, whereby the signal is divided into more than two outputs of either equal or unequal power level. Nearly all fiber optics coupler applications occur in military data bus systems, primarily airborne.

8

SYSTEMS APPLICATIONS

Jeff D. Montgomery

Fiber optics cable systems may be used to replace any signal transmission medium in virtually all applications. Most optical fibers, however, are more expensive than most signal wires on a per meter basis. Fiber optics cable systems also have the added cost of the transmitter and receiver modules. Therefore, the initial penetration of fiber optics has been in applications involving significant cost factors beyond the basic wire cost. Some of these added costs, often hidden or unrecognized, include

Isolation from conducted interference (use of transformers or optical couplers)

Isolation from radiated interference (use of shielding)

Protection from explosion initiation (use of explosion-proof, pressure-tight conduit and junction boxes)

Reduction of weight (worth several hundred dollars per pound in aircraft, and valuable in automotive and other applications)

Need for greater bandwidth (requiring coaxial cable or waveguide, which is much more expensive than wire)

Need for much greater bandwidth in restricted volume (limited duct space in large metropolitan telecommunications trunks)

No detectable radiated signal (military secure communications)

Voltage isolation required (industrial instrumentation and controls; ground loop elimination)

Protection from lightning strikes

These and numerous other factors have led to the establishment of fiber optics cable systems as the most economical choice in numerous applications in the late 1970s. The growing use of fiber optics has resulted in major price reductions, due to the high economy of large-scale production of components. These price reductions, in turn, have opened up additional applications.

The type of fiber optics cable system needed varies widely, depending on the application. The requirements for the various significant applications are discussed on the following pages.

Business and Office Electronics

The major demand for fiber optics cable systems in office electronics will be in the complex "office of the future," combining a wide range of office electronics into a single system. Such a system will interconnect typically a large number of individual offices, often involving several neighboring buildings, as in an office or industrial park, or involving several floors of an office building. Facsimile reproduction, document storage with remote retrieval, and conference video phones are major consumers of bandwidth, and many of these units will be included in the typical future office complex. Word processing and data processing systems, with numerous remote terminals, are also components of this system.

The basic data distribution approach for such a complex is exemplified by the Xerox-patented Ethernet concept, which incorporates message storage and forwarding. The field operational prototype of this concept was installed in 1978 in the U.S. Presidential Executive Office Building, using coaxial cable. Many future systems, however, will need greater bandwidth than can be accommodated by coaxial cable. Also, the fiber optics cable system can be less expensive and easier to install than coaxial cable, particularly in view of building codes that require conductive lines to be installed in conduit in some areas.

The best fiber optics cable system for the office of the future will use graded-index fiber with data rates exceeding 500 Mb/s and loss below 6 dB/km. High-flexibility cable with minimal external protection, and without strength members, will be used. Data distribution will be parallel rather than serial, with individual fiber lines connecting from the various terminals to the central processing units. This approach affords greater flexibility for future system expansion, contraction, modification, and relocation, at lower cost, compared to multiplexing into serial lines with couplers. Injection laser diode transmitters will be used for a few of the highest data rate applications, but most systems will be handled by high-radiance LED transmitters. Most receivers will be of the APD type. Multipin connectors will be commonly used at the peripheral end of the cables, with single-pin connectors interfacing to backplane connectors at the central processing units.

Office electronic equipment, such as copiers and word processors will also be substantial users of short plastic fiber lengths within the equipment. These will be applications that require a very low data rate, with the fiber optics link providing signal isolation. Copiers, for example, generate a very high noise level from electromechanical relays, solenoids, switches, motors, and flash lamps. These noise spikes must be isolated from the microcomputer controller, and the low-cost fiber optics isolator is a good solution.

Communications Systems

Commercial communications systems, providing long-distance transmission of both voice and digital data, constitute the largest single long-term application for fiber optics. The high reliability, ease of installation, and low cost–bandwidth ratio are major advantages of fiber optics in this field. Communication services have been opened to competition from several different suppliers, particularly in the United States. This competition, coupled with the very strong business and consumer demand for increased communications services, is pushing all contenders into providing greater bandwidth. In transnational and intercontinental business communications, new vendors such as SBS (Satellite Business Systems) are competing strongly with the established telecommunications companies.

Most of the new business data communications systems propose to use satellites, in some cases combined with terrestrial microwave links, for long-haul transmission. After the high-bandwidth signal is received at the antenna of a business operation, however, it must be distributed throughout the business offices, often to several buildings or several floors of a single building. Fiber optics cable is a strong favorite for this local distribution, since existing telephone lines will not have the required bandwidth.

Telecommunications public utilities are recognizing the threats to their future business, as well as the huge overall business potential, and are moving rapidly into fiber optics use. Fiber optics field trial installations have been operational since 1977 in Japan, the United States, Canada, and several European countries. The experience from these field trials has been highly encouraging, leading telecommunications operating companies to accelerate their acceptance of fiber optics.

The major near-term use of fiber optics cable systems in telecommunications, through the mid-1980s, will be in interexchange trunk applications. Apart from publicity value, there are two major conditions pushing operating companies into greater fiber optics use:

Demand for greatly increased channel capacity (bandwidth) with limited duct space, such as in the downtown areas of New York, Chicago, and Los Angeles.

Microwave transmission problems in metropolitan areas, owing to building interference with line of sight or to spectrum congestion.

The initial operating experience with fiber optics in telecommunication installations has shown very high reliability for the optical components. Also, the lower loss of optical fiber, compared to coaxial cable or twisted wire pairs, will greatly reduce the number of repeaters required, thus decreasing cost and increasing reliability. Over 90% of interexchange trunk lines can be installed with no repeaters when fiber optics are used, compared to the need for several repeaters for conventional cable. For new installations in particular, elimination of repeaters also will eliminate the need for access manholes, which are very expensive to install and maintain.

Nearly all interexchange trunk transmission takes place at rates of less than 50 Mb/s, and most at rates below 10 Mb/s. System designers, however, are projecting ahead to much greater bandwidth requirements in planning their fiber optics cable systems. Graded-index fiber will be used with ILD transmitters and APD receivers. Precision low-loss connectors and splices will be used throughout the system. Therefore, new fiber optics systems that are installed will accommodate major expansion of signal inputs. Telephone operating companies foresee drastic competition for broadband transmission capabilities, coming from operations such as Satellite Business Systems Corporation in the business data field, and various cable television operators in the home broadband services field. They are planning to meet this competition, and fiber optics will play a major role in expanding their bandwidth capacity.

Long-line transcontinental telecommunications transmission will continue to be handled by coaxial cable, terrestrial microwave, and satellite microwave up to the mid-1980s. Beyond that point, however, monomode fiber will be used. The microwave spectrum is becoming increasingly crowded in metropolitan areas. Also, many large customers are becoming increasingly concerned about the lack of security of microwave transmission. Monomode fiber will be capable of transmitting over 1 Gb/s per fiber, with tens of kilometers between repeaters. Operation will be in the range 1.2–1.3 μm, using quaternary solid-state laser diodes and detectors.

Much of the local subscriber interconnection, after 1985, will also be accomplished in fiber optics. Although step-index fibers will be adequate for nearly all such applications, graded-index fibers will be used to minimize supply and interconnection problems. Hybrid LED transmitters and PIN receivers will be used with simplified, low-cost connectors.

Cable television, as an industry, is now beginning to achieve the expansion in subscribers and services which was forecast over five years ago. Subscribers for conventional and special television programs are being added rapidly to existing systems, and interest in two-way systems is increasing. Considerable controversy has developed between advocates of digital-format video transmission (such as Harris Communications) and advocates of analog transmission (such as Times Communication). In either case, low-loss fiber is needed with trunk distribution links of several kilometers. A data rate of 1 Mb/s is adequate for most analog systems, but a rate of several hundred Mb/s will be needed for planned digital systems. Graded-index fibers will be used most commonly with high-radiance LED transmitters and PIN receivers.

Computer Systems

The application of electronic data processing has multiplied geometrically during the 1970s, and strong growth will continue. The costs per function of logic, memory, and input–output circuits will continue their downward trend, making electronic data processing (EDP) feasible for a rapidly widening range of new applications. Former applications of EDP, such as business information and scientific calculation, will use processing power more intensively. The need for greater transmission line bandwidth has become apparent and will be more pressing in the future. The increasing bandwidth needs are based mainly on two established trends:

> The trend to greater use of distributed systems, with portions of the system computing power located at various points throughout the system, rather than all concentrated in a central processing unit.
>
> The trend to more data processing per unit time at any one location, with a need to transfer that information to other points within the system.

Many computer systems now have outgrown the use of conventional wire for signal transmission and have adopted coaxial cable. Coaxial cable is large, heavy, more difficult to install, and much more expensive than wire. Fiber optics cable assemblies are an excellent choice for data transmission between peripherals and mainframes, and between different processing units in the same system.

Computer systems interconnection cables are relatively short. Less than 5% of these cables are over 100 m in length, and over half are less than 20 m long. Nearly all interconnections require less than 50 Mb/s data rate, and most need less than 1 Mb/s, although some future applications will require rates of several hundred Mb/s. Fiber loss and dispersion are therefore of minor concern to most computer transmission systems designers. Plastic fiber with loss as low as 200 dB/km is being incorporated into many low-bandwidth applications such as connection of data entry terminals to mainframes. Step-index fibers, with losses in the range 10–20 dB/km, will be adequate for most other applications. Some designers, however, prefer to use fibers of lower loss and of higher bandwidth than needed for immediate applications in order to accommodate less expensive emitter and detector diodes of lower performance but greater reliability. There are wide differences in design approaches taken by transmission planners of the several leading worldwide computer manufacturers.

LED transmitters and PIN receivers are used almost exclusively for computer transmission systems. The high-radiance LED has sufficient power output and modulation bandwidth capability to meet essentially all computer systems needs; most can be met by conventional LEDs. There is little interest in the use of avalanche photodetectors. Their gain is not needed, and the higher operating voltage they require is a handicap with general low-voltage computer circuitry. There is no need for repeaters in computer systems.

Fiber optics transmission systems layouts in computers are almost entirely point to point, with a separate fiber connecting each peripheral to a mainframe, rather than a data bus system with drop-offs from a single line. Therefore, couplers find little use in commercial computer systems. As the cost of integrated multiplexing and demultiplexing circuits decreases in the future, the data bus concept will be more widely used. Fiber optics line costs, however, will be dropping rapidly at the same time, so that the data bus concept may not make major inroads.

Connectors used in computer systems will have medium-performance capability. Loss is not as important a consideration as it is in long-distance communications lines. Also, there is greater price sensitivity. Nearly all connectors will be attached at the factory, making feasible the use of relatively expensive semiautomated connector assembly machinery but simplifying connector design. Miniaturized connectors are preferred in computer systems so that they are not in the way during cable installation and so that they can be accommodated within limited space at the backplane entrance to the mainframe.

Fiber optics cables in computer systems, in addition to providing the needed greater bandwidth with less loss, also serve the useful function of eliminating ground loops and conducted or radiated interference. The small size and high flexibility of fiber optics cable, in contrast to conventional cables, which have diameters as great as 3 cm, greatly simplify installation. Also, since the fiber is nonconductive, the installer can avoid the use of expensive conduit, which is required by building codes for conventional cable in many areas.

Consumer Electronics

The main consumer application for fiber optics will be in automotive electronics. The automobile industry is rapidly changing over to the use of microcomputers for engine control, transmission control, and convenience features such as environmental control. Microcomputers can provide better fuel economy, reduced pollution, greater safety, and more pleasant driving. In automotive use, fiber optics links provide two major advantages: (a) reduction of size and weight; and (b) elimination of conducted and radiated interference.

The major appeal of fiber optics lies in its ability to reduce size and weight of cable harnesses. In luxury automobiles fully equipped with accessories, fiber optics with simple multiplexing can eliminate over 25 kg of weight from the car. Weight reduction, even in family and compact automobiles, will be significant and is therefore of top priority for automotive designers wanting to increase gasoline mileage. The space required by conventional wiring also is often of concern; there is strong interest in mounting more controls on the steering column rather than on the instrument panel, but space for control cables within the column is severely limited. Fiber optics links are a convenient solution to this problem.

Elimination of interference also is a major fiber optics advantage. Most signal line use is for interconnecting remote sensors to the microcomputers. The automotive environment includes high noise spikes from opening and closing of relay and solenoid contacts, ignition, and other sources. The fiber optics link is an excellent isolator.

Fiber optics signal lines were used in prototype automobiles in 1977 and in a few thousand trial production units in 1978. Fiber optics use is directly tied to incorporation of microcomputers. By 1985, over 90% of the automobiles produced in the United States will incorporate one to three microcomputers, and most microcomputers will have two to five fiber optics interconnecting links. This represents a demand for over 40 million individual links per year.

Automotive fiber optics interconnections will use plastic fiber for most applications. However, in the high-temperature environment in the engine compartment, step-index glass fiber will be used. Since most links will be less than 2 m and all less than 10 m long, fiber loss is not a significant consideration. Since very low data rates are used, the simplest, least expensive, conventional LEDs and PIN detectors can be used, without additional signal processing circuitry in the link. Fibers will be bonded directly to the diode chips, using epoxy followed by molded encapsulation. Any required installation interconnection will be by low-cost permanent splices, such as fusion, epoxy plus grooved block, or compression sleeve. Cost of components is of major concern in automotive use, and such link prices will be reduced to below $1.00 for each link by 1985.

Trucks, buses, and recreational and off-highway vehicles also will be major users of fiber optics links to microcomputers. The application of fibers and the associated components in these large vehicles will be similar to that in smaller cars, although some of the environmental conditions will be even more severe.

Microcomputers also will be used in other home appliances, as they are already used in microwave ovens. As automotive high-volume use drives the cost of fiber optics links down to an acceptable level, these links will start to phase into other consumer uses, such as appliance controls and home environmental control by the late 1980s.

Industrial Electronics

Fiber optics systems fit a wide range of specialized niches in the industrial electronics field. Most usage is in process control and machine control systems, connecting remote sensors to a central computer. Process control systems, such as in petroleum refineries, often use many kilometers of transmission line. Electrical interference to such transmission lines often is a problem, requiring isolation from the computer. In many systems, also, fire or explosion is a major concern, so that conductive wire must be installed in conduit. Fiber optics lines solve the problems, often at lower cost. Step-index fiber 10–20 dB/km is adequate for most applications. Low-loss graded-index fiber (less than 6 dB/km) is used for long-distance installations. LED transmitters and PIN receivers are used. Process control transmission line systems typically are laid out point to point rather than as a single data bus line with coupler taps.

Automatic machines typically have dozens of sensors and switches which must be interconnected to the computer, at distances of less than 10 m, most below 5 m. The noise isolation of fiber optics is a major advantage in this environment. Plastic fiber meets most needs, with conventional LED transmitters and PIN receivers interconnected with low-position, low-cost connectors. This fiber optics application has much in common with automotive electronics, except that quantities are much smaller and price pressure is not severe.

The capability of fiber optics to transmit data without transferring voltage is extremely useful in many industrial applications. Fiber optics lines are used to transfer sensor data across high voltage barriers, as in power generation equipment. Fiber optics lines also are being substituted in controls of equipment used in high-voltage areas, to protect personnel. The capability of optical fiber to sense temperature change, without transferring voltage, also is being used to advantage by transformer and motor manufacturers. By winding a fiber optics data link among the copper windings, any excessive "hot spot" within the unit can be detected and the unit shut down before severe damage occurs.

Fiber optics transmission lines are becoming widely used for instrumentation data transmission in nuclear power plants. Safety is of paramount concern in these plants. The fiber optics lines often are installed as redundant systems, parallel to wire lines, with the logic that an unexpected occurrence which might cause interruption of the wire line, such as a short circuit or fire burning the insulation, would not affect the fiber optics line.

Instrumentation

While most fiber optics cable systems for instrumentation are used in industry, especially in the industrial electronics field, they are also used in laboratory test, analytical, and medical instrumentation. Isolation of the electrical or electronic phenomena being tested, from the instrument, often is important, and a fiber optics transmission link is a good solution to this problem.

Advanced, automated instrumentation is increasingly making use of a computer, with data input from a wide range of sensors; fiber optics lines can simplify this interconnection. In medical applications, voltage isolation between patient and instrument can be accomplished with fiber optics systems. Instrument systems applications require short links, typically less than 5 m, and nearly all less than 10 m. Data rates are low, nearly all below 1 Mb/s. Plastic fiber, with conventional low-cost LED transmitters and PIN receivers, is adequate for most instrument applications. Some designers are using step-index glass fiber in preference to plastic.

Military Electronics

Military applications led to the early usage of fiber optics cable systems and have continued to contribute significantly to component development. In a wide range of military systems, fiber optics provides certain very useful advantages which are difficult or impossible to duplicate in other transmission systems. The early use of fiber optics by military systems designs in the late 1960s and early 1970s has been a major force in the continued component performance up-grading and price improvements. Military needs encompass all types of fibers, transmitters, receivers, and other components.

Airborne electronics systems are leading candidates for fiber optics use. Because an advanced fighter or small bomber aircraft has several hundred kilograms of copper cable for signal and power transmission, size and weight reduction is extremely important for avionic cables. Signal cable requirements can be drastically reduced by multiplexing. Further reductions can be achieved by converting from coaxial to fiber cable. Electrical interference also is a major problem among the closely packed conductor cables in an aircraft; this problem can also be solved by fiber cable.

Early aircraft installations used bundled fiber to gain greater reliability from the redundancy of hundreds of fibers carrying the same signal. As fiber strength has improved, however, and as cabling design and connectors have evolved which put less strain on the fiber, it has become apparent that much redundancy can be avoided. Future systems will employ typically single-fiber lines, with parallel redundant lines. Early connectors were low-precision devices, simple modifications of SMA coaxial connectors or military standard multipin cylindrical connectors. These are much larger than necessary. Future single-fiber connectors for airborne use will be precision machined and will be an order of magnitude smaller than bundled-fiber connectors.

Avionic fiber optics systems use LED transmitters and PIN receivers. Nearly all lines are less than 10 m in length, most less than 5 m, so that high emitter power output or detector gain are not needed. Most data rate requirements are modest, although some systems are now operating at up to 100 Mb/s. Reliability in environments up to 100°C, however, is very important. Early avionic fiber optics systems used point-to-point interconnections, but design is evolving to increased data bus use with signal couplers.

Fiber optics cables are being adopted for interconnecting satellite ground station receivers to a remote data center, typically at distances of 1 to 3 km. Security of transmission is of major concern for many of these installations, and the much greater difficulty of tapping into a fiber optics line without detection is an important advantage. The low loss of the graded-index fiber used for these systems, less than 6 dB/km, eliminates the need for repeaters. Lightning, which has caused severe damage to conventional cables and equipment in some receiving stations, is not a problem with fiber optics. Data transmission rates are high in these systems, with a long-term need for a capability of several hundred Mb/s per fiber. Low-loss graded-index fiber is used almost exclusively, with precision connectors. ILD transmitters are normally used, mainly because of their higher modulation rate capability and higher power output. Receivers use APD detectors. These modules operate in a benign environment, and size–weight restrictions are not severe. Redundant hot standby units are provided to achieve high reliability.

Army field communication historically has been accomplished by analog transmission on wire line laid out on the ground, or by radio. There are increasing problems with radio use, such as susceptibility to monitoring, many forms of electronic interference with effective transmission in battle conditions, and ability to pinpoint transmitter locations. Wire line analog communication also has numerous problems, including inability to handle a large number of channels, and need for amplifiers when long distances are involved. These problems are leading to conversion of field wire communications to digital format. A typical system is the U.S. Army TTC-39 switching system. This system incorporates a digital switch at the company level, accommodating about 20 subscribers. Some subscribers, such as data terminals, are inherently digital, and voice subscriber transmission is digitized at this point. Transmission, then, is digital to a larger division level switching system, which accommodates 600 channels. Division level nodes, in turn, may be interconnected by digital video or by transmission line. Fiber optics cables are excellent for interconnection from division to company level switches, for interconnection among equipment vans at the division level center, and for interconnection between major division level nodes. The fiber optics cable is much smaller and of lighter weight than the coaxial cable bundles otherwise required. The use of fiber optics in this system permits elimination of several trucks and associated personnel required to transport and install the larger, heavier coaxial cable. The lightweight fiber optics cable may be laid by helicopter rather than by truck. A major reduction in the number of repeaters is possible, simplifying maintenance and increasing reliability. Finally, fiber optics transmission is much more secure. These field communications systems use low-loss graded-index fiber (i.e., less than 6 dB/km) with precision multipin connectors. Long lines use ILD transmitters and APD receivers, with up to 10 km between repeaters.

The principle of using a paid-out trailing transmission line to transmit data for guidance of a launched missile has been well established over the past decade, with missiles such as the U.S. Army TOW and Dragon. Optical fiber is a promising substitute for wire in these guidance systems. The lighter weight and lower loss of fiber are significant advantages. Recently developed higher-strength step-index fiber will be used. A similar guidance technique for torpedoes is being evaluated. These applications use high-radiance LED sources and PIN detectors.

The loss in optical fiber varies as a function of fiber bending, temperature, nuclear radiation, and other environmental changes. When long fiber sections are used, such as 1 km, very minute changes in the environment can produce a measurable change in loss. This characteristic makes optical fiber a good choice as a sensor element, as well as a data transmission line, in numerous military applications. A hydrophone system for detection of submarines and surface craft is being developed by the U.S. Navy, for example, using monomode optical fiber. Fiber optics systems have been used for measurement of nuclear radiation rates in spacecraft. The use of fiber optics in sonar arrays and various other sensing applications is being evaluated.

9

SYSTEMS ASPECTS

Helmut F. Wolf

The successful implementation of an optical communications system requires the solution of a number of engineering problems that are primarily a consequence of its unique properties. Because optical fiber cables offer considerable advantages compared to metal wires, they will increasingly tend to be a substitute for the more conventional metal interconnection approaches. Contrary to the previous analysis which dealt mainly with the maximization of component performance, a discussion of systems performance has to consider the compatibility of all components and hence the optimization of the overall performance (1). Furthermore, and frequently most important, the economic aspects of system implementation and the various resultant trade-offs must be taken into account. This chapter will be concerned with the interface of technological and economic factors that have to be considered in practical systems.

GENERAL SYSTEMS PERFORMANCE

The principal components of an optical communications system are shown in Figure 9.1. Its design involves a number of options, each having advantages and disadvantages. Each component in the link must function not only by itself but in cooperation with all other components in the system (2,3). For example, the source cannot be selected independently of the detector, since the characteristics of each depend on wavelength. Some of the otherwise useful sources may be efficient emitters of adequate optical power at a given wavelength, but the detector characteristics may be unacceptable at this wavelength. Also, the absorption characteristics of the fiber must be considered in selecting the wavelength.

Aside from the criterion of component compatibility, some other system parameters can fortunately be specified arbitrarily; however, it is necessary to be aware of the consequences of that choice (4). For example, there is a cost increase associated with an increased data transfer rate. This increased cost is due to enhanced switching time and multiplexing demands, both of which have a significant impact on system cost. Since the switching time demand increases with pulse rate, the repeater cost varies with data transfer rate. This

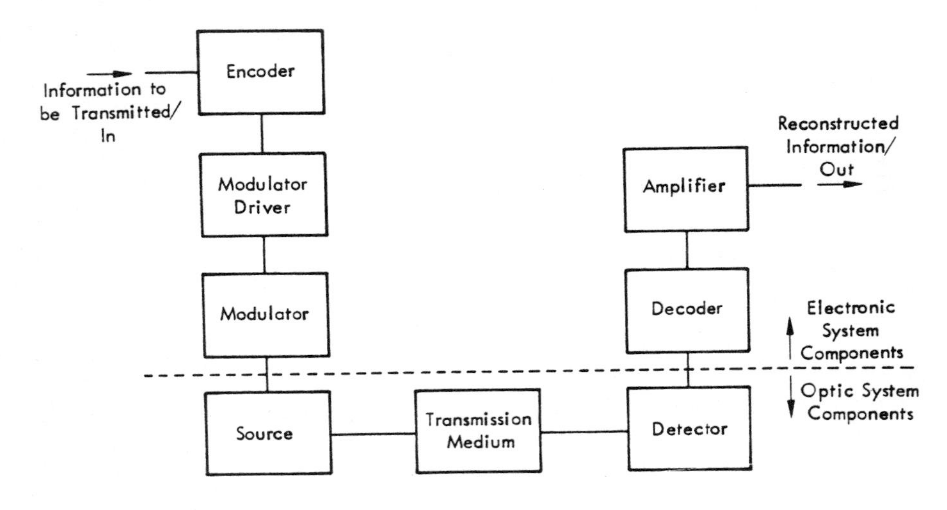

Figure 9.1 Principal components of optical communications systems.

results in an optimum pulse rate and an optimum repeater spacing. To obtain an optimum system design, we must have an understanding of the factors that affect data transfer rate limitations and of the consequential aspects. In metal wire cable systems, attenuation is usually the dominant factor; the repeater spacing, and hence the cost, can be adjusted only by changing the size of the cable, which results in high systems cost and hence often is unacceptable. In fiber optics cable systems, by contrast, entirely different factors cause data transfer rate limitations, and a corresponding change in fiber size will not produce an equivalent transfer rate improvement.

For the purpose of this discussion, fiber optics (FO) components of present design have been divided into the categories shown in Figure 9.2. This categorization will facilitate the following analysis. Optical systems can be divided into communications and data bus, with a further division of communications into digital and analog systems; data bus systems are usually digital.

In comparing the advantages and disadvantages of an optical communications system and a more conventional metal wire system, the following factors must be considered:

Electromagnetic Compatibility. Optical data transmission depends on the use of a dielectric waveguide. Hence, it is not susceptible to radio frequency interference and does not emit RF energy. Hence, optical communications offers inherent electromagnetic compatibility. Because in electrical cables electromagnetic interference may cause random burst errors, error detection and correction techniques as well as retransmission techniques must

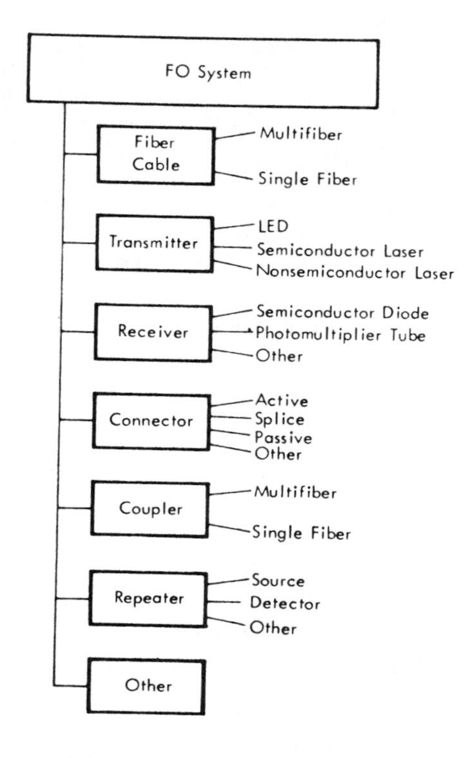

Figure 9.2 Categories of fiber optics systems components.

be employed to ensure data integrity; this adds hardware cost and decreases the effective data rate. Optical communication, on the other hand, reduces this problem and permits the use of less sophisticated and hence less expensive error detection techniques.

Ground-Level Independence. In an optical waveguide, there is no need for a return line, and the signal is not referenced to a particular ground potential. Because the two system parts that are interconnected are electrically independent, data transmission can be achieved without regard to ground potential shifts, which in electrical systems detrimentally affect the reliability of the data transmission, add to system cost, and affect performance and reliability.

Size and Weight Reductions. Both size and weight of an optical cable are significantly less than those of a metal cable. Typically, the diameter of a coated optical fiber is about 0.5 mm, compared to about 10.0 mm for an equivalent coaxial cable. This size difference of more than an order of magnitude is significant already, but the full advantage of an optical communications system becomes even more apparent when the optical cable is designed to contain several data channels through the use of space division multiplexing.

Such a multifiber cable allows parallel data transmission with a minimum increase in cable size. A substantial size reduction is also possible for the cable connectors because of the smaller optical cable size. Furthermore, cable weight is substantially reduced by the use of fibers because of the elimination of metals. For example, a fiber of 1 km length typically weighs about 50 g.

Cost and Performance Trends. Optical cables will experience considerable price reductions in the future, contrary to the price trends of metal cables. At the same time, the performance of optical cables will make significant advances, whereas coaxial cables, because of their maturity, will not display substantial performance improvements. As a result, the cost–performance ratio of optical cables will show dramatic reductions, whereas the cost–performance ratio of coaxial cables will show a slight increase. However, the cost of the fiber in an optical communications system represents only a small portion of the total system cost.

Other Fiber Advantages. In the future, optical cable will display a significantly higher data transmission rate than coaxial cables. Although the potentially high data rates of fibers will not be a reality in the next few years, the upgrading of existing data lines will allow improvements in systems interconnection techniques. Other advantages of optical cables related to the absence of signal ringing, line capacitance, and disabling short circuits, although of only secondary importance, will also make fiber optics superior to coaxial cables.

Receiver Time Delay. The signal delay associated with the receiver gain is one of the drawbacks of optical communications links in comparison with coaxial systems. The needed receiver gain is a result of the inefficiency of most source–fiber interfaces where the ineffective coupling of the optical source to the optical waveguide is associated with a high insertion loss which must be compensated by additional receiver gain; the insertion loss is typically 30 to 40 dB for an LED and 20 to 30 dB for an injection laser. The receiver gain causes signal delay and hence affects the frequency response detrimentally. This is, however, important only in short communications links, where a higher gain is required to return the signal to a logic level. In a long communications link, where the fiber loss is significantly lower than that of a coaxial cable, the total gain may be less than that of a corresponding coaxial system.

DIGITAL VERSUS ANALOG OPERATION

In communications there is a general trend from analog to digital data transmission. Although digital data transfer offers many advantages, its general acceptance is limited by the operational mode of all telecommunications systems which are presently still built around the need for analog voice communications.

The major requirement of a digital system is an increase in bandwidth, whereas that of an analog system is an improvement in linearity. Hence, in the future the relative advancement of both systems types depends on the relative improvements in switching speed and linearity, but it is expected that digital advances will surpass analog advances, so that ultimately digital techniques will dominate.

Digital operation offers the following advantages over analog operation (5):

Noise Immunity. Digital systems offer the promise of nearly unlimited transmission distance. The quality of the transmission is independent of path length through the inclusion of regenerating low-noise repeaters. Analog communication, by contrast, does not allow the elimination of noise once introduced into the system, and the signal is degraded accordingly. Hence, the limit on systems performance of an analog system is set by the accumulated signal degradation, which cannot be overcome by the use of repeaters. Both the noise and the signal are amplified in the repeaters, with the repeaters adding additional noise.

Bandwidth Increase. Digital systems allow a better utilization of the bandwidth available from components. Because in digital systems a conversion of the analog signal into digital form at the transmitter and the reconversion from the digital to the analog form at the receiver is required, an important parameter is the number of bits attributed to a word. With an n-bit word, the amplitude at a particular time can be represented as any one of 2^n distinct levels; for example, an 8-bit word corresponds to $2^8 = 256$ possible levels. An increase in word length, however, tends to degrade the bandwidth of the system, so that for faithful analog operation long words and reduced bandwidth are common. In digital systems, transmission encoding and redundancy removal, however, increase the bit rate requirements.

Cost Reduction. Digital systems allow the use of low-cost digital integrated circuits. In analog systems that use, for example, frequency-division multiplexing (FDM), precise and complex bandpass filters are required in the terminal devices. Analog FDM systems utilize multiplexing and demultiplexing functions, which can be handled only with expensive filters. Digital systems, on the other hand, need only inexpensive low-pass filters. Because of the decreasing cost of digital integrated circuits in the future and the absence of expensive filters, the cost per channel of a digital system will continue to be lower than that of an equivalent analog system.

User Flexibility. Digital systems allow the use of a common language. Hence, network design can be optimized. Also, with digital channels, voice, video, and facsimile transmission by the same network is possible, since the signals appear the same whether originally they were voice, video, facsimile, or data.

IMPLEMENTATION

The practical implementation of the combination of the various available individual components to a fiber optics communications system tends to present several difficulties (6). These difficulties are caused by source, transmission medium, and detector. Table 9.1 summarizes the principal problem areas and their suggested solutions. Unfortunately, the normally suggested solutions to primary problems cause other problems, listed here as secondary problems, together with possible solutions. Most of the secondary problems, however, have been common with conventional coaxial or pair cable systems. For example, the modulation nonlinearity of all lasers is a disadvantage mainly in analog operation, whereas in digital operation a two-level transmission line offers a theoretical solution. However, in practice there are several potential problems. For example, the spectral line of the timing signal sometimes disappears. Also, the interruption of the transmission for error monitoring can cause difficulties. In avalanche photodiodes, bias instability problems have been overcome theoretically by the use of ac-coupled amplifiers, but in practice this has resulted in the problem of dc wander in the received pulse train.

The introduction of redundancy by line coding may be a solution to these difficulties (7). For example, a two-level coding plan can provide an abundant timing component. It represents a useful concept if its decision threshold is insensitive to the received signal level variation. Such a line coding plan takes advantage of the large bandwidth of the fiber and has the following additional desirable features: (a) error detection is possible without interrupting transmission, utilizing the redundancy offered by the plan; (b) reduction of the signal-to-noise ratio due to the existence of redundancy is possible; (c) a timing tank circuit of only a relatively small Q value is required because the rectified pulse train contains an abundant timing frequency component; (d)

TABLE 9.1. Summary of Possible Problem Solutions in Optical Communications Systems

System component	Problem	Solution
Source (laser)	Primary: nonlinearity	Two-level transmission
	Secondary: occasional timing disappearance; on-line error detection	Zero code suppression; scrambling
Transmission medium (fiber)	Primary: pulse dispersion	Monomode operation; light source of narrower spectral width
	Secondary: splicing	Improved connection schemes
Detector (APD)	Primary: instability in biasing; high voltage bias	AC coupling; high voltage generation
	Secondary: baseline wander; powering efficiency	Clamping; scrambling

automatic gain control is not required, since the decision threshold of the received waveform is insensitive to level variations; and (e) a more efficient repeater powering scheme is possible. In this bias scheme a high ac voltage (several hundred volts) is fed to each repeater, instead of the more conventional approach in which a low dc voltage (of the order of 10 V) is used in conjunction with a high-voltage generator for each APD. On the other hand, the ac voltage may cause photodiode gain ripple due to imperfect A/D conversions.

We saw earlier that detector noise limitations arise from the loss of pulse amplitude due to waveguide attenuation, whereas dispersion limitations arise from the loss of pulse sharpness as a result of waveguide and material dispersion. Both types of system limitation display a different dependence of the maximum data transfer rate upon system length. The dominant portion of the noise is associated with the photodetector. In a nonavalanche detector the major noise contribution comes from thermal noise in the detector load resistor. A higher load resistance is therefore more desirable. In an avalanche detector quantum or shot noise usually dominates because of the detector gain. In general, avalanche detectors are preferred in systems where low optical power is available, but their higher sensitivity compared to nonavalanche detectors is offset by higher voltage requirements, higher cost, and the need for temperature stabilization networks. Thus, when the received optical signal is adequate, nonavalanche devices are preferable. A higher optical power can be obtained from injection lasers than from light-emitting diodes. Lasers also allow a higher modulation frequency and cause a smaller pulse dispersion. On the other hand, their threshold characteristics require careful temperature stabilization, resulting in additional circuits, higher power dissipation, and higher cost.

Systems performance not only depends upon source and detector but is also a strong function of the characteristics of the fiber. Its choice depends mainly upon the attenuation and dispersion characteristics, in addition to economic considerations. Furthermore, information on the transfer functions, applicable to the source properties and the coupling methods used, must be considered, as well as other factors, such as radiation resistance.

As implied by the earlier discussion, considerations of noise and distortion apply mainly to analog systems, where faithful reproduction of the input signal by the output has to be optimized and where nonlinearities are intolerable. In digital systems, where signal distortion can be tolerated until it begins to cause decision errors at the receiver or repeater, these criteria also apply, but they are much less critical. In digital systems, other considerations are therefore important, such as intersymbol interference between succeeding symbols and "eye diagrams." Pulse distortion depends critically upon the detailed frequency spectrum of the digital symbols transmitted and upon their phase relationships. Consequently, that set of symbols is usually chosen which, when convoluted with the full amplitude and phase modulation transfer function, will result in the least intersymbol interference (8,9).

TABLE 9.2. Illustration of the Total Loss of an Optical Communications System Consisting of Source, Detector, and Two Fibers (Total System Loss Given for Three Different System Lengths)

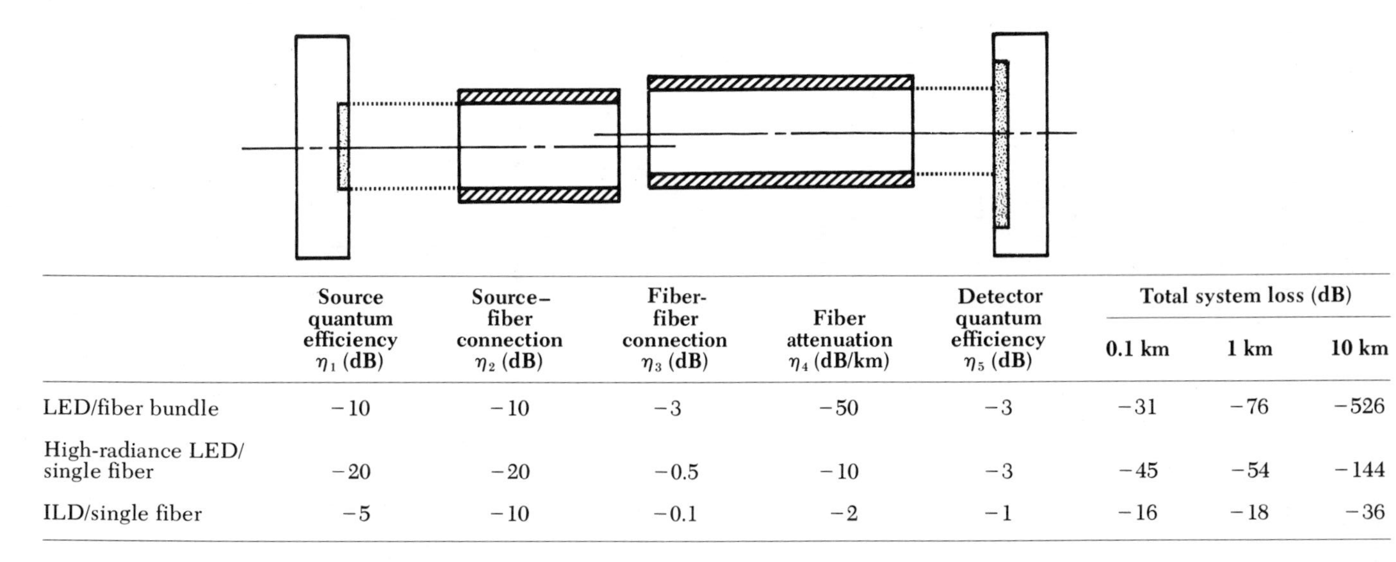

	Source quantum efficiency η_1 (dB)	Source–fiber connection η_2 (dB)	Fiber-fiber connection η_3 (dB)	Fiber attenuation η_4 (dB/km)	Detector quantum efficiency η_5 (dB)	Total system loss (dB)		
						0.1 km	1 km	10 km
LED/fiber bundle	−10	−10	−3	−50	−3	−31	−76	−526
High-radiance LED/ single fiber	−20	−20	−0.5	−10	−3	−45	−54	−144
ILD/single fiber	−5	−10	−0.1	−2	−1	−16	−18	−36

Frequency and attenuation are important systems performance and design criteria. A summarizing review follows below.

System Attenuation. The total optical loss of an optical communication system is composed of material and coupling contributions. Table 9.2 illustrates the composition of the total system loss for arrangements consisting of single fibers or fiber bundles and different light sources. The indicated individual losses are only approximate examples and are intended to serve as guides. For reasons described earlier, any system loss in excess of about 50 dB is unacceptable; hence, system lengths shown here which exceed this value do not offer a practical solution. In general, sources with high frequency response correspond to small optical power, and fibers with large numerical aperture have a higher attenuation but also a smaller fiber–fiber coupling loss. These are not principal but practical trade-offs which may be overcome by appropriate designs.

The source quantum efficiency η_1 expresses the conversion of electric power to optical power. The source–fiber connection efficiency η_2 is, under the assumption that source and fiber core diameters agree and coincide:

$$\eta_2 = \tfrac{1}{2}\,(\mathrm{NA})^2\,(\sigma + 1) \tag{9.1}$$

where σ is the directivity coefficient which is discussed in Chapter 3. For a Lambertian source, where $\sigma = 1$, $\eta_2 = (\mathrm{NA})^2$, whereas for a laser ($\sigma = 10$) the coupling efficiency is significantly higher, assuming the same numerical aperture. The fiber–fiber coupling efficiency η_3 given here assumes a fiber displacement $<\frac{1}{5}a_1$ and a fiber–fiber separation $<4a_1$, where a_1 is the core radius; this indicates that the displacement tolerance is much less critical than the separation tolerance, which requires alignment on the order of 1 μm. The fiber attenuation η_4 is independent of frequency. The detector quantum efficiency η_5 is rather high and typically approaches 100% because the translation of optical power into electrical power can be achieved without appreciable loss. Furthermore, Fresnel reflections can be minimized, and with avalanche photodiodes internal detector gain may result in $\eta_5 > 1$.

System Bandwidth. The frequency response and hence the bandwidth of the total system depends upon the frequency response of source and detector and the dispersion characteristics of the waveguide. Obviously, the system bandwidth is limited by the weakest link of the arrangement. Typically, source limitations exist above about 100 MHz for LEDs and about 1000 MHz for ILDs.

Fiber dispersion is caused by the dependence of the light velocity upon refractive index (mode dispersion) and light wavelength (chromatic dispersion). Both mode and chromatic dispersion determine the upper frequency of operation of the waveguide. For mode dispersion:

$$\tau/L = 50\,\Delta n/n \qquad \text{(SI fiber)} \tag{9.2a}$$

$$\tau/L = \Delta n/n \qquad \text{(GI fiber)} \tag{9.2b}$$

For chromatic dispersion:

$$\tau/L = \Delta\lambda/\lambda \tag{9.3}$$

In these expressions τ/L is given in ns/km if $\Delta n/n$ and $\Delta\lambda/\lambda$ are expressed in percent. For typical fibers it can be assumed that $\Delta n/n = 1\%$, whereas for typical LEDs $\Delta\lambda/\lambda = 4\%$ and for typical ILDs $\Delta\lambda/\lambda = 0.1\%$. Because $\Delta n/n$ is proportional to $(\text{NA})^2$, mode dispersion is proportional to $(\text{NA})^2$; hence, a small numerical aperture is desirable to achieve small dispersion or high bandwidth. Generally,

$$\text{NA} = 0.2\,(\Delta n/n)^{1/2} \tag{9.4}$$

With the above expressions for mode and chromatic dispersion the limiting frequency of the fiber is approximately

$$f_{\text{max}} = 1/2\tau \tag{9.5}$$

Thus, for a multimode step-index fiber the dispersion is about 50 ns/km, corresponding to a limiting frequency of 10 MHz, whereas for a graded-index fiber the mode dispersion is significantly smaller, and for the monomode step-index fiber it is completely absent. At higher frequencies chromatic dispersion rather than mode dispersion determines the upper frequency limit. Numerical data indicate that the upper frequency limit of incoherent sources combined with step-index fibers is determined mainly by mode dispersion, whereas coherent sources with monomode step-index fibers are limited mainly by chromatic dispersion. With graded-index fibers both mode dispersion and chromatic dispersion are comparable.

Because, as we have stressed in earlier chapters, the performance of an optical communications system depends upon the performance of each of its components, in Tables 9.3 through 9.8 we review the state-of-the-art of reported components and systems. These lists may not be complete but are intended to indicate the mainstream of developments. The following is a summary of the principal optical components and systems and their present (1978) status.

Optical Waveguides. In systems, signal amplitude loss and frequency limitations are of overriding concern in selecting fibers of adequate performance.

Attenuation. Rayleigh scattering represents the absolute lower limit of attenuation. Because it decreases inversely proportional to the fourth power of wavelength, the trend is toward transmission at larger wavelengths. Very pure fibers exhibit an inherent loss of about 1.0 dB/km at 0.85 μm and of about 0.5 dB/km at 1.2 μm, whereas at about 1.25 μm a small absorption peak and at 1.40 μm a large absorption peak, caused by OH^- radicals, can be observed. Hence, fibers used in systems should ideally be operated in the range 1.2 to 1.4 μm. Most optical waveguides consist of GeO_2–B_2O_3–SiO_2, GeO_2–SiO_2, or P_2O_5–SiO_2. The lowest loss has been observed in fibers made of thallium brome iodide, with a loss of about 0.01 dB/km between 4 and 5 μm.

Dispersion. Pulse broadening, which determines the system data rate capability in multimode step-index fibers, is caused predominantly by mode dispersion. The highest transmission bandwidth (limiting frequency at 6 dB) has been found to range from 20 to 60 MHz at 0.85 μm, corresponding to a pulse broadening in excess of 30 ns/km, measured in multimode step-index fibers. Hence, these waveguides can be used only in systems of low data transfer rates.

In graded-index fibers mode dispersion is minimized by a continuously changing refractive index, so that the data transfer rate is potentially much higher. The optimum refractive index profile depends upon, among others, the transmitted wavelength, the spectral bandwidth of the source, and the optical dispersion ($\partial^2 n/\partial\lambda^2$). In present graded-index fibers, mode dispersion may be smaller or larger than material dispersion. Also, disturbances such as splices, connectors, or couplers may cause mode coupling in fibers that have a high mode content, so that a dependence of pulse broadening on fiber length proportional to L^m has been observed, where $0.5 \leq m \leq 0.7$. Hence, pulse widening may increase linearly with fiber length up to a critical length, known as the mode coupling length, then display an increase proportional to $L^{1/2}$ because of mode coupling. Ultimately, the linear dependence dominates again, now because of material dispersion.

Waveguide dispersion, on the other hand, which depends on the spectral width of the source, is usually about an order of magnitude smaller. The best reported pulse-broadening data for graded-index fibers are 0.2 ns/km at 0.90 μm, with a source spectral width of 3 nm, corresponding to 2 GHz km. In monomode step-index fibers the bandwidth is limited by both material and waveguide dispersion. In the wavelength range from 1.2 to 1.4 μm, material dispersion of quartz is negligible or negative, so that the positive waveguide dispersion of these fibers can be partially compensated. At a certain wavelength, therefore, no dispersion at all can be anticipated for monomode step-index fibers, whereas only a very small amount of mode dispersion is found in properly designed graded-index fibers.

The above review implies that the most desirable wavelength range extends from 1.2 to 1.4 μm because of the favorable attenuation and dispersion characteristics of the available waveguides. To utilize this range, however, optical sources and detectors have to be developed. Tables 9.3 and 9.4 give the characteristics important for optical communications systems of present state-of-the-art waveguides. The fabrication processes listed are CVD (chemical vapor deposition), PCVD (plasma-activated CVD), MCVD (modified CVD), and POD (plasma outside diffusion). The pulse broadening data apply to a zero-dimensional light source and may vary considerably depending upon the coupling conditions. The cabling process may cause an attenuation increase by several dB/km, whereas pulse dispersion is not significantly affected by cabling, at least as long as there are no splices or other connections contained in the system.

Optical Sources. Sources for optical communications systems are either light-emitting diodes, semiconductor injection lasers, or solid-state nonsemiconductor lasers. Semiconductor lasers can be directly modulated up to very high data rates, and they allow long repeater spacings. LEDs are deployed principally in short-distance systems because of their relatively small optical output power, large spectral width, and small modulation bandwidth. Of future systems interest will be semiconductor and nonsemiconductor lasers.

Semiconductor Lasers. Double-heterostructure (DH) devices have achieved prominence so far because they can be modulated directly by variation of the injection current up to several GHz, because their emitting area corresponds to the dimensions of most fiber cores, and because their electro-optical conversion efficiency is high. GaAsAl lasers with DH structure and stripe geometry currently have the most desirable characteristics in regard to modulation capability, magnitude of the threshold current, and lifetime. Additional features, such as improvements in crystal homogeneity, reduction and optimization of stripe width and thickness to achieve a lower current density, optimization of waveguiding properties of the stripes, and strong attenuation

TABLE 9.3. Principal Characteristics of State-of-the-Art Optical Waveguides

Characteristic	SI–MM		SI–SM	GI–MM	
Material	Quartz	Multi-component	Quartz	Quartz	Multi-component
Core diameter (μm)	60	80	5	60	60
Cladding diameter (μm)	125	125	125	125	150
Attenuation at 0.85 μm (dB/km)	2–5	5–10	2–5	3–5	5–10
Maximum bandwidth for 1 km (MHz)	65	30	3000	1200	1000

TABLE 9.4. Principal Characteristics of State-of-the-Art Graded-Index Waveguides

Characteristic	GeO_2-doped quartz glass CVD	GeO_2-doped quartz glass PCVD	Multi-component glass MCVD	F-doped quartz glass POD
Core material	SiO_2–GeO_2	SiO_2–GeO_2	GeO_2, P_2O_5, Sb_2O_3	SiO_2
Cladding material	SiO_2	SiO_2	SiO_2/B_2O_3	SiO_2–F
Numerical aperture	>0.16	>0.20	>0.18	>0.19
Maximum index difference (%)	1.5	1.0	2.0	0.8
Core diameter (μm)	63	50	40	50
Attenuation at 0.85 μm (dB/km)				
Naked fiber	<6	<4	<8	<5
Fiber cable	<7	<9	—	—
Pulse broadening at 0.85 μm and $\Delta\Lambda$ = 3 nm (ns/km)	0.2–1.0	0.2–1.0	1.0–5.0	2.0–4.0

outside of the stripes to achieve improved light confinement, tend to increase the usefulness of this source type.

Examples of these optimized semiconductor lasers are the transverse junction stripe (TJS) device, the channeled substrate planar (CSP) device, and the buried heterostructure (BHS) device. They are able to reduce the number of modes to a single transverse and almost a single longitudinal mode, to eliminate nonlinearities (kinks) in the electrical–optical relationship, to reduce the threshold current (to about 5 mA at 0.7 μm stripe width and to about 30 mA at 4 μm stripe width), to decrease relaxation oscillations in the range of the threshold current, and to reduce the operational temperature and thus to increase the device lifetime.

The elimination of kinks and of relaxation oscillations effects a reduction of the variations in pulse peak power (pattern effect) that are less than 5% if the laser bias current does not deviate more than 10% from the threshold current. Aging tests of these sources at elevated temperature (>60°C) and increased light output (>10 mW) indicate an expected lifetime of about 10^5 hr at 25°C, compared to 10^6 hr for GaAsAl LEDs.

In order to utilize the characteristics of fibers at longer wavelengths, attempts are being made to apply the experience gained with GaAsAl to other semiconductor alloys, such as AlGa–AsSb and InGa–AsP. Lasers made from these materials have yielded an output wavelength in the range 1.0 to 1.2 μm, whereas GaAsInP–InP lasers theoretically allow emission in the range 0.95 to 1.7 μm. Experimentally, the range 1.05 to 1.28 μm has been verified.

Compared to GaAsAl lasers, these devices are superior because of a simpler structure (smaller number of layers), which allows a better thermal dissipation; experimental buried heterostructure devices with 5 μm stripe width have yielded more than 15 mW at 1.25 μm wavelength. Because the lifetime of this laser appears to be comparable to the GaAsAl laser, GaAsInP lasers presently offer promise. The spectral width of these devices, however, is still several nm. Its further reduction leads to the distributed feedback (DFB) laser. In this device a periodic structure on the laser stripe is used to filter out a single spectral line (of about 0.05 nm width) so that a very narrow spectral bandwidth can be achieved. Because the device is easily damaged during fabrication, a distributed Bragg reflector at the laser ends is added which controls the laser wavelength and keeps it constant over a certain temperature range. Characteristics of semiconductor lasers useful for application in optical systems are compared in Table 9.5.

Nonsemiconductor Lasers. Miniaturized solid-state lasers have been developed for the 1.05 μm wavelength (NbP_5O_{14}, $LiNbP_4O_{12}$, and Nd:YAG) and for the 1.30 μm wavelength ($LiNbP_4O_{12}$). These lasers have to be pumped by a semiconductor laser or light-emitting diode, and they have to be modulated by an external waveguide modulator, such as one made of $LiNbO_3$. These nonsemiconductor lasers have been developed mainly for potential use

TABLE 9.5. Principal Characteristics of State-of-the-Art Semiconductor Optical Sources

	LED		ILD
Characteristic	GaAsAl	GaAsIn	GaAsAl
Wavelength at peak intensity (μm)	0.82	1.06	0.82
Spectral bandwidth (nm)	40–60	50–60	2–5
CW optical power (mW)			
Total	0.1–8.0	0.1–0.5	5–20
Into fiber	0.01–0.5	—	1–5
Rise time (10%–90%) (ns)	3–20	5–10	0.1–1.0
Bandwidth (MHz)	30–150	—	100–1000
Threshold current (mA)	—	—	75–350
Operating current (mA)	200–300	40–60	100–400
Emitting area (μm^2)	1500–2000	2000–3000	4–80
Lifetime (hr)	10^6	10^6	10^5

TABLE 9.6. Principal Characteristics of State-of-the-Art Nonsemiconductor Optical Sources

Characteristic	NbP_5O_{14} (NDP)	$LiNbP_4O_{12}$ (LNP)	$Y_{2.97}Nd_{0.03}Al_5O_{12}$ (Nd:YAG)
Wavelength of peak intensity (μm)	1.051	1.048	1.064
Spectral bandwidth (nm)	3.0	1.7	0.5
Nd concentration (10^{21} cm^{-3})	4.0	4.4	0.14
Refractive index at 0.63 μm	1.62	1.58	1.83

in integrated optics systems and presently offer only limited promise for discrete optical systems. For integrated optics they must be fabricated as a thin film deposited on a substrate of lower refractive index, and their Nd concentration must be adequate for the laser effect. Table 9.6 gives the main characteristics of presently known nonsemiconductor lasers that are of potential benefit in optical systems.

Optical Detectors. To be useful in optical systems applications, detectors must have a high quantum efficiency, small excess noise, small dark current, and fast response time. Most detectors are based on silicon and are either of the PIN or of the APD type. Assuming the same quantum efficiency, an avalanche photodetector has a sensitivity that is higher by the device gain compared to the PIN diode. Silicon detectors, however, can be used only up to about 1.1 μm.

For longer wavelengths, other semiconductor materials must be employed. The Ge APD is applicable to the range 1.0 to 1.5 μm, although in most other requirements it is inferior to comparable Si devices. The GaAs detector, although useful in the same range, has a shorter response time. Other materials that show sufficient promise to permit further investigations are GaAlSb and GaAsIn. Furthermore, $CuInSe_2$–CdS has been investigated as a possible material, with a maximum device sensitivity at 1.1 μm.

Avalanche photodetectors are usually fabricated in either of two basic structures: mesa and planar diode. In the mesa APD, the p-n junction is directly illuminated by the light to provide a long interaction path for the photons and a short drift path for the carriers. In the planar diode the p-n junction is vertically illuminated by the impinging light. In both mesa and planar APD a gain–bandwidth product of about 200 to 300 GHz can be achieved by the appropriate choice of the dimensions of the avalanche region and of the electric field distribution.

In contrast to avalanche photodiodes, the PIN diode has a much inferior sensitivity; hence, it is useful mainly for systems of low data transfer rate and of short length or for those systems where cost is of overriding concern. Most PIN diodes are made of silicon. Some of the characteristics of state-of-the-art

TABLE 9.7. Principal Characteristics of State-of-the-Art Optical Detectors

Characteristic	PIN (Si)	APD (Si)	APD (Ge)
Detection wavelength range (μm)	0.4–1.1	0.4–1.1	0.4–1.6
Wavelength of peak sensitivity (μm)	0.8–0.9	0.5–0.9	1.1–1.2
Quantum efficiency (%)	60–80 (0.82 μm)	40–80 (0.82 μm)	40–50 (1.15 μm)
Dark current (nA)	0.1–5.0	0.02–15	10–200
Rise time (ns)	0.5–3	0.1–0.5	0.1–0.2
Detecting area (10^4 μm^2)	4–6	1–15	50–70

photodetectors useful in optical communications systems applications are summarized in Table 9.7.

Optical Connections. This category encompasses fixed and engagable connections, such as fiber–fiber, source–fiber, and fiber–detector joints. All of the connections affect attenuation and dispersion of signals in an optical system, with nonmatching cross-sectional dimensions, numerical apertures, ellipticity and excentricity, and lateral and longitudinal displacement being among the most critical causes of connection-induced signal deterioration. Engagable fiber–fiber connections typically have losses in the range 0.1 to 1.0 dB, whereas fixed connections (splices) typically allow lower losses, ranging from about 0.01 to 0.5 dB.

Splices cause mode conversion in graded-index fibers of high mode content, so that splices may alter the system bandwidth. The mode conversion in the splice connections is believed to be larger than that in the fiber. The source–fiber and fiber–detector connection losses are usually significantly higher than the fiber–fiber connection losses; for source–fiber connections about 20 dB and for fiber–detector connections about 1 dB can be assumed.

The source–fiber connection loss can be considerably reduced by the insertion of a small lens or by melting of the fiber end to form a microlens together with a laser diode or an etched-well emitter with an LED. With this method the LED–fiber loss can be reduced to about 12 dB and the ILD–fiber loss to about 2 dB. These techniques allow the achievement of an optical power of up to about 5 mW from an injection laser in cw operation at the entrance of a graded-index fiber with 50 μm core diameter and an NA of 0.2, and of up to about 0.5 mW from a high-radiance (Burrus) LED at the entrance of a fiber with an NA of 0.3.

The coupling of optical energy out of fibers can be accomplished by the use of transparent Si or GaAsAl–GaAs PIN diodes, which yield a radiation loss of about 15% by reflection and recombination in the 0.8 to 1.1 μm region. Their coupling loss is on the order of about 2.5 dB.

Optical Systems. The combination of the above-reviewed components in an optical communications system involves several compromises, mainly because most of the components can be optimized for a specific property which may not necessarily match the optimization of other components. Both analog and digital data transmission systems are used, although there is a trend toward increased application of digital techniques.

Analog Systems. The analog transmission of audio and video signals has been achieved in several experimental systems. One of them is described at the end of this chapter. In these trials, normally short distances (less than 1 km) and small bandwidths (less than 20 MHz) have been realized. It is expected that in the interim, until digital systems are fully established, further progress in analog systems in terms of system length and bandwidth will be achieved.

Digital Systems. Various experimental systems using digital transmission have been established. These systems allow the transfer of data, control, and measurement signals over short distances (less than 1 km) at data rates of less than about 100 Mb/s and error rates of up to 10^{-7}. Other systems that allow information transfer at several hundred Mb/s are in various phases of development, and the achievement of 1.12 Gb/s transmission over a cable length of several kilometers is being investigated at several laboratories in Japan and in Germany (at the Heinrich-Hertz-Institut in Berlin), supported by 140, 280, and 560 Mb/s system experiments. In the short-term future, repeater spacings of 30 km for wideband long-distance transmission of audio and video signals over monomode step-index fibers at several hundred Mb/s, as well as repeater spacings in excess of 50 km for transmission at 100 Mb/s using injection lasers at 1.0 to 1.4 μm wavelength, can be expected.

A few experimental optical transmission systems are compared in Table 9.8. Other systems not listed here are in various stages of development and may exceed in some cases the characteristics given here. The required system repeater spacings can be derived from an analysis of the transfer function, $S(f)$. For a coaxial cable it is

$$S(f) = -L(2.35\sqrt{f} + 0.019f) \tag{9.6a}$$

whereas for a fiber optics cable it is

$$S(f) = -\alpha L - 15.46f^2\left[(\tau_{\text{mat}}L)^2 + (\tau_{\text{mod}}L^m)^2\right] \tag{9.6b}$$

where τ_{mat} and τ_{mod} are the pulse broadening components per km caused by material and mode dispersion, m is the dimensionless fiber length exponent, f is the transmission frequency (in MHz), L is the fiber length (in km), and α is the fiber loss per km. The variation of $S(f)$ with frequency is illustrated in Figure 9.3A. It implies that in coaxial cable the transmission of broadband signals

TABLE 9.8. Principal Characteristics of Experimental Short-Range Optical Communications Systems

	Analog			Digital		
Characteristic	AEG–Telefunken	Siemens	ITT	Siecor	AEG–Telefunken	ITT
Bandwidth (MHz)	5	5	20	—	—	—
Data rate (Mb/s)	—	—	—	1	10	20
Source	LED	LED, B-LED	B-LED	LED	LED	B-LED
Detector	PIN	PIN, APD	PIN, APD	APD	PIN	PIN
Fiber attenuation (dB/km)	10–20	5–100	10–30	10–30	10–20	10–30
Input impedance (Ω)	1000	75	75		1000	50
Output impedance (Ω)	20	75	75, 600		20	50

requires considerable efforts to compensate for distortion. Contrary to the case for coaxial cable, the optical fiber displays a fundamental attenuation that is independent of frequency. At higher frequencies a decrease in $S(f)$ with the square of frequency is observed. Hence, broadband transmission by the use of optical fibers requires the employment of complex techniques to compensate

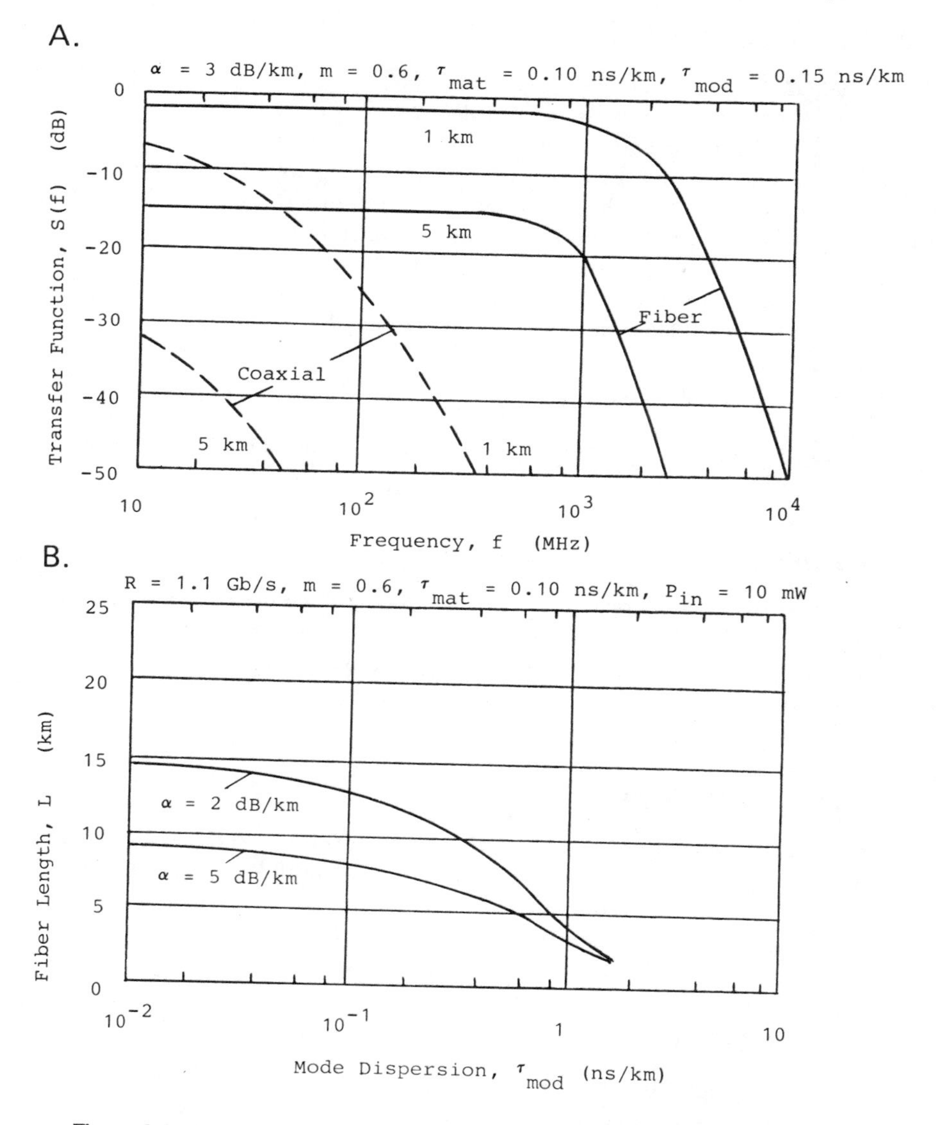

Figure 9.3 Variation of the data transfer function with frequency (A) and variation of the corresponding repeater spacing as a function of mode dispersion (B).

for distortion only at frequencies that are several orders of magnitude higher than those appropriate for coaxial systems. Figure 9.3B shows the resulting repeater distances for an optical communications system that is operated at 1.12 Gb/s at an error rate of $\epsilon = 10^{-9}$.

NOISE LIMITATIONS

In analogy to other communication systems, the presence of noise in an optical system is of major concern to designers. It is usually not possible to eliminate noise in an optical communications system once it has been introduced. All subsequent regenerative processes in repeaters tend to amplify the present noise as well as the signal. Hence, all efforts to improve the ratio of signal power to noise power must concentrate on a reduction or elimination of all noise sources and an increase in signal level. As in all-electronic systems, the signal-to-noise ratio (S/N) is defined as the ratio of the mean square signal current to the total mean square noise current,

$$S/N = \langle i_S^2 \rangle / \langle i_N^2 \rangle \tag{9.7}$$

Noise in an optical communication system is generated mainly by the following factors:

Transmitter

The only noise generator within the transmitter is the optical source, assuming that the electronic circuits in front of the optical source are noiseless (which is usually a reasonable assumption, if the electric signal at the source is at a high level compared to the associated electric noise sources). The source noise depends upon the source type.

LIGHT-EMITTING DIODE

In an LED the major noise contribution originates from random-phase generators. An LED can be considered as consisting of a multitude of random-phase noise sources that cover the entire frequency spectrum (up to about 2×10^{14} Hz) and that are simultaneously intensity modulated by variations in the drive current.

INJECTION LASER

In a semiconductor feedback laser, noise is caused not by a multitude of random-phase noise generators but by noise resonance at high modulation frequency. A rise in noise output is observed near threshold that is quenched by saturation effects as the strength of the oscillation is increased. In addition, under certain excitation levels, anomalously high resonant peaks have been observed in the noise spectrum, usually, however, at frequencies near 1 GHz or above. Thus, at modulation frequencies up to a few MHz, noise introduced by an injection laser into an optical data link can be neglected. This is an additional reason for the superiority of injection lasers over light-emitting diodes.

Receiver

The photodetector is the major noise source of an optical system; hence, it limits system performance. There are three major detector noise sources.

QUANTUM OR SHOT NOISE

It is associated with the quantization of the light upon detection and is a consequence of the statistical nature of the process by which the incoming light generates electrical carriers in the detector.

THERMAL OR JOHNSON NOISE

It is caused by thermal noise fluctuations in the equivalent input resistor of the amplifier required to elevate the detected signal level and in the amplifier itself.

DARK OR LEAKAGE CURRENT NOISE

This noise contribution is caused by leakage current, even in the absence of any illumination. It can be divided into noise originating in the photocarrier generation region that experiences the full avalanche gain and noise originating past the avalanche region so that it effectively has bypassed the gain mechanism. In a PIN diode without gain, the first of these two contributions is negligible, whereas in an Si APD operating at a high gain, the second contribution is negligible. In a well-designed system, the signal current is large compared to the dark current, so that only quantum and thermal noise have to be considered.

Repeater

Repeaters that are necessary in long systems to regenerate the amplitude and pulse shape of the optical signal also introduce system deficiencies. In a digital system, a repeater increases mainly the bit error rate, and in an analog system it introduces noise to the signal. Although digital repeaters also introduce noise, the noise increase is less critical than in analog systems. In an analog system containing a chain of N identical repeaters, the signal-to-noise ratio at the output is related to that of the input of the chain by

$$\frac{(S/N)_{out}}{(S/N)_{in}} = -10 \log(N + 1) \tag{9.8}$$

This shows that each repeater decreases the signal-to-noise ratio by about 3 dB. In practice, the cable length is therefore chosen such that it is based on $(S/N)_{out}$ or third-harmonic distortion, whichever limit is reached first.

Spurious Ambient Light

Light not associated with the signal may reach the detector if precautions are not carefully taken. There are several possible sources for such undesirable light, which occurs as electrical noise at the detector output: ambient light, light multiply reflected at discontinuities, and luminescence generated by the fiber itself. Normally, however, light resulting from these sources is negligible. For example, fiber ends can be shielded from unwanted light sources, and proper cladding and jacketing of individual fibers in a multifiber bundle cable will reduce optical crosstalk to insignificant levels. Also, multiple reflection effects have been found to be small. Luminescent effects are usually also small, except under irradiation conditions. In normal commercial applications such irradiation does not occur, but in military applications radiation effects (X rays, γ rays, or neutrons) caused by nuclear explosions may be a serious problem. Also, in satellite and other space applications, this effect has to be considered because it may seriously affect the noise characteristics of the system.

Detector Noise

In the following, we will give a more detailed analysis of detector noise because of the critical importance of the contribution to the system noise by the detector. A more system-oriented approach will be taken. Generally, the product of the signal-to-noise ratio and the bandwidth B is

$$(S/N)B = \frac{m^2 I^2}{4(qIF_0 + 2kTF_1/R_e G^2)} \tag{9.9}$$

where m is the modulation index ($m = 1$ is a reasonable assumption), I is the generated signal photocurrent, F_0 is the equivalent noise factor associated with the gain mechanism (F_0 is gain dependent such that $F_0 = G^{1/2}$—so that in the absence of gain, as in the p-n or PIN photodiode, $F_0 = 1$; in an avalanche photodiode, where $G = 50$, it is about $F_0 = 7$), F_1 is the noise factor of the amplifier (typically $F_1 = 4$), G is the current gain within the photodetector ($G = 1$ for p-n and PIN diodes, $G = 40$–150, typically 100, for avalanche photodiodes), R_e is the equivalent detector load resistance (typically ranging from less than 100Ω to several MΩ), and T is the absolute temperature. Most of these terms depend upon the operating bias. The above relationship shows that for a given photodetector operated at a given bias level, there is a trade-off between S/N and bandwidth. If it is desirable to improve S/N, the term $\langle i_N^2 \rangle$ must be minimized and the term $\langle i_S^2 \rangle$ must be maximized. The mean square noise current is

$$\langle i_N^2 \rangle = \langle i_Q^2 \rangle + \langle i_T^2 \rangle = 2qIBG^2F_0 + 4kTBF_1/R_e \tag{9.10}$$

where i_Q and i_T represent the quantum and thermal noise current contributions. The mean square signal current, on the other hand, is

$$\langle i_S^2 \rangle = \tfrac{1}{2}(mIG)^2 \tag{9.11}$$

The generated signal photocurrent is

$$I = (q/hc)\lambda\eta P \tag{9.12}$$

where q and h are electron charge and Planck's constant, respectively, c and λ are velocity and wavelength of light, respectively, η is the detector quantum efficiency, and P is the optical power arriving at the photodetector. For $\lambda = 0.85$ μm and $\eta = 0.85$, the current–power relationship is $I = 0.57P$. Thus, the generated photocurrent per received unit of optical power $I/P \approx 0.8\lambda\eta$, which indicates that the available current per watt of optical power is proportional to the wavelength and the quantum efficiency of the detector; a wavelength increase from 0.85 μm to 1.05 μm hence results in a detection conversion efficiency enhancement of about 25%. A quantum efficiency of about 85% can be considered to be the maximum achievable. Any improvements obtained are marginal and are gained at a high penalty. A low load resistance R_e (about 100 Ω) allows the use of a high (several hundred MHz) bandwidth without the need for equalization. A high load resistance (about 1 MΩ) allows the detection of digital signals with an adequate S/N, but the pulse shape is badly distorted and equalization is required at high data rates.

Several conclusions can be drawn from the above discussions. At high signal level, shot noise dominates; at low signal level, thermal noise is more significant. At low received light level the gain of an avalanche detector makes it superior to a p-n or PIN device, while at high light level the noise associated with the avalanche amplification process results in the superiority of the PIN diode. The variation of the (S/N)B product of avalanching and nonavalanching photodetectors with the optical power received at the detector based on the previous discussion is shown in Figure 9.4 for a typical situation at room temperature with low, medium, and high load resistance. At high light level the influence of the load resistance is insignificant and the (S/N)B prod-

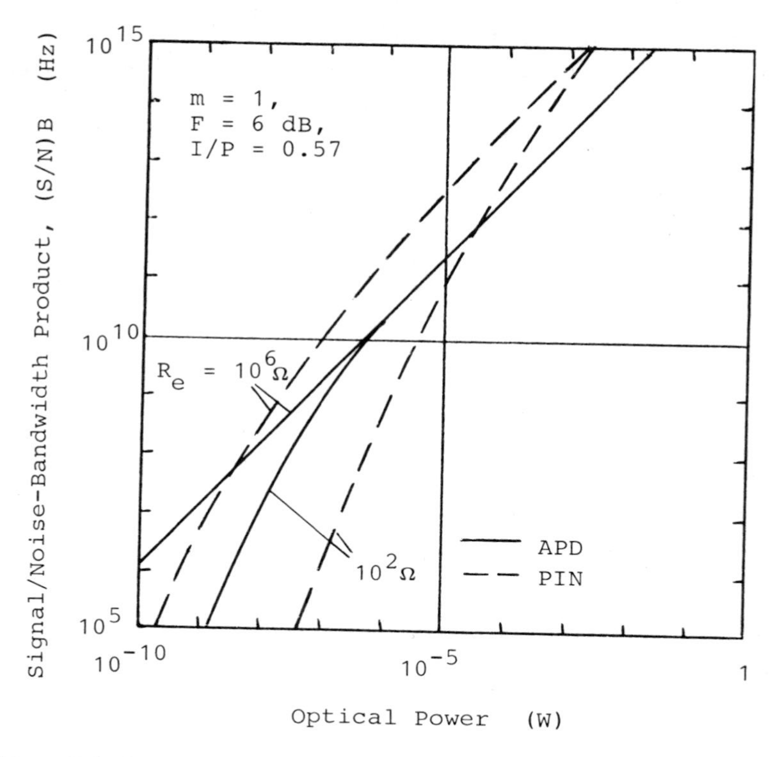

Figure 9.4 Signal-to-noise ratio–bandwidth product as a function of optical power, load resistance, and detector type.

uct is directly proportional to the light level. At low light level the (S/N)*B* product increases with the square of the optical power and proportional to the load resistance. The (S/N)*B* of a p-n or PIN diode is about one order of magnitude higher than that of an APD at high optical power, and it is about two orders of magnitude lower than that of an APD at very low optical power. For example, at an available optical power of 100 mW at a bandwidth of 100 MHz and for a load resistance of 100 Ω, the (S/N)*B* product of a p-n or PIN photodetector is about 5×10^{16} Hz, corresponding to a permissible S/N of 87 dB; however, that of an avalanche photodetector under the same operating conditions is S/N = 77 dB. On the other hand, at an available optical power of 100 mW and a load resistance of 100 Ω, the S/N of a p-n or PIN diode at 100 MHz is only 4 dB and that of an APD is 36 dB.

From this analysis we appreciate that the reduction of the signal-to-noise ratio in an optical communications link is of considerable importance. The tolerable S/N is a function of the transmission mode of the system and hence differs for digital and analog systems. Television transmission provides a representative example of noise requirements. Binary digital pulse transmission of good cable TV quality requires S/N of better than 22 dB; this corresponds to a bit error rate (probability of error per transmitted bit) of 10^{-10}, assuming Gaussian noise that is independent of signal level. If S/N is reduced to 20 dB, the bit error rate increases to 10^{-6}, which becomes unacceptable. Analog television transmission requires a minimum S/N of 45 dB for good cable TV quality; S/N of 30 dB represents the lower limit of marginally acceptable TV quality. This leads to the following more general conclusions in regard to noise performance: Satisfactory noise performance is considered to be equivalent to the condition that the noise current does not exceed 1% of the signal current. In digital data transmission a minimum S/N = 25 dB is required in order to provide error-free data transfer, whereas in analog data transmission a minimum S/N = 40 dB is required.

The signal-to-noise ratio of an analog communications system can be improved by pulse-position modulation (PM), compared to intensity modulation (IM). This can be achieved at the expense of increased system complexity and reduced transmission bandwidth. A comparison of the two modulation techniques follows.

INTENSITY MODULATION (IM)

In an intensity-modulated system, information is transmitted by intensity variations of the source output with the modulating electrical signal. Demodulation is accomplished through the use of a photodetector whose electrical output is proportional to the power of the optical input. The IM transmitter and receiver are relatively simple, and transmission is achieved in the signal bandwidth. On the other hand, the lack of efficiency in S/N is due to the fact that the optical source cannot be modulated more than 100% without severe distortion (in practice, the modulation level may be less than 100% because of source or modulator nonlinearities).

PULSE-POSITION MODULATION (PM)

In a PM system, information is transmitted via a sampled version of the analog signal, whereby each sample is assigned a time slot and the sample amplitude is represented by the location of a narrow optical pulse within the time slot (7). Thus, the time shift of the pulse relative to a reference (usually the center of the time slot) is proportional to the sample amplitude. The duration of the pulse is usually small compared to the length of the time slot, resulting in a low duty cycle. The use of narrow pulses expands the required bandwidth. Since the information is conveyed by the location of the pulses rather than by their detailed shape, source and modulator nonlinearities do not restrict the PM system operation as much as they restrict IM systems. On the other hand, PM transmitters and receivers are more complex than those required in an IM system. This is a consequence of the need to sample the signal, to generate the optical pulse train, to detect the pulses, to extract the sample values, and to reconstruct the signal waveform. Also, synchronization must be maintained.

DISPERSION LIMITATIONS

In addition to system performance limitation by noise, system degradation as a result of signal dispersion has important economic implications. A detailed discussion of dispersion was given in Chapter 2. Here the important aspects of signal dispersion are reviewed as they affect system performance.

As evident from our previous discussion, waveguide and material dispersion significantly affect the system data transfer rate through influences on the pulse shape (10, 11). Depending on the fiber index profile and the number of modes the fiber can support, we can distinguish three cases:

Multimode Step-Index Fiber. In this fiber the maximum pulse rate depends on the power distribution among the various modes and on the index difference between core and cladding. Assuming equal modal attenuation, it is approximately

$$R_{\max} \cong \frac{c/n_1\Delta}{L} \tag{9.13}$$

where the terms are defined as above. The factor $c/n_1\Delta$ takes into account the mode power distribution. It is, however, pessimistic because higher-order modes display stronger attenuation than low-order modes. Furthermore, mode mixing will reduce modal dispersion so that the maximum pulse rate of a step-index fiber exhibits an $L^{-1/2}$ rather than an L^{-1} dependence on fiber length, according to

$$R_{\max} \approx \frac{3c/n_1\Delta}{2\sqrt{L}} \tag{9.14}$$

Multimode Graded-Index Fiber. Basically, the same considerations apply to a graded-index fiber as to a step-index fiber, except that pulse dispersion is significantly smaller in the graded-index fiber because of light-speed enhancements resulting from the index profile. Again, the power distribution among the modes affects the maximum pulse rate, which can be approximated by

$$R_{\max} \approx \frac{2c/n_1\Delta^2}{L} \tag{9.15}$$

A small improvement over this value can be achieved if mode mixing is taken into account.

Monomode Step-Index Fiber. Because in a single-mode fiber only one mode can propagate, pulse dispersion is minimized and the above consideration of mode mixing does not apply. In this case, the maximum pulse rate is approximately inversely proportional to the square root of fiber length:

$$R_{\max} \approx A/\sqrt{L} \tag{9.16}$$

where A is a factor that includes the maximum allowable pulse overlap and the second derivative of the axial wave number; typically $A = 23 \times 10^9$ pulses/s, assuming a pulse overlap $\delta = 0.01$.

The maximum distance between repeaters, L, as a function of a given pulse rate or of the maximum pulse rate for a given fiber length, is shown in Figure 9.5 for the three major fiber types with and without mode mixing. Also, the cable length is plotted against data transfer rate for typical coaxial cables. Some variations are frequently observed, depending on the coaxial cable dimensions and attenuation. The data transfer rates of the Bell System, ranging from T1 through T4 and representing the spectrum from 1.5 to 281 Mb/s, are indicated. The illustration shows the significant increase in pulse rate for a multimode step-index fiber and the substantially less apparent increase for a graded-index fiber. A considerable improvement in either permissible fiber length or data transfer rate is evident for a monomode fiber compared to multimode fibers, but even a graded-index fiber exhibits a considerable improvement over a multimode step-index fiber. A comparison with present coaxial

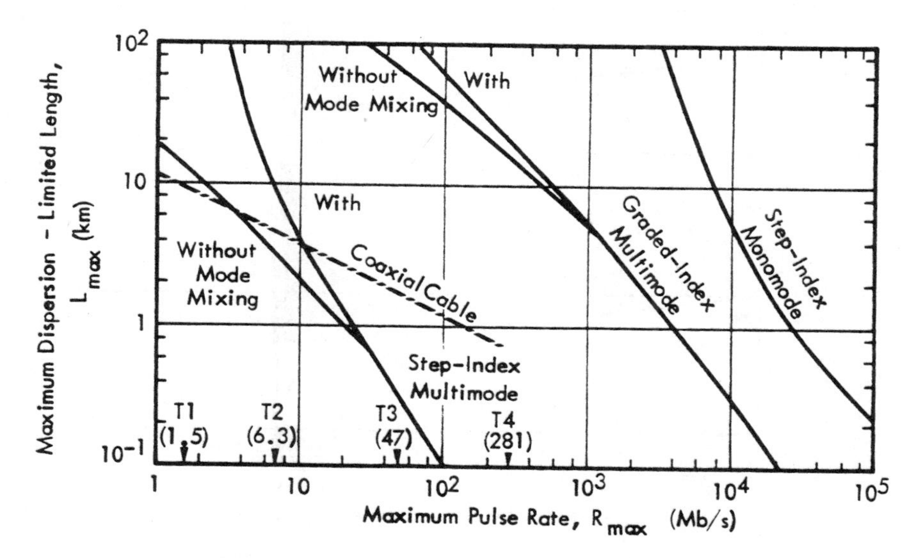

Figure 9.5 Relationship between fiber length and maximum data transmission rate.

cables points out the signficiant technical advantages that can be obtained, at least in theory, by optical cables.

In contrast to an ideal optical waveguide (in which a perfect index profile, constant cross-sectional fiber area over the entire fiber length, and the absence of structural imperfections are assumed), a real waveguide exhibits dispersion characteristics that are modified by the structural imperfections within the fiber and the stress-induced discontinuities through cabling. The resulting modifications cause mode mixing, radiative losses, and differential mode losses. These, in turn, result (a) in dispersion characteristics that are not linearly proportional to fiber length and (b) in a high sensitivity of the transmission properties to the launching conditions. Consequently, in system design the modified dispersion characteristics, as defined by waveguide, material, and multimode dispersion, form practical design limits.

In a multimode fiber, waveguide dispersion is negligible for all modes not close to the cutoff point. Even for those near cutoff, the dispersive effects can be ignored because they represent a small fraction of the total energy and generally suffer a higher loss. Material dispersion is typically about 4 ns/km for a light-emitting diode of a spectral width of 50 nm and about 0.1 ns/km for a laser of 2 nm width, as shown in Figure 9.6. Modal dispersion depends on the fiber index profile. For a step-index fiber it is given by

$$\tau_{SI} = (NA)^2 L / 2n_1 c \tag{9.17}$$

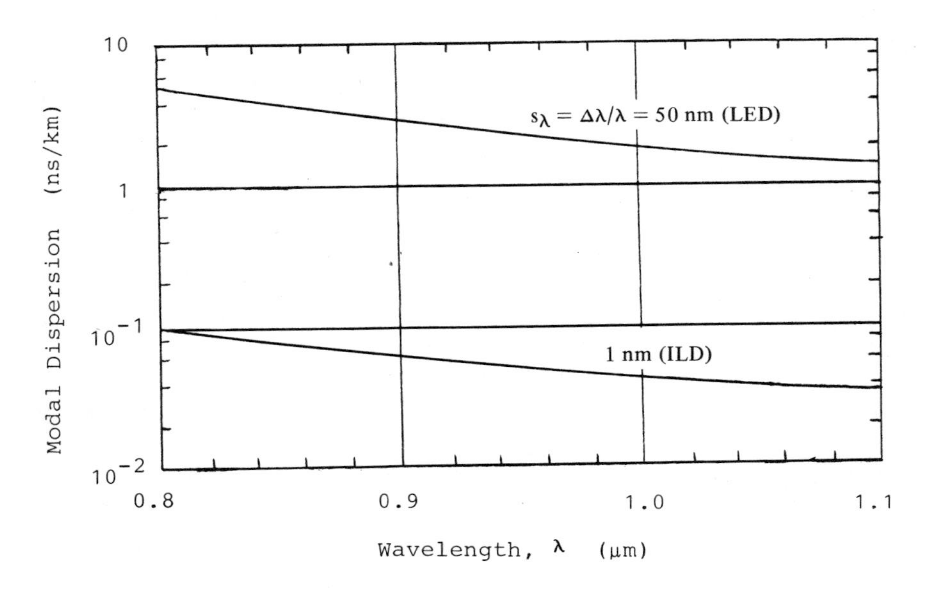

Figure 9.6 Example of the decrease of dispersion with wavelength and spectral bandwidth of the source for different sources.

and for a graded-index fiber by

$$\tau_{GI} = (NA)^2 L\Delta / 4n_1 c \tag{9.18}$$

Differential loss and mode coupling cause dispersion modifications that prevent encountering the optimum pulse delay characteristics in practice. This, however, is primarily of concern in step-index fibers, where there is a substantial difference in modal velocities. In graded-index fibers, modal velocities are almost equal, so that mode mixing and differential loss are usually negligible.

This leads to the determination of the repeater spacing as a function of fiber loss and spectral width of the laser source.

The optimum repeater spacing in an optical communications system as a function of the signal-to-noise ratio of the system, independent of the type of fiber as long as its bandwidth is sufficiently large, is

$$L = \tfrac{1}{2}\alpha \ \ln(P_0/0.6h\nu RF) \tag{9.19}$$

where α is the fiber attenuation, P_0 is the average input light intensity, $h\nu$ is the light energy, R is the data rate, and F is the noise factor given by $F = (S/N)_{opt}/(S/N)_{el}$ (the subscripts refer to the optical and electrical noise contributions).

Analysis of total systems performance requires a knowledge of the detector characteristics. The detector voltage level depends on a threshold value, which is a function of the type of photodetector used, and also on the employed detection method. In the case of a small noise level, one is interested in a high impedance level, which allows an improvement in the compensation threshold level. Hence, at low bit rates (below 10 Mb/s) one employs PIN photodiodes. For an error of 10^{-9}, a minimum detection level of -45 dB can be assumed for PIN diodes and -55 dB for avalanche photodiodes. Above 10 Mb/s, classical detection methods are employed, which make the use of PIN diodes difficult. Around 100 Mb/s and above, a detection threshold of -45 dB can be achieved with avalanche photodiodes of the pπpn type. This is illustrated in Figure 9.7.

DISTORTION LIMITATIONS

Signal distortion is another major cause of bandwidth and system length limitations. There are several potential causes of distortion that occur in a digital or analog signal during transmission from the system input to the system output. Often signal distortion is related to pulse amplitude and width degradations, but it is also a function of source and detector linearity. The signal nonlinearities of optical systems elements can generally be classified as those

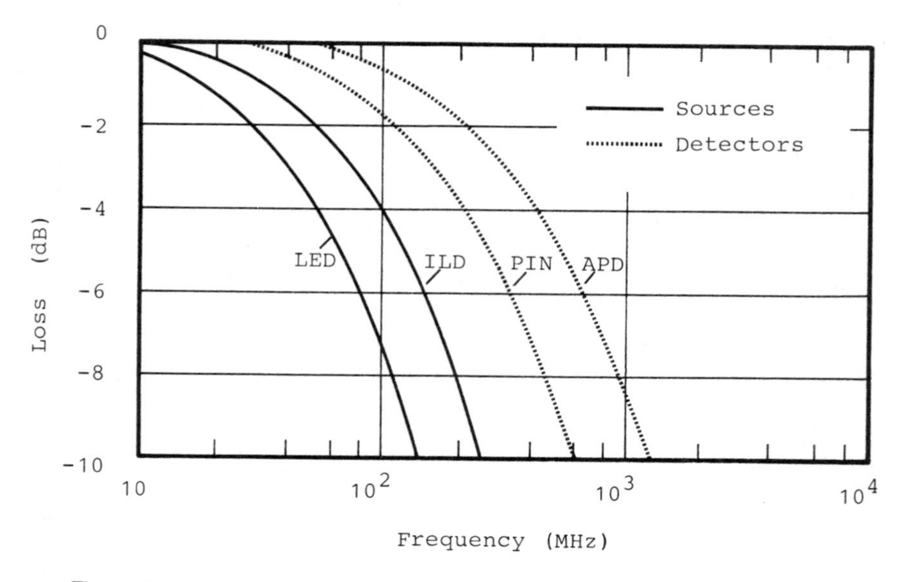

Figure 9.7 Device loss as a function of modulation frequency of typical sources and detectors.

resulting from source, detector, or fiber. Distortion caused by the electrical components will not be discussed here, since it is common to all communications systems.

Source Nonlinearities. In a light-emitting diode, signal deterioration due to source inadequacies may be either amplitude or frequency dependent. In practical applications, amplitude and frequency limitations are usually simultaneously observed.

Amplitude-dependent distortion. The most prominent amplitude-dependent signal distortion associated with an LED depends on the drive current level. At low current, nonradiative processes that saturate with increasing drive current introduce nonlinearities unless the device is operated in the linear portion of the optical power–current curve. At high current, the resultant temperature increase causes a reduction in the radiation efficiency and a consequent nonlinearity. These effects are illustrated schematically in Figure 9.8. At high current level, pulsed operation and/or adequate heat sinking tends to result in improved linearity.

Frequency-dependent distortion. The frequency response of an LED is a function of the recombination time of the injected carriers. The shortest observed recombination time in an experimental LED is about 1 ns, corresponding to a frequency of 1 GHz, with typical values of the order of 20 ns. An example of the frequency limitation of state-of-the-art LEDs is also shown in

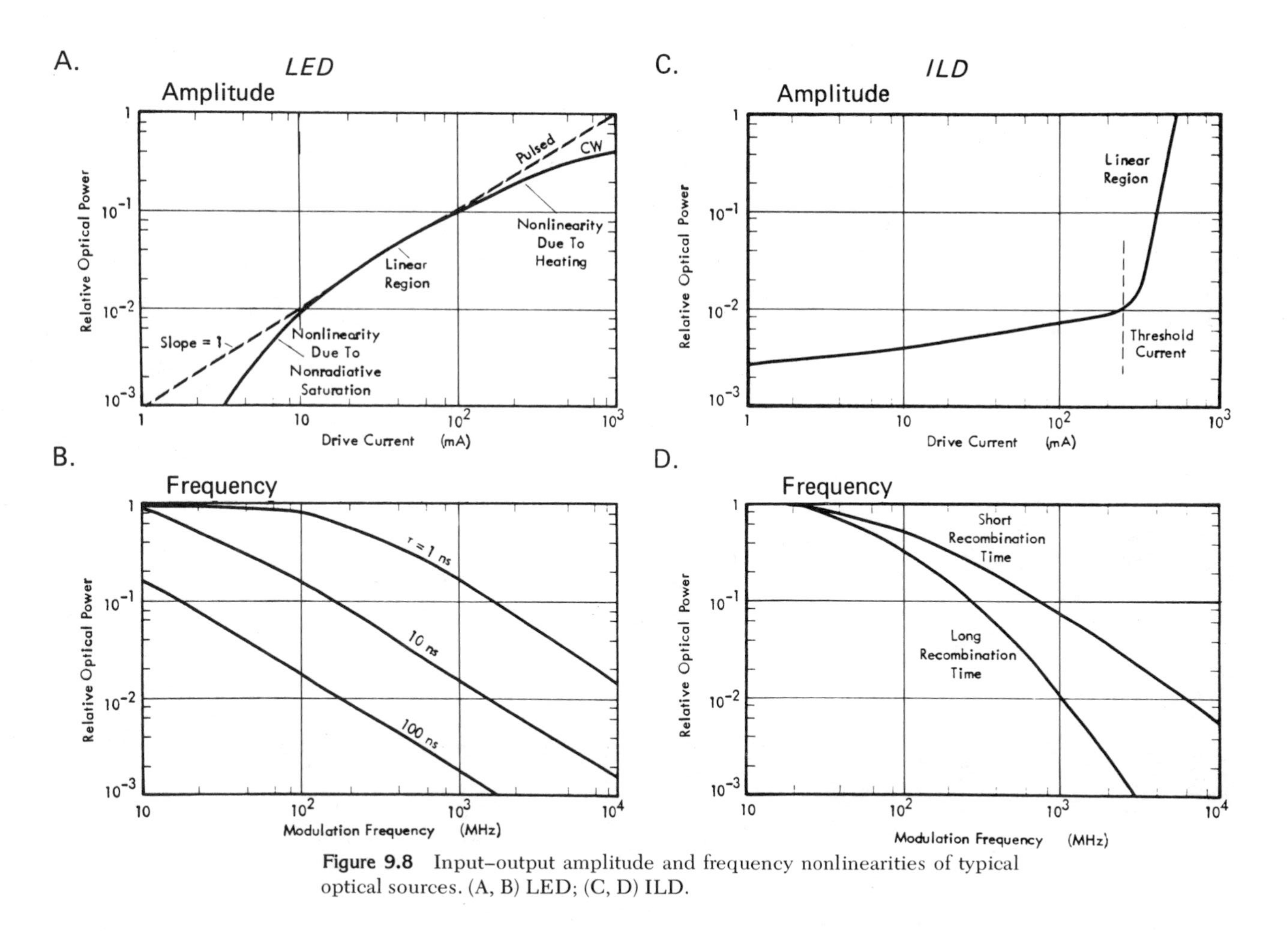

Figure 9.8 Input–output amplitude and frequency nonlinearities of typical optical sources. (A, B) LED; (C, D) ILD.

Figure 10.8, where the relative frequency response of the optical output power has been calculated from

$$P(f,\tau)/P_0 = \left[1 + (2\pi f\tau)^2\right]^{-1/2} \tag{9.20}$$

where f is the modulation frequency, τ is the LED recombination time, and P and P_0 are the high-frequency and low-frequency output powers, respectively. It can be seen from this illustration that the output power decreases linearly with increasing modulation frequency above a critical frequency; the output power decreases also linearly with an increase in recombination time.

In an injection laser, nonlinearities have a different origin. As in the LED, however, amplitude and frequency nonlinearities limit the laser performance. The output of a typical injection laser as a function of amplitude and frequency is also shown in Figure 9.8, with the understanding that considerable differences may apply to different lasers.

Amplitude-dependent distortion. Because a semiconductor laser is a threshold device, a linear output–input relationship can be expected only above the threshold current. Above threshold the laser optical output is an almost linear function of the electrical input. If the drive current drops below its threshold value, signal clipping can occur.

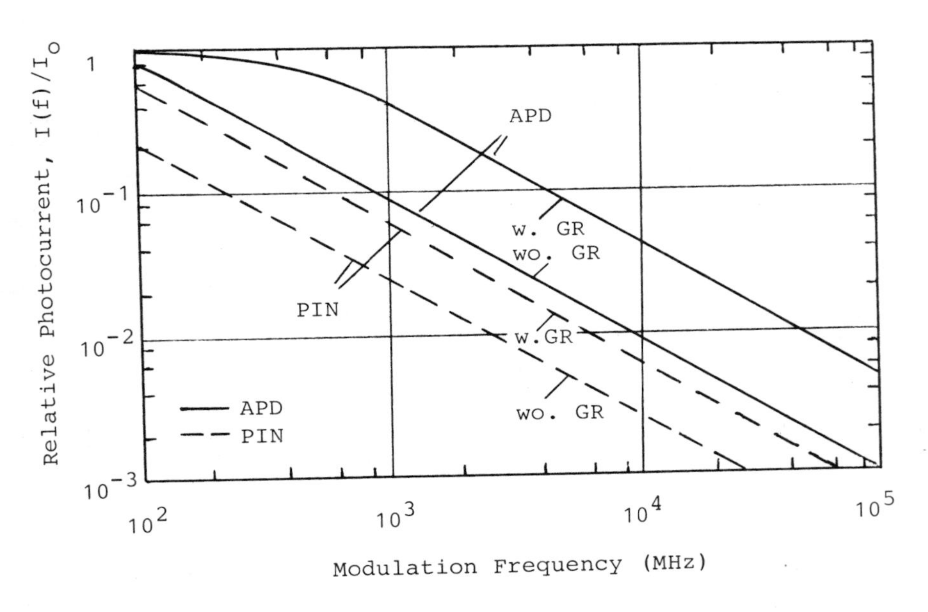

Figure 9.9 Relative decrease in photocurrent with change in frequency for typical optical detectors with (w.) and without (wo.) guard ring (GR).

Frequency-dependent distortion. In a laser, light emission is stimulated, as opposed to the spontaneous emission of light in an LED. Hence, the optical output is not a simple function of the recombination time as in the case of an LED. Modulation frequencies in excess of 1 GHz have been applied to injection lasers. In addition, laser frequency response is limited by internal response resonance.

Detector Nonlinearities. Photodetectors are generally not amplitude limited. Their frequency limitation, however, may be significant, particularly in the case of avalanche photodiodes. The frequency limitation of all photodetectors is determined by the time or the spread of times required to collect the photogenerated carriers. This is shown for typical detectors in Figure 9.9.

Nonavalanching detectors. The response times of p-n and PIN diodes usually range from about 0.1 to 10 ns, which corresponds to an upper frequency range of about 0.1 to 10 GHz. The effective data-transfer rate is usually lower by about a factor of 3. Thus, nonavalanche diodes do not present a practical limitation of available systems. In future systems, however, in which both source and waveguide allow higher bandwidth than presently achievable, improved photodiode response times will be required. Generally, speed can be increased by considering optimization of geometry, impurity levels, and capacitance reductions.

Avalanching detectors. Avalanche photodiodes are typically slower than nonavalanche devices, usually by at least one order of magnitude, unless the structures are enhanced by guard rings or other devices. Their response time usually is about 2 to 3 ns; consequently, their use can lead to substantial signal distortion at higher frequency. In order to enhance the response time, frequently a guard ring is added to the structure, which results in a significant improvement of the detector bandwidth capabilities. With this added feature, the frequency response of an APD is superior to that of a nonavalanche detector.

Fiber Nonlinearities. Signal distortion caused by the optical fiber is potentially the most important dispersive effect. Generally, there are two major dispersion mechanisms in an optical waveguide, as discussed earlier.

Material dispersion. When the source has a relatively large spectral width, dispersion in the fiber material causes a chromatic spread of propagation times, thereby distorting the received signal. The magnitude of the distortion depends on the material dispersion and the spectral width of the source. Typically, in silica fibers this spread is about 4 ns/km for LEDs and about 0.1 ns/km for injection lasers. In borosilicate and other fibers it is about 2.5 ns/km with LEDs and about 0.05 ns/km with injection lasers. Hence, with lasers material dispersion is negligible, but with light-emitting diodes it may be significant, particularly at high modulation frequencies.

Mode dispersion. Because different propagation modes in an optical fiber have different velocities, the ultimate fiber bandwidth is achieved if only one mode is carried by the waveguide. This, however, results in core diameters of only a few micrometers, which leads to practical fabrication and coupling difficulties in addition to potential reliability and reproducibility problems. In practice, if a preliminary system design indicates that it is power limited (inadequate S/N), fibers with a larger numerical aperture must be chosen. If, on the other hand, the system is dispersion limited, then fibers with a smaller numerical aperture are more advantageous.

The dependence of the fiber signal amplitude on the modulation frequency is illustrated in Figure 9.10 for fibers 1 km long. It is important to remember that dispersive effects are usually not linearly proportional to fiber length owing to mode mixing and other effects. Also, the measured fiber dispersion is a function of the efficiency of source–fiber coupling. The details, therefore, of the fiber construction, particularly in graded-index fibers, may deviate considerably from predicted values.

MODULATION AND MULTIPLEXING

Signal modulation in a communications system can be achieved generally by any of four techniques: amplitude or intensity modulation (AM), frequency modulation (FM), pulse amplitude modulation (PAM), and pulse-coded modulation (PCM). All of these schemes have found applications in telecommunications in general. The nature of transmission through optical media, however, allows only amplitude and pulse-coded modulation techniques. This is more a function of the characteristics of the available light sources than of the optical fibers, although the transmission spectrum of fibers does not allow a large degree of freedom in this respect. Frequency modulation is also not at all suitable in conjunction with the optical transmission of intelligent information. Pulse-coded modulation is the most promising technique applicable to optical communications systems. As the previous analysis has indicated, existing light sources operate only at a fixed frequency or wavelength which cannot be altered by modulation; hence, light intensity rather than light frequency can be advantageously modulated. Pulse-coded modulation allows noise-independent transmission because the signal consists of a rapid sequence of pulses of constant amplitude arranged in binary code groups (sequences of 0's and 1's). They correspond to numerical values and represent the amplitude values needed to recreate the slope of the signal, or the pulses can be used as a code to represent data bits. The significant advantage of PCM is that trains of pulses can be regenerated almost perfectly by any number of repeaters over any distance, since the information is not related to the amplitude of the pulses, as long as pulse dispersion does not limit pulse recognition.

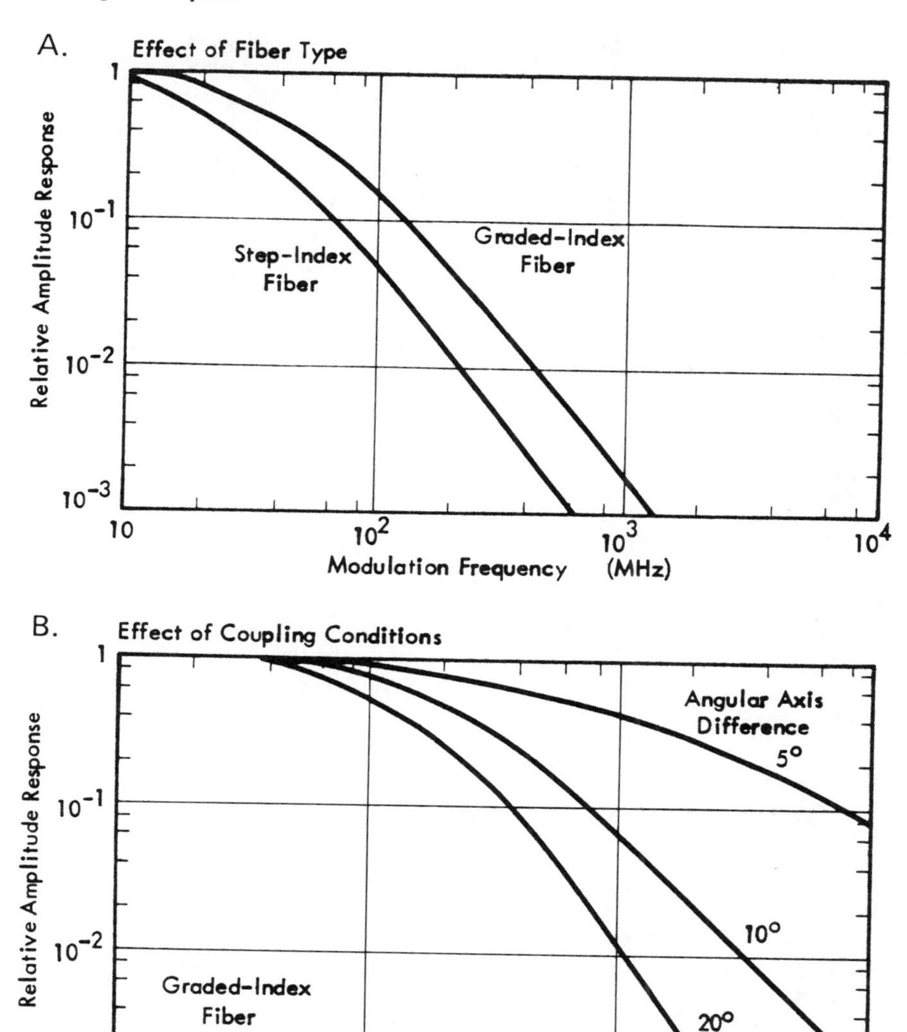

Figure 9.10 Influence of fiber type and coupling on frequency response of fiber output signal (NA = 0.15).

The implementation of practical optical systems suitable for transmission of high-transfer-rate signals over long distances implies, in addition to the use of electronics for signal processing, the operation of sources (especially lasers) in the pulsed mode. Several modulation techniques have been suggested which usually can be reduced to either analog or digital methods. Direct analog modulation is applicable mainly to low-capacity links (e.g., voice or video channels) through the modulation of light-emitting diodes, whereas pulsed analog modulation applies chiefly to systems of medium capacity and may involve pulse-position modulation. Digital modulation is best utilized in high-capacity links. Its main advantage is that it involves only the distinction and recognition of 0's and 1's.

Major Multiplexing Techniques

In order to achieve maximum utilization of the optical fiber capabilities, it is usually desirable to employ multiplexing of several signals. Three major types of multiplexing techniques applicable to optical communications can be distinguished.

Time-Division Multiplexing (TDM). In time-division multiplexing the sequential transmission of information is utilized. This technique takes full advantage of the optimization of the fiber for a given pulse. The various signals are sampled repetitively and their instantaneous amplitudes (or the digital equivalents of their amplitudes) are transmitted sequentially. In this way, a large number of signals can be transmitted simultaneously, with negligible interference among them. The main disadvantage of time-division multiplexing is the concept of sampling, which does not necessarily allow for the interrogation of all signal pulses.

Space-Division Multiplexing (SDM). In space-division multiplexing, each fiber of a bundle carries different information and is thus associated with a particular channel. Because of the near-complete optical isolation, there is negligible cross coupling between data paths. The expected economic advantages of such a bundle, however, are partially offset by the need for high-precision connections which allow for the accurate alignment of individual fibers.

Frequency- or Wavelength-Division Multiplexing (FDM or WDM). In wavelength- or frequency-division multiplexing, several optical sources (each of a distinct wavelength and each avoiding high-absorption regions) are coupled into a single fiber. Dichroic filters, which allow transmission of only one wavelength while reflecting all others, can be used to provide wavelength separation at the receiving end. The wavelength-division multiplexing technique requires the utilization of a relatively complex assembly of circuits in order to achieve the economic implementation of the multiplexing and demultiplexing operations. Alternately, FDM can be used with a single laser source which feeds multiplexed carriers, with the carrier spacing being determined by the length of the laser cavity; the spacing can be maintained through mode locking.

COMPARISON OF TECHNIQUES

These three multiplexing schemes are compared in Table 9.9. The choice of the multiplexing technique to be employed usually depends on a variety of factors. Space-division multiplexing depends on the existence of fiber bundles with their inherent packing fraction limitations and their need for small fiber diameters and correspondingly small light source diameters. Furthermore, the system cost is a function of the multiplexing scheme used. It must also be recognized that when signals of the same bandwidth or of different

TABLE 9.9. Principal Multiplexing Schemes of Potential Use in Optical Communications Systems

Multiplexing technique	Advantages	Disadvantages
Time division (TDM)	Allows transmission of a large number of signals (>1000) Negligible interference between signals Efficient use of optical fiber Suitable for long cables	Not all signal pulses can be interrogated Restriction on maximum bandwidth of system Complex circuitry, hence expensive Requirement of channel separation before pulse regeneration in repeater
Space division (SDM)	Negligible interference between signals Allows transmission of a large number of signals (>100) Relatively economical No need for channel separation before pulse regeneration in repeater Continuous identification of each signal; channel dropping filters are not required Total system capability is the sum of the capabilities of all individual fibers of the bundle	Need for high-precision connectors Fiber splicing difficulties Requirement of coherent bundles, limited to short cable lengths
Frequency or wavelength division (FDM, WDM)	Can be used with a single laser source or with several sources Carrier spacing determined by laser cavity dimensions Suitable for long cables	Limited to small number of signals (<10) Requires use of dichroic filters Requires low fiber loss over wide wavelength spectrum Complex circuitry, hence expensive Requirement of channel separation before pulse regeneration in repeater

bandwidths are multiplexed by either method, the total bandwidth or frequency required must be in excess of the sum of the individual bandwidths plus some margin to provide channel separation.

The WDM system offers advantages only in systems which do not demand a large number of different channels, since the sharpness and influence of the fiber loss minima as well as the unavailability of filters of sufficiently narrow spectral width prevent the simultaneous use of more than a few different wavelengths. Whereas the FDM system is composed of a combination of sources and modulators and the signals maintain their identity in the frequency domain, in the TDM system the component signals are identifiable in the time domain but occupy a common frequency domain. The multiplexed signals must be separated at the receiver by using appropriate filters in the frequency or time domains. In wavelength- or frequency-division multiplexing, thousands of voice signals can be assembled over a single medium. This is the most widely used multiplexing method in twisted pairs and coaxial cables, but it presents technical difficulties in optical waveguides. Hence, at present, time-division multiplexing is the most economic and technologically most desirable multiplexing method. It is capable of offering adequate transmission characteristics and hence is preferred over other techniques. It lends itself mainly to digital data transmission, although in principle it is also applicable to analog transmission.

SYSTEM ECONOMY

System operation at the highest data transfer rate is usually desirable from a technical point of view. From an economic standpoint, however, this is not necessarily the case. Consequently, the design of an actual optical communications system represents a compromise between technical and economic considerations. The following discussion is devoted to the economic aspects of system design.

In our previous analysis we have seen that an increased pulse rate corresponds to more stringent requirements on the spectral width of the source and the dispersion characteristics of the fiber, causing a reduction in the maximum length of the waveguide and an increase in the number of required repeaters. Furthermore, higher pulse rates necessitate more sophisticated repeater and receiver designs.

It is apparent that under usually encountered conditions, system cost increases with data transfer rate. Generally, the system cost C per kilometer can be shown to vary with pulse rate R according to the relationship

$$C = c_0 R^b \tag{9.21}$$

where the terms c_0 and b are numerical factors that depend upon achievable transfer rate and system complexity. For very low values of R the term b may become negative, meaning that an increase in data rate will lead to a system cost reduction. However, because of increased system complexity, the reverse will become true if the pulse rate is increased further. In this case, the value of b will become positive, indicating an increase in system cost with data transfer rate. At intermediate pulse rate it is reasonable to assume that system cost increases as the square root of pulse rate, corresponding to $b = 0.5$. As system data rates are pushed to ever higher rates, the exponent b will ultimately approach or exceed unity, at which point a further increase in data transfer rate will become uneconomic.

Figure 9.11 illustrates this situation in schematic form, first in a general way to show the cost contributions, and then to indicate cost–performance trends as the state-of-the-art advances. On the other hand, systems requirements on the various components depend upon the overall communications needs and usually involve a variety of compromises. As the technology matures, there will be a tendency to increase the pulse rate to the maximum allowable value. However, it will be necessary to consider the economic aspects of such an advance, together with the considerations brought about by the transfer from a dispersion-limited to a noise-limited operational mode as pulse rate is increased.

To establish the total system cost it is necessary to analyze the limitations imposed by performance factors, such as system length L and specified maximum data transfer rate R. In addition, limitations resulting from economic factors, such as fiber and repeater costs, have to be included. The performance-limiting cases can be divided into data rate restrictions due to signal dispersion, detector noise, and auxiliary circuits. Signal dispersion due

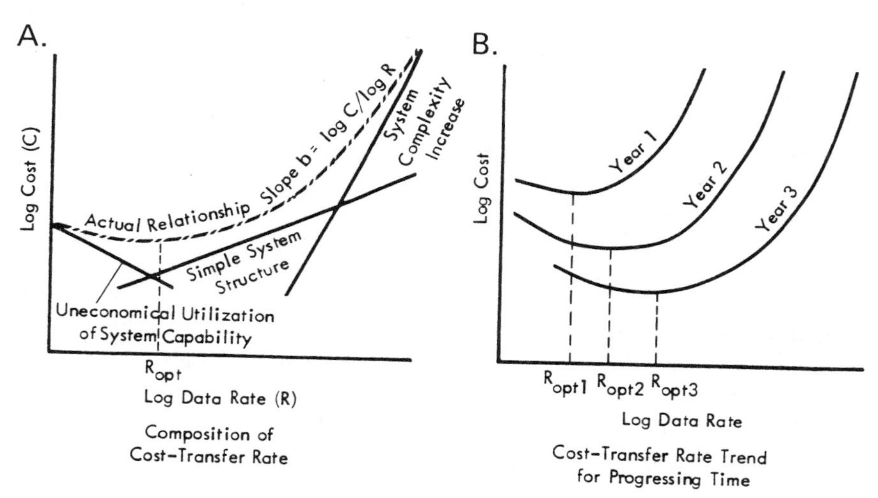

Figure 9.11 Schematic diagram showing the relationship between systems cost and systems data rate.

to the transmission medium results in an increasing error rate; knowledge of the relation between L and R is required when operation is in this regime. In a lossy transmission medium where the output signal power is small, the system length is limited by detector noise, yielding an unacceptably high probability of error in correctly detecting the presence or absence of a pulse. For a given system length the maximum pulse rate may also be limited by the associated digital circuitry which is not fast enough to handle the high data rates. A digital circuit constraint of 0.5 Gb/s can be assumed for 1978 and of 3.0 Gb/s for 1990. For a very short system the limitation by circuits prevails, whereas for longer systems either dispersion or noise may be dominant restrictions. Dispersion and noise limitations are interdependent because the signal peak value decreases as dispersion increases. The point at which the two limitations are equally important is a crossover point beyond which the economic picture changes drastically.

The difference between system limitation due to dispersion and that due to noise is of considerable economic importance. While a system may be economically viable in the case of a dispersion-limited operational mode because of a relatively simple structure, it may become uneconomic in the noise-limited mode. This is a result of unavoidable increases in system complexity in order to improve the signal-to-noise ratio, either through more sensitive detectors or a larger number of repeaters. We will shortly derive the total system cost, but let us first review the achievable maximum pulse rates for each of the above cases.

Dispersion Limitation. This case has been analyzed previously, so that no detailed discussion is required here. For a single-mode fiber the maximum data transfer rate is approximately

$$R_{\max} \cong 23 \times 10^9 L^{-1/2} \tag{9.22}$$

For an allowed pulse overlap $\delta = 0.01$ and for a multimode fiber it is

$$R_{\max} = (1/2s_0)L^{-1/2}\left[(1 - 2\Delta t^*/\alpha^* s_0)\delta\right] \tag{9.23}$$

where s_0 is the inverse of intramodal and intermodal bandwidth, Δ is the index difference, and t^* and α^* are the signal delay difference and the attenuation difference between the fastest and slowest mode, respectively. The multimode value of $R_{\max}$ is proportional to $1/L$ for the large values of L and narrow transmitted pulse widths.

Noise Limitation. Because the detector responds to individual photons and not to the total incident optical energy over a wide band, two cases must be distinguished.

Direct photodetection. In this process frequency and phase information is lost, leading to a relationship between the modulation scheme and the nature of the detector. Only the amplitude modulation of the carrier is detected, whereas the detector does not respond to frequency or pulse modulation.

Indirect photodetection. In this process the incoming optical signal is mixed at the detector with a coherent local oscillator frequency to produce a difference frequency which contains the transmitted information. Photomixing generally increases the detector sensitivity.

There are obvious inherent performance advantages in the indirect photomixing technique. The disadvantages lie in the increased receiver complexity and resultant higher cost. The maximum data transfer rate in the case of shot noise dominance is given by

$$R_{\text{max}} = 2.5 \times 10^{18}\, P_i e^{-\alpha L} / \,|\ln(2\epsilon)| \tag{9.24a}$$

where ϵ is the error bit rate and P_i is the optical signal power at the transmitter. It is relatively insensitive to the error rate. Assuming $\epsilon = 10^{-9}$, it is

$$R_{\text{max}} = 1.3 \times 10^{17} P_i e^{-\alpha L} \tag{9.24b}$$

where P_i and R are expressed in W and Hz, respectively. In this discussion it is assumed that the relationship between system length and maximum pulse rate is independent of material dispersion, but it depends upon signal attenuation, detector noise power, modulation scheme, and the nature of the detector.

Circuit Limitation. The case in which the electric rather than the optic system components determine the upper limit of the data transfer rate is of lesser interest in this context because for a given state of development it is fixed, although economic reasons may also influence the choice of a particular circuit configuration. Also, the limitations imposed by the digital or analog circuits will change with time. Therefore, in the following discussion principal attention will be focused on the economic aspects of the optical system components rather than those of the electrical components.

COMPARISON

Figure 9.12 illustrates the maximum pulse rate for the three limiting cases. The use of multimode fibers presents a more significant pulse rate limit than that caused by the associated circuits. If a monomode fiber is used, the circuit frequency response will continue to be the limiting factor, although it will be one to two orders of magnitude less severe than the constraints resulting from fiber dispersion in multimode fibers. For dispersion-limited operation the maximum pulse rate is independent of the fiber loss and of optical power. This is a reasonable conclusion, because it is assumed that there is adequate power at the detector but the pulses have overlapped because of dispersion, precluding an intelligent decision on the presence or absence of a pulse. For noise-limited operation, however, the maximum pulse rate is a strong function of both fiber loss and length. The slopes of the various curves given in Figure 9.12 are important. For dispersion-limited operation the variation of the data rate with fiber length is algebraic, varying with L^{-1} or $L^{-1/2}$; the monomode fiber has the more desirable $L^{-1/2}$ dependence and the multimode fibers show the L^{-1} dependence. For noise-limited operation the variation of the maximum data transfer rate with length is exponential, since the pulse rate varies with optical power or its square. An increase in attenuation causes a change in the slope and the location of the curves. For comparison, in a coaxial cable the variation of maximum pulse rate with cable length displays an L^{-2} dependence, which is much less desirable than the L^{-1} and $L^{-1/2}$ variations of dispersion-limited systems.

A more detailed evaluation indicates that the intersection of corresponding curves represents the change from dispersion-limited to noise-limited operation. Generally, operation in the noise-limited regime should be avoided. As the data transfer rate demand increases, one must design for reduced spacing between repeaters or between transmitter and receiver. The reduction of fiber length for increased data rate is rather gradual in dispersion-limited operation, but it becomes severe in noise-limited operation. This has an effect on the economics of the system. Furthermore, it is advantageous to allow for a gradual rather than an abrupt degradation of operating conditions.

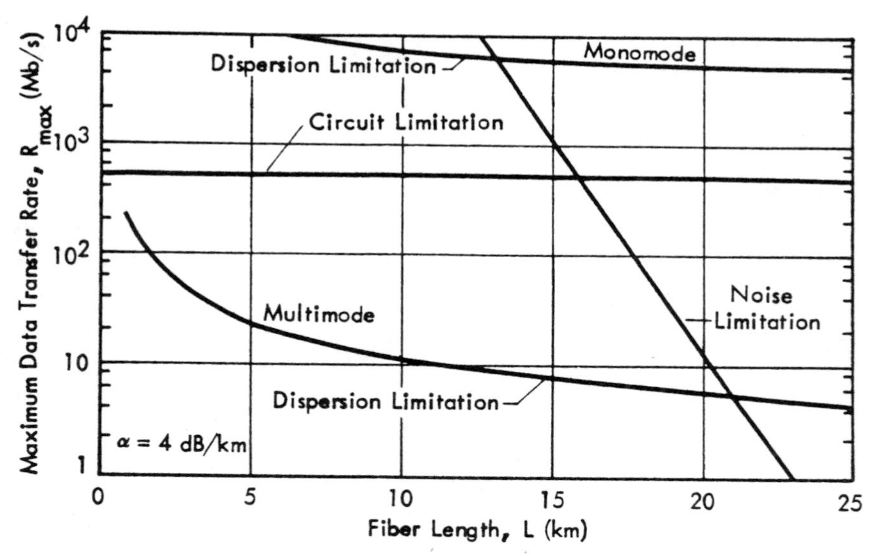

Figure 9.12 Dependence of transfer rate on fiber length for limiting cases (1978).

Hence, taking into account fiber and source aging and other performance deteriorations, it is advisable to design the system for a fiber length adequately below the dispersion–noise crossover point.

System Cost

We turn our attention now to the analysis of the system cost. Not only is it a function of the various costs of the individual components, but it is also related to the optimization of system performance. It is usually advisable to operate an optical communications system considerably below the maximum achievable data transfer rate. Thus, it is advantageous to employ an optimum data rate and an optimum fiber length rather than the maximum the state-of-the-art allows (12). In addition to the performance constraints that enter the analysis of systems cost, the economic constraints imposed by fiber cost and repeater cost are important design criteria. Although they are usually beyond the control of the systems designer, the variation of fiber and repeater cost with data rate has to be considered. In a first approximation, the fiber cost (per unit length) can be considered to be independent of the data rate. However, with maturing manufacturing process the fiber cost will be a function of the data rate, since the more expensive monomode fiber is capable of a higher data rate than the less expensive multimode step-index fiber; furthermore, a graded-index fiber of low dispersion is more expensive than a multimode step-index fiber of much higher signal dispersion.

The total cost of the optical components of the system can be expressed as

$$C = n_R C_R + (n_R + 1) C_F L \tag{9.25}$$

where C_F is the fiber cost per unit length ($/km), C_R is the cost per repeater (dollars), and n_R is the number of repeater stations per system (11, 12). This allows the derivation of a normalized system cost (given in km^{-1}) as follows:

$$\bar{C} = C/C_{R_0}L = n_R (R/R_0)^b/L + (n_R + 1)g \tag{9.26}$$

The normalized system cost is a function of the data transfer rate, the multiplexing technique used, and the total number of repeaters present in the system. The expression for the normalized cost holds for all three limiting cases up to their respective maximum values with proper use of the applicable terms. The term g takes care of the multiplexing technique used. It depends only on economic factors. For time-division multiplexing $g = n^{b-1}\ C_F/C_{R_0}$, and for frequency-division multiplexing $g = C_F/nC_{R_0}$. For example, if the fiber cost is assumed to be $C_F = \$1000/\text{km}$ and if the reference repeater cost is $C_{R_0} = \$10{,}000$, then the ratio of fiber cost to repeater cost is $C_F/\overline{C}_{R_0} = 0.1/\text{km}$; various values of the number of repeaters per station (n) will then be reflected in g. Increasing the value of n has the same effect as decreasing C_F in the case of frequency-division multiplexing or decreasing C_{R_0} in the case of time-division multiplexing.

The repeater cost depends upon data rate, complexity, and several other factors. Also, its regenerating circuitry is subject to the same constraints imposed by digital or analog circuits as the transmitter and receiver are. Furthermore, repeaters are usually specifically designed for a particular data rate and are not optimized for substantially deviating data rates. The following model is used for estimating the cost per system repeater, assuming time-division multiplexing:

$$C_R = C_{R_0} (R/R_0 n)^b \tag{9.27}$$

where C_{R_0} is a reference cost, R and R_0 are the actual and reference data rates, respectively (R_0 is assumed to be 100 Mb/s), n is the number of repeaters at each station (where the value of n is determined by the multiplexing scheme employed), and b is a dimensionless exponent, ranging from about 1.0 at 10^2 Mb/s to about 2.5 at 10^4 Mb/s. Thus, for reference purposes it is assumed that a repeater capable of a data transfer rate of 100 Mb/s costs $10,000.

The normalized system cost as a function of fiber length is shown in Figure 9.13 for the three limiting cases considered, where the fiber length L corresponds to the spacing between repeaters. It is assumed that the data transfer rate is optimized for maximum economy but that it is not necessarily at its maximum R_{max}. The system cost is lowest in the circuit-limited case, assuming optimized data transfer rate, whereas the dispersion-limited case is associated with a relatively high system cost. Different numerical values are obtained if b differs from 2 and if the maximum rather than the optimum data transfer rate is used. As the transfer rate increases, the cost of the repeater increases, so that the maximum system cost corresponds to the maximum transfer rate. Obviously, the most economic operation is achieved if the data transfer rate is optimized rather than maximized. Generally, however, little difference in system cost for maximum and optimum transfer rates is observed in the case of dispersion limitation, but in the case of noise limitation the differences are substantial.

A further example of the system cost as a function of fiber length and data transfer rate will serve to illustrate the previous analysis in more detail by estimating both maximum and optimum transfer rates. The system economics are evaluated for two different fiber lengths; in all cases $b = 2$ is assumed. First, fiber length of $L = 2.5$ km is given. In this case one can operate at the economic optimum of $R_{opt} = 15$ Mb/s at a normalized cost slightly above $\bar{C} = 1 \times 10^{-9}$ or, if a higher transfer rate is required, at $R_{max} = 45$ Mb/s at a normalized cost of $\bar{C} = 2 \times 10^{-9}$. Thus, an increase in data rate by a factor of 3 results in a cost increase by a factor of 2. Second, for comparison it is assumed that the fiber length is allowed to range from $L = 15$ to $L = 20$ km. It is evi-

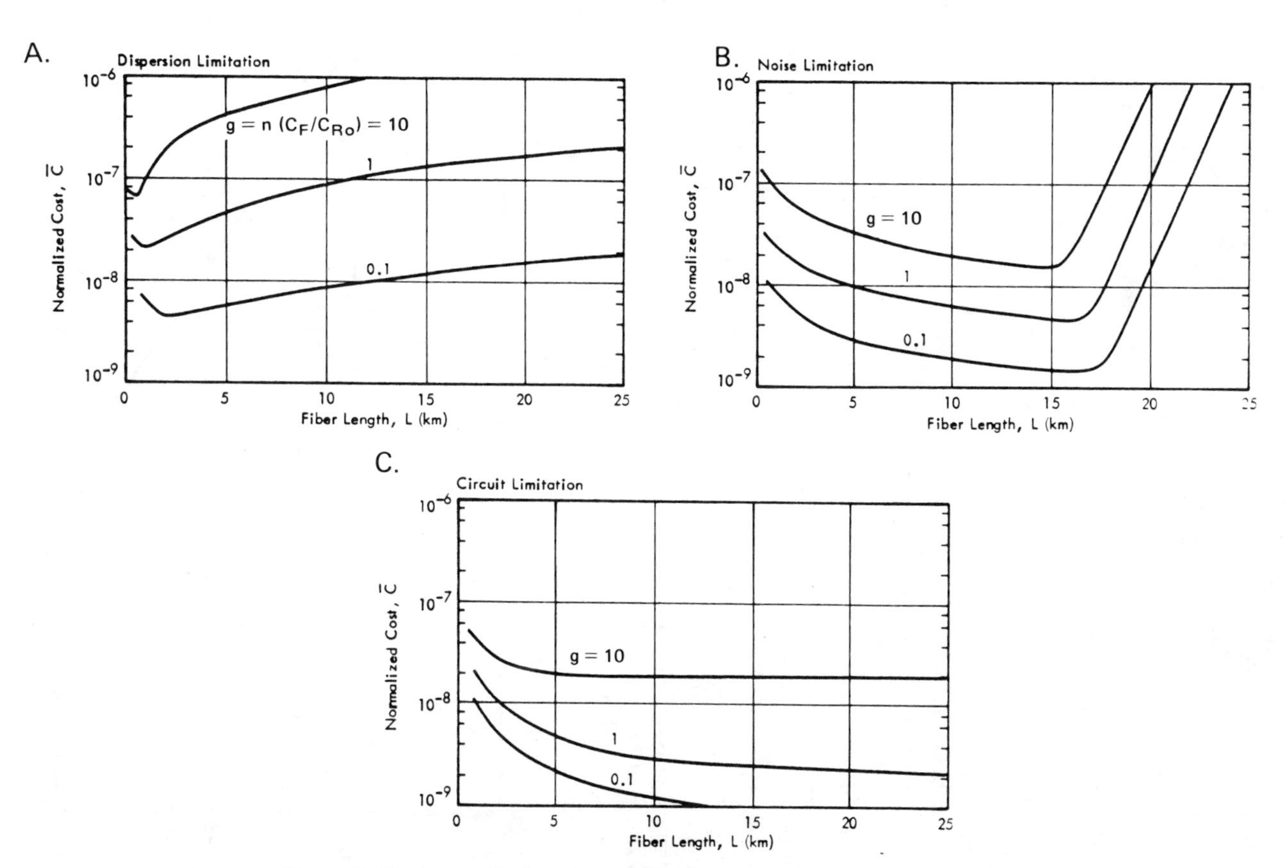

Figure 9.13 Dependence of normalized systems cost on limitations by dispersion, noise, and circuit at optimized transfer rate ($b = 2$).

dent in this case that the noise-limited case is applicable if a monomode fiber is used, so that the data rate ranges from approximately 40 Mb/s to 4 Gb/s. At $L = 20$ km, the optimum and maximum transfer rates are identical, $R_{opt} = R_{max} = 40$ Mb/s, and $\bar{C}$ is about 5×10^{-10}, which is a relatively economic situation; if $b = 1$ were assumed, the operation would be less economic, but the same data transfer rate is possible. At $L = 15$ km, the optimum cost at about the same data rate $R_{max} = 40$ Mb/s, with about the same normalized cost $\bar{C} = 5 \times 10^{-10}$; but if the system is operated at the maximum transfer rate $R_{max} = 4$ Gb/s, the cost effectiveness is substantially reduced, yielding a normalized cost of $\bar{C} = 7 \times 10^{-10}$. Thus, the cost is higher by a factor of about 40 if the data rate is increased by a factor of 100. However, if $b = 1$, the cost for operation at the maximum data rate 4 Gb/s and assuming $L = 15$ km is only $\bar{C} = 7 \times 10^{-10}$.

We can illustrate this in a more general way, as shown in Figure 9.14. This graph is derived from the above analysis of cost as a function of fiber length for various limiting conditions, with the cost given as a function of the data transfer rate for the noise-limited case. From the slopes of the curves we see that operation at a data transfer rate that differs from the optimum is not economic in most cases of practical concern. This allows us to arrive at the following conclusion: For simplicity of the discussion we assume again that the repeater cost per station $C_R = \$10,000$. Then a system cost $\bar{C} = 10^{-9}$ corresponds to an actual cost of \$0.65 per channel kilometer for a channel transfer rate of 65 kb/s; this value is encountered even when the fiber cost is

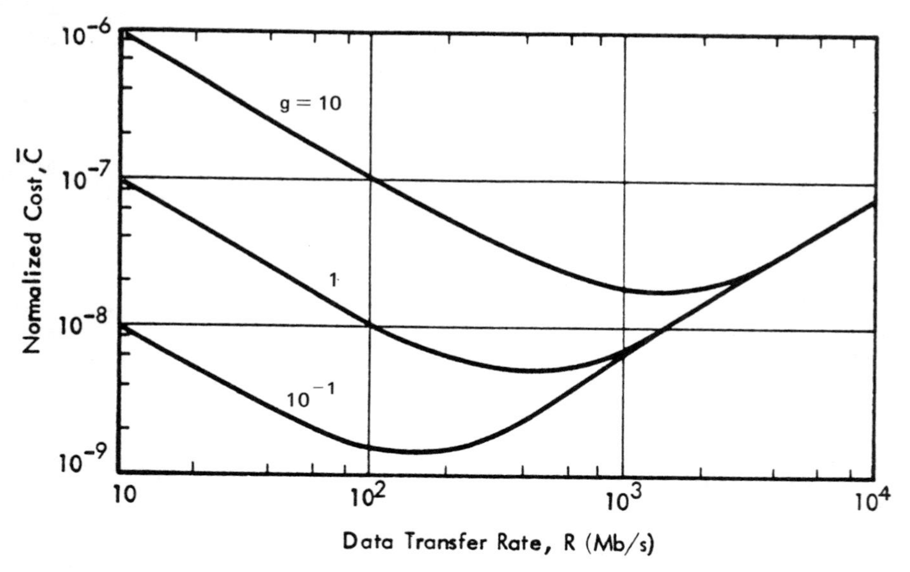

Figure 9.14 Dependence of systems cost on data transfer rate in the noise-limited case ($b = 2$).

$1000/km. An increase in the number of repeaters per station, assumed to be unity in this analysis, results in a reduced sensitivity of system cost on fiber cost. The signal distortion introduced by a monomode fiber is insignificant over a substantial operating range, and only fiber attenuation affects the system economics; hence, this fiber is an excellent transmission medium at high data rates and large fiber lengths. Less expensive and slower repeaters used in conjunction with appropriate multiplexing techniques have the capability to reduce the system cost relative to that of a system consisting of high-speed repeaters.

From an economic standpoint it is not necessarily advisable to operate the system at the highest possible data transfer rate, since the cost of the repeaters increases with the data rate. If it is assumed that the fiber cost increases with the maximum allowable data rate, operation at the optimum rather than the maximum transfer rate is even more important. Furthermore, the relationship between repeater cost and data transfer rate will change with time, as will the fiber cost. Hence, a system cost projection that takes all of these factors into account is rather complex, and a model that gives an accurate representation of the various trade-offs has to be derived on an individual basis.

Economy requires that the system operate in the dispersion-limited mode rather than in the noise-limited mode. This necessitates low-loss fibers and high-intensity light sources, at least for long communications links. For short data paths, high-loss fibers may be acceptable if moderate data rates are to be transmitted. In the noise-limited case, the maximum possible pulse rate decreases approximately inverse exponentially with fiber length. Thus, the rapid degradation of system capability with system length makes the operation in the noise-limited regime undesirable, since it may potentially lead to severe economic consequences. These can be overcome partly by the reduction or elimination of photodetector limitations due to thermal noise. If the optical pulse width of the signal is reduced in favor of increased amplitude, thereby retaining a fixed pulse energy, it is possible to reduce thermal noise so that over a wider range shot noise will be the principal noise contribution. This procedure will facilitate the economic implementation of long-haul systems at high data rates.

Because the design of an optical communications link for a specific application necessitates the evaluation of all system components as a combined network (since their compatibility and interaction determine the overall system performance and economics), it is useful to analyze the bandwidth of the major components of a system as a function of link length, as shown in Figure 9.15. The bandwidth given here refers to a single channel or fiber; bandwidth increases are possible through employment of several fibers in parallel. Depending upon a particular bandwidth or length specified, different system components may determine the system bandwidth or length limit. Also, system limitations will shift in time, since further advances are being

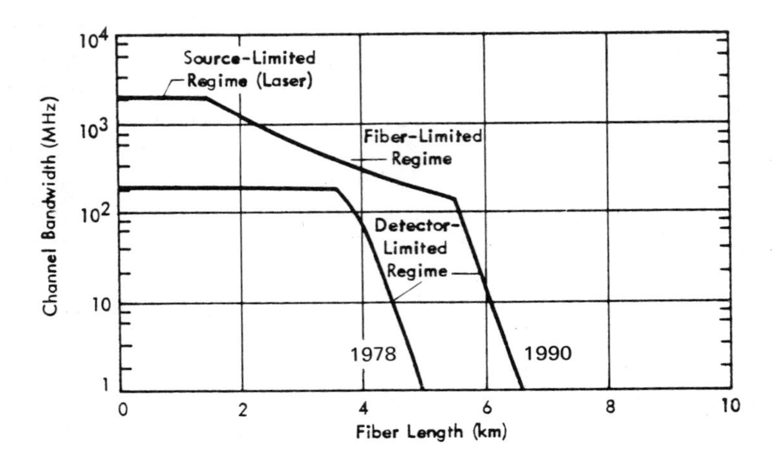

Figure 9.15 Expected progress in the variation of bandwidth with fiber length.

made by individual components at different rates. Generally, in short communications links the upper bandwidth is determined by source frequency response, whereas in long links the system bandwidth is restricted by detector limitations. In medium-length links fiber dispersion or attenuation may affect the bandwidth, although fiber limitations are usually overridden by other components.

This has important implications with regard to the application of optical fibers to various types of electronic equipment. Thus, systems that require a high data transfer rate are limited to very short distances, whereas long systems can use only a smaller bandwidth. Hence, long-link applications of fiber optics are found in military and telecommunications equipment, whereas short-link applications are mainly found in high-speed computers. Instrumentation and industrial control equipment also require generally only short optical cables, but in these applications cost usually overrides demands on bandwidth. A summary of the principal advantages of fiber optics in specific applications is given in Table 9.10. These advantages allow the practical or potential realization of a number of functions that are largely unique to optical systems.

Two examples of digital transmission systems using optical waveguides can be used to demonstrate the capabilities of optical communications systems. The first example provides low-data-rate transmission of four-bit data words, whereas the second example allows high-data-rate color TV transmission.

TABLE 9.10. Advantages of Optical Communications Systems in Major Applications Areas

Application area	Principal advantages	Principal functions	Needed advances
Communications	High bandwidth Low cost/channel km Large repeater spacing Small size Simple transmission line characteristics Lack of crosstalk and interference	Intercity trunks Interoffice loops Wired city Dedicated networks Submarine cables CATV/CCTV	Higher radiance sources Longer source lifetime Lower system cost Better connectors Improved cable structures Improved bandwidth
Computer	High bandwidth Immunity to interference Small size Equalized data transfer rate	CPU–peripherals connection Distributed systems CPU architectural improvements	Better source efficiency and bandwidth Higher source radiance Improved multiplexing techniques
Consumer	Small size and weight Fiber flexibility	Automotive use throughout cars and trucks TV hookups	Lower cost–performance ratio More reliable connectors
Government/military	Small size and weight Immunity to interception Reduced use of nonstrategic materials	Military communications Weapons and surveillance instrumentation Internal wiring of weapons systems	Improved cable structures Better connectors Better source reliability
Industrial control	Immunity to interference Lack of electrical input–output connection High temperature range	Computer-based process control Interference-free communications in power plants Deployment in explosion-, radiation-, and fire-prone environments	Compatible input–output transducers Lower systems cost Improved cable structures Higher radiance sources
Instrumentation	Immunity to interference Small size and weight Fiber flexibility	Use in small and low-weight instruments Medical applications Industrial use in areas of difficult access	Improved bandwidth Better connectors Lower systems cost

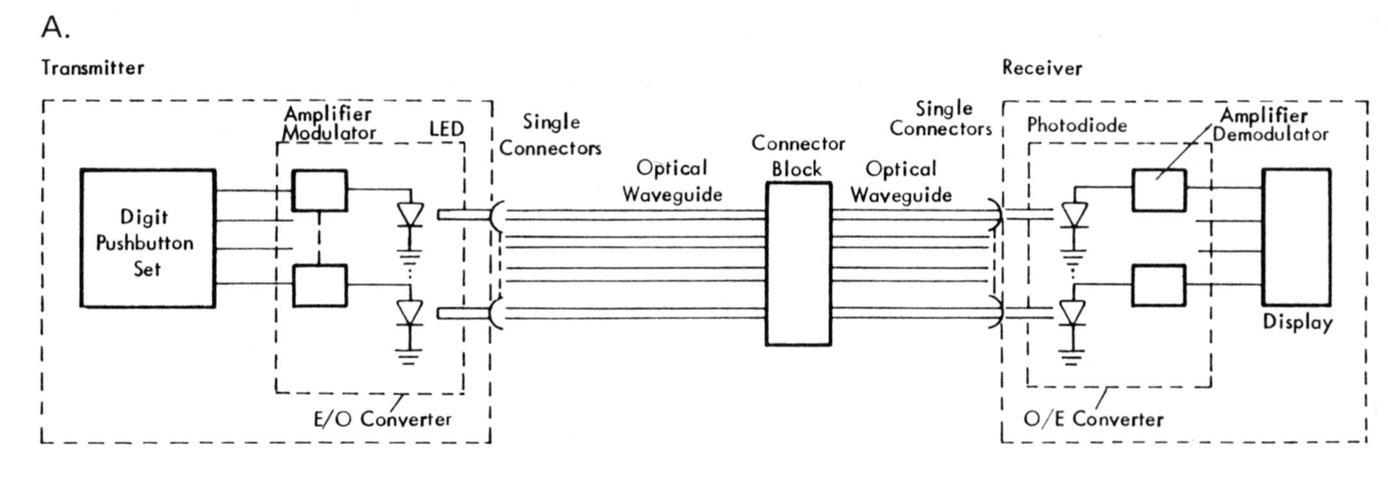

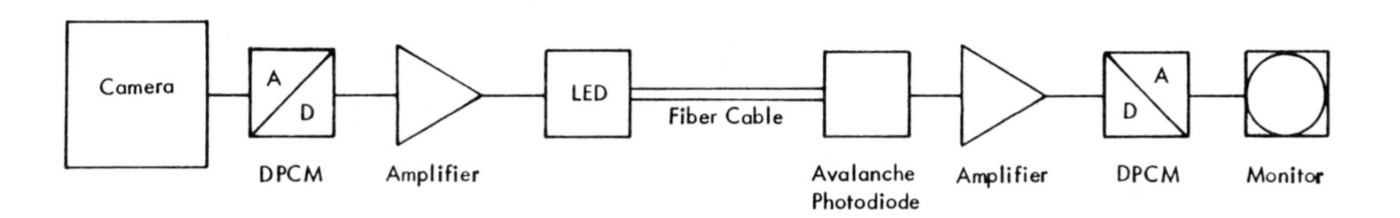

Figure 9.16 Examples of digital transmission systems. (A) Short distance; (B) long distance.

Short-Distance System. The short-distance data transmission system, shown in Figure 9.16A, is capable of delivering four-bit data words (the base-ten digits 0 through 9 given in binary code) over a distance of 20 m. The information is transmitted on four signal channels in parallel-by-bit and serial-by-character configuration. Pulse repetition rate is 1 MHz, system attenuation is 28 dB (including connectors), data rate per channel is 1 Mb/s, signal delay is 4.5 ns/km, and S/N is $\geqslant$ 30 dB.

Long-Distance System. The long-distance example of an optical communications link is shown schematically in Figure 9.16B. This digital video transmission system has a length of 2 km and provides data transfer at a rate of 80 Mb/s. The system allows the transmission of one-channel color television pictures of high quality.

REFERENCES

1. Tingye Li. Optical Transmission Research Moves Ahead. *Bell Lab. Rec.* **53**:333–339, 1975.

2. A. Jacobsen. Optical Transmission by Fiber Optics. *Feinwerktech. Messtech.* 3:117–121, 1975.

3. J.E. Mazo & J. Salz. On Optical Data Communication via Direct Detection of Light Pulses. *Bell Sys. Tech. J.* **55**:347–369, 1976.

4. R.L. Gallawa et al. *Optical Fiber Links for Telecommunications.* NTIS Rep. AD-767–544, Washington, D.C., 1973.

5. O.B. Medved. New Light on Fiber Optic Communications. *Opt. Spectra* **9**:34–36, 1975.

6. J. Samson. The Telecommunications Challenge. *Elect. Weekly,* pp. 13–17, Sept. 24, 1975.

7. D.E.N. Davies & S. Kingsley. Method of Phase-modulating Signals in Optical Fibres: Application to Optical-Telemetry Systems. *Electron. Lett.* **10**:21–22, 1974.

8. G.J. Foschini et al. Optimum Direct Detection for Digital Fiber-Optic Communication Systems. *Bell Sys. Tech. J.* **54**:1389–1430, 1975.

9. Y. Takasaki & M. Tanaka. Line Coding Plan for Fiber Optic Communication Systems. *Proc. IEEE* **63**:1081–1082, 1975.

10. J. Fulenwider & G. Killinger. Advantages of Optical T-Carrier Systems on Glass-Fiber Cable. International Wire & Cable Symposium, Cherry Hill, N.J., November 1975.

11. R.L. Gallawa & M. Kayoma. *A Cost Model for Optical Waveguide Communication Systems.* NTIS Rep. COM-75–10941, Washington, D.C., 1974.

12. R.L. Gallawa et al. *Telecommunication Alternatives with Emphasis on Optical Waveguide Systems.* Office of Telecommunications Rep. 75–72, Washington, D.C., U.S. GPO, 1975.

10

ENDOSCOPY

Tadashi Morokuma

The concept of nonmodulated optical image transmission through fiber bundles, generally known as endoscopy, was described as early as 1929 by Hansell in a British patent disclosure (1). Since then, many researchers have attempted to advance fiber bundles for practical, medical or industrial uses. In 1930, Lamm reported on experiments concerning the transfer of images of simple objects through a fiber bundle produced by means of combs. The images were very poor, however (2). Important progress was achieved by van Heel when he described a glass fiber coated with a transparent plastic material and the transmission of images with a flexible fiber bundle (3).

This waveguide, now called a step-index fiber, is composed of a core through which light energy is transmitted and a cladding whose refractive index is lower than that of the core. In present fibers, the cladding consists of a low-refractive-index glass which is more durable than most plastic materials. Since the core surface is protected by the cladding, transmission characteristics are, in general, not significantly affected by stain, dirt, and foreign materials attached to the surface of the fiber or by scratches made on the surface. This implies that for a long period of time during normal use hardly any image deterioration is observed in a properly made step-index clad fiber. In the case of fibers that lack a cladding layer, loss in light energy may increase abruptly after a short period owing to slight scratches developed during the use of the flexible bundle. In addition, very large amounts of the energy may be lost from the surfaces where the fibers contact each other.

Progress in the development of the step-index fiber made possible the practical realization of the concept of fiberscopes. For example, the work of van Heel stimulated activities in the development of various fiber optics image guides and their applications throughout the world. Among the workers engaged in the field of fiber optics, Kapany published and presented a number of scientific papers that contributed to progress in fiber bundle techniques, flexible fiberscopes, fiber optics faceplates, field flatteners, and so on (4–13).

In 1957, the first attempt was made toward medical endoscopy in the U.S., where research and development were most active, covering a wide range of the aspects of fiber optics. Hirschowitz, Peters, and Curtiss developed a flexible fiberscope for the examination of stomach and duodenum and established a foundation for medical flexible endoscopy (14,15).

Since then, fiber optics endoscopy has been accepted by medical doctors throughout the world as a successful diagnostic tool. It now plays a very important role and is indispensable for the diagnosis of diseases within the human body. For this application, the quality of the fiber bundle itself does not require any further improvement. Fiberscopes, however, still need to be improved, although medical treatment is already possible with the use of endoscopes, thereby eliminating the need for hospitalization in conventional cases.

In mechanical and chemical engineering, conventional borescopes are being replaced by flexible fiberscopes, although the demand is limited to uses in such areas as the inspection of turbine blades, engine bores, the insides of pipes, or furnaces. The structure of the industrial fiberscope is essentially the same as that of the medical scope. Improvement may be unnecessary except as required for increased length, easier operation in a cavity with a complicated shape or narrow path, or higher heat resistance.

The graded-index fiber and rod lens were first introduced by Kita and Uchida in 1968. They attempted at first to develop a waveguide for optical communications by providing focusing capability to the fiber (16,17). However, the fiber was also found to have image transfer capability, so graded-index rod lenses were developed and applied to endoscopy (18,19). Although the rod lens is conceptually different from the fiber, it will be treated in this chapter because it offers the potential of a new field of endoscopy.

TRANSMISSION THROUGH A BUNDLE

When a light energy pattern impinges on one end surface of a fiber bundle, the image of the pattern is obtained on another end surface, either deformed or undeformed, depending upon the type of the bundle. Fiber bundles may be classified as follows (illustrated in Figure 10.1):

Classification of Fiber Bundles

Uniform Coherent Bundle. Each fiber is arranged on one end surface so as to occupy the same relative position as another surface. A light energy pattern is, therefore, transmitted through the bundle, forming an image at unity magnification without distortion. This type of bundle has been widely used for rigid fiberscopes as well as flexible fiberscopes, and as optical elements in optical and electrooptical instruments. Although there are three types (i.e., rigid image conduit as shown in Figure 10.1A, flexible image guide as shown in Figure 10.1B, and faceplate as shown in Figure 10.1C), only the former two are applicable to endoscopy.

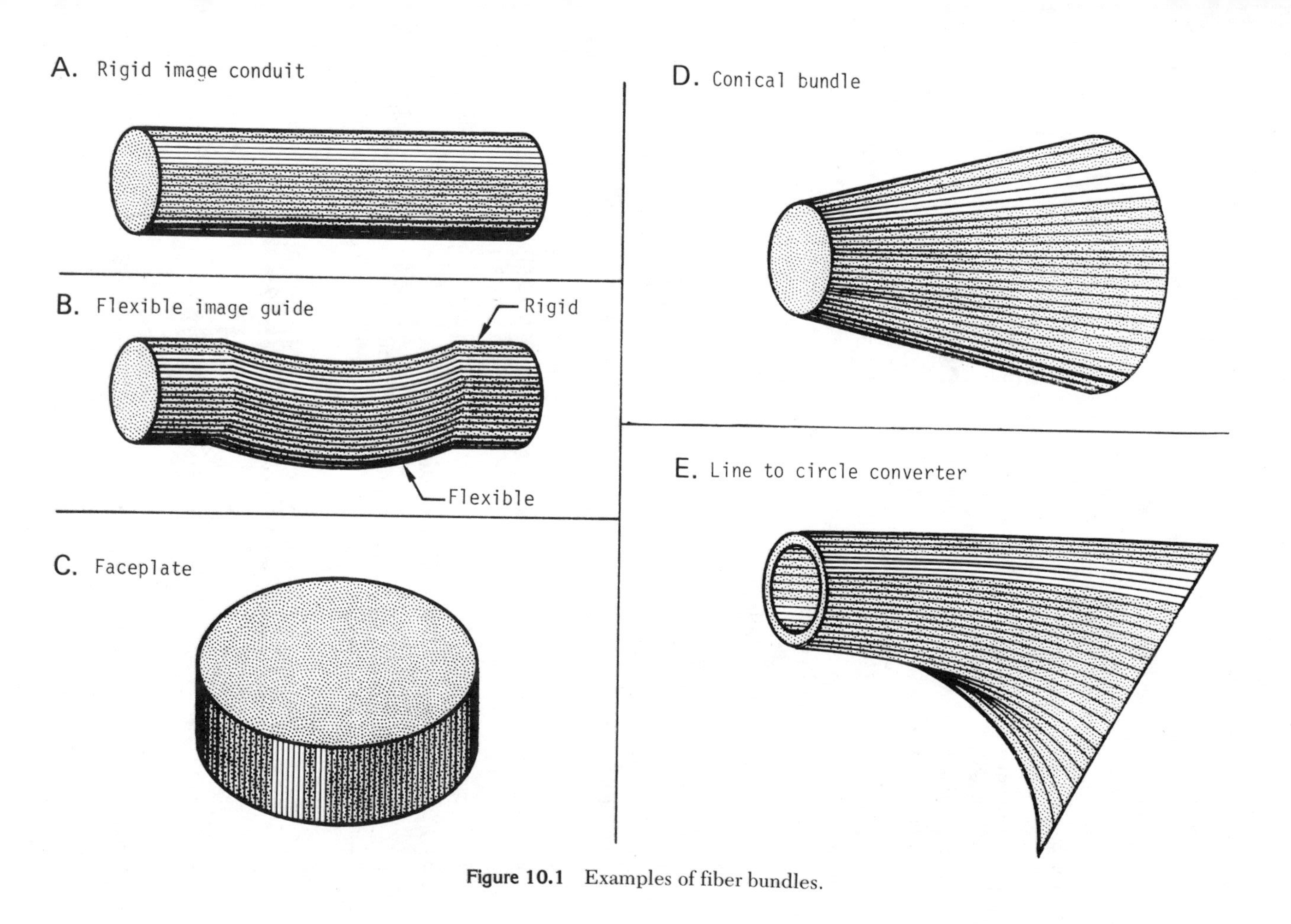

Figure 10.1 Examples of fiber bundles.

Conical Coherent Bundle. If the diameter of one end of a bundle is made larger (smaller) than the diameter of the other end without changing the relative positions of the individual fibers (as illustrated in Figure 10.1D), an enlarged (reduced) image can be obtained at the larger (smaller) end corresponding to the light energy pattern incident on the end. This bundle arrangement has the significant advantage that magnified or reduced images can be obtained in an object-to-image distance that is much shorter than that required in the case of a lens system (20). The conical bundle may be either rigid or flexible. However, a rigid one is more useful because of the short object-to-image distance and because of easy alignment when it is to be installed within a very small space. If it becomes available at relatively low cost as compared with the conventional lens system, new and interesting applications may be found for endoscopy.

Converter Bundle. This type is also referred to as an image dissector. A line-to-circle converter is a typical example of this bundle type, as depicted in Figure 10.1E. It picks up a linear array of picture elements and rearranges them into a circle. By scanning the light energy that emerges from the circular end, the intensity pattern corresponding to the picture elements can be detected. This bundle type is applicable to information processing, picture scanning, and other areas.

Incoherent Bundle. This is a fiber bundle in which each fiber at one end is randomly positioned with respect to another end. It is used as a light guide which transmits light energy from one position to another, for the uniform illumination of objects, or for the detection of light intensity. In practical light guides, the fiber configuration of the bundle cannot be made perfectly random, but it possesses a partly regular pattern. Since high transmittance is desirable, fibers of large diameter are usually employed in light guides of this type.

Requirements on Fiber and Bundle

It is important for fiber optics in general and for endoscopy in particular that images be bright enough for photographic recording as well as for visual

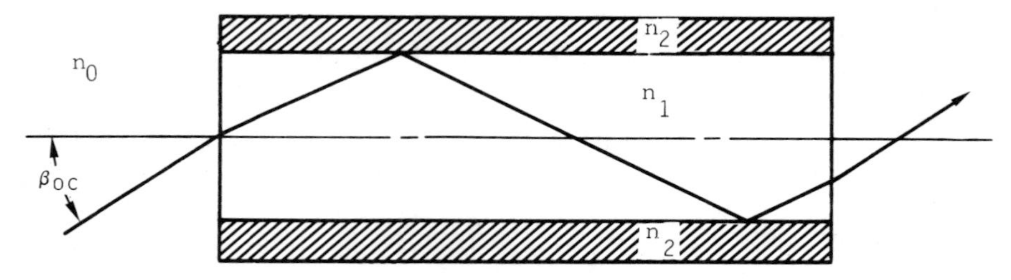

Figure 10.2 Transmission of a light ray at the critical angle through a fiber.

observation. Therefore, the transmittance of the fibers should be kept as high as possible. The factors to be considered in this respect are numerical aperture, Fresnel reflection loss, packing fraction, spectral transmittance and scattering by the fiber materials, and loss at total reflection. Image quality is also to be considered in the case of image guides. Factors related to quality are resolution, defects in fiber bundle, and nonuniformity in transmittance. Although these criteria were discussed in previous chapters, a brief review is given here to summarize the considerations most important to endoscopy.

NUMERICAL APERTURE

The numerical aperture of a step-index fiber is defined by

$$\mathrm{NA} = (n_1^2 - n_2^2)^{1/2} \tag{10.1}$$

where n_1 and n_2 are the refractive indexes of fiber core and cladding, respectively. The beam of light incident on the end surface of a fiber is totally reflected when it impinges on the surface at an angle smaller than a critical angle β_{oc}, as shown in Figure 10.2. The critical angle is

$$\begin{aligned}\beta_{oc} &= \sin^{-1}(\mathrm{NA}/n_0)\\ &= \sin^{-1}\left[(n_1^2 - n_2^2)^{1/2}/n_0\right]\end{aligned} \tag{10.2}$$

where n_0 is the refractive index of the medium surrounding the fiber. If the incident angle is larger than the critical angle, light energy is refracted out of the interface of the core and cladding and cannot be transmitted through the fiber owing to multiple reflection at the interface. On the other hand, energy can be transmitted without loss when the beam of light is incident at an angle smaller than β_{oc}, assuming the fiber to be straight and its material free of both absorption and scattering. Therefore, the higher the numerical aperture, the more energy can be transmitted without loss. The numerical aperture of the fiber and the critical angle are given in Table 10.1.

FRESNEL REFLECTION LOSS

On the surface of a glass plate, light energy is partly reflected (Fresnel reflection). The reflectance is given by

$$R_f = \left[(n_1 - n_0)/(n_1 + n_0)\right]^2 \tag{10.3}$$

For example, assuming $n_0 = 1$ and $n_1 = 1.62$, 5.6% of the incident energy is reflected. Since reflection takes place at both entrance and exit surfaces of a fiber bundle, 10.9% of the energy is lost. This loss can be reduced to almost zero by applying antireflection multilayer coatings on both sides.

TABLE 10.1. Numerical Aperture and Critical Angle for Various Values of Refractive Indexes n_1 and n_2, Assuming $n_0 = 1$

n_1 \ n_2		1.45	1.46	1.47	1.48	1.49	1.50	1.51	1.52	1.53
1.55	NA	0.548	0.520	0.492	0.461	0.427	0.391	0.350	0.303	0.248
	β_{oc}	33.2°	31.4°	29.4°	27.4°	25.3°	23.0°	20.5°	17.7°	14.4°
1.56	NA	0.575	0.550	0.522	0.493	0.462	0.428	0.391	0.351	0.304
	β_{oc}	35.1°	33.3°	31.5°	29.5°	27.5°	25.4°	23.1°	20.5°	17.7°
1.57	NA	0.602	0.577	0.551	0.524	0.495	0.464	0.430	0.393	0.352
	β_{oc}	37.0°	35.3°	33.5°	31.6°	29.7°	27.6°	25.5°	23.1°	20.6°
1.58	NA	0.628	0.604	0.579	0.553	0.526	0.496	0.465	0.431	0.394
	β_{oc}	38.9°	37.2°	35.4°	33.6°	31.7°	29.8°	27.7°	25.5°	23.2°
1.59	NA	0.652	0.630	0.606	0.581	0.555	0.527	0.498	0.467	0.433
	β_{oc}	40.7°	39.0°	37.3°	35.5°	33.7°	31.8°	29.9°	27.8°	25.6°
1.60	NA	0.676	0.655	0.632	0.608	0.583	0.557	0.529	0.500	0.468
	β_{oc}	42.6°	40.9°	39.2°	37.4°	36.7°	33.8°	31.9°	30.0°	27.9°
1.61	NA	0.700	0.679	0.657	0.634	0.610	0.585	0.559	0.531	0.501
	β_{oc}	44.4°	42.7°	41.0°	39.3°	37.6°	35.8°	34.0°	32.1°	30.1°
1.62	NA	0.722	0.702	0.681	0.659	0.636	0.612	0.587	0.560	0.532
	β_{oc}	46.3°	44.6°	42.9°	41.2°	39.5°	37.7°	35.9°	34.1°	32.2°
1.63	NA	0.744	0.725	0.704	0.683	0.661	0.638	0.614	0.589	0.562
	β_{oc}	48.1°	46.5°	44.8°	43.1°	41.4°	39.6°	37.9°	36.1°	34.2°

PACKING FRACTION

Figure 10.3 shows typical examples of possible packing configurations. Only the light energy impinging on the area of the core fibers can be transmitted and contributes to image formation. The packing fraction, i.e., the ratio of the core area to the total area, is given by

$$f_p = \tfrac{1}{4}\,\pi(a_1/a_2)^2 \qquad \text{for square packing} \tag{10.4a}$$

$$f_p = \frac{\pi}{2\sqrt{3}}\,(a_1/a_2)^2 \qquad \text{for triangular packing} \tag{10.4b}$$

$$f_p = (a_1/a_2)^2 \qquad \text{for hexagonal packing} \tag{10.4c}$$

where a_1 and a_2 are the dimensions, as indicated in Figure 10.3.

For commercially available fiberscopes, the thickness of the cladding ranges from 1.0 to 2.5 μm, which according to waveguide theory is larger than the value required for light to be guided without loss. Figure 10.4 gives an example of the packing fraction calculated for a cladding thickness of 1.5 μm as a function of $2a_1$. When high resolution is required, the diameter of the fiber must be decreased while its cladding thickness is kept constant, thus resulting in a decrease in the packing fraction and therefore in a decrease in the transmittance of the bundle.

SPECTRAL TRANSMITTANCE

Recently, there has been an increasing need for longer fiberscopes in medical as well as industrial applications. For example, the colonofiberscope is required to be about 2 m long with a light guide of 1.5 m, and the industrial fiberscope must be longer than 3 m, with a light guide of about 2.5 m, for the inspection of aircraft engines. In contrast to optical communications uses, the numerical aperture should be as large as possible, as previously mentioned, to obtain bright images. Therefore, the core glass should have as high a refractive index as possible and the lowest absorption. In general, the higher the index, the higher is the absorption loss or the less uniform is the spectral response, resulting in dark and colored (in most cases yellowish) images.

A typical spectral transmittance is shown in Figure 10.5 for a commercially available fiber bundle 2 m long. The absorption in the green to blue region results from impurities such as Cr, Fe, and others in the raw materials, especially for core glass. Although the elimination of these impurities allows the attainment of low-loss fibers, pure materials necessary for the fibers are still too expensive to use for commercial fiber bundles. Since multicomponent glasses have been intensively studied for the use in optical communications, it is expected that improvements will be made in the near future also in low-

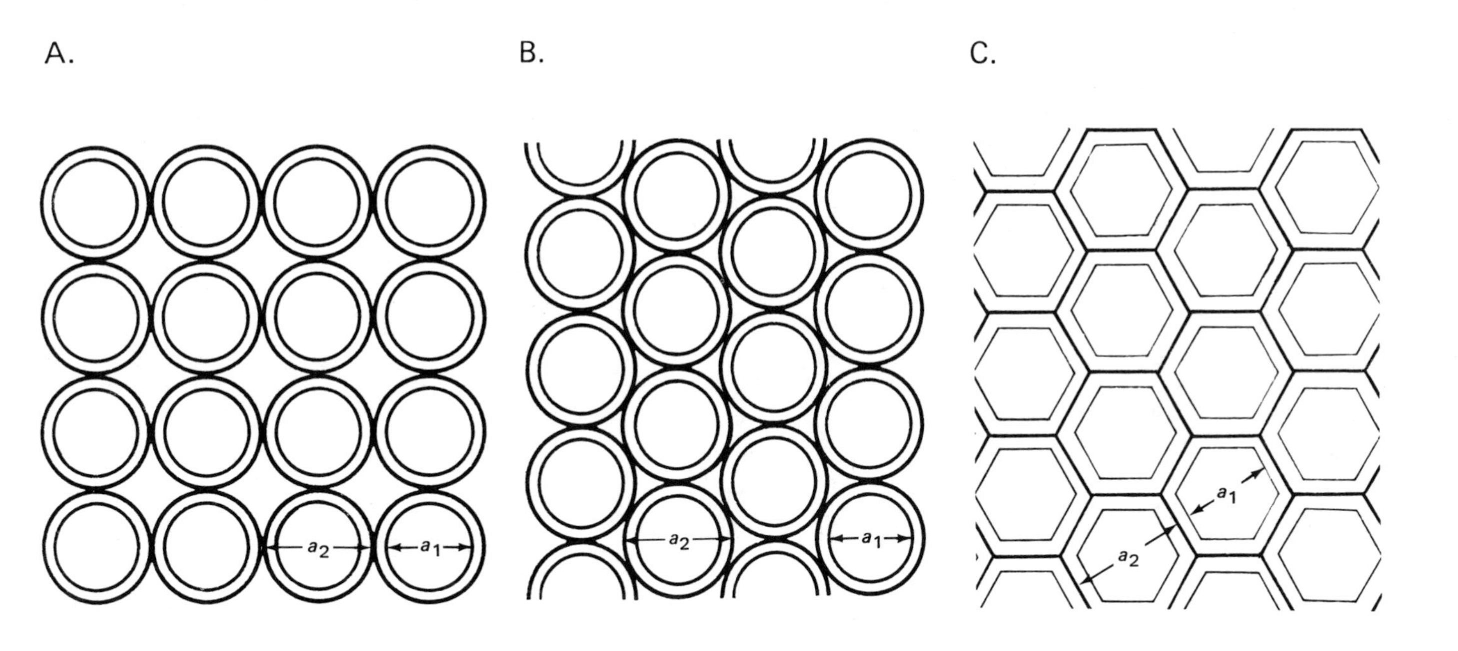

Figure 10.3 Examples of packing configurations; (A) square, (B) triangular, (C) hexagonal.

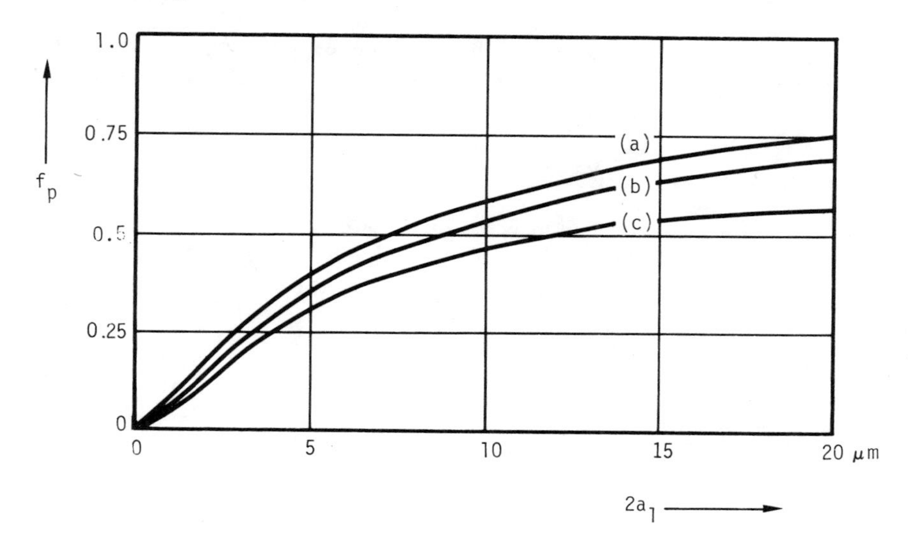

Figure 10.4 Variation of packing fraction with the diameter of the fiber core. Thickness of cladding 1.5 μm. Packing: a, hexagonal; b, triangular; c, square.

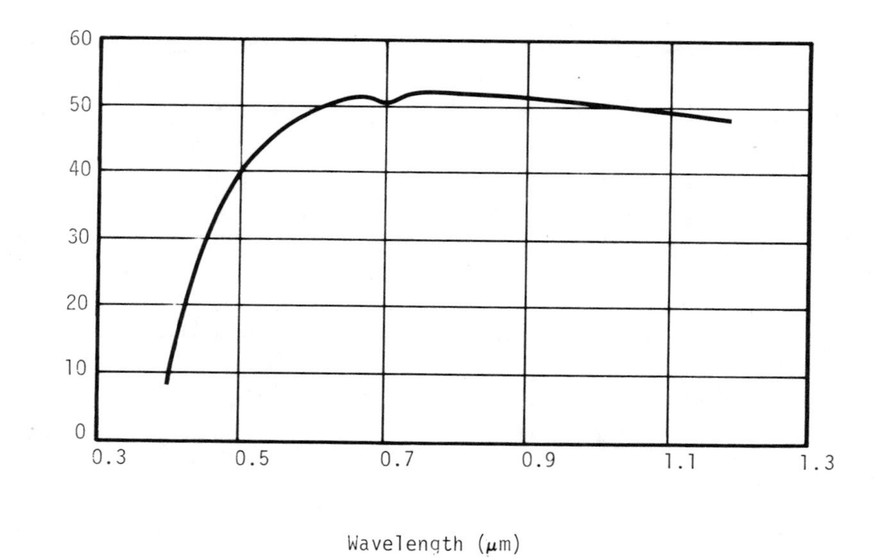

Figure 10.5 Spectral transmittance of a fiber bundle 2 m long of NA = 0.55.

loss fiberscopes. Furthermore, high-index glasses with very low loss should be investigated concerning development of fibers with higher numerical aperture. These are necessary for wide-angle fiberscopes.

LOSS AT TOTAL REFLECTION

At the interface between fiber core and cladding, energy may be lost due to absorption and scattering after repeated reflections. The number of the total reflections for a meridional ray incident at the critical angle is given by

$$N = \frac{L}{2a_1} \frac{(n_1^2 - n_2^2)^{1/2}}{n_2} = \frac{L}{2a_1} \frac{\mathrm{NA}}{n_2} \tag{10.5}$$

where L and $2a_1$ are the length of the fiber and the diameter of the core, respectively. The intensity decrease after N reflections can be expressed

$$T_r = R_t^N \tag{10.6}$$

where R_t is the reflectance of the interface. Even if the loss at the interface is very small, a considerable loss is expected over the fiber length because the number of reflections becomes very large. The intensity decrease as a function of the fiber length is shown in Figure 10.6 with R_t as a parameter. If the

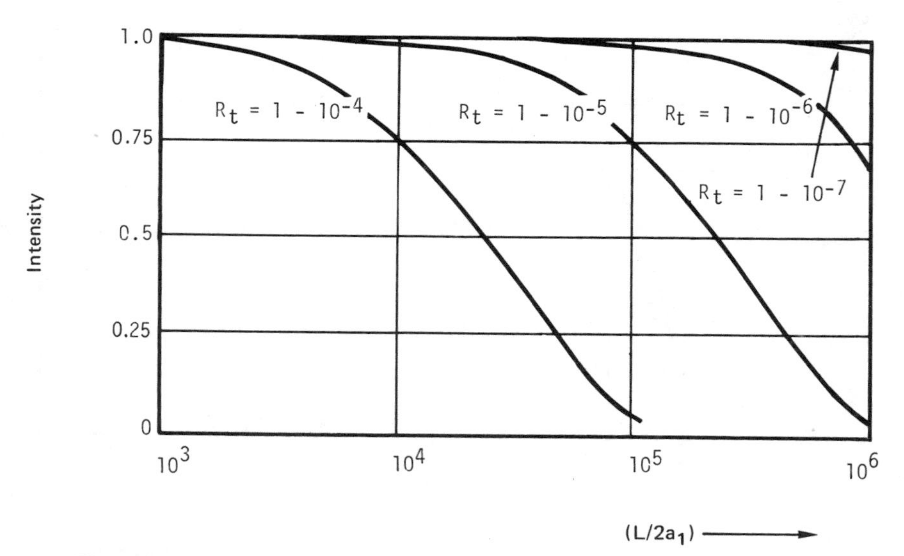

Figure 10.6 Decrease of intensity with fiber length. The decrease in intensity is caused by repeated reflections at the interface.

Figure 10.7 Image obtained with a fiber bundle of triangular packing.

thickness of the cladding becomes small, energy leaks out of the fiber, giving rise to loss. If the fiber diameter becomes smaller, the number of reflections and hence the energy loss tend to become larger.

IMAGE QUALITY

As the image area of the fiber bundle is divided into elementary areas determined by the packing configuration of individual fibers, the images are of a discrete nature that is governed by the configuration. Figure 10.7 is an example of the images obtained with a fiber bundle of triangular packing. The image quality is also determined by the total number of the fibers included in the image area. When the diameter of a bundle is required to be small, as in the case of the bronchoscope, the total number is limited to the extent at which image smoothness is apparently lost. However, sufficient information can be obtained from the image. The image quality is affected by three major factors: resolution, defects, and nonuniformities.

Resolution. For ordinary optical image-forming systems, such as lenses or reflectors, a point spread function can be defined as an invariant with respect to position within the isoplanatic region where aberrations stay constant. On the other hand, the point spread function depends on the position in the case of the fiber image guide. The point spread function is characterized as follows: There is a uniform intensity distribution within the cross section of a fiber core when a point light source is located in the corresponding area in the incident surface, whereas no light energy is transmitted when the light source is located outside this area; this is illustrated in Figure 10.8. Therefore, resolution cannot be defined in the same sense as the Rayleigh criterion. In order to understand resolution, the following factors should be considered:

Spot resolution. Assuming that the incident beam has a circular cross section, the diameter $2a_s$ of the beam should be larger than $2(2a_2/\sqrt{3} - a_1)$ in order for the light beam to be always detectable regardless of its position for a bundle of triangular packing, as illustrated in Figure 10.9. The light energy cannot be transmitted if the diameter is smaller than the above value and the beam is located outside the core area. Two beams of diameter $2a_s$ should be separated by a distance S_p larger than $2(a_1+a_2+a_s)$ so that the beams can be distinguished. For example, when the core diameter is 10 μm and the cladding diameter is 13 μm, $2a_s > 5$ μm and $S_p > 28$ μm. This implies that in this example a circular object can be recognized if its diameter is larger than 5 μm, and two objects of the same diameter can be distinguished if they are separated by a distance larger than 28 μm in terms of center-to-center distance. Hence, the resolution is 5 μm or 28 μm, respectively.

Line resolution. A line object is recognized as an array of spots located at the cross section of fiber cores. The image varies as shown in Figure 10.10,

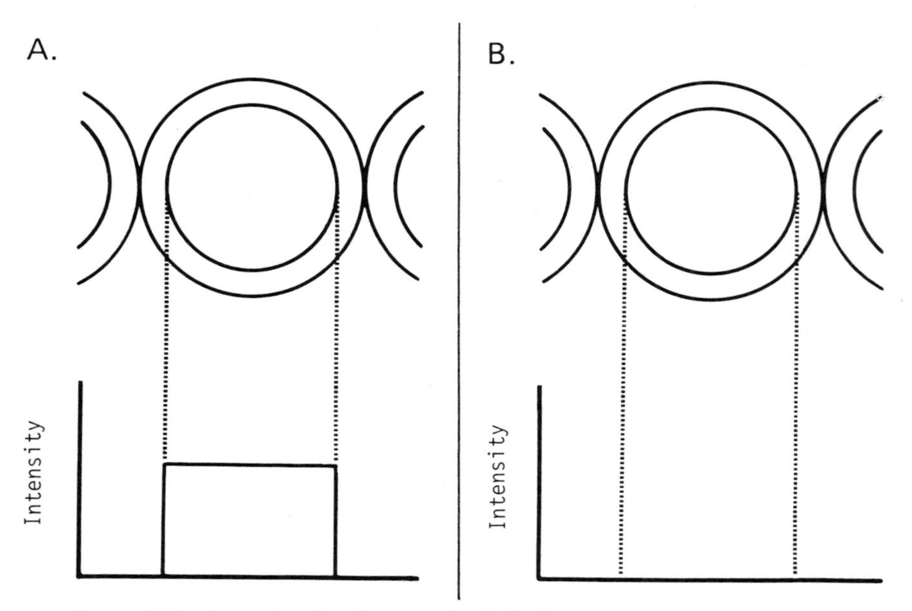

Figure 10.8 Intensity distribution of a light beam emerging from a fiber bundle as a function of the position of the incident point light source, located within (A) and outside (B) fiber core.

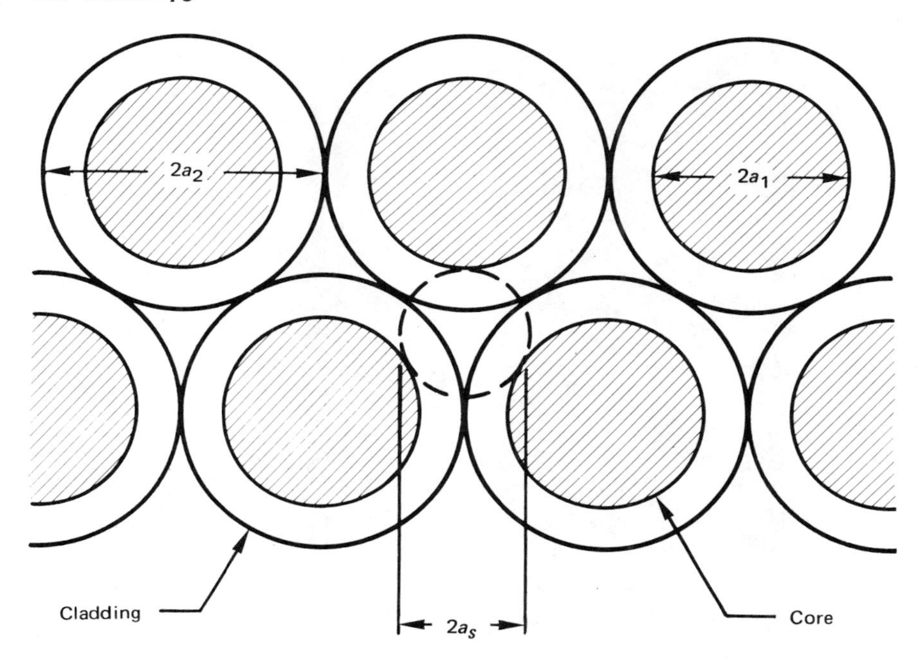

Figure 10.9 Spot resolution for triangular packing. Dashed circle, incident light beam diameter, $2a_s = 2\,(2a_2/\sqrt{3} - a_1)$.

when the azimuthal orientation of the object changes. The line width should be larger than $\sqrt{3}\,a_2 - 2a_1$ in order for the line to be detected regardless of its position and orientation. Two lines of width a_l should be separated by a distance S_l larger than $2a_1 + 2a_2 + a_l$ so that the lines can be distinguished. For example, under the same conditions as in the example above, $a_l > 1.26\ \mu m$ and $S_l > 24.26\ \mu m$; this implies that a line object can always be detected if its width is larger than 1.26 μm, and that two objects can be distinguished if they are separated by a distance larger than 24.26 μm.

The above examples indicate that the resolution of a fiber bundle can be defined as the center-to-center separation of two circular objects each having a nonnegligable diameter. The resolution can be increased if the fiber bundle is vibrated randomly in the transverse directions by keeping both end surfaces geometrically fixed. The image obtained in this manner is called the dynamic image (9), whereas the static image is defined as that obtained without vibration. The resolution of a dynamic image in general is higher than that of an equivalent static image. Although a better image can be obtained, dynamic images have not been utilized in endoscopy because of the technical complexity involved in their practical implementation.

Defects. Since not all fibers are manufactured with dimensional and optical uniformity, some defects are noticeable when the fiberscope is illumi-

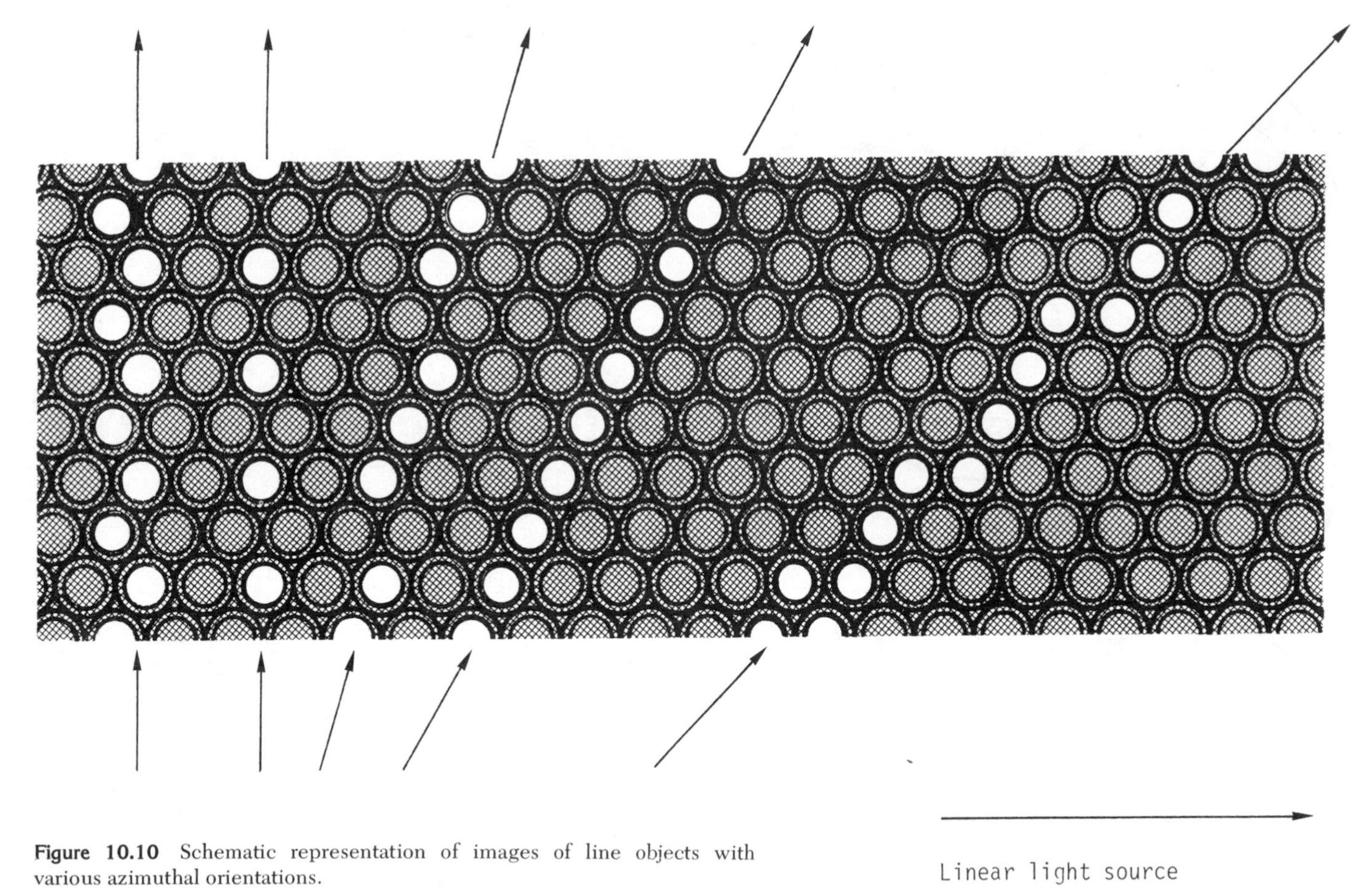

Figure 10.10 Schematic representation of images of line objects with various azimuthal orientations.

nated uniformly with a white light source. Some of them originate from the manufacturing process and others from repeated uses of the scope. The most prominent defects are as follows:

Dark spot. This is the most noticeable defect that is caused by fracture or crack in a fiber of the bundle. It appears as a black spot because no light is transmitted through the defective fiber when illuminated at the entrance surface. Since the bundle usually consists of 20,000 to 40,000 fibers, it is very difficult to prevent any of the fibers from fracturing in the process of manufacturing, and several dark spots often exist even in a virgin fiber bundle. The number of dark spots usually increases as the bundle is repeatedly used. Although the spots are clearly seen, 20 to 30 spots may be tolerated in actual applications. Spots may not be completely dark in some cases because of a slight crack in the fiber, uneven thickness of cladding layer (core positioned eccentrically or nonuniform thickness in longitudinal direction), or inclusion of opaque or semitransparent material. In other cases, the dark spot may appear as a group of semitransparent fibers in fiber bundles manufactured through a process in which individual fibers are fused together. When a fiber includes a solid particle somewhere along its length, the diameter of surrounding fibers may be squeezed at that location, causing a flowerlike dark spot.

Void. A flowerlike spot is also formed when a fiber is not packed at the position where it should be located. This is recognized as an irregularity in packing. If the spot size is less than the resolving power of the eyepiece, this defect is hardly detectable.

Dark line. A dark line may be seen at a location where an array of fibers is not contacted closely by its adjacent array. This is often seen in fiber bundles made by means of foil stacking.

Nonuniformity. A group of fibers is often recognized as a tinged area, slightly bluish or yellowish, or slightly darker or brighter. The fibers in the tinged group may, for example, have a thinner cladding layer, resulting in a loss in the longer-wavelength region, hence giving a bluish tinge. The difference in spectral absorption of the core material may also cause a bluish or yellowish tinge. The fiber diameter is also related to nonuniformity in transmittance. The change in diameter, however, does not cause a change in spectral transmittance or a change in color if the cladding thickness is kept larger than $\pi\lambda$. A 10% diameter may cause 20% transmittance change. In medical examinations, diagnosis of tissue disease is frequently based on the observation of color change and aerial shape. Since very delicate color changes must be detected, nonuniformities in transmittance may detrimentally affect the capability of fiberscopes.

TRANSMISSION THROUGH A GLASS ROD

If a glass rod has a refractive index profile expressed by

$$n(r) = n(0)\left(1 - \tfrac{1}{2} Ar^2\right) \tag{10.7}$$

a ray of light travels along a sinusoidal path at a period $p = 2\pi/\sqrt{A}$ within the rod, exhibiting light-focusing properties needed for light guidance. Here, r is the radial distance from the center of the rod, $n(0)$ is the refractive index at the center, and A is a parameter related to the focusing properties. Since the graded-index glass rod possesses a capability equivalent to a system of lenses, it may be called a rod lens. As is well known in conventional lens systems, image position and magnification can be calculated using the equations given in elementary books of geometric optics (21) if the focal length and the positions of the principal planes are known.

The focal length of the rod lens is given by

$$f = \left[n(0)\sqrt{A}\ \sin(\sqrt{A}L)\right]^{-1} \tag{10.8}$$

where L is the length of the rod lens. It is understood that the focal power becomes either positive or negative depending on L. In other words, a lens of desired focal length can be obtained simply by cutting and polishing the end surfaces until the rod has a length that is determined by the above equation. The focal length lies within the range between $1/n(0)\sqrt{A}$ and infinity. The principal planes are located at a distance d from each end,

$$d = (1/n(0)\sqrt{A})\ \tan\left(\tfrac{1}{2}\sqrt{A}L\right) \tag{10.9}$$

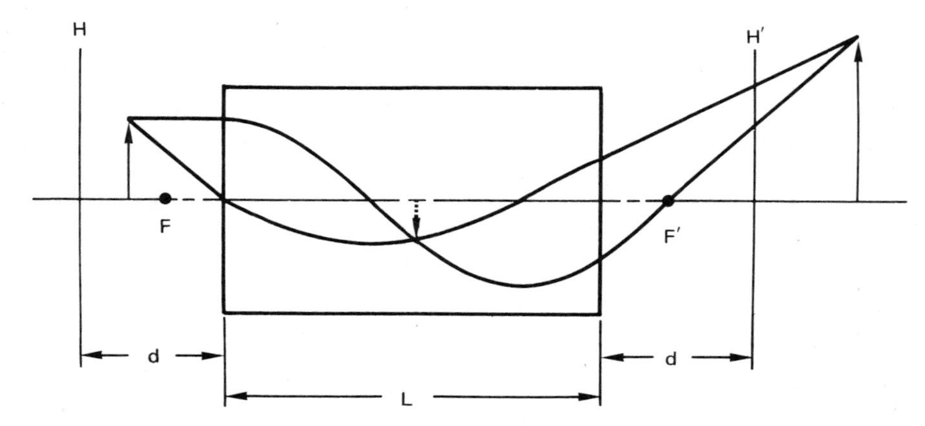

Figure 10.11 Image formation by a rod lens. F and F' represent focal points, H and H' principal planes; $f < 0$ and $d < 0$.

This is illustrated in Figure 10.11. The position of the principal plane can be either positive or negative, depending on the length L. From the above equations, characteristics of image formation can be classified as illustrated in Table 10.2.

The resolution of the rod lens depends on the object position and the quality of the lens. Assuming a lens free of aberration and an object located at infinity, the resolution is given by

$$s = 1.22\lambda/n(0)\sqrt{A}b \tag{10.10}$$

according to Rayleigh's theory, where b is the radius of the rod and λ is the wavelength.

For endoscopy, chromatic aberration may cause a serious problem in the case of the rod lens because images are transferred from one end to the other by periodic image formation. The aberration can be expressed

$$\frac{\Delta p}{p} = \frac{1}{2}\frac{[1/\nu(b) - 1/\nu(0)] - [1/n(b)\nu(b) - 1/n(0)\nu(0)]}{n(0)/n(b) - 1} \tag{10.11}$$

where Δp is the difference between values for $p = 2\pi/\sqrt{A}$ at the F and C lines and $\nu(0)$ and $\nu(b)$ are Abbe's numbers at the center and the periphery, respectively. Ikeda et al. (22) reported a rod lens whose chromatic aberration was reduced to 0.004 by employing glass containing Cs_2O.

The rod lens has the advantage over a conventional lens that very short focal lengths or small diameters can be easily obtained. Some examples are shown in Table 10.3 (23). The rod lens can be applied to either the objective lens or the image guide of an endoscope.

FIBER OPTICS ENDOSCOPES

Since the fiberscope is suitable for the observation of the inside of a cavity which otherwise cannot be observed, it has found applications in the examination of the interior of the human body or the inside of mechanical parts such as an aircraft engine bore. Although many applications have been proposed or conceived, medical applications have found the widest use. In other words, medical endoscopes represent the largest quantity of fiberscopes used throughout the world.

Structure of a Flexible Endoscope

Fiber optics endoscopes can generally be divided into six sections: distal end, bending section, insertion tube, control section, connecting cable, and light supply section, as shown schematically in Figure 10.12.

TABLE 10.2. Characteristics of Image Formation by Rod Lenses (P, Object; Q, Image)

Length of rod lens	Positions of object and image	Classification
$0 < L < \frac{1}{4}\frac{2\pi}{\sqrt{A}}$	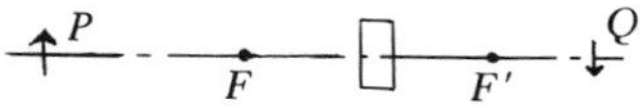	Inverted real image
	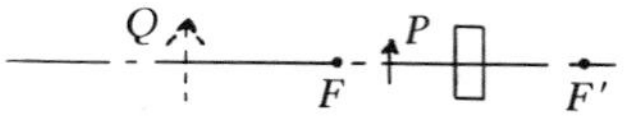	Erect virtual image
$L = \frac{1}{4}\frac{2\pi}{\sqrt{A}}$	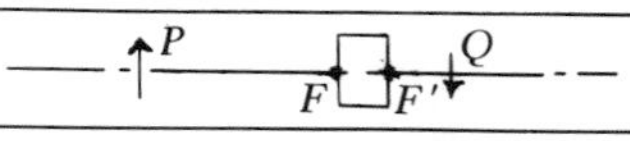	Inverted real image
$\frac{1}{4}\frac{2\pi}{\sqrt{A}} < L < \frac{1}{2}\frac{2\pi}{\sqrt{A}}$	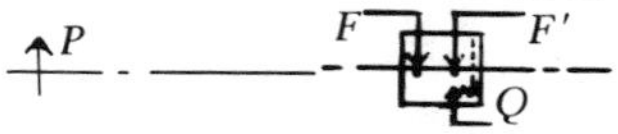	Inverted virtual image
	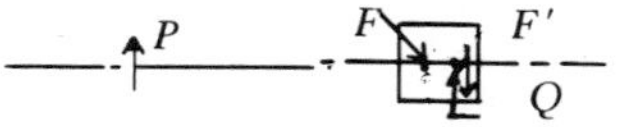	Inverted image at end surface
	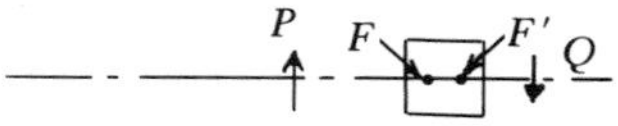	Inverted real image
$L = \frac{1}{2}\frac{2\pi}{\sqrt{A}}$	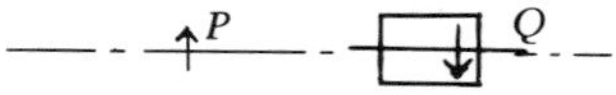	Inverted virtual image with unit magnification
	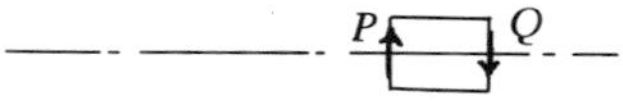	Inverted image at end surface with unit magnification
$\frac{1}{2}\frac{2\pi}{\sqrt{A}} < L < \frac{3}{4}\frac{2\pi}{\sqrt{A}}$	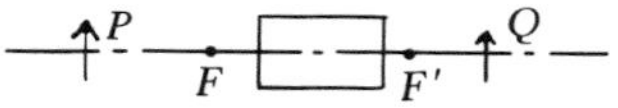	Erect real image
	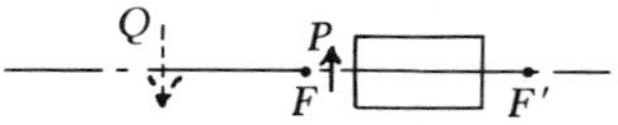	Inverted virtual image
$L = \frac{3}{4}\frac{2\pi}{\sqrt{A}}$	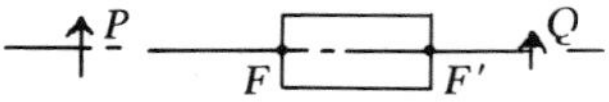	Erect real image
$\frac{3}{4}\frac{2\pi}{\sqrt{A}} < L < \frac{2\pi}{\sqrt{A}}$	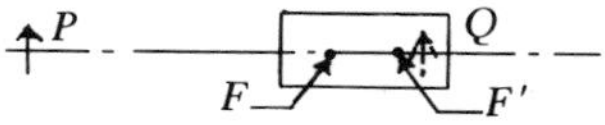	Erect virtual image
	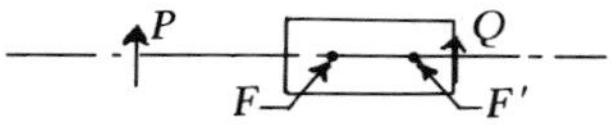	Erect image at end surface
	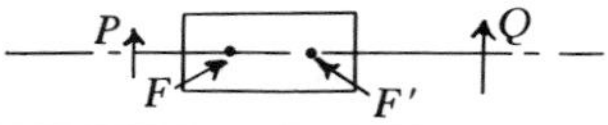	Erect real image
$L = \frac{2\pi}{\sqrt{A}}$	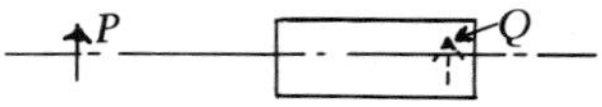	Erect virtual image with unit magnification
	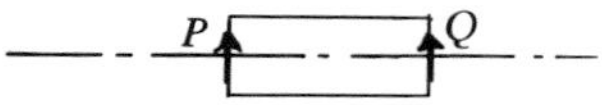	Erect image at end surface with unit magnification

TABLE 10.3. Characteristics of Typical Rod Lenses

Lens type	$2b$(mm)	A(mm^{-2})	p (mm)	$n(0)$ (λ = 0.589 μm)	NA	Beam angle (degrees)	$\Delta p/p$	Cutoff (μm)
Standard lens	1.0	0.15	16.2	1.553	0.30	35	0.07	320
	1.5	0.07	23.8					
	2.0	0.04	31.4					
Wide-field angle lens	0.35	3.41	3.4	1.616	0.50	60	0.05	380
	0.50	1.46	5.2					
	1.0	0.36	10.5					
	1.5	0.16	15.7					
	1.8	0.12	18.1					
	2.0	0.09	20.9					
Low chromatic aberration lens	1.0	0.020	44.4	1.525	0.10	12	0.004	260
	1.0	0.012	57.4	1.525	0.08	9.5	0.004	260
	1.0	0.009	66.2	1.525	0.07	8	0.004	260

Figure 10.12 Structure of a flexible endoscope (A) and views of the distal end (B); a, objective lens; b, light guide (illumination optics); c, suction and forceps channel; d, air/water feeding port; e, forceps raiser.

A.

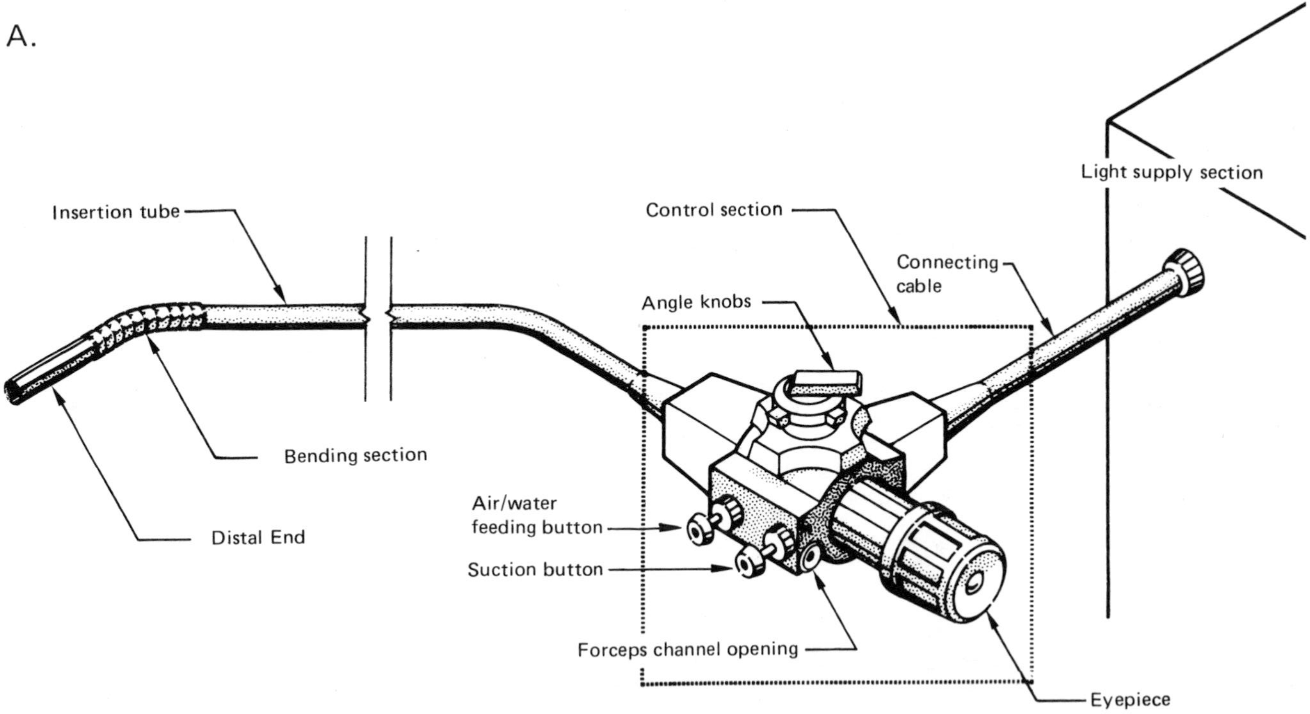

B. Distal end

Side viewing system

Forward-oblique viewing system

Forward viewing system

The distal end consists of an objective lens system, illumination optics, outlets for a variety of forceps, diathermic snares or cytology brushes, outlets for air and water feeding and water suction, a photosensitive element for automatic exposure, a flash lamp, and a small camera. Except for the objective lens and illumination optics, these components are installed selectively depending on the type of fiberscope. The insertion tube contains an image guide fiber bundle, a light guide fiber bundle, guide tubes for forceps, snares, or brushes, and tubing for air and water feeding and water suction. These components are placed inside a flexible sheath. The insertion tube is connected with the distal end through a bending section whose flexure is controlled by wires so that the distal end can deflect in two or four directions. The wires are actuated by turning angle knobs in the control section. The main portion of the control section is an eyepiece with a diopter adjustment ring and angle knobs for deflection adjustment. Depending on the demands of the endoscopist, other parts are added, such as air and water feeding control and water suction control, as well as inlets for forceps, snares, or brushes. The light supply section consists of a light source, power supplies for the source, an automatic exposure control, and an air compressor. It is connected with the

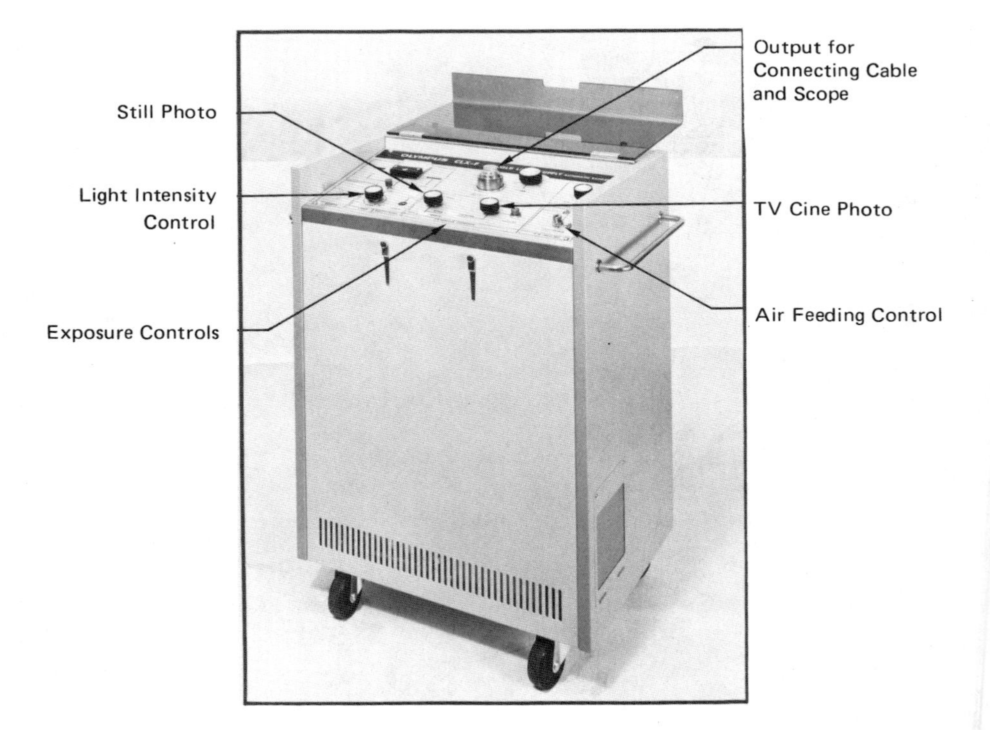

Figure 10.13 Photograph of cold light supply.

control section by means of the connecting cable, which consists of a sheath, a light guide, and power supply and control signal lines. Intense but cold light from a high-power xenon short-arc or halogen lamp is generally used for illumination. Figure 10.13 shows a photograph of a cold light supply. A 500W xenon short-arc lamp is used as a continuous wave or flash lamp at a color temperature of 6000°K. The shutter speed can be automatically controlled by means of a TTL (through-the-lens) system from 1/15 to 1/500 s. An "electric eye" system is also provided for cinematography in which the amount of light reaching the film is automatically adjusted by the servocontrol of a diaphragm opening. In addition, air and water pumping systems are provided for inflating the organ to be examined and cleaning the objective lens during endoscopy. An automatic heat ray control system is built into the light supply which prevents the light guide surface from burning the internal wall of the organ in case it approaches the wall too closely.

Medical Flexible Endoscope

Gastrocameras were used for the examination of the stomach or the early detection of stomach cancer before fiber gastroscopes became available. The gastrocameras developed by the Olympus Optical Company have been used widely since 1949. They were preceded by ten years of camera improvement as well as progress in the ability to interpret stomach photographs in terms of disease detection. The gastrocamera has a small camera together with a flash lamp for illumination in its distal end. It is manipulated by holding its control section and adjusting the deflection angle of the distal end during insertion. Since no observation optics are provided, it requires careful and skillful operation. In spite of this difficulty, the gastrocamera still plays an important role in the examination of the stomach.

The first gastrofiberscope, developed by Hirschowitz et al., opened the possibility for relatively easy examination of the interior of the human body and triggered the development of new examination methods, such as biopsy, endoscopic treatment, and surgery. Since then, various types of fiber optics endoscopes have been developed, and it has become possible to examine or treat a variety of parts of the human body, as illustrated in Figure 10.14. In medical practice various fiberscopes have been developed. Their main features are highlighted below.

GASTROCAMERA WITH FIBERSCOPE

This type of endoscope is provided with a camera in its distal end and viewing optics composed of an objective lens, a fiber optics image guide, and an eyepiece. Illumination is made with a small light source installed at the distal end. Its resolution is far better in a photograph taken directly with the camera at the distal end than in one taken with a camera at the proximal end after

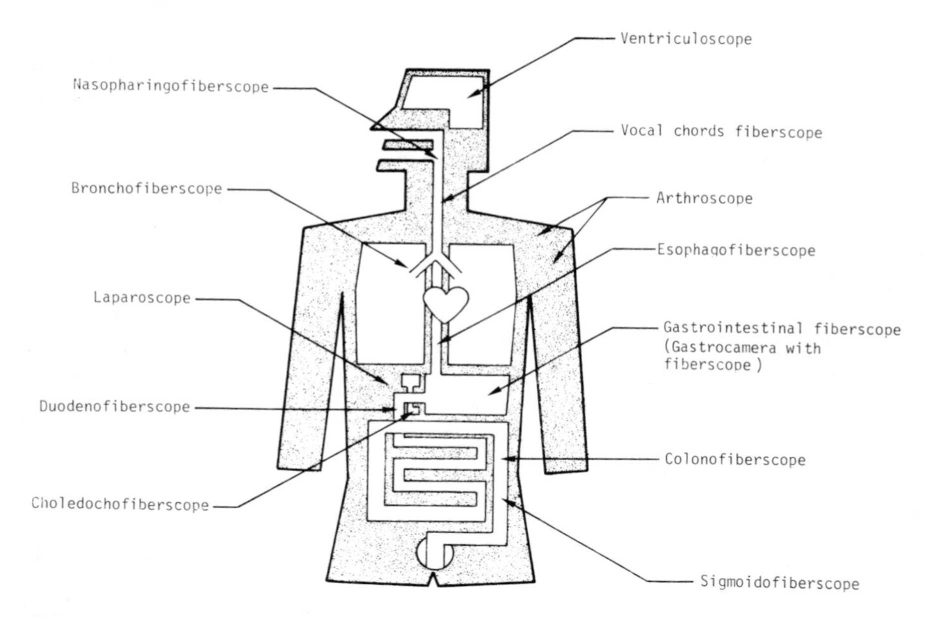

Figure 10.14 Portions of the human body which can be observed or treated by fiber optics endoscopes.

transmission through a fiber optics image guide. The distal end can be operated so that the viewing element can be directed toward the target. It is important that the dead angle be made as small as possible.

Figure 10.15 shows an example of a gastrocamera with fiberscope. Single-lens reflex (SLR) optics, in which the same lens is used for both picture-taking and observation, is installed under a side window of a rigid cylindrical tip at the distal end. The portion to be examined is illuminated with a light beam from a lamp located adjacent to the SLR camera. Exposure is automatically controlled by detecting the intensity of light reflected from the wall of the stomach. The intensity is detected by a photosensitive element located next to the observation–photo lens. A strip of photographic film loaded in a small cartridge is placed in the tip from its top end.

The SLR system is best suited to obtaining clear pictures; it can eliminate the possibility of taking a hazy picture due to lens clouding. The operation of the camera is made with an angle knob, a shutter release lever, and a rewind lever.

It is important for the doctor to operate the gastroscope in such a way that the control section is held snugly in the left hand; thus air and water feeding can be easily carried out by a simple push button operated singlehandedly. Suction should also be done easily by a lever control in this instrument. A

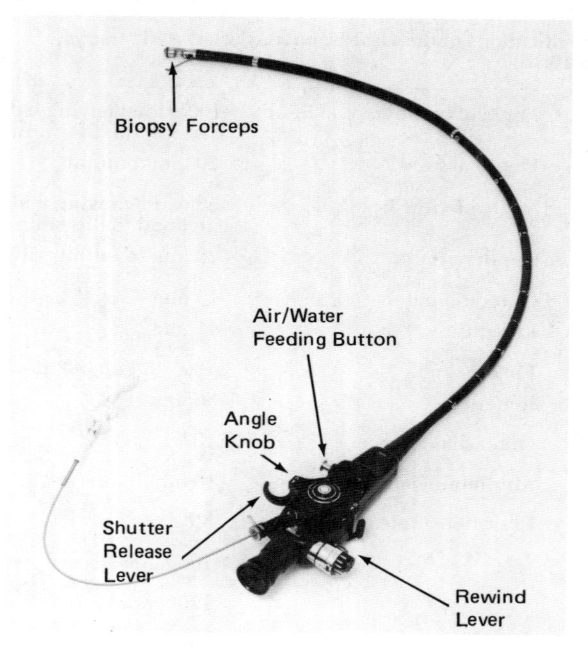

Figure 10.15 Gastrocamera with fiberscope (side viewing scope).

biopsy forceps can be inserted through a channel, and tissue samples can be collected efficiently by a forceps-elevation mechanism (forceps raiser) for cytology and histodiagnosis. Table 10.4 lists the principal specifications of the instrument to illustrate the general concept of the gastrocamera–fiberscope system.

GASTROINTESTINAL FIBERSCOPE

In this type of fiberscope, photographic recording is made with a camera attached to its control section. Although the quality of the photographs is slightly poorer than that taken with a distal camera, the gastrointestinal fiberscope has been widely accepted throughout the world because of its versatility. Employing a forward or forward-oblique viewing optical system, it is possible to examine, in addition to the stomach, the esophagus, duodenal bulb, upper flexure, and descending limb. Figure 10.16 shows a gastrointestinal fiberscope of small diameter. It was developed for the observation and photographic recording of the esophagus, stomach, and duodenum sequentially at one insertion. The insertion tube is 8.8 mm in diameter and 1260 mm in total length as a fiberscope; it can be introduced into the upper gastrointestinal tract with little premedication and discomfort to patients. Further im-

TABLE 10.4. Specifications of the Gastrocamera With the Fiberscope (shown in Figure 10.15)

Optical system for photography	Angle of photographic field	100° side viewing with optical axis inclined forward by 5°
	Depth of field	20 mm to infinity
Optical system for observation	Angle of view field	53° side viewing with optical axis inclined to the same direction
	Depth of field	20 mm to infinity
Camera section (distal end)	Outer diameter	13 mm
	Length	41 mm
Bending section	Flexure angle	180° (90° up, 90° down)
	Length	56 mm
Insertion tube	Outer diameter	10.7 mm
Biopsy forceps	Minimum visible distance	10 mm from camera section
	Elevation angle	85°
Working length		1000 mm
Total length		1200 mm
Film	Width	4 mm
	Picture size	4 mm × 5 mm
	Number of frames	26

provements of the endoscope itself are necessary to ease its operation and thus to enhance the effectiveness of premedication and technical expertise.

There are several types of fiberscopes commercially available, so that the doctor can select and use at least one most suitable for the portions of the gastrointestinal tract under examination. In Figure 10.16, an objective lens is located at the top of the distal end, allowing forward viewing over an 85° viewing angle with a fixed focus. The distal end can be deflected by a maximum of 200° by means of a four-way wide deflection mechanism, with which 180° up, 60° down, and 100° left and right deflections are achievable.

Forceps can be used for biopsy sampling, although the tube is very thin. This type of scope is being used for screening, routine, and emergency endoscopy. It is also suitable for the examination and extraction of a foreign body from pediatric and geriatric patients and applicable to hypersensitive patients and those with esophageal stenosis.

COLONOFIBERSCOPE

Unlike the upper gastrointestinal tract, bowels are necessarily curved in complicated forms or folded tightly. Special considerations must be given to the design of the colonofiberscope so that it can be introduced deep into the colon

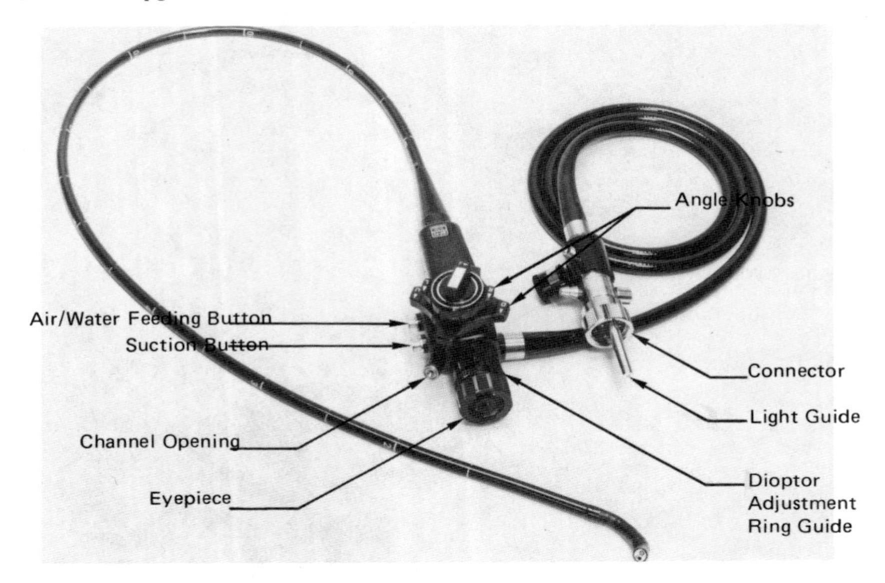

Figure 10.16 Gastrointestinal fiberscope (forward viewing scope).

without giving pain and discomfort to patients. The scope should have front viewing optics for the purpose of steering the scope as well as the examination of the wall of the colon, which forms an internal surface of a tubular tract. The distal end must be deflectable in any direction up to almost 180° so that the scope can be advanced smoothly in the colon.

The colonofiberscopes in Figure 10.17 are 2020 and 1270 mm long and their diameters at the insertion tubes are each 13 mm. The objective lens is of fixed focus and wide angle of view (85°). The insertion of the scope can be achieved very smoothly because the distal half of the thin insertion tube is made softer than its proximal half and because of the four-way deflection control, 180° up and down and 160° left to right. The scopes shown in Figure 10.17 enable examination up to the cecum through sigmoid, descending, transverse, and ascending colons.

BRONCHOFIBERSCOPE

The routine inspection of the bronchial tree became feasible after the first development of a bronchofiberscope, reported by Ikeda in 1965. The flexible fiberscope has made it possible for endoscopists to examine the bronchial tree better than with the rigid bronchoscope.

A currently available bronchofiberscope shown in Figure 10.18 has a diameter of 5.9 mm at its insertion tube; it is highly flexible at the distal portion, which ensures the easiest insertion and the minimum discomfort under local anesthetic. This instrument has an automatic aspiration adapter,

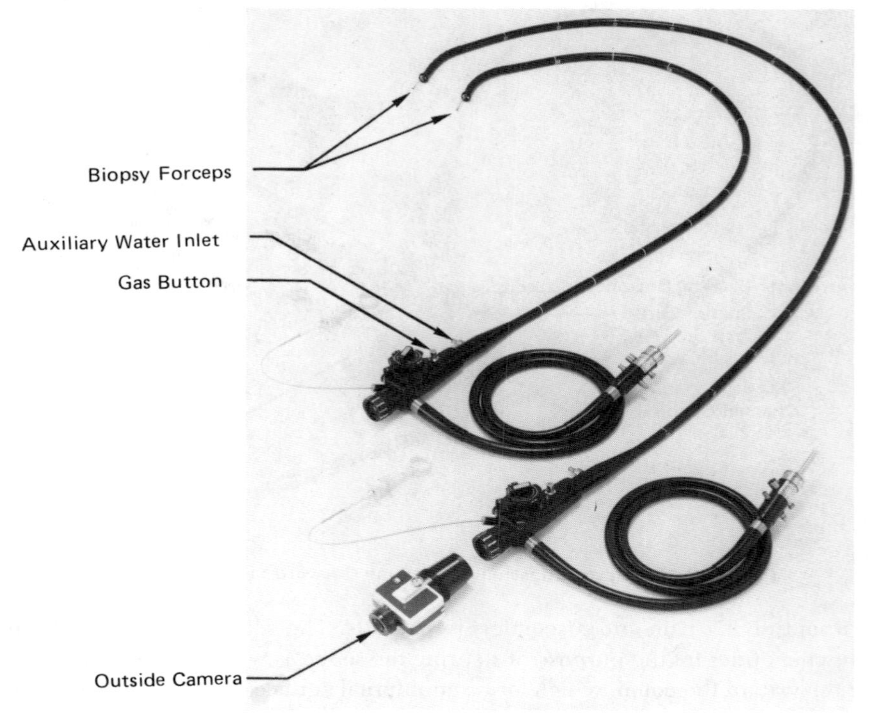

Figure 10.17 Colonofiberscopes of different lengths (forward viewing scopes).

permitting lignocaine instillation and a suction adapter for the removal of the sputum. Since the bronchofiberscope can eliminate the great pain experienced by the patient during the observation by the rigid bronchoscope, it allows examination as far as the subsegmental bronchi. It can sample tissues as large as those obtained with a rigid bronchoscope, permitting identification of the histological type.

EXAMPLES OF TREATMENT

In the following we want to give a few examples that illustrate practical treatment with the fiberscope. Since the time of institution of endoscopic examination, the concept of treatment using a fiberscope has been envisioned, and the removal of polyps has been practiced to a limited extent using the sigmoidscope. It was hoped that the endoscopic treatment would obviate the need for laparotomy in certain surgical situations. A revolutionary development in endoscopic treatment took place when Shinya developed a technique for the surgery of colonic polyps, colonoscopic polypectomy, which made hospitalization almost unnecessary for many patients. Since this technique is

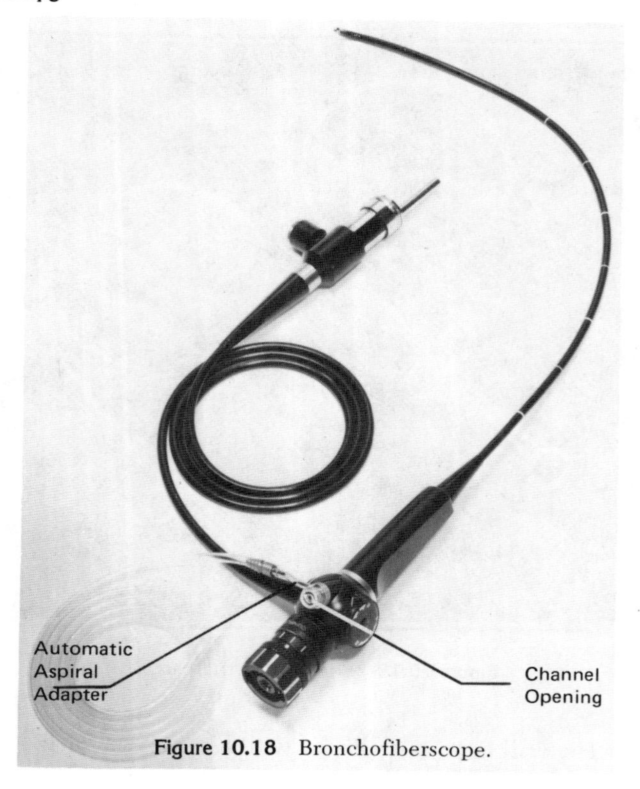

Figure 10.18 Bronchofiberscope.

relatively new, in the medical profession there was some skepticism about whether it represented a significant improvement over existing methods or whether it might be too risky. In spite of this, the colonoscopic polypectomy has been accepted by many doctors who have already acquired considerable experience and expertise in diagnosis by colonofiberscopes (24).

The removal of polyps is carried out with a diathermic snare, shown in Figure 10.19, energized by high-frequency electric surgery power. The snare is introduced through a channel of the colonofiberscope and manipulated to pass over the head of the polyp and wind around the stalk, as illustrated in the inset of Figure 10.20. The cautery power is applied to the wire so that the polyp can be removed without bleeding. After the polyp is removed, it is extracted with a grasping forceps or by suction. Figure 10.20 shows a grasping forceps and a diathermic snare both extending from a colonofiberscope that has two channels for the simultaneous handling of the forceps and the snare to ease the surgery. The diathermic procedure can be applied to other portions of the body when removal or coagulation is required.

For several years attempts have been made to coagulate bleeding or to control ulcers on the wall of the stomach, liver, esophagus, small and large intestines, etc. by the use of a coherent light beam from a high-power

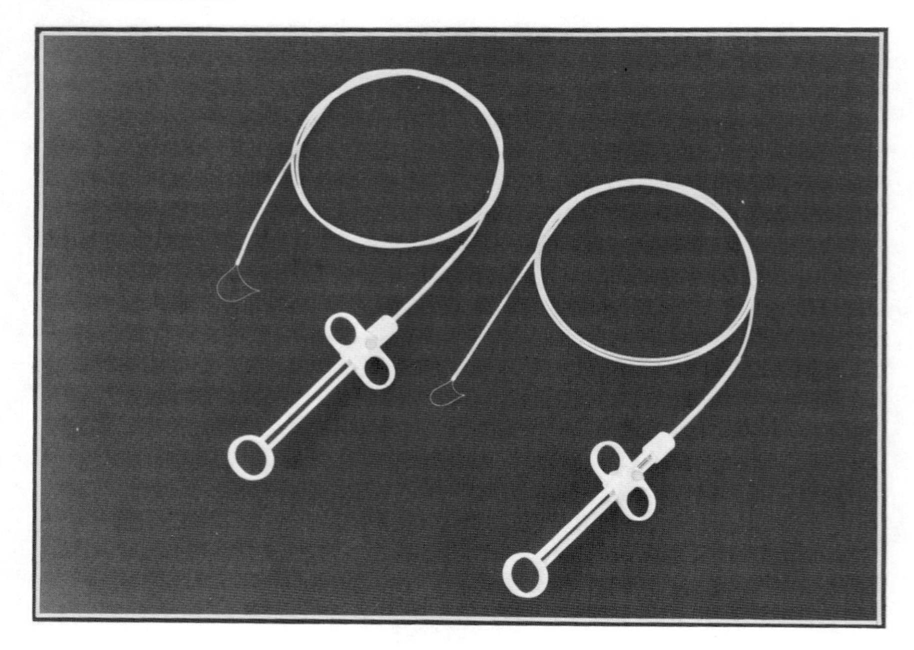

Figure 10.19 Diathermic snares.

laser (25–28). The wall may be irradiated with a light beam of an argon, CO_2, or Nd:YAG laser through a fiber or a bundle of fibers. The effect of the high-power laser beam on tissues and the effectiveness of the treatment are not fully understood and still remain to be investigated. In spite of this, these attempts may open a new field in endoscopic surgery. Although the light beam from a high-power CO_2 laser has been used as a surgical knife, the development of far-infrared fibers that transmit the laser beam with very low loss is necessary before lasers can be used for endoscopic surgery.

Medical Rigid Endoscope

Following the development of fiberscopes, many rigid endoscopes have been replaced by flexible endoscopes. Those replaced include the esophagascope, gastroscope, laryngoscope, bronchoscope, choledochoscope, and rectoscope. Flexibility is not required, on the other hand, in the case of the venticuloscope, thoracoscope, laparoscope, cystoscope, hysteroscope, and arthroscope. In order to obtain thin endoscopes, instruments that use fiber optics bundles or graded-index rod lenses have been developed which can be made thinner than conventional scopes. These new scopes are therefore more suitable for the venticuloscope, thoracoscope, cystoscope, laparoscope, and arthroscope, in which a small diameter is desirable. The new laparoscope and the arthroscope are briefly described.

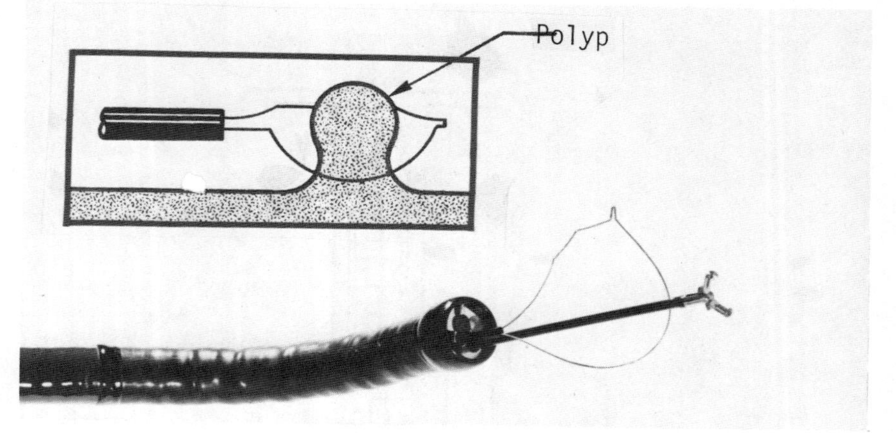

Figure 10.20 Diathermic snare and grasping forceps applied to a two-channel colonofiberscope.

LAPAROSCOPE

The laparoscope has been used for the examination of the surface of internal organs such as the stomach, intestines, and liver by puncturing a portion of the abdomen. The conventional laparoscope that is still widely used employs a series of relay lenses as its image guide. If the relay lenses are replaced by a fiber optics image guide or by a graded-index rod lens, it is possible to build a laparoscope with a thinner insertion tube. Making the diameter as small as possible can minimize pain to the patient and eliminate suturing the puncture after the examination, which enables the laparoscopist to examine his patient outside the operating room, i.e., at bedside. As illustrated in Figure 10.21, the fiber optics laparoscope consists of an objective lens, a rigid image conduit, an eyepiece, a light guide fiber bundle that is placed around the image conduit, and a metal sheath to protect optical elements. The diameter of the insertion tube can be made as small as 3.4 mm, permitting broader selection of puncture sites, more comfortable positioning of the patient, and less pre-operative preparation and postoperative care than with conventional scopes.

ARTHROSCOPE

The graded-index glass rod lens can transfer images, as previously described, even when its diameter is much smaller than the rigid image conduit. For example, a rod lens of 1 mm diameter and 150 mm length can be used as an image guide. An extraordinarily thin arthroscope using the lens, called the Selfoc arthroscope, was developed by Watanabe in 1970 (29). Since then, the scope has been improved so that the outer diameter of the sheath, as exemplified by the Selfoscope shown in Figure 10.22, becomes as small as 1.7 mm, with a working length of 105 mm. A deep view from 1 mm to infinity can be obtained with a 55° viewing angle in the case of a perpendicular end

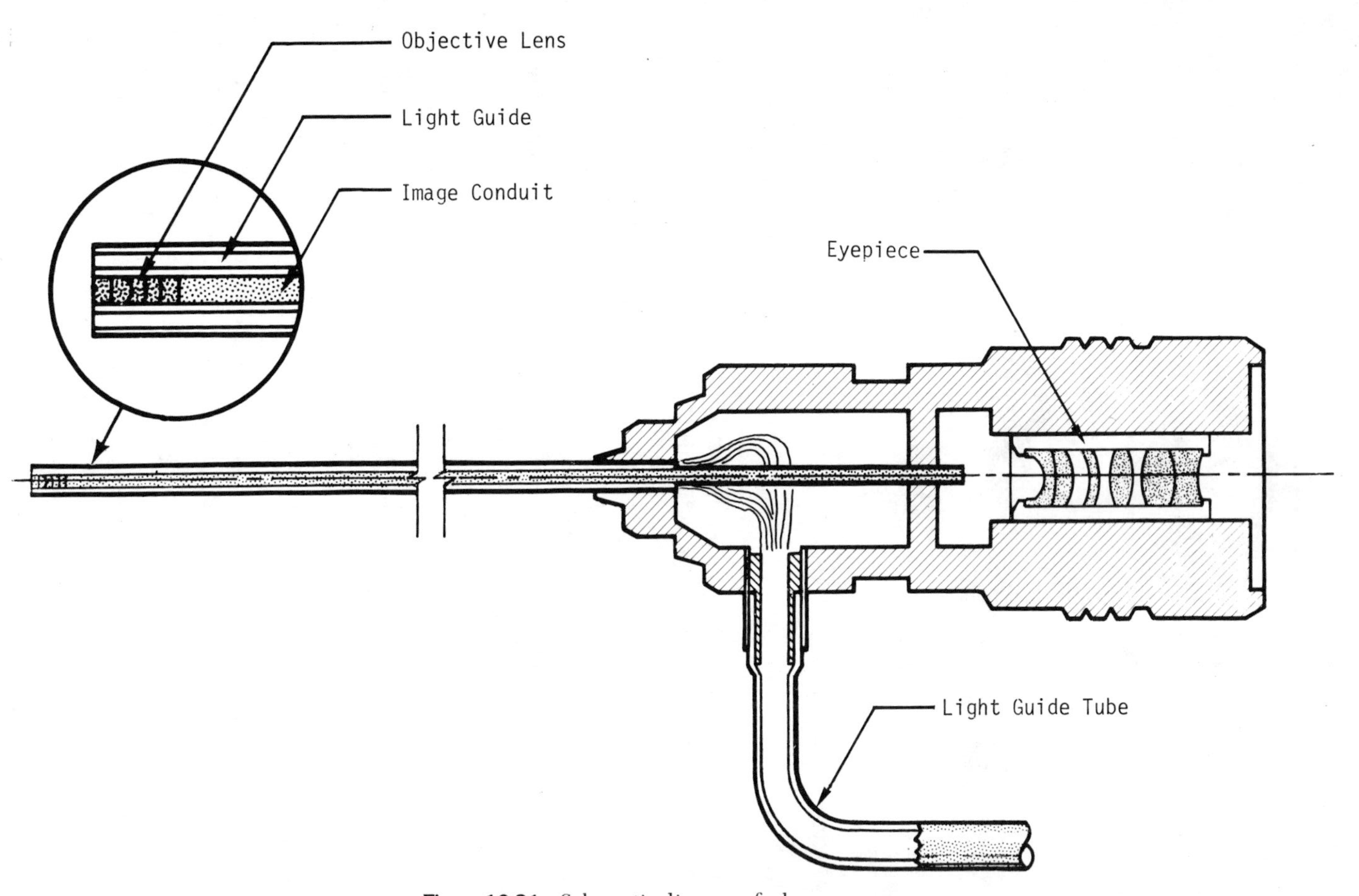

Figure 10.21 Schematic diagram of a laparoscope.

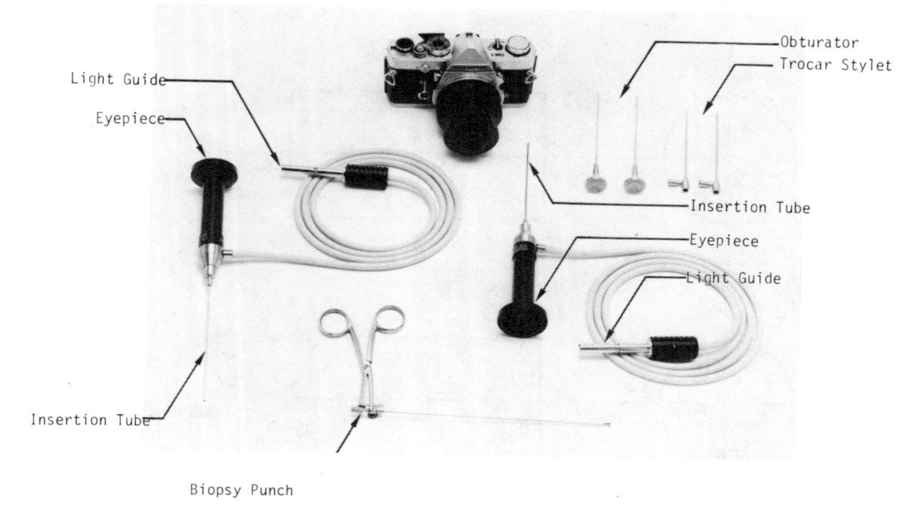

Figure 10.22 Example of an arthroscope using a graded-index rod lens.

surface at its distal end or 75° in the case of an obliquely cut end surface, as shown in Figure 10.23. This instrument has been used for the examination of shoulders, child's elbow joint fracture, congenital hip dislocation, early stage of coxarthrosis, child's joint disorder, osteochondral fracture, and others. With the biopsy punch introduced through the sheath, tissue samples can be collected for histodiagnosis.

Industrial Endoscope

The industrial fiberscope does not differ substantially from the medical scope in structure. It differs in its diameter and the length or the materials forming the insertion tube. Channels for various forceps, air and water feeding, and suction are not needed. The outermost surface of the insertion tube is not necessarily coated with elastic organic material as in the case of medical scopes. The insertion tubes of the scopes shown in Figure 10.24 are made of stainless-steel braided wires. The control section is provided with two-way deflection, 120° up and down, for the manipulation of the scope. Objects are illuminated with the light beam transmitted through light guides from a light source and viewed by means of an objective lens, image guide, and eyepiece. The first scope shown in Figure 10.24 has forward viewing optics suitable for the inspection of the internal surface of a tubular object; the second scope has side viewing optics suitable for the inspection of a cavity of relatively large volume. SLR cameras can be attached to the eyepieces for photographic recording.

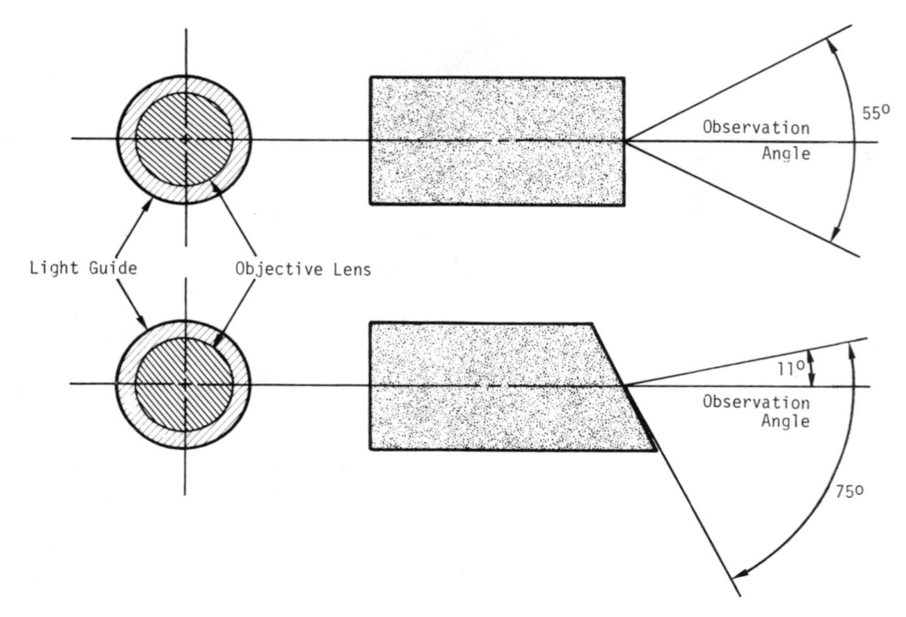

Figure 10.23 Form of the optical element at the distal end, and the observation angle.

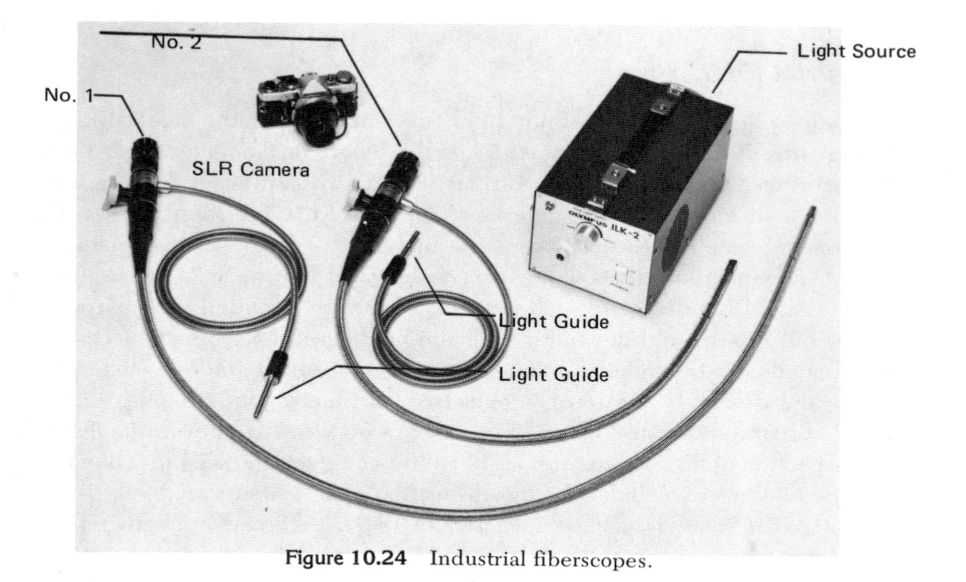

Figure 10.24 Industrial fiberscopes.

Figure 10.25 Inspection of an aircraft engine.

The conventional borescope, which employs a series of relay lenses, has been used for the inspection of the internal surface of a tube, pipe, or bore. However, the use is limited to the case where the surface is of simpler form and easily accessible with a straight and rigid scope. On the other hand, the flexible fiberscope makes it possible to examine structures of very complicated form which are not accessible by other means.

Typical applications are the examination of turbine blades or the inside surface of engines or the inspection for foreign matter which may affect the normal operation of aircraft. This is illustrated in Figure 10.25. For example, the combustion chamber can be examined with a fiberscope inserted through the igniter plug hole. Other applications are the inspection of engines or transmission boxes in automobiles, the inspection of molding machines, the detection of uneven inner surfaces of the machine in the plastic and casting industries, the inspection of the interior of pipes, boiler, etc. in the chemical industry, and the inspection of atomic reactors.

REFERENCES

1. C.W. Hansell. Improvements in or Relating to Means for Transmitting Radiant Energy Such as Light, and to Apparatus for Use Therewith. Br. Patent 295601, Feb. 21, 1929.

2. H. Lamm. Flexible Optical Instruments. Z. *Instrumentenkd.* **50**:579, 1930.

3. A.C.S. van Heel. A New Method of Transporting Optical Images Without Aberrations. *Nature* **173**:39, 1954.

4. H.H. Hopkins & N.S. Kapany. A Flexible Fiberscope Using Static Scanning. *Nature* **173**:39, 1954.

5. H.H. Hopkins & N.S. Kapany. Transparent Fibers for the Transmission of Optical Images. *Opt. Acta,* **1**:164, 1955.

6. R.E. Hopkins & N.S. Kapany. Field Flatteners Made of Glass Fibers. *J. Opt. Soc. Am.* **47**:117, 1957.

7. N.S. Kapany. An Introduction to Fiber Optics. *J. Opt. Soc. Am.* **47**:117, 1957.

8. N.S. Kapany. Fiber Optics, Part I: Optical Properties of Certain Dielectric Cylinders. *J. Opt. Soc. Am.* **47**:413, 1957.

9. N.S. Kapany et al. Fiber Optics, Part II: Image Transfer on Static and Dynamic Scanning With Fiber Bundles. *J. Opt. Soc. Am.* **47**:423, 1957.

10. N.S. Kapany & R.E. Hopkins. Fiber Optics, Part III: Field Flatteners. *J. Opt. Soc. Am.* **47**:594, 1957.

11. N.S. Kapany & J.N. Pike. Fiber Optics, Part IV: A Photorefractometer. *J. Opt. Soc. Am.* **47**:1109, 1957.

12. N.S. Kapany. Fiber Optics, Part V: Light Leakage Due to Frustrated Total Reflection. *J. Opt. Soc. Am.* **49**:770, 1959.

13. N.S. Kapany. Fiber Optics, Part VI: Image Quality and Optical Insulation. *J. Opt. Soc. Am.* **49**:779, 1959.

14. L.E. Curtiss et al. A Long Fiberscope for Internal Medical Examinations. *J. Opt. Soc. Am.* **47**:117, 1957.

15. B.J. Hirschowitz et al. Preliminary Report on a Long Fiberscope for Examination of Stomach and Duodenum. Univ. Mich. Med. Bull. **23**(5):178, 1957.

16. H. Kita & T. Uchida. Light Focusing Glass Fiber and Rod. *1970 SPIE Seminar Proc.* **21**:117, 1970.

17. T. Uchida et al. Optical Characteristics of a Light Focusing Fiber Guide and its Applications. *IEEE J. Quantum Electron.* **QE-6**(10):606, 1970.

18. H. Kita et al. Light Focusing Glass Fiber and Rods. *J. Am. Ceram. Soc.* **54**(7):321, 1971.

19. K. Matsushita & K. Ikeda. Newly Developed Glass Devices for Image Transmission. *SPIE Semin. Proc.* **31**:23, 1972.

20. E. Glazer. Taper Measurement Techniques. *SPIE Semin. Proc.* **31**:13, 1972.

21. F.A. Jenkins & H.E. White. Thick Lenses, in *Fundamentals of Optics.* New York, McGraw-Hill Book Co., Chapter 5, 1957.

22. K. Ikeda et al. Study on the Low Chromatic Aberration of Light-Focusing Glass Rods With a Parabolic Distribution of the Refractive Index. 10th I.C.G. Proc., 6, 1974.

23. Data for Selfoc Single Microlens taken from company brochure on materials manufactured by Nippon Sheet Glass Co., Ltd.

24. H. Shinya & W.I. Wolff. Colonoscopic Polypectomy: Technique and Safety. *Hosp. Pract.* **10**(9):71, 1975.

25. R.L. Goodale et al. Rapid Endoscopic Control of Bleeding Gastric Erosions by Laser Radiation. *Arch. Surg.* **101**:211, 1970.

26. G. Nath et al. First Laser Endoscopy via a Fiberoptic Transmission System. *Endoscopy* **5**:208, 1973.

27. G. Nath et al. Transmission of a Powerful Argon Laser Beam Through a Fiberoptic Flexible Gastroscope for Operative Gastroscopy. *Endoscopy* **5**:213, 1973.

28. P. Fruhmorgen et al. Experimental Examinations on Laser Endoscopy. *Endoscopy* **6**:116, 1974.

29. M. Watanabe. Arthroscope. Present and Future. *Surg. Ther.* **26**(1):73, 1972.

11 OPTICAL COMMUNICATIONS ACTIVITIES

Optical communications is a rapidly maturing industry. This is a result of the development efforts on fiber optics components and systems in a large number of research laboratories throughout the world. Many of the goals sought earlier have been achieved, and more economic means for their practical implementation are now being investigated. The outlook for the application of fiber optics systems has been substantially improved by the major advances that have been made during the past years, and most of the practical difficulties that have inhibited the use of optical communications technology have been overcome.

In order to describe the progress that has accompanied the development of fiber optics components and systems, this chapter reviews the present and anticipated optical communications activities of the major industrial, academic, and Government organizations involved in research, design, and implementation. The discussion is restricted to work in the United States, Western Europe, and Japan, although work is also being performed in Canada, Australia, the Soviet Union, and other countries.

OVERVIEW

Helmut F. Wolf

Because of diligent efforts of various organizations in the United States, Japan, and Western Europe, as well as some work in Canada and Australia, the state-of-the-art of fiber optics has progressed from a laboratory curiosity to that of preproduction and field trials. Thus, most of the major technological breakthroughs that had been expected have been achieved, and many present activities are directed toward the economic implementation of practical systems. The recent emphasis on cost effectiveness has led to a feedback from enhanced fabrication concepts to technological improvements whose end is not yet in sight. Particularly, the compatibility of the various independently developed system components is being improved in order to provide simultaneous system optimization in both economy and performance.

As a result of the varied efforts by the multitude of companies engaged in preparing the market for fiber optics components and systems, systems and circuit designs necessary for economic and practical implementation of fiber optics systems are being developed. Useful fibers of adequate characteristics, and sources and detectors of nearly optimal features, are becoming available. These efforts have led to the development of prototype fiber optics transmission systems which are presently being tested by using existing components. Preliminary test data indicate that most of these prototype systems meet the early expectations.

In the 1960s the applications of optical fibers were limited mainly to light guide uses or short-distance uses in instruments, such as faceplates for cathode ray tubes, image intensifiers, and similar applications. These uses led to the initial growth of the fiber optics industry and provided a financial basis for the further development of more sophisticated applications.

The decorative market in particular was the stimulus for the emphatic development of low-loss fibers, although initially requirements on these fibers were restricted to their aesthetic effects. In addition, the card reader market provided a good financial base for the development of improved fibers. The automotive industry has also been a market for optical fibers and associated components. These and many other nonintelligent applications are still growing, but the majority of all future uses of optical fibers and systems will be in intelligent data transmission. Hence, almost all research efforts in optical fibers and other components are concerned with improvements in transmission characteristics for the transfer of intelligent information.

Major organizations in the United States, Japan, and Western Europe are equally heavily engaged in the further development of fiber optics components and systems. Most companies are pursuing research in the visible and near-infrared wavelength regions, whereas ultraviolet data transmission is being investigated by only a few companies or organizations, and far-infrared transmission is being studied mainly in conjunction with integrated optics.

Generally, glass is considered to be the most favored transmission medium, whereas plastic and liquid waveguides are receiving only very limited attention at this time. No major change is anticipated for the future, although the use of plastic as a fiber material will increase, mainly in automotive and other short-length applications. Mediumless transmission is generally investigated only for specialty uses.

About equal amounts of developmental efforts are being devoted to research and development of LEDs and semiconductor lasers. On the other hand, nonsemiconductor laser work is presently a subject of research at somewhat fewer companies, although in the future a considerable increase in the developmental activities on these optical sources is expected to lead to substantially improved lasers.

The development activities relative to photodetectors are concentrating mainly on enhancements of the frequency response and of the detection efficiency at 1.05 μm and at longer wavelengths. Compared to technological developments in sources, the efforts devoted to the improvement of detectors are relatively minimal.

Connectors are facing a relatively active future. Many companies are working on improved connectors that provide a lower coupling loss than previously achievable. Considerable emphasis on connector standardization can be anticipated. Similar considerations apply to couplers. Again, standardization is an important future task, and considerable future activity will be found in this area. Further development of the T and star couplers will be pursued.

The area of repeaters is being investigated mainly by those companies primarily engaged in work on telecommunications. Also, the interconnection of computer terminals with CPUs requires additional repeater development work by those companies that serve the computer market.

Finally, integrated optics presently is mainly a subject of research activities. Several organizations are vigorously pursuing research in this field because of its potential for future applications. Considerable advances in integrated optics research have been made, and most companies and research laboratories are utilizing the experiences gained by the semiconductor integrated circuit industry.

Generally, research and development (R&D) on fiber optics components and systems is about equally advanced in the United States, Western Europe, and Japan. Differences are found, however, in the relative emphasis on various component and system aspects. For example, in the United States the prevalent interest lies in telecommunications, computers, and military applications, whereas in Japan the dominant applications area is telecommunications. In Western Europe also, telecommunications is of primary concern, except in Great Britain, where military applications are also vigorously pursued.

The United States is probably furthest along in the practical implementation of fiber optics systems, but the R&D efforts pursued in the other countries are similarly advanced. Although the United States may not always be leading in terms of innovations and inventions, the translation of laboratory devices and concepts into practical hardware has been a superior characteristic of all fiber optics systems made in the U.S. compared to those that have been made in other countries.

Furthermore, the highly advanced state of the U.S. semiconductor industry, particularly in the area of advanced integrated circuits, continues to have important beneficial repercussions on the advancement of cost-efficient fabrication techniques that can hardly be matched by the other countries engaged in optical communications work, except perhaps in Japan. This is expected to be even more significant in the development of practical integrated optics systems, which demand highly sophisticated and economic production concepts.

In Germany in particular, consumer applications utilizing partly unguided transmission are also of considerable interest. For example, the use of light to transmit the audio portion of television and stereo sound to wireless headphones is receiving attention. Also, optical rather than acoustic remote TV channel selection and implementation of other TV control functions by unguided optical means are areas of considerable interest. In addition, several fiber optics communications experiments for telecommunications applications are being pursued by the Heinrich-Hertz-Institut in Berlin.

In Great Britain many advances in fiber optics have been made. The quality and quantity of the innovations that have originated in Great Britain have been above average, and many of the practical advances achieved throughout the world have been accomplished by British organizations.

In France, most optical communications activity is devoted to telephone and other data transmission applications. As in the other leading research centers in the world, substantial efforts are also being devoted to the actual utilization of integrated optics, although a considerable amount of work remains to be done.

In Japan, almost all companies involved in fiber optics research are engaged in work for the practical and reliable implementation of optical systems to replace conventional telephone cables. This is being done in view of the highly congested telephone ducts and the relatively wide penetration of the densely populated country by telephones. Most companies are actively involved in telecommunications applications in direct or indirect cooperation with Nippon Telephone and Telegraph (NTT). Lately, considerable activity in integrated optics has led to a surge of publications on this subject.

As an example of international cooperation, Corning and Siemens are jointly operating Siecor, a company established to provide a line of fiber cables and related hardware as well as continuous engineering support.

UNITED STATES ACTIVITIES

Helmut F. Wolf

Although much of the original work in optical waveguides and their associated components was originally conceived and developed in Western Europe, a significant amount of research and development effort has been concentrated in the United States, more recently also in integrated optics. A few organizations in particular are responsible for the substantial activity in both discrete and integrated optical communications that has led to several practical discrete systems, e.g., Bell Labs, Corning, ITT (International Telephone and Telegraph), RCA and a few others. Each of these laboratories has contributed a large number of papers on various subjects relating to either fiber optics components or systems. Some of the most important or most representative of these are reviewed in this chapter; only a brief selection has been considered due to the extent of the published work. A comprehensive review of the status of fiber optics up to 1973 was given by Miller et al. (1). A few examples of installed fiber optics systems are given below. The selection is not intended to be a complete description of all installed systems but gives only an indication of potential applications.

In 1977 and 1978, the Bell Telephone System conducted one of the first optical communications system evaluations. This experimental system was the first to provide a wide range of telecommunications services, carrying voice, data, and video signals on light pulses over 2.5 km of underground cable. Other companies in the United States and abroad conducted earlier commercial tests of optical communications systems carrying voice signals. For example, in 1977 GTE (General Telephone and Electronics) announced the world's first optical communications system to provide regular telephone service to the public in California. This test system transmits voice conversations between two telephone offices. For comparison, also in 1977, Fujitsu installed an optical transmission system to provide telephone conversations between telephone offices in Singapore.

In 1978, Valtec and BART (Bay Area Rapid Transit) installed a demonstration fiber optics full-color television monitoring system connecting two subsidiary railroad (subway) stations in the San Francisco area. The fiber cable is located next to a 34 kV power line without receiving any interference from the high voltage. Eventually all 24 stations are expected to be connected by fibers for such a monitoring task. The center of the fiber cable is constructed from a 2.4 cm standard steel strength member used to pull the cable into position. The steel-reinforced cable was tested for 660 kg pull over a length of 100 m without fiber breakage; the individual fibers alone can withstand a 275 kg pull without damage. Although repeaters are not necessary for this application, copper wires surrounding the six fiber subcables allow an electrical supply if repeaters are required for ultrahigh bandwidth or complex video systems.

Another system that relies on optical fibers for information transfer and that uses fibers of similar characteristics was installed in Las Vegas in early 1978 to transmit several thousand telephone calls or a number of videotape transmissions simultaneously over a 4.2 km distance. The system, installed by Valtec, is capable of connecting a hotel and the Central Telephone Company's main switching office.

The longest and highest-capacity fiber optics telephone link in the world to date is being installed by Harris in Canada. It is the first operating 274 Mb/s telephone link and extends over a distance of 32 km; it consists of a six-pair, 1.3 cm diameter cable that at full capacity is able to carry 20,160 conversations simultaneously.

Optical Fibers

Work on optical waveguides commands the largest amount of effort, followed by activities on advanced optical sources, as indicated by the number of papers on these subjects. Work on optical fibers is largely confined to studies of improved attenuation and dispersion, more economic fabrication methods, better matching of interface characteristics with those of other system components, and superior cabling techniques. In addition, a considerable amount of work is being pursued in the field of environmental testing. Fibers with the lowest values of attenuation have been fabricated by both Corning and Bell Labs. MacChesney et al. (2) described a method to produce the lowest-loss waveguides of high silica composition and prepared by chemical vapor deposition (CVD). This graded-index fiber consists of a GeO_2-doped SiO_2 core and SiO_2 cladding, and it combines low loss with a relatively large index difference between core and cladding. Although a large index difference normally corresponds to a large dispersion, in this fiber the index profile has been chosen such that dispersion has been minimized; the index profile parameter of this fiber was selected to be $i = 1.6$.

An adaptation of this technique was described by Tasker and French (3). They report fiber attenuation as low as 1.9 dB at 0.86 μm and 1.1 dB/km at 1.02 μm. Throughout the range from 0.69 to 1.10 μm the attenuation of this fiber is less than 5 dB/km.

Kaiser et al. (4) reported an easy-to-fabricate fiber that consists of a pure fused silica core surrounded with a loosely fitting, extruded cladding tube. This fiber has been produced for different ranges of NA values. A large numerical aperture that allows the efficient transmission of incoherent light from a light-emitting diode displays a loss of 7.6 dB/km at 0.80 μm, increasing to 14 dB/km at small numerical aperture.

A number of papers have dealt with loss mechanisms and the fundamental limitations on fiber attenuation. For example, Tynes et al. (5) gave a description of the various loss contributions and reported loss measurements as a function of fiber diameter and fiber length. The dependence of attenuation on fiber dimensions has been recognized; for example, the reduction of the core diameter from 20 to 3.2 μm on a specific group of samples led to an increase in the sum of extinction coefficients due to scattering and absorption by about 15% if measured in cm^{-1}.

Pinnow et al. (6) also reviewed the various loss mechanisms. They identified the fundamental optical scattering and absorption mechanisms and used these for comparing different waveguide materials. They concluded that pure fused silica is a preferable material, having ultimate total losses of about 1.2 dB/km at 1.06 μm (Nd:YAG laser wavelength), 3.0 dB/km at 0.80 μm (GaAsAl LED and laser wavelength), and 4.8 dB/km at 0.70 μm (GaP:ZnO emission wavelength).

Other types of glass have been investigated by Schroeder et al. (7). Particularly, experiments relating the K_2O mole content to various fiber properties, such as attenuation, mechanical characteristics, and fictive temperature, as described.

Measurements of fiber losses are also reported in a paper by Kaiser et al. (8). The authors give details of experimental data on various vitreous silica and soda-lime-silicate glasses in the wavelength range between 0.5 and 1.1 μm. Agreement between theory and experiment is found to be excellent in most cases.

Most of the theoretical aspects of fiber propagation have been investigated by researchers from Bell Labs. Among the most important publications coming from this research group are those by Gloge (9–13), Marcuse (14, 15), Personick (16), DiDomenico (17), and Smith (18). Most of these publications discuss the fundamental aspects of fiber propagation and are therefore largely theoretical in nature. Each of these provides further literature references. Although most of these papers fall into the 1972–1975 period, the validity of their treatment remains unchallenged. Among the many other papers of equal importance, a few have been selected as representative: those by Keck (19) of Corning and Gallawa (20) of the U.S. Department of Commerce.

Much work has been done in the area of environmental test investigations. Lebduska (21) presents test results obtained at the Naval Electronics Laboratory Center on more than 200 cables. Measurements described in this paper include photometric, fiber break detection, and performance (such as mechanical strength, bending radius, tensile, torque, mandrel, cyclic flexibility, vibration, and shock tests), as well as environmental and chemical tests. Lebduska concluded that all of the fibers and cables evaluated are able to withstand the physical and environmental conditions imposed by Navy usage.

Sources and Detectors

As in the area of optical fibers, a considerable amount of research activity has been reported by authors from several companies. Whereas the early work dealt mainly with light-emitting diodes, later papers are concerned principally with various types of semiconductor and nonsemiconductor lasers.

Among the first reports on an optical source specifically designed for use in conjunction with optical fibers were papers by Burrus and Miller (22) and Dawson and Burrus (23). In these publications detailed descriptions of the "Burrus diode" are given, which refer to a special LED that allows efficient coupling between source and fiber.

Panish (24) describes work on a variety of semiconductor lasers. Among the device types reviewed are single-heterostructure (SH) devices, where one heterojunction confines both light and carriers to one side of the structure, double-heterostructure (DH) devices, where both light and carriers are confined on both sides to the same region, and separate-confinement–heterostructure (SCH) devices, where the carriers are separately confined to a narrow region within the optical cavity. The paper focuses on the reduction of strain from mismatch and bonding of contacts at the GaAs and AlGaAsP interface and reports low current threshold and high lifetime (in excess of 10^5 hr). Detailed studies of the physical and chemical properties have led to the capability for the growth of complex structures and the understanding of the behavior of these structures as lasers.

Melchior (25) gives a comprehensive review of various types of photodetectors suitable for optical communications detectors. The paper describes the operation of photomultipliers, p-n photodiodes, and avalanche photodiodes. Devices both with and without gain provision are discussed.

An optical repeater with high-impedance input amplifier operating at 6.3 Mb/s is described by Goell (26). One of the major objectives of this work was the reduction of thermal noise, which is achieved by utilizing an input circuit with a time constant that is long compared to the bit interval and by equalizing after the signal has been sufficiently amplified to set the signal-to-noise ratio.

Goell also reported work on a 274 Mb/s repeater (27) that employs a directly modulated source. Such a repeater is considered to be substantially less complex compared to one that utilizes an external modulator. Because of its lower cost, it is potentially advantageous for optical communications. The high data transfer rate requires the use of a single-mode source, hence a GaAs laser, which is not a necessary requirement at 6.3 Mb/s.

A detailed analysis of receiver design has been provided by Personnick (28). It gives a systematic approach to the design of repeaters and is concerned with properly choosing preamplifier and biasing for the photodetector and establishing the relationship between the desired error rate and the bit rate. Then, the results of the theoretical analysis are applied to obtain numerical results on specific receiver types.

Integrated Optics

Recently, most leading organizations involved in optical communications research have shifted their primary interest to integrated optics, with most of the fundamental aspects of fiber optics having been explained. Bell Laboratories, Texas Instruments, United Technology, Rockwell, RCA, and several other laboratories are among the most active organizations involved in these activities.

For example, Cross and Kogelnick (29) investigated corrugated-waveguide filters for multiplexing circuits. To be useful for this application, these filters must have very low side lobe levels and larger bandwidths. The work described examines several specific taper functions and a linear chirp and concentrates on physically realizable gratings.

Marcatili (30) examined the most important reflection properties from tapered and chirped gratings, such as maximum reflection bandwidth and out-of-band reflection level. He found that Bragg-like reflections from a grating depend on the area under the coupling coefficient.

Stolen (31) reported a novel time dispersion tuned-fiber Raman oscillator that is tuned by using the fiber's group velocity dispersion. Such a tunable oscillator can be used to demonstrate the usefulness of fibers as active optical devices.

Further work on integrated optic waveguides by Tsai et al. (32) indicates that a large Bragg bandwidth can be obtained by the use of optimized anisotropic Bragg diffraction in $LiNbO_3$ waveguides. In the past, most work was concentrated on isotropic Bragg diffraction. Anisotropic Bragg diffraction, however, provides inherent advantages, such as a large Bragg bandwidth and small background noise.

Also concerned with optical waveguide modulators and switches is work by Hammer of RCA (33). The paper gives a review of possible applications of waveguide modulators and switches and discusses the still unresolved problem of coupling of the modulator switches to monomode fibers.

A review of recent work on optical waveguide lenses, reflectors, and lens systems, together with surface elastooptic spatial modulators and waveguide-coupled detectors, is given by Anderson and August (34). They concluded that a thin-film Luneburg lens yields the most useful planar approach based on the use of oxides. Further Rayleigh scattering reduction is required, because it limits the resolution and dynamic range of the Fourier transformation.

Comerford et al. (35) of IBM have received a patent on a compact integrated optics transmitter consisting of a laser array, a cylindrical lens, and a waveguide, all mounted on a grooved silicon wafer. The array may contain more than ten lasers of relatively high power; it is fabricated in a single bar. The patent also covers a similar receiver module that can likewise potentially be used in integrated optic systems.

EUROPEAN ACTIVITIES

Oskar Krumpholz, Stefan Maslowski, and Klaus Petermann

The foundations for today's worldwide activities in fiber optics communications were laid in Europe in 1966, when Kao and Hockham of the Standard Telecommunication Laboratories (STL) in England published a paper proposing the use of modulated light guided in glass fibers as a wideband transmission medium (36). Their proposal was based on the expectation that the prohibitively high attenuation of fibers at that time could be lowered to useful values. In the same year, Börner at the Research Institute of AEG-Telefunken in Germany took the view that economic fiber optics communications systems should also be feasible with relatively high fiber attenuation, provided that semiconductor light emitters modulated by the injection current were used as sources and semiconductor photodiodes as receivers (37). Since then, an increasing amount of research effort has been expended on fiber optics communications by about 100 institutions throughout Europe. The number of publications resulting from these efforts in the field already exceeds 1000.

As a consequence of these extensive and widespread activities, the need arose for a special European Conference on Optical Communication. It has been held annually since 1975 and has been attended by more than 500 participants each time. The distribution of accepted papers among the European countries, as shown in Table 11.1, may be regarded as an (admittedly very rough) estimate of the corresponding activities.

It should be mentioned in this connection that a considerable interest in fiber optics communications, and in some instances noteworthy activity as well, also exists or is beginning to awaken in other European countries not yet included in the table. For example, Norway and Finland are engaged in a joint Scandinavian research program on optical communications together with Denmark and Sweden.

In view of this background, it is a small wonder that European researchers have contributed considerably to the rapid progress in this young field. We shall try to outline some of these contributions, without claiming completeness. In some instances, the references given will not represent the original

TABLE 11.1 Distribution of Papers Held at the First and Third European Conferences on Optical Communication

	London 1975		Paris 1976		Munich 1977		Total	
United Kingdom	29		11		21		61	
Federal Republic of Germany	11		8		21		40	
France	7		18		6		31	
Italy	—		7		8		15	
The Netherlands	1		1		1		3	
Denmark	—		1		2		3	
U.S.S.R.	1		1		—		2	
Sweden	—		—		1		1	
Subtotal of European countries	49	49	47	47	60	60	156	156
Japan	15		13		11		39	
U.S.	14		14		4		32	
Canada	2		2		—		4	
Australia	1		—		—		1	
Subtotal of other countries	32	32	29	29	15	15	76	76
Total		81		76		75		252

paper on a special topic but rather subsequent, more comprehensive work by the same authors on the same subject. In this way we will try to cover the extensive literature a little more efficiently by a limited number of references.

Fiber Technology

In Europe the work on fabrication of low-attenuation fibers was at first aimed mainly at the production of high-purity, low-loss glasses. Despite the fact that Jones and Kao (38) had already found in 1969 that commercially available fused silica could have losses below 5 dB/km, the activity concentrated to a larger extent on low-melting-temperature glass systems rather than on materials with high fused silica content. After the successful applications of CVD and MCVD processes for low-loss silica-based fiber production had been reported by Corning Glass Works and Bell Laboratories in the U.S., European groups soon started corresponding activities. As a result, numerous laboratories in nearly all of the European countries included in Table 11.1 and additionally those in Norway have now obtained attenuation values as low as a few dB/km at the wavelength of GaAlAs emitters.

In the course of their work on fiber technology, European researchers contributed to an appreciable extent to the understanding of the investigated glass compounds and glass-forming mechanisms involved and to the enrichment of the fiber production processes by new variants and materials.

Various low-melting-temperature glass systems have been investigated with respect to their potential use for low-loss fibers; e.g., sodium borosilicate (39–41), flint glasses (42), and alkali germanosilicate (43). The purest raw materials have been used, and special care was taken to minimize the contamination of the glass melts by impurities from the crucibles. For example, Scott and Rawson (40) introduced the radio frequency (rf) induction heating of the glass melt within a crucible kept at a lower temperature than the melt itself for this purpose. The main technique employed for converting the glasses to fibers has been the double-crucible method for pulling, either with interdiffusion of core and cladding glass for graded-index fibers or without for step-index fibers. In this way, fibers with typical medium attenuation losses of more than 30 dB/km at best have been obtained. An exception is the result of the work carried out by the group at the British Post Office (BPO) (44); here sodium borosilicate fibers with losses down to 3.4 dB/km at 0.84 μm wavelength and with pulse dispersion in the range of 1 to 5 ns/km have been recently drawn using the double-crucible method.

In the CVD and MCVD techniques for producing high-silica-content fibers, the refractive index of fused silica has to be modified by suitable dopants in either the core or the cladding of the fibers or in both. To obtain a higher refractive index than that of fused silica, Payne and Gambling (45) were the first to use phosphorus as a dopant. For a lower refractive index, Mühlich et al. introduced fluorine as a very effective dopant (46) and used it for the development of a plasma outside deposition technique, which leads to very large vitreous preforms, sufficient for a fiber length of at least 40 km (47).

During their studies on silica-based fibers prepared by CVD within fused silica tubes, Payne and Gambling (48,49) recognized that OH diffusion from the supporting tube to the deposited layers can cause OH impurity absorption in the fiber attenuation characteristics. They found that the OH diffusion to the fiber core could be avoided by depositing buffer layers or borosilicate cladding layers on the tube wall beneath the core layers, resulting in almost OH absorption-free fiber attenuation characteristics even with OH-containing supporting silica tubes. The mechanism of OH diffusion was investigated in detail recently by Ainslie et al. (50).

A significant modification of the CVD technique has been introduced by Küppers and co-workers (51,52) by using a microwave discharge to stimulate the reactions for deposition of oxides. It may lead to an excellent control of the refractive index profiles in fibers.

Fiber Theory

Preliminary theoretical investigations on round dielectric waveguides were carried out in Europe as early as in 1910 by Hondros and Debye (53). The use of optical fibers in optical communications systems however, was first discussed more than 50 years later in 1966 by Kao and Hockham (36). Since the publication of this paper, which is now already historical, a vast amount of material has been published and a book on optical waveguide theory written by Unger is already available (54).

With the advent of low-loss optical fibers, the theoretical activity increased sharply. The graded-index fiber in particular has attracted more and more interest. Important contributions to this area with respect to realistic graded-index fibers with a truncated index profile were made by Clarricoats and Chan (55) and later by Kirchhoff (56). While these calculations were carried out for only a few modes, fibers with hundreds or thousands of modes are often of more practical importance. Here the main effort was first devoted to the search for an optimum profile with respect to minimum delay spread. Timmermann (57) succeeded in finding an optimum profile by adding fourth-power terms to the square-law profile. He found an optimum profile with a delay spread which is smaller by a factor of 3 compared to that of the square-law profile. More recent numerical calculations (58) which also take profile dispersion into account may yield still less delay spread.

Modal attenuation in optical fibers deserves theoretical interest in addition to dispersion. In this respect, it has been shown that even in fibers composed of lossless material there are weakly attenuated modes of the leaky mode type (59), which must be taken into account when calculating transmission properties. Fibers may also be characterized by using these leaky modes (60). Leaky modes must also be considered when applying the near-field scanning technique to the determination of the refractive index profile (61), although more recent theoretical investigations show (62) that leaky-mode correction becomes uncertain for imperfect fibers.

Fibers which deviate from ideal straightness are another point of interest to European theoreticians. The calculation of microbending effects leads to the formalism of statistical modes, which had been derived in 1969 (63) for optical rays in hollow tubes. By use of such a formalism, Jeunhomme et al. (64) succeeded in predicting the type of perturbation from measurement of the stationary power distribution among modes.

The theoretical prediction of the transmission properties of fibers which are joined together represents another challenge to theoreticians. Eve has shown (65) that the use of fibers with undercompensated as well as overcompensated profiles leads to a strong modification of the transmission behavior.

In the case of single-mode fibers with microbending, a simple analytic formula has been found (66) which may facilitate the design of transmission routes. As far as uniformly bent single-mode fibers are concerned, Gambling et al. (67) have observed the interesting phenomenon that bent single-mode fibers exhibit radiation in terms of discrete beams rather than a continuum. In a theoretical treatment, Sammut (68) explained this astonishing fact by the interference between the fundamental mode and the first-order leaky mode.

Fiber Characterization

The investigation of the properties of optical fiber waveguides in Europe increased rapidly at the end of the 1960s. Since at that time the fiber loss was very high, measurement techniques for the bulk loss in the glass were of prime interest in order to permit comparison of materials suitable for fiber manufacture. Considerable effort has been expended on the determination of total loss (38), absorption loss (69), and scattering loss (70) in bulk glasses, especially at STL, the BPO, and the University of Southampton in Great Britain.

The same techniques were also employed to examine fibers once these became available (71, 72). Measurement of the total loss in fibers, however, is much easier than in bulk material, primarily because very long samples can be used. Since the two-point technique commonly applied to transmission loss measurements in fibers is inconvenient with cabled fibers, Hillerich proposed derivation of the fiber loss from the reduced amplitude of a pulse reflected by an external mirror at the fiber end (73). As shown previously (74), this echo pulse technique can also be used to determine the fiber length and to locate fiber imperfections (e.g., breaks).

Pulse response measurements on short lengths of high-loss step-index multimode fibers were initially made by Williams and Kao (75) and by Heyke (76). They used either a pulsed GaAs laser or a mode-locked HeNe laser as the light source. From the observed pulse broadening, they estimated that a transmission rate of a few MHz might be possible over a distance of 1 km. However, their results differed considerably from those reported later by others. A group at Southampton (77) showed experimentally that this is due to differing launching conditions. In order to permit launch-independent measurements, Eve et al. suggested using mode scrambling over a short length near the fiber input (78). The influence of mode coupling on the pulse response (and loss) due to induced microbendings was studied by Geckeler and Schicketanz (79). Methods for the determination of mode conversion coefficients in fibers from measured fiber parameters are described in references 80 and 64.

Since the pulse response of multimode fibers is largely determined by the refractive index profile across the core diameter, a need for a direct and accurate profile measurement technique exists. Various methods have been proposed by European researchers. Eickhoff and Weidel obtain the refractive index profile by focusing a laser beam onto a polished fiber endface and measuring the reflected optical power (81). Costa and Sordo demonstrated that the sensitivity of this method can be improved considerably if the fiber end is immersed in a liquid with a refractive index close to that of the core (82). Unfortunately, the reflection technique requires careful and time-consuming sample preparation. This is not the case for the near-field technique, which derives the refractive index profile from the intensity distribution on the endface of a fully excited fiber. However, as shown by researchers at Southampton, the results have to be corrected for the effect of leaky modes (61). The same is true for the method proposed by Summer (83), which is the reciprocal of the near-field technique. This method derives the profile from the variation of the power coupled into a (graded-index) fiber as a function of the radial position of a focused laser beam at the input. Stewart has suggested a direct profile measurement technique which does not require any leaky-mode correction (84). As in the case for the near-field method, there are two possibilities. Only one case will be mentioned here, in which a small spot is scanned across the fiber endface. However, as opposed to the near-field measurement, the power guided in the fiber is not measured, but rather that part of the refracted power which escapes from the core.

The refractive index difference between core and cladding and thus the numerical aperture, can be deduced from the measured profile. In the case where this profile is not known, the numerical aperture can be determined by scanning the angular far-field pattern of the fiber when it is either fully excited at the input or when it is illuminated with a laser which is weakly focused onto the fiber input at a variable launching angle. The second alternative method was proposed by Jeunhomme and Pocholle (85). From measurements of the dependence of the numerical aperture on the wavelength, Sladen et al. determined the profile dispersion parameter, which is of importance for the design of optimum profiles with respect to equalization of transit time differences between modes (86). The bandwidth of optical fibers, however, is limited not only by transit time differences but also by material dispersion, particularly if broad-linewidth sources are used. Payne and Hartog have described a pulse delay measurement technique to determine the material dispersion of an optical fiber over a wide wavelength range which included the zero of material dispersion around 1.3 μm (87).

The excitation and propagation of individual modes in multimode fibers were studied by several European researchers. Midwinter used a prism and tapered fiber end to access any of the bound modes in small-core step-index fibers (88). Thus, it became possible to investigate the loss by tunneling from the modes closest to cutoff into the surrounding material. Stewart also succeeded in launching light into the fiber through the cladding using a prism coupler (60). In this way a considerable range of leaky modes can be excited in large-core multimode graded-index fibers. Several fiber parameters can be derived from measurement of the characteristics of these modes. Eve described a novel technique for the determination of differential mode attenuation in graded-index fibers (89). The method is based on the measurement of the output intensity as a function of angle and position using parallel beam injection at the fiber input.

Initial investigations on monomode glass fibers were done in 1970. Krumpholz found that the near-field and far-field intensity distributions of the fundamental mode are affected because of mutual diffusion of the core and cladding glasses (90). This has a strong influence on the launching efficiency of this mode. More recently, researchers from the University of Southampton described a method for determining the core diameter and the refractive index difference between core and cladding from the far-field pattern of single-mode fibers (91).

Coupling Techniques

Transmission length and repeater spacing of optical systems depend to a large extent on the efficient coupling of the various components. The active dimensions of the optical elements, however, are extremely small, usually less than 100 μm. This requires precise alignment of the components relative to one another. Coupling between two fibers and of the laser to the fibers turned out to be most critical, and European researchers have made significant contributions to overcome the inherent difficulties.

FIBER–FIBER COUPLING

Detachable fiber–fiber connectors are necessary to facilitate assembly, testing, and maintenance of optical systems. The first device reported was an adjustable connector (92), in which the fibers were eccentrically mounted in offset pins. Adjustment of the fibers is obtained by rotating the pins and maximizing the optical signal transmitted through the coupled fibers. An improvement has recently been introduced in which the connector loss is minimized by observing the scattered light by means of a built-in monitor (93). Insertion losses of less than 0.4 dB have been obtained even for monomode fibers. For multimode fibers, losses as low as 0.1 dB with immersion liquid have been reported. However, it is considered that this principle would be used only where more severe demands on alignment accuracy exist, such as those made by monomode fibers.

Nonadjustable connectors are preferable for multimode fibers. Various concepts have been reported in order to achieve the required alignment accuracy. The most commonly used technique is to mount the fibers centrally in precision plugs, which are aligned in close-fitting sleeves. ITT, for example, achieves centering of the fiber by a watch jewel which fits into a precisely machined recess in one end of the plug (94). Coupling losses of 1.0 to 1.5 dB with an 85 μm diameter core fiber without index matching have been reported. However, several high-precision components are necessary for this type of connector. This is not the case for those connectors in which fiber location and alignment are performed simultaneously. In the AEG-Telefunken v-groove concept (95), exact positioning of the fibers relative to one another is achieved using a flat-sided pin and one surface of the v groove. The angle of the v groove, which can be machined to ordinary workshop tolerances, is the only precision section. In the triple-ball connector described by Hansel (96), the diameter of the balls is the critical dimension. These balls are available with extremely small tolerances. To locate and align the fibers, two interlocking sets of three balls are used. Losses as low as 0.2 dB have been obtained with both connector types.

Since practical optical cables are likely to contain more than one fiber, it may also be advantageous to have multiway fiber–fiber connectors, which can be used to couple several single fibers simultaneously. ITT has developed a multipin bayonet type connector for use in military aircraft (94). Losses of about 3 dB per connection have been achieved with fibers of 50 μm core diameter. In the AEG-Telefunken multiple connector, a series of v grooves in which the fibers are positioned are cut into the circumference of two cylinders (97). Insertion losses of between 0.5 and 2 dB with graded-index fibers of 30 μm core diameter are reported. Siemens, on the other hand, described a flat multiple-fiber connector using a standard photolithographic process to facilitate alignment grooves in which the fibers are held in place (98). Coupling losses less than 0.22 dB with immersion liquid between the fiber ends have been obtained using fibers with a 92 μm core diameter.

In addition to remateable fiber connections, permanent connections, so-called splices, are needed to join cables into lengths which exceed the fabrication length and to repair damaged fibers. Various techniques are being pursued by European groups. These include the normal v groove (99), collapsed sleeve (100), and fusion splices. In reference 101 a microplasma torch and an oxyhydric microburner are described as the heating source instead of an electric arc to weld fibers. Low-melting-point glass solder as the connecting material for splices has successfully been tried (see ref. 102). A completely different method has been proposed by Hensel (103). His joint is formed by tightly wrapping the fiber ends in a single turn of a plastic tape. Reported splice losses for all splice connections are in the range of a few tenths of a dB.

For efficient jointing, perfectly flat fiber ends normal to the fiber axis must be used. Cumbersome and time-consuming polishing can be avoided by one of the proposed fiber fracture techniques. Alternative methods to the original one proposed by Gloge et al. have been suggested by European researchers. They use either spark erosion (104) or a hot wire (105) to produce mirrorlike fiber endfaces.

LASER–FIBER COUPLING

The investigations being conducted to improve the coupling efficiency between the light source and the transmission fiber by means of various light beam transformers are also manifold.

In the case of an LED light source, lenses cannot increase the coupling efficiency to a multimode fiber when the light-emitting area is larger than or equal to the fiber core. If the light-emitting area is smaller, however, a certain degree of coupling improvement can be achieved with a microlens. In reference 106 a method of contact bonding an epoxy resin lens to the fiber endface is outlined. If the cladding of the fiber is etched to reduce the fiber diameter almost to that of the fiber core before forming the lens, much better results are obtained (107). Nevertheless, coupling efficiencies of only a few percent for LEDs are typical.

Much higher figures are obtained with an injection laser light source due to the small light-emitting area, much higher brightness, and the directivity of the radiation pattern. Thus, coupling efficiencies as high as 80% could be achieved with microlenses using a graded-index fiber with a low NA value (0.13) (107). A coupling efficiency of 65% was measured with an arrangement in which a tiny glass bead is glued to the fiber endface (108). The laser's noncircular geometry, on the other hand, suggests the use of a cylindrical rather than a spherical lens for coupling. Weidel (109) described an arrangement with a thin fiber perpendicular to the axis of the transmission fiber. The first fiber acts as a cylindrical lens. Coupling efficiencies of more than 70% have been obtained, even when using a monomode fiber with a core diameter of 4.5 μm.

Sources

The light-emitting diode (LED) and the semiconductor injection laser appear to be the most important sources for use in optical communications systems. Although most of the European activities are devoted to these semiconductor devices, the ultraphosphate laser might also be a suitable emitter (110–112). These lasers may be excited by light-emitting diodes and are followed by an external modulator. In the following discussion, however, we will confine our attention to the activities on LEDs and semiconductor lasers.

LIGHT-EMITTING DIODES

Many European laboratories are concerned with LEDs, but the most significant results have been obtained at Plessey, U.K. and at the Institut für Allgemeine Elektrotechnik, Technical University of Munich.

Light-emitting diodes can be made with a radiance up to 100 W/sr cm^2 (113). These diodes have a bandwidth of about 30 MHz. Large bandwidths may be achieved with higher doping of the active layer, which results in bandwidths of up to 500 MHz, as reported by Goodfellow and Mabbit (114). Bandwidths in excess of 1 GHz may be achieved (115) if in addition to the high doping a small active layer width is chosen. Diodes with these extreme bandwidths still have the remarkable radiance of 3.5 W/sr cm^2.

The high reliability of light-emitting diodes is their main advantage compared to the semiconductor laser. At Plessey, the reliability of LEDs has been tested at elevated temperatures (113), and from these measurements lifetimes in excess of 10^7 hr (more than 1000 yr) are expected.

The light-emitting diodes mentioned above are GaAs or GaAlAs devices, which emit at around 0.90 μm. Since optical fibers exhibit lower loss and less dispersion at larger wavelengths, GaInAs LEDs emitting at 1.06 μm have been built (116) with a radiance as large as 15 W/sr cm^2. The quaternary GaInAsP system also looks promising for realizing longer wavelengths.

Light-emitting diodes have the disadvantage that only a small portion of the emitted power can be launched into optical fibers. Superluminescent devices, which make use of the stimulated gain, emit in a much narrower angle and may therefore be coupled more efficiently to optical fibers. Superluminescent diodes have been built, and it has been shown that they also exhibit a better dynamic behavior (117), since the rise time decreases with increasing stimulated emission. However, in order to make full use of this behavior, the diode must be operated at relatively high current densities.

INJECTION LASER DIODES

The semiconductor injection laser appears to be the most attractive source of optical communications, and many European researchers are working on its improvement. Today (1978) the main activity is concentrated on the GaAlAs heterostructure injection laser, but in the future the development of lasers based on GaInAsP which emit at longer wavelengths may become more important.

A large variety of laser structures are still of interest, but at the moment lasers with an oxide or a proton-implanted stripe as well as the low-mesa-stripe laser are the most important. Some work has recently been devoted to the improvement of proton-implanted devices. It has been demonstrated by Bakker (118) that the depth of implantation should stop 1 μm above the active layer in order to avoid defect centers in the active layer. The stability of the lasing operation is improved when proton-implanted lasers of the narrow-stripe type are used (119). Other investigations which have been carried out by Steventon et al. (120) have shown that the threshold current may be reduced to 10 to 50 mA if the laser is optimized with respect to the cavity length, stripe width, and the depth of implantation.

European researchers have also investigated laser structures with stabilized light output characteristics. Such stabilized lasing behavior may be achieved by using an embedded structure, as proposed by Kirkby and Thompson (121). A laser structure with separate confinement for light and injected carriers has also been built by the Standard Telecommunication Laboratories (122). This laser has a narrow angular emission both parallel and perpendicular to the active layer. It has been observed at Philips Laboratories that a more stable laser characteristic may be achieved simply by using a tilted stripe or by changing the geometry of the stripe contact along the stripe length (123).

Degradation is one of the main problems in connection with semiconductor injection lasers. Degradation tests are being carried out at the British Post Office at elevated temperatures (124) in order to predict the actual laser lifetime after a short observation time. In this test, oxide-stripe lasers supplied by Standard Telecommunication Laboratories have been used and room-temperature lifetimes of 50,000 to 100,000 hours are expected.

At AEG-Telefunken, lifetime tests are performed at an elevated temperature of 40°C and at a constant light output power of 5 mW (125). Lifetimes of more than 20,000 hr have been achieved at this elevated temperature. Similar lifetimes at room temperature have been obtained at STL (126). Continuous-wave lasers are already commercially available from both companies.

The modulation behavior of semiconductor lasers is of great importance for use in optical communications systems. It is well known from theory that semiconductor lasers exhibit relaxation oscillations (127). Theoretical investigations carried out by Harth (128) indicate that the relaxation resonance frequency decreases with increasing modulation amplitude, a fact which has also been observed experimentally (129).

Despite these relaxation oscillations, Russer and Schulz succeeded in modulating a semiconductor injection laser with a bit rate of 2.3 Gb/s at an early stage (130). This high bit rate is made possible by proper adjustment of the bias current.

Many other laboratories are also concerned with the modulation of semiconductor lasers; in particular, the work carried out at the Technical University of Denmark deserves some attention. The actual modulation behavior of a semiconductor laser has been described theoretically to a high degree of accuracy by taking into account transverse diffusion in the active region (131).

Besides simply characterizing the modulation behavior, experiments have also been conducted to improve the modulation characteristics by damping the relaxation oscillations. Such a damping is possible by using small-cavity volume lasers yielding an increase of the spontaneous emission rate, which also leads to damping of the relaxation oscillations (132). Another possibility consists of injecting coherent light into the modulated laser. Relaxation oscillation damping has been demonstrated theoretically (133) as well as experimentally (134).

In addition to the activities mentioned above, some European research groups are also concerned with more special problems. Researchers at the University of Bern, for example, are devoting their activities to semiconductor lasers coupled to external resonators (135). An improved insight into laser characteristics is obtained by using such an arrangement. Another group, at the Swiss Federal Institute of Technology, is concerned with the noise properties of semiconductor lasers, which have been investigated at high microwave frequencies (136) as well as at low frequencies (137).

Detectors

From the beginning, there was no doubt that semiconductor photodiodes would be the most favorable detectors for fiber optics communications. In particular, Si avalanche photodiodes are most attractive because of their excellent demodulation and noise properties.

Detector activity in Europe is confined to a small number of institutions. AEG-Telefunken has introduced a new type of detector (138) based on early work on planar photodiodes. By irradiating the light into the depletion layer of a mesa diode from the circumference, carrier diffusion can be avoided; in addition, the sensitivity is extended to longer wavelengths. PIN as well as avalanche photodiodes have been realized using this mode of operation. In particular, avalanche photodiodes with a high gain–bandwidth product have been achieved (139) with a planar front-illuminated reach-through structure. This was made possible by proper design of the carrier multiplication region and by employing ion implantation in order to achieve the desired doping profile. The measured gain–bandwidth product could be increased to about 300 GHz. The same figure is reported by Müller and Ataman from the Technical University of Braunschweig (140). They developed a thin silicon structure which uses a highly reflecting back contact to improve the internal quantum efficiency, which otherwise would be low because of the small depletion layer thickness. Extremely low pulse rise times, less than 100 ps, have been obtained even for a multiplication factor of 100. With PIN diodes of the same kind, values as low as 65 ps are reported (141). If an additional window is provided in the back contact of these thin-film diodes, a transparent photodiode results (141). When supplied with fiber tails on both sides, this new and promising device can be inserted into the optical path of a fiber route. Thus, it becomes possible to extract a small amount of the optical power, which can be used for monitoring and data distribution. AEG-Telefunken has realized a diode of this type on the basis of the GaAs/GaAlAs double heterostructure (142). Further detector studies are being done at Philips (143) and Marcoussis (144) laboratories.

It is expected that detectors around the 1.3 μm wavelength will be of interest in the near future because of the outstanding properties of optical fibers in this wavelength region. A group at Thomson-CSF recently succeeded in developing a GaInAs homojunction avalanche photodiode with a high sensitivity over the wavelength range between 1.1 and 1.7 μm (145). A quantum efficiency of 75% was reported at 1.3 μm. Near the breakthrough voltage of 22 V, a multiplication factor of 100 and a pulse rise time of 0.1 ns were measured. The dark current at 10 V reverse bias is as low as 2 nA. $CuInSe_2$–CdS heterodiodes may also be useful (146). So far, however, it is unknown whether or not carrier multiplication can be obtained in this material.

Integrated Optics and Optical Components

Research work in Europe has been devoted to both single-mode and multimode optical devices. In the past, the main emphasis was laid on single-mode integrated optics, since single-mode devices lead to many more functional elements than do multimode devices. Integrated optics, however, must always be considered in conjunction with optical fiber systems. Since at the moment multimode fiber systems are preferred in Europe, the interest in multimode devices is growing.

In single-mode integrated optics, optical waves are guided in thin dielectric films. A number of materials for these dielectric films have been proposed by European researchers. Besides passive waveguides, active waveguides have also been realized; in particular waveguides based on lithium niobate films have been widely employed. Such films have been prepared by outdiffusion as well as by rf sputtering techniques (147). At Siemens, phase modulators based on lithium niobate have been fabricated (148). At Thomson-CSF considerable efforts have been made to realize electrooptical couplers. These single-mode directional couplers consist of two parallel waveguides in which the interaction between the waveguides depends strongly on the phase match. Papuchon et al. (149) have proposed realization of these parallel waveguides with lithium niobate films of such a crystal orientation that an applied voltage of only 6 V suffices for a phase mismatch and therefore for an effective switching action.

The coupling between a fiber and an integrated optical device is another important problem. Butt-joint coupling between the fiber end and the planar waveguide of the integrated optical device represents the simplest method but requires good field matching between the fiber and waveguide. Distributed coupling, as proposed by Bulmer and Wilson (150), might be more effective. Their method consists of depositing a grating on the film waveguide in order to match the propagation constants. The fiber with reduced dimensions is then pressed on the grating. Slight tapering of the fiber is proposed in order to relax the tolerances, so that this coupler then represents a tapered velocity coupler (151).

Additional efforts have been made in order to provide components for fiber–fiber coupling. Soares et al. (152) have proposed holographic elements for coupling single-mode fibers. Branching elements which distribute the light from one incoming fiber to several outgoing fibers have also been investigated (153).

As mentioned above, there is a growing interest in multimode devices that can be used in multimode optical communications systems. Such multimode devices have been realized in planar technology (154), but devices which can be better connected to optical fibers are also of interest.

The light from an incoming fiber often has to be distributed to several outgoing fibers, corresponding to the star-coupler arrangement. Such a coupler has been described by Auracher (155). He designed a planar branching network by using light-sensitive materials prepared by a subsequent photolithographic process. Another approach has been proposed by Ulrich (156). He makes use of the multiple-imaging properties of slab and rectangular waveguides. A self-imaging double-image 3 dB coupler has been shown to exhibit a loss of only 1 dB (157). A star coupler may also be built by using a thick clad core fiber, which acts as a mixer rod (158). Such a mixer rod, however, suffers from the packing fraction loss.

The design of directional multimode couplers has also attracted considerable interest. Jeunhomme and Pocholle (159) and Stewart and Stewart (160) have realized such couplers without cutting the fiber link. They make use of mode conversion induced by a grating, where the radiated light is collected by a photodetector. Such elements may also be used as variable attenuators for measuring purposes as proposed by Stewart and Stewart (160).

Witte (161) has made another proposal for tapping light from a fiber link. The fiber is cut and the fiber ends are then mounted with a slight offset with respect to each other so that a tapping element can collect light which is coupled to another fiber.

The devices described above are designed to be used in systems with single fibers. Thomson-CSF is also concerned with the development of systems using fiber bundles, and several types of couplers have been made (162). A straightforward solution to the problem of how to launch and how to detect light in a fiber optics link is that of using transparent light-emitting diodes and transparent photodiodes. Such devices have been proposed by Heinen and Proebster (163) and by Müller and Eickhoff et al. (141, 142), respectively. These transparent devices are simply inserted into the fiber transmission route, allowing a simple access to the guided light.

The field of integrated optics is still expanding, and new developments are expected in the near future. Future trends in this field are therefore difficult to foresee.

Field Trials

Fiber optics data links for industrial applications are commercially available from several European companies. Military projects are also being realized in several countries. Furthermore, some PTT administrations and governmental organizations are performing or planning field trials with fiber optics transmission systems in order to prepare for future incorporation into the communication networks.

In the United Kingdom, the British Post Office (BPO) has been performing several feasibility trials with fiber optics transmission systems since 1977 for future application in the communications network (164). PCM systems operating at 8 Mb/s are intended for use in the junction network and systems operating at 140 Mb/s are intended for main-line applications. A system operating at 140 Mb/s has been supplied by STL (165). Here a 9 km long STL cable (166) has been installed between exchange buildings at Hitchin and Stevenage on a route provided by the BPO. Two dependent repeaters are housed in manholes. BPO systems at 8 Mb/s as well as at 140 Mb/s are being tested with about 20 km of cable from BICC (167, 168) laid in existing ducts between the BPO Research Centre at Martlesham and Ipswich Telephone Exchange (169, 170). One 8 Mb/s system on this route (14 km) with a surface repeater housed in the Kesgrave telephone exchange is intended to be formally incorporated in the Post Office Eastern Telecommunications Region network in the future.

In Germany the Deutsche Bundespost (DBP), with the sponsorship of the Ministry of Research and Technology (BMFT), is conducting a field trial of fiber optics PCM systems operating at 34 Mb/s (171). Four systems from different suppliers (AEG-Telefunken, SEL, Siemens, and Tekade-Felten und Guilleaume) have been installed in parallel in West Berlin between two local exchanges about 5 km apart. Another project sponsored by the BMFT is underway at the Heinrich-Hertz-Institut (HHI) in West Berlin (172). Here the feasibility for fiber optics links for the simultaneous transmission of data, telephone, picturephone, stereo sound, and video signals is being studied at bit rates of 140, 280, and 560 Mb/s.

In France the Centre National d'Etudes des Telecommunication (CNET) has installed two optical fiber cables underground at Lannion (173). The cables were supplied by Les Cables de Lyon (174). One of them (750 m, 19 fibers) was laid in 1976 and is used for a system operating at 34 Mb/s, the other (1 km, 6 fibers, laid in 1977) for a system operating at 140 Mb/s. For long-range transmission, several fibers within the cables are spliced in tandem.

In Italy the Centro Studie Laboratori Telecomunicazioni (CSELT) started the first field test, called the COS 1 experiment, in 1976. A cable (1 km) manufactured by Pirelli was laid by SIRTI in a trench within the CSELT premises (175), and the transmission properties were monitored over one year (176). On the basis of the results, a fiber cable of about 4 km is now being installed in existing ducts in Torino. It is intended for systems operating at 34 and 140 Mb/s (experiment COS 2).

So far, no field trials have been performed by the other European PTT administrations. However, the progress of fiber optics communications is being followed closely and its introduction is envisaged for the future in most instances.

JAPANESE ACTIVITIES

Tsuneo Nakahara

Optical communications activities in Japan are being carried out rather emphatically. The vigor of these activities can be explained by the unique features that can be offered by optical fiber communications systems when these are considered in relation to the Japanese environment, as discussed below. It must be recognized that all of these are directly or indirectly related to the restrictions and constrictions of available space.

Higher Transmission Capacity. Optical fiber communications systems can provide much larger transmission capacity per unit cross-sectional area than can conventional copper cables. The space required for the installation of an optical fiber transmission system is small.

Lower Interference Problems. The high congestion in Japan in comparison with other major industrialized countries necessitates that many different facilities be located next to each other. This creates difficult problems from the viewpoint of communications network installation, since electromagnetic noise exists almost everywhere, resulting in severe problems. Optical fiber communications systems can provide a solution to the difficulties created by electromagnetic interference.

High Information Transfer. The high population density and the high ratio of urban areas to the total land area of Japan leads to a continuous large amount of information of various kinds to be exchanged throughout the country. The forms of such information transfer can be video signals, audio, digital data, etc. Optical fiber communications is a near-ideal medium for the transmission of such information throughout Japan.

Convenient Electrical Compatibility. Optical fiber communications networks can be installed in parallel with the electric power transmission lines because the optical fiber communications network are immune to the electromagnetic fields generated by the electric power transmission lines. This feature becomes extremely useful particularly in Japan. The land space required for the installation of communications networks for new information services is small.

Optical communications activities in Japan are considered to be among the most advanced in almost all areas (177). Many field trials have been carried out in various applications of optical fiber communications, and some new field trials are being prepared. Examples of field trials include two major experiments by NTT and an optical CATV project by MITI (Ministry International Trade and Industry), an optical fiber IF (intermediate frequency) transmission system by NHK (Nippon Hohso Kyokai) at the Kawai satellite broadcasting station, an optical fiber transmission system for practical PCM relays for electric power control at three substations by TEPCO (Tokyo Electric Power Company), similar systems by KEPCO (Kansai Electric Power Company) and Kyushu Electric Power Company, an optical fiber lightning counter and polarity-determining system by KEPCO, and several others.

In the area of research and development of the components required for the optical communications systems, Japan has been quite active and made significant progress in almost every field. Examples include new fabrication processes of fiber preforms (178) such as the inside-coated CVD (chemical vapor deposition) method, the rod-in-tube method, and VAD (vapor-phase axial deposition) method, as well as considerable progress in the areas of optical sources, connections, and systems.

Features of Optical Fiber Communications

One of the most important features of optical fiber communications is the potential large transmission capacity. As discussed in Chapter 2, the typical average transmission loss as a function of the wavelength of multimode step-index fibers implies that the transmission capacity of an optical fiber can be extremely large in comparison with that of a conventional copper cable. For example, attenuation is approximately 0.6 dB/km at a wavelength of about 1.3 μm, much lower than that of conventional copper cable. Consequently, the cost per channel-kilometer of an optical fiber cable could also be much less, since there would be fewer repeaters required for long-distance systems. An estimated potential transmission capacity per unit cross section of this type of optical fiber is 10^3–10^4 times larger than that of conventional coaxial cables (the bandwidth of a conventional coaxial cable for trunk lines would be about 300 MHz).

Figure 11.1 shows the cross section of typical fiber structures. Distinction is made between double- and triple-coated fibers of equal outer diameter and equal fiber diameter.

Another outstanding feature of optical fibers is that the optical signals transmitted through the optical fiber cables are not influenced by external electromagnetic noise and that they are highly secured in comparison with those transmitted through conventional copper cables. This feature can be very useful also for the data lines for various security systems, in addition to the applications mentioned above.

The raw material of optical fiber cables is usually silica. The abundance of the raw material supply implies that the main portion of the optical fiber cost would be manufacturing cost. Therefore, it would be reasonable to expect that in the future the cost of optical fiber cables will substantially decrease as mass production increases.

The main features of optical fiber cables which can provide us with advantages for designing optical fiber video systems have been discussed in the previous chapters. These features include wide bandwidth, small size and light weight, low transmission loss, electromagnetic noise immunity, no crosstalk and no signal leakage, potential low cost, and total electrical isolation.

Major Advances in Optical Fiber Communications in Japan

The major activities in optical fiber communications in Japan are concentrated in mainly two areas, digital communications and analog CATV systems.

DIGITAL COMMUNICATIONS

As of this writing, field trials of digital communications media have been performed in two major areas, i.e., in telephone networks and digital data transmission for electric power control circuits. Research activities in optical fiber communications at NTT (Nippon Telegraph and Telephone Public Corporation) were reportedly initiated in June 1966. Early research emphasis was placed on the development of fiber manufacturing, fiber splicing, cabling technology, optical sources, transmitters and receivers, repeaters, splitters, optical switches, and submarine optical fiber cables.

It was reported on several occasions in Japanese newspapers that there are two major field trials. In December 1976, 20 km transmission without repeaters was successfully achieved by NTT. In early 1978, Japanese newspapers reported that NTT had a commercial test project of digital transmission of telephone voice signals at 100 Mb/s (1440 telephone voice channels equivalent per system) using a cable consisting of 48 graded-index fibers over a distance of 20 km between Karagasaki, Ohtemachi, and the Kuramae telephone exchange. This commercial test project was designed to evaluate

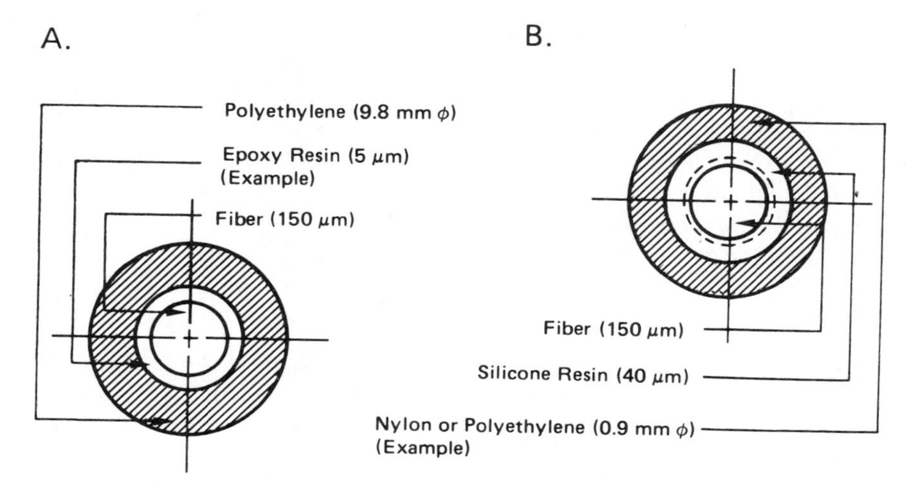

Figure 11.1 Cross-sectional views of typical (A) double- and (B) triple-coated fiber structures.

the overall performance of the digital transmission of telephone voice signals through a 48-fiber cable. It is the first of its kind in terms of data rate and distance. The cable installation was initiated during the second half of 1978, and the transmission of the 100 Mb/s digital signals will be initiated in early 1979.

In addition to the above two major projects, in the middle of 1978 NTT replaced the conventional copper cables with optical fiber cables for the computer internal transmission network of DIPS (Dendenkosha Information Processing System).

The TEPCO (Tokyo Electric Power Company) field trial using an optical fiber data transmission system for electric power control has the following characteristics: (a) Four-fiber cables were installed alongside 275 kV underground high-voltage power cables. (b) The transmission capacity is 6.312 Mb/s, transmitting 24 channels of signals with bit rates of 54 or 42 kb/s. (c) DPPM (differential pulse-position modulation) signals are used to intensity modulate the semiconductor laser output.

The 54 kb/s signals from the multiplexer telecommunications equipment are multiplexed in the PCM data terminal equipment and the 6.3 Mb/s digital multiplexer equipment. The 6.3 Mb/s signal is transmitted through the optical fiber link. The object of the optical fiber cable communications is to protect the power systems which the Tokyo Electric Power Company will introduce in the future. The terminal equipment is installed at the Shinjuku substation, and the optical fiber cable is laid in a tunnel and conduit. The total length of the optical fiber cable is approximately 3 km. The optical repeater is installed in a manhole approximately 1 km from the Shinjuku substation.

The IF optical fiber transmission system at the NHK Kawai Satellite broadcasting station consists of an optical transmitter and receiver and a two-fiber cable of 430 ms, with one fiber included as a spare. The frequency band of the IF signals is 16.5 to 22.5 MHz. One LED was used as the optical source and one PIN photodiode was used as the detector. The optical fiber cable is nonmetallic with a cable outer diameter of 12.2 mm, a cladding diameter of 400 μm, a core diameter of 200 μm, and an NA of 0.25. The major specifications are listed in Table 11.2.

ANALOG COMMUNICATIONS

Optical fiber is one of the most suitable transmission media for analog CATV systems. At present, analog CATV systems utilizing optical fiber cables can be described by citing the example of the Hi-OVIS project (Higashi-ikomo Optical Visual Information System), which was established by VISDA (Visual Information System Development Association) in 1972 (179). This project is entirely sponsored by MITI (Ministry of International Trade and Industry), which is part of the Japanese Government. The main objectives of this project are threefold: first, to stimulate research in the optical fiber communications industry in general; second, to perform a practical field trial of an optical fiber cable video system in the form of a two-way interactive CATV system which is applied to the model town of Higashi-ikoma, a suburb of Osaka; and third, to evaluate the social impact of such systems on modern societies. The basic transmission scheme of the Hi-OVIS signals, consisting of analog video, FM audio, and digital data, is baseband transmission using one fiber for each channel signal.

The Hi-OVIS project consists of two phases. Phase I was completed in November 1976. The main objective of Phase I was to study the feasibility of a large optical fiber video transmission system for wideband video signals consisting of analog video, FM audio, and digital data signals. The Phase I system included three main subsystems: the center control equipment with a minicomputer, the optical fiber cable transmission system with a 6 × 16 video

TABLE 11.2. Outline of System Specifications of the NHK Optical Fiber Satellite Broadcasting System

Parameter	Specification
Input signal level	104 dB/μV
Output signal level	80 dB/μV
920 kHz beat	$<$ 45 dB
Noise figure	$<$ 50 dB
DG	±5%
DP	±5%

switch, 500 m of 12-fiber cable, and 500 m of 18-fiber cable, and the home terminal equipment with a TV monitor and a keyboard.

The successful completion of the feasibility study and the subsequent demonstration of the Phase I system convinced VISDA and MITI to proceed with Phase II as planned, on the scale of 160 home subscribers and eight local studio extension terminals located on various public premises, such as City Hall and in schools, where many public activities were expected to take place. The center equipment consists of a data control computer, VCR (video cassette recorder), transmitters, and the transmitters and receivers for TV signal rebroadcasting, etc. The transmission of signals consisting of TV video, FM audio, and digital data is baseband transmission from the center to each home subscriber. A distribution cable approximately 6 km long of 36 fibers, 400 m of 24-fiber cable, 500 m of 18-fiber cable, 500 m of 12-fiber cable, approximately 5.5 km of 6-fiber cable, approximately 1.5 km of 4-fiber cable, and approximately 31 km of 2-fiber cable are to be installed.

The total fiber length is approximately 400 km for the entire Hi-OVIS project. The longest transmission distance of the composite video signals (analog video and FM audio) is approximately 4 km from the UHF receiving station at the top of Ikoma mountain to the control center at Higashi Ikoma station. No repeaters are used throughout the entire system. The Hi-OVIS system was designed with 30 channels available for transmission, allocated as shown in Table 11.3. The schematic diagram of the Hi-OVIS Phase II system is shown in Figure 11.2. It is envisioned that 160 home subscribers and eight local subscribers using presently only one subcenter will take part in the field trial. The system depicted in this illustration could handle up to 504 (168 × 3) subscribers if all three subcenters were in operation. The subscriber drop is comprised of a two-fiber cable, one each for upstream and downstream transmission. The 16 pairs of fiber cables lead from an optical junction box, where they are connected to the distribution cable (180).

TABLE 11.3 Hi-OVIS Channel Allocation

Purpose	Channels
Retransmission of TV broadcasts	6 VHF plus 3 UHF
Video cassette recorders (dedicated)	3 dedicated
Local studio broadcasting	1
Still picture projection	1
Video cassette recorders (selection)	3 for 6 programs
Maintenance	1
Telop	4
Automatic cassette control	1
Character generator	1
Character display	6

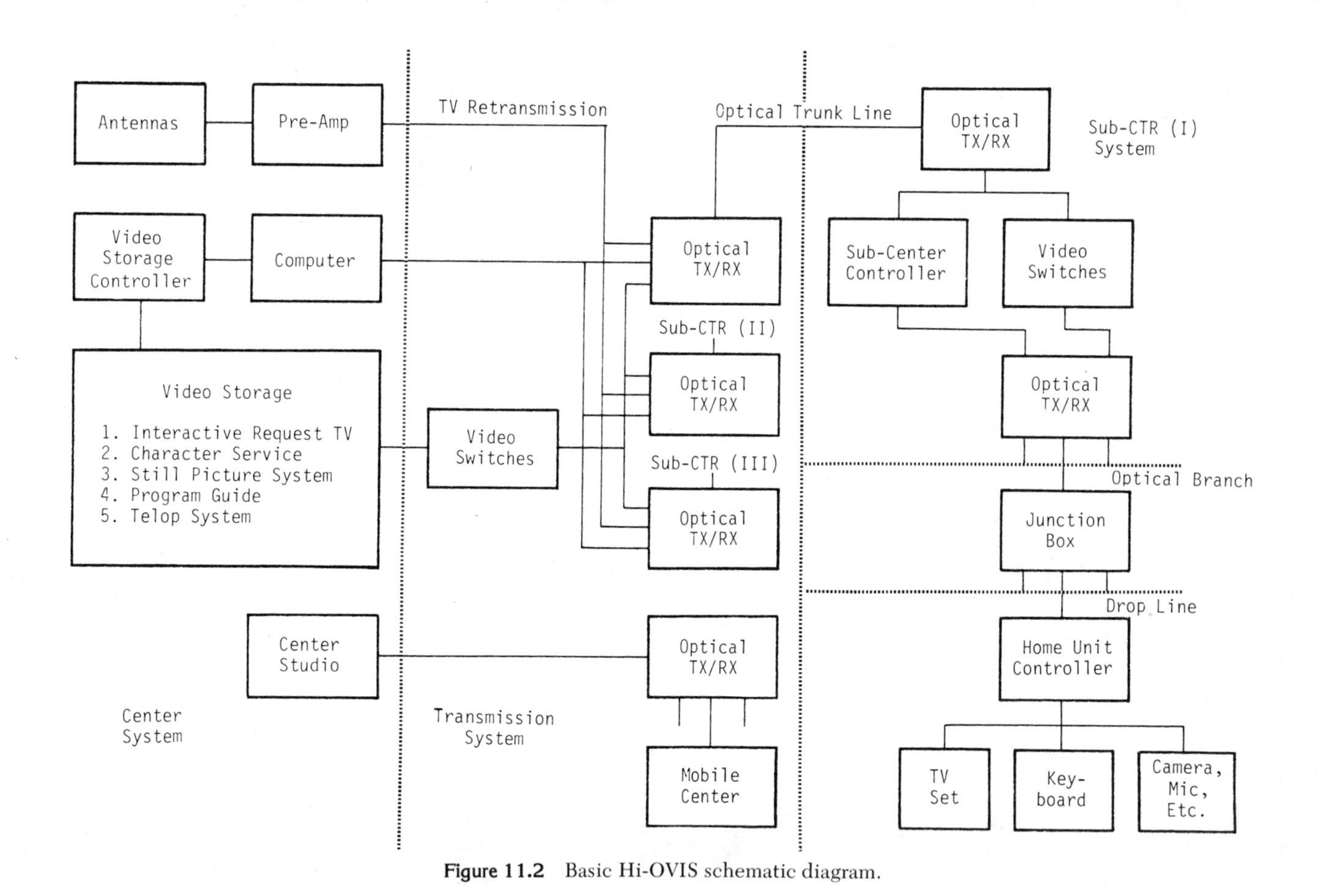

Figure 11.2 Basic Hi-OVIS schematic diagram.

Fiber Innovations

In Japan two recent innovations in the area of waveguide fabrication have achieved some degree of prominence, vapor-phase axial deposition (VAD) and the rod-in-tube method for CVD fabrication.

VAPOR-PHASE AXIAL DEPOSITION METHOD

One of the most suitable fiber fabrication techniques is vapor-phase axial deposition. It offers advantages of better fiber uniformity and greater fiber length. Because the transmission loss of optical fibers can be less than 1 dB/km, the repeater spacing can be large. It is therefore reasonable to expect the unit length of optical fiber cable to be larger than that of conventional copper cable. It is also necessary to look into recent developments in the fabrication of optical fiber preforms.

NIT has developed a new method to fabricate optical fiber preforms. Similar efforts are also being pursued by Sumitomo Electric Industries, Fujikura Cable Works, and Furukawa Electric. Some examples of recent developments in the fabrication of optical fiber preforms include a modified CVD (chemical vapor deposition) and a double-crucible compound glass structure.

Figure 11.3 shows a schematic diagram of the vapor-phase axial deposition method for continuous fabrication of optical fiber preforms reported by

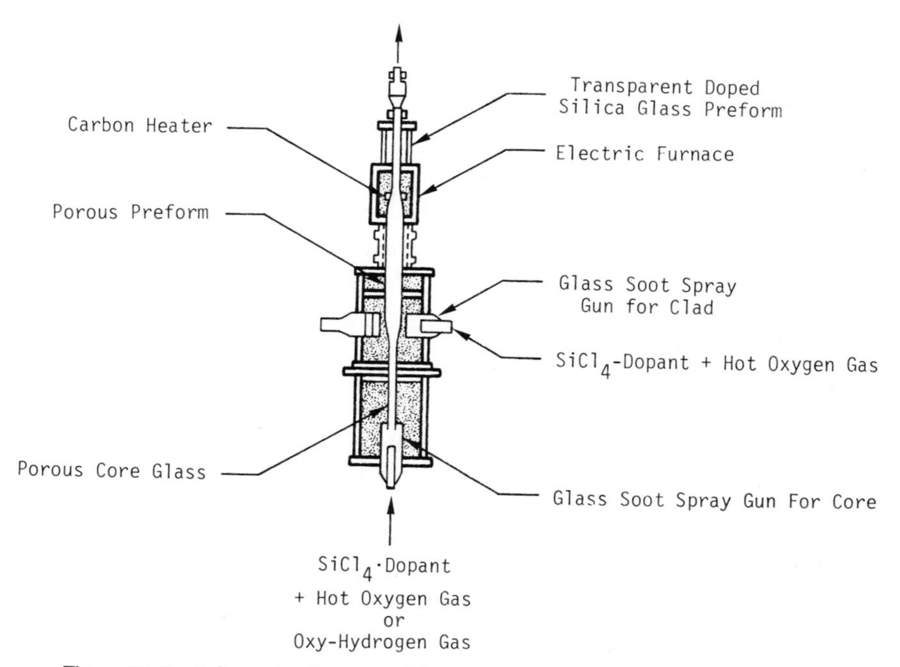

Figure 11.3 Schematic diagram of the vapor-phase axial deposition method.

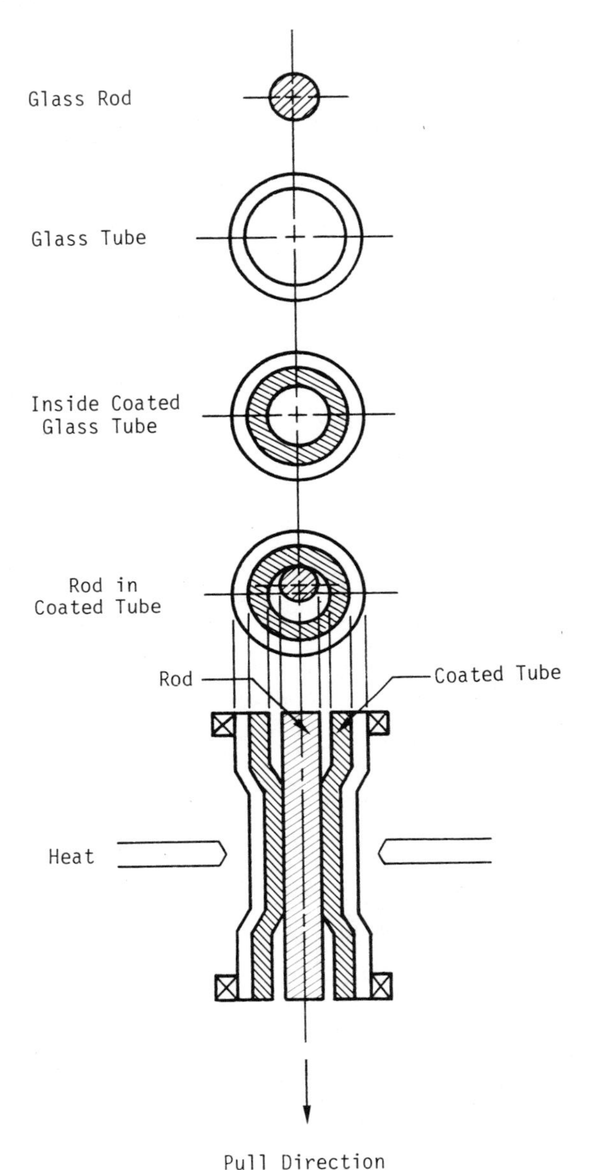

Figure 11.4 Illustration of the rod-in-tube CVD fiber manufacturing method.

the Electrical Communication Laboratory of NTT and Sumitomo Electric Industries. Such fabrication methods allow almost unrestricted fiber lengths. If optical fiber preforms are fabricated by this method, at least in principle the unit length of optical fiber cables could be as long as the system design requires.

ROD-IN-TUBE METHOD FOR CVD FIBER FABRICATION

The rod-in-tube method adapted for CVD fiber manufacture is a new method of making silica preforms for optical fiber production. It has been reported by several organizations, such as NTT, Sumitomo Electric Industries, Hitachi Cable, Dainichi-Nippon Cable, and others. The new method is a combination of the inside-coated CVD method and the rod-in-tube method (181). The CVD method uses the Bernoulli vapor deposition of dopants onto the inside surface of a silica tube. The rod-in-tube method uses a silica tube which is collapsed around a glass rod in differing chemical composition. The combination of the two methods is shown schematically in Figure 11.4. In the CVD step of the new process, B_2O_3–SiO_2 is coated onto the inside glass surface of the tube. The combined rod-in-coated-tube structure is collapsed to a single preform rod. By using this method, the subsequently drawn fiber has three layers, the middle layer of which is boron. The resulting refractive index profile is a classical w-shaped profile, as shown in Figure 11.5. The introduction of the boron layer has the following effects:

The effective numerical aperture is increased.

By adjusting the layer thickness, the fiber transmission characteristics can be controlled.

An effective barrier against the OH^- absorption loss is obtained.

A protective layer against microbending losses from external mechanical stresses during manufacture and cabling is provided.

In view of the above-mentioned merits, it is advantageous to use the new preform fabrication method for fiber manufacture. One difficulty with this fabrication method is that purity and index profile uniformity must be maintained under stringent conditions, similar to the CVD method. Nonuniformities in the silica rod circumference result in core circumference irregularities; thus, the concentricity of the core and middle layer becomes critical.

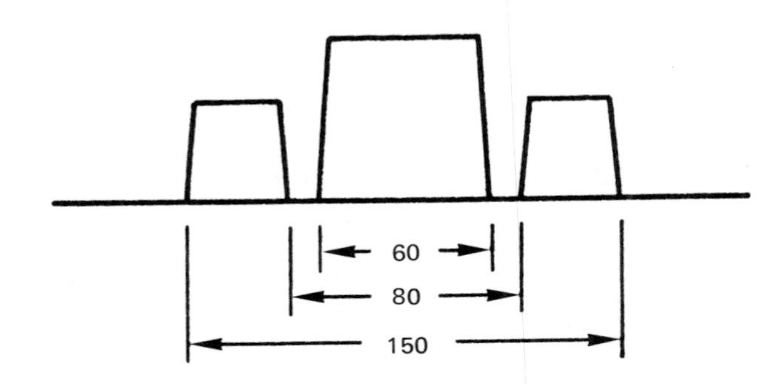

Figure 11.5 Refractive index profile of a fiber made by the rod-in-tube method for CVD fiber manufacture.

The transmission loss of trial fibers manufactured by the new method is not exactly as good as that of CVD at the present time. However, the overall transmission characteristics of the CVD rod-in-tube fiber are suitable for general use in optical fiber communications. The transmission characteristics of the new fiber will be improved soon by better control on all system variables to approach the quality of CVD fibers (182).

Developments in Optical Devices

Research and development activities on various optical communications systems components generally fall into the categories of discrete devices, such as sources, detectors, fibers, and connecting elements, and of integrated optics. In all of these areas considerable progress has recently been made (183–188). The most important of the nonfiber activities in Japan will be described below.

OPTICAL SOURCES

The light source (either light-emitting diode [LED] or injection laser diode [ILD]) is one of the most important devices in optical communications systems. Examples of optical sources used in optical communications systems in Japan are shown in Table 11.4. In Japan, a large number of experimental optical systems have been installed. Usually, a LED is used for analog transmission systems (such as for video signals) or for computer links, whereas an ILD is used for high-bit-rate digital data communication systems.

Considerable efforts have been made, especially in the area of analog communications systems, using light-emitting diodes. These systems have many desirable characteristics, such as narrow transmission bandwidth, simple structure of transmitter, receiver, and repeater, relatively low cost, and high reliability. There are, however, some minor technical problems in using an LED in analog communication systems, such as a relatively low coupling

TABLE 11.4. Examples of Japanese Optical Communications Systems (1978)

Source	System	Signal	Fiber maximum span
LED	NHK Kawai satellite station	VIDEO-IF (16.5, 22.5 MHz)	Plastic clad fiber (430 m)
	Hi-OVIS CATV system	Video baseband FM-audio, data	Plastic clad fiber (1.5 km)
	MITI computer link	8 Mb/s	Plastic clad fiber (50 m)
ILD	NTT	32 Mb/s	Step-type CVD (8 km)
	TEPCO	6.3 Mb/s PCM	Step-type CVD (2 km)

efficiency of the LED to the fiber and the nonlinear distortion of the LED radiance versus the driving current response.

Various methods are used in Japan for the optical coupling of LED and fiber. Examples include (a) a spherical lens is attached to the emitting surface which narrows the output radiance angle and thus achieves a higher coupling efficiency, and (b) by melting the fiber end, a spherical lens can be made which enlarges the effective NA of the fiber and thereby achieves a higher coupling efficiency. By using these methods, a 10% coupling efficiency (for a fiber whose NA = 0.2 and whose core diameter $2a_1 = 60\ \mu m$) has been obtained.

When multirepeater systems for analog video transmission systems are designed, the nonlinear distortion of the LED typically presents problems. To avoid these, electrical compensation of the distortion is usually employed. The most practical technique is the predistortion method; a schematic diagram of this technique (189) is shown in Figure 11.6. Considerable improvement in system performance is achieved by its use. For example, the DG distortion is reduced to 0.3% from about 4.2% without compensation. This is equal to an improvement of 28 and 31 dB of the second and third harmonics, respectively.

Recent research on laser diodes in Japan has led to several advances, such as the following:

> A reduction in the required threshold current has been achieved, with the minimum now being about 20 mA. Such ILDs have a small stripe width and a high stability of the radiant mode, such that the coupling to the fiber is both efficient and stable.
>
> The frequency response of ILDs is now greater than 1.3 GHz, and in experimental systems a data rate of 800 Mb/s has been successfully obtained.
>
> InGaAsP laser diodes that radiate at wavelengths ranging from 1.0 to 1.6 μm have been investigated. Figure 11.7 shows the structure of such a device.
>
> Direct modulation experiments at 1.13 GHz have been performed with these lasers (at $I_{th} = 200$ mA, $\lambda = 1.3\ \mu m$), and optical communications systems which use InGaAsP laser diodes have been installed.

Although a large amount of research and development efforts are still needed to improve the performance of light sources and particularly of light detectors in the wavelength range from 1.0 to 1.6 μm, the future for these semiconductor devices at long wavelengths appears to be promising.

Substantial research and development efforts have been conducted in Japan on the improvement of the lifetime of both injection lasers and light-emitting diodes. Hayashi (190) estimates the achievable lifetime of LEDs to be on the order of 10^6 hr and that of ILDs approaching 10^5 hr. Operating life of GaAsAl lasers or LEDs has been demonstrated to be more than 10^4 hr. Accelerated life tests of GaAsAl lasers or GaAs LEDs have been carried out by increasing environmental temperature up to a few hundred degrees.

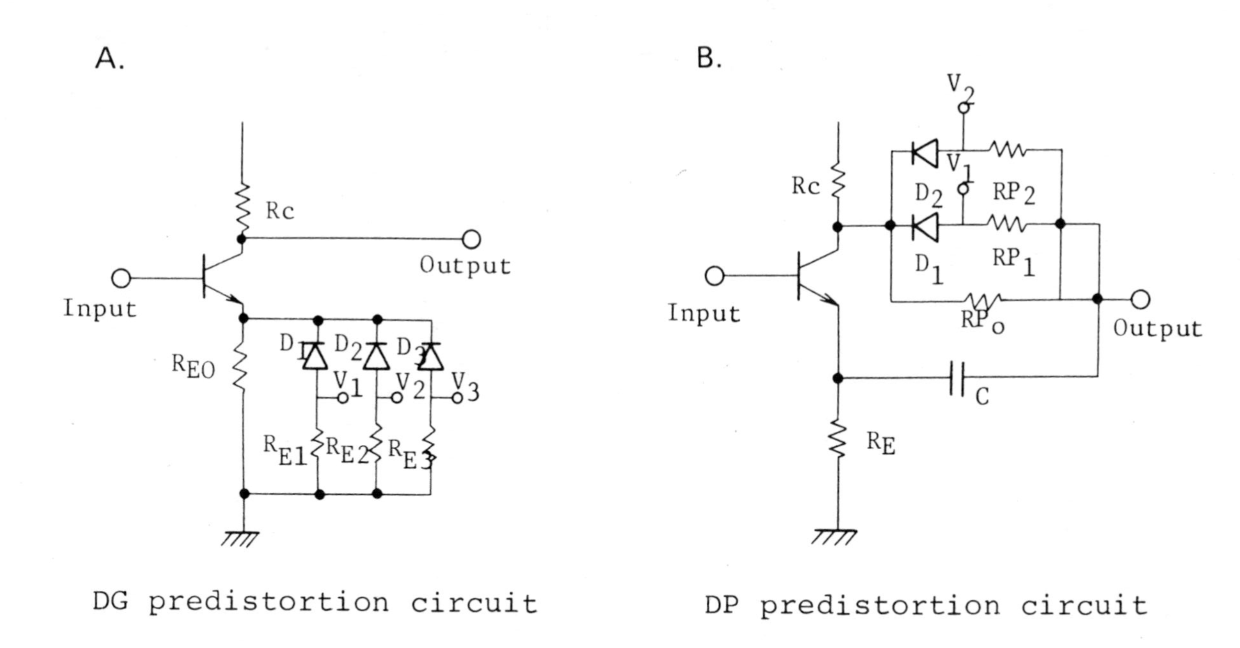

Figure 11.6 Examples of predistortion circuits used in Japan.

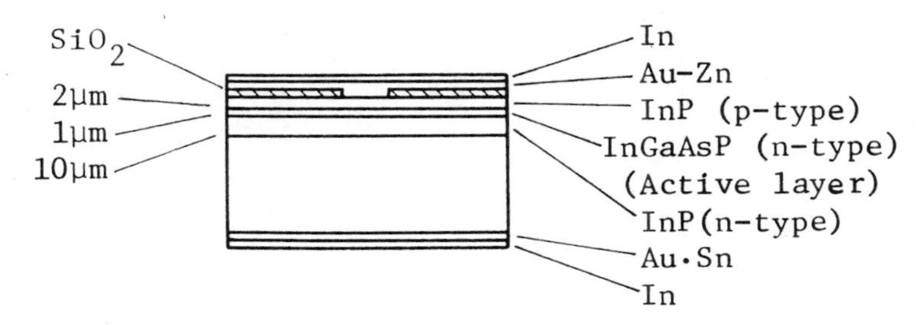

Figure 11.7 Schematic diagram of the structure of an InGaAsP injection laser.

Based on accelerated life tests, the operating life time of GaAsAl lasers and GaAs LEDs ranges from 10^5 to 10^6 hr (191).

OTHER OPTICAL COMPONENTS

The many passive optical components which have been developed for applications in optical communications systems include optical switches, modulators, directional couplers, mixers, splitters, and other devices.

The optical switch is one of the most important components in optical communications systems because it allows simple point-to-point transmission to be expanded to multipoint systems (such as telecommunications networks) using the traffic exchange function of these devices. There are several methods of optical switching. Examples include the mechanical movement of a fiber or optical lens and the use of the electrooptic properties of special semiconductor and other crystalline materials. The former method has a low switching speed and a low insertion loss; switches made in this way are already in practical application in Japan. The latter method has a high switching speed and at present a relatively high insertion loss. While such electrooptical switches are not yet in practical use in Japan, these components are highly desirable because they are suitable for mass production and thus can be fabricated in an integrated optical IC package, coupling light sources, optical modulators, and other components in small units.

An optically activated optical switch has been demonstrated in both planar and grooved branched waveguide structures using epitaxial single-crystal films (192). Figure 11.8 shows the device; it consists of an epitaxial layer 40 μm thick on a $\langle 100 \rangle$ BSO substrate of 700 μm thickness and 4 mm total length. When the direction of the propagating light in the waveguide is in the $\langle 100 \rangle$ direction, the refractive index change due to electrooptic birefringence in the region of the higher electric field rotates the polarization of an incident beam (45° to $\langle 001 \rangle$ axis) and maximizes the light intensity passing the polarizer at the guide exit. When the excited light is turned off, the electric field in the layer becomes about 5% of its value in the on state and the guided beam is switched off. The on–off ratio thus obtained is about 25 to 30 dB at 400 V for 0.633 μm He–Ne laser beams.

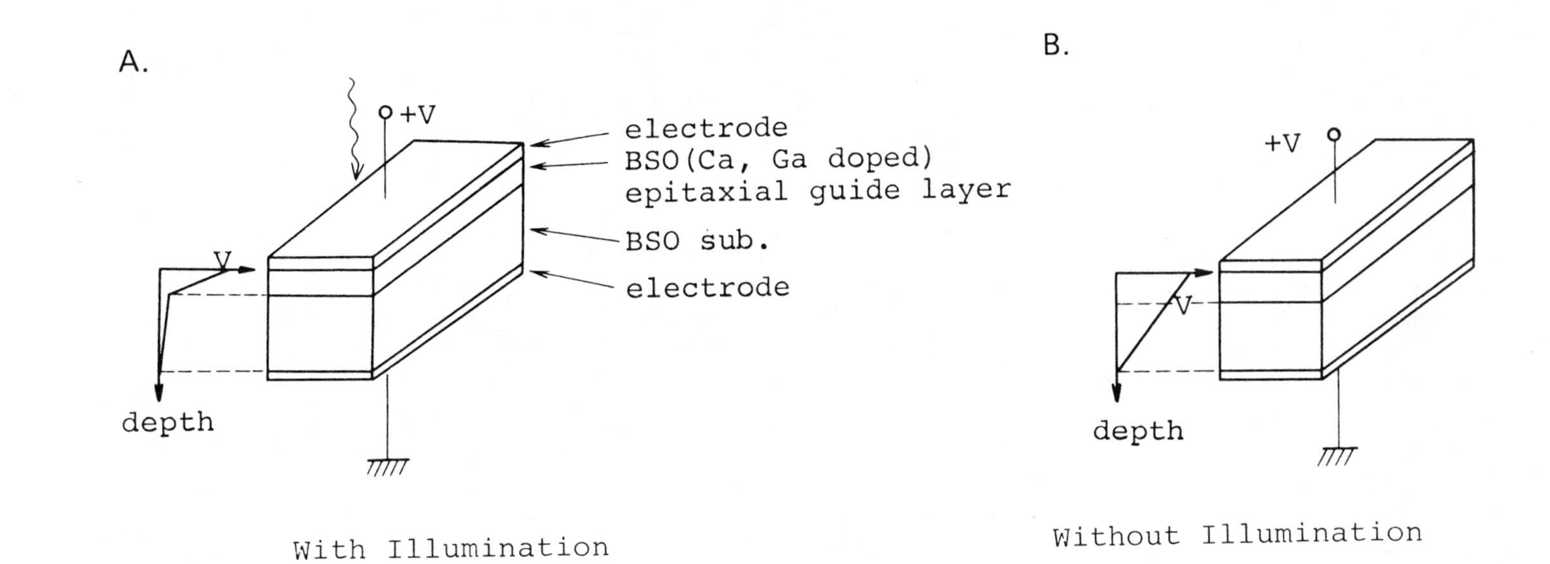

Figure 11.8 Optically activated switch with (A) and without (B) illumination.

INTEGRATED OPTICS

Distributed-feedback (DFB) semiconductor lasers, integrated on a single chip to realize a frequency-multiplexing light source (193), were successfully operated at room temperature. The thin-film device utilizes light diffraction in a periodic structure. Other devices include a thin-film filter, distributed Bragg reflector lasers, and grating couplers for dielectric waveguides.

The first experiments on DFB semiconductor lasers in Japan were performed with GaAs surfaces corrugated at wavelengths of 0.115 μm (first order) and 0.345 μm (third order); holographic photolithography and ion milling were used in the fabrication. The threshold pumping intensity was on the order of 10^4 W/cm^2 at 77°K for lasing to occur at room temperature.

Figure 11.9 shows the lasing spectrum of a typical SCH (separate-confinement heterostructure) DFB laser. The diode lases in a single longitudinal mode with a spectral bandwidth of 0.5Å. SCH DFB lasers described in

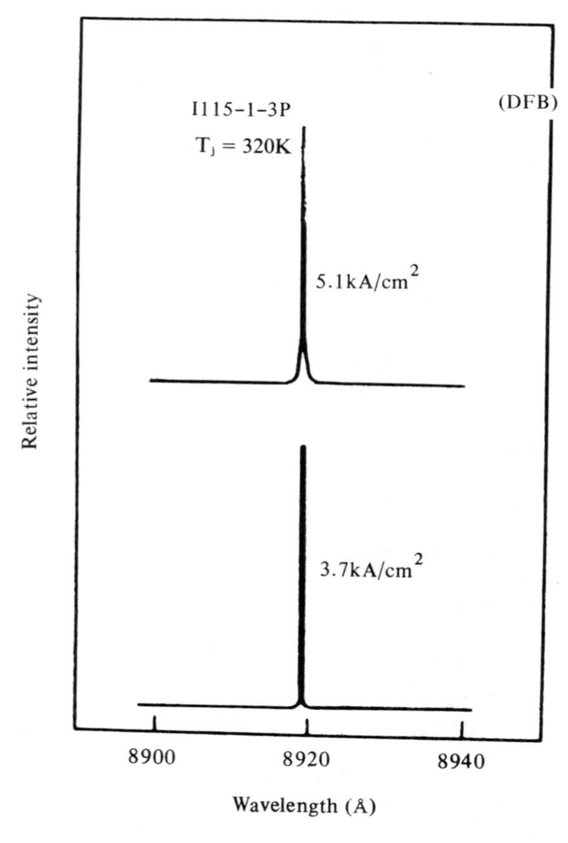

Figure 11.9 Lasing spectrum of a typical distributed-feedback laser (J_{th} = 3.4 kA/cm^2).

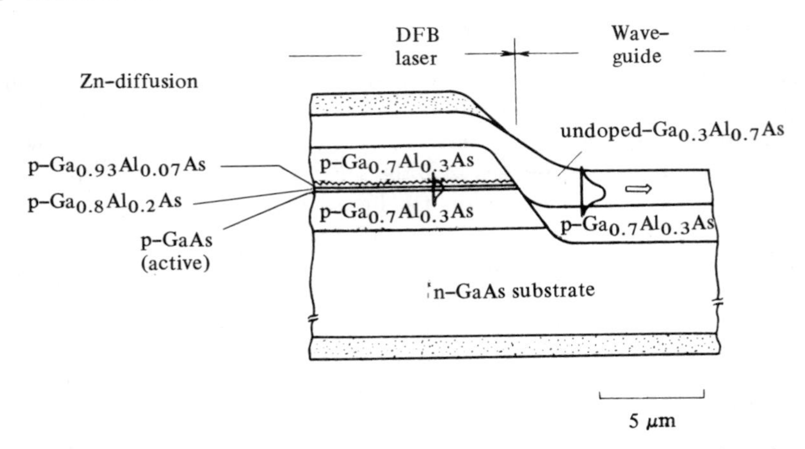

Figure 11.10 Cross-sectional view of the coupling region of a distributed-feedback laser.

the literature have operated continuously at room temperature. Figure 11.10 shows a cross-sectional view of the coupling region.

REFERENCES

United States

1. S.E. Miller et al. Toward Optical Fiber Transmission Systems. *Proc. IEEE* **61**:1703, 1973.
2. J.B. MacChesney et al. A New Technique for the Preparation of Low-Loss and Graded-Index Fibers. *Proc. IEEE* **62**:1280, 1974.
3. G.W. Tasker & W.G. French. Low-Loss Optical Waveguides With Pure Fused SiO_2 Cores. *Proc. IEEE* **62**:1281, 1974.
4. P. Kaiser et al. Low-Loss FEP-clad Silica Fibers. *Appl. Opt.* **14**:156, 1975.
5. A.R. Tynes et al. Loss Mechanisms and Measurements in Clad Glass Fibers and Bulk Glass. *J. Opt. Soc. Am.* **61**:143, 1971.
6. D.A. Pinnow et al. Fundamental Optical Attenuation Limits in the Liquid and Glassy State With Application to Fiber Optical Waveguide Materials. *Appl. Phys. Lett.* **22**:527, 1973.
7. J. Schroeder et al. Rayleigh and Brillouin Scattering in K_2O-SiO_2 Glasses. *J. Am. Ceram. Soc.* **56**:510, 1973.
8. P. Kaiser et al. Loss Mechanisms in Various Glasses Useful for Optical Fibers. *J. Opt. Soc. Am.* **63**:1141, 1973.
9. D. Gloge. Propagation Effects of Optical Fibers. *IEEE Trans. Microwave Theory Tech.* **MTT-23**:106, 1975.
10. D. Gloge. Weakly Guiding Fibers. *App. Opt.* **10**:2252, 1971.
11. D. Gloge & E.A.J. Marcatili. Multimode Theory of Graded-Core Fibers. *Bell Sys. Tech. J.* **52**:1563, 1973.
12. D. Gloge. Optical Power Flow in Multimode Fibers. *Bell Sys. Tech. J.* **51**:1767, 1972.
13. D. Gloge. Impulse Response of Clad Optical Multimode Fibers. *Bell Sys. Tech. J.* **52**:801, 1973.
14. D. Marcuse. Radiation Losses of Dielectric Waveguides in Terms of the Power Spectrum of the Wall Distortion Function. *Bell Sys. Tech. J.* **48**:3233, 1969.

15. D. Marcuse. Coupled Mode Theory of Round Optical Fibers. *Bell Sys. Tech. J.* **52**:817, 1973.

16. S.D. Personnick. Time Dispersion in Dielectric Waveguides. *Bell Sys. Tech. J.* **50**:843, 1971.

17. M. DiDomenico Jr. Material Dispersion in Optical Fiber Waveguides. *Appl. Opt.* **11**:652, 1972.

18. R.G. Smith. Optical Power Handling Capability of Low Loss Optical Fibers as Determined by Stimulated Raman and Brillouin Scattering. *Appl. Opt.* **11**:2489, 1972.

19. D.B. Keck. Spatial and Temporal Power Transfer Measurements on a Low-Loss Optical Waveguide. *Appl. Opt.* **13**:1882, 1974.

20. R.L. Gallawa. Optical Waveguide Technology. World Telecommunication Forum, Geneva, Oct. 1975.

21. R.L. Lebduska. Fiber Optic Cable Test Evaluation. *Opt. Eng.* **13**:49, 1974.

22. C.A. Burrus & B.I. Miller. Small-Area, Double-Heterostructure Aluminum-Gallium Arsenide Electroluminescent Diode Sources for Optical-Fiber Transmission Lines. *Opt. Commun.* **4**:307, 1971.

23. R.W. Dawson & C.A. Burrus. Pulse Behavior of High-Radiance Small-Area Electroluminescent Diodes. *Appl. Opt.* **10**:2367, 1971.

24. M.B. Panish. Heterostructure Injection Lasers. *IEEE Trans. Microwave Theory Tech.* **MTT-23**:20, 1975.

25. H. Melchior. Sensitive High Speed Photodetectors for the Demodulation of Visible and Near Infrared Light. *J. Lumin.* **7**:390, 1973.

26. J.E. Goell. An Optical Repeater with High-Impedance Input Amplifier. *Bell Sys. Tech. J.* **53**:629, 1974.

27. J.E. Goell. A 274 Mb Optical-Repeater Experiment Employing a GaAs Laser. *Proc. IEEE* **61**:1504, 1973.

28. S.D. Personnick. Receiver Design for Digital Fiber Optic Communication Systems. Part I. *Bell Sys. Tech. J.* **52**:843, 1973; Part II. *Bell Sys. Tech. J.* **52**:875, 1973.

29. P.S. Cross & H. Kogelnick. Sidelobe Suppression and Band-Broadening in Corrugated-Waveguide Filters. IOOC 77, p. 17, Tokyo, July 1977.

30. E.A.J. Marcatili. Reflection from Tapered and Chirped Gratings. IOOC 77, p. 21, Tokyo, July 1977.

31. R.H. Stolen. Tunable Fiber Raman Oscillators. IOOC 77, p. 45, Tokyo, July 1977.

32. C.S. Tsai et al. Wideband Guided-Wave Anisotropic Acoustooptic Bragg Diffraction in $LiNbO_3$ Waveguides. IOOC 77, p. 57, Tokyo, July 1977.

33. J.M. Hammer. Progress in Optical Waveguide Modulators and Switches. IOOC 77, p. 133, Tokyo, July 1977.

34. D.B. Anderson & R.R. August. Progress in Waveguide Lenses for Integrated Optics. IOOC 77, p. 231, Tokyo, July 1977.

35. L.D. Comerford et al. U.S. Patent 4,079,404, April 1978.

Western Europe

36. K.C. Kao & G.A. Hockham. Dielectric-Fibre Surface Waveguides for Optical Frequencies. *Proc. IEEE,* **13**(7):1151, 1966.

37. M. Börner. Mehrstufiges Übertragungssystem für in PCM dargestellte Nachrichten. German Patent 1,254,513, Dec. 21, 1966; Brit. Patent 1,202,418; French Patent 1,548,972; U.S. Patent 3,845,293.

38. M.W. Jones & K.C. Kao. Spectrophotometric Studies of Ultra Low Loss Optical Glasses, Part 2: Double Beam Method. *J. Sci. Instrum. (J. Phys. E), Ser. 2* **2**(4):331, 1969.

39. K.J. Beales et al. Preparation of Sodium Borosilicate Glass Fibre for Optical Communication. *Proc. IEEE* **123**(6):591, 1976.

40. B. Scott & H. Rawson. Preparation of Low Loss Glasses for Optical Fibre Communication Systems. *Opto-electronics* **5**:285, 1973.

41. G.R. Newns et al. Absorption Losses in Glasses and Glass Fibre Waveguides, *Opto-electronics* **5**:289, 1973.

42. M. Faulstich et al. Highly Transparent Glass for Producing Optical Fibers for Telecommunication. Topical Meeting on Optical Fiber Transmission, Williamsburg, Va., January 1975.

43. H.M.J.M. van Ass et al. Preparation of Graded-Index Optical Glass Fibres in the Alkali Germanosilicate System. *Electron. Lett.* **12**(15):369, 1976.

44. K.J. Beales et al. Low-Loss Compound-Glass Optical Fibre. *Electron. Lett.* **13**:755, 1977.

45. D.N. Payne & W.A. Gambling. New Silica-based Low-Loss Optical Fibre. *Electron. Lett.* **10**:289, 1974.

46. A. Mühlich et al. A New Doped Synthetic Fused Silica as Bulk Materials for Low-Loss Optical Fibres. *IEE Conf. Proc.* **132**, 1975.

47. A. Mühlich et al. Preparation of Fluorine-doped Silica Preforms by Plasma Chemical Technique. Third European Conference on Optical Communication, Munich, September 1977.

48. D.N. Payne & W.A. Gambling. Preparation of Water-free Silica-based Optical-Fibre Waveguide. *Electron. Lett.* **10**:335, 1974.

49. D.N. Payne & W.A. Gambling. A Borosilicate-cladded Phosphosilicate-Core Optical Fibre. *Opt. Commun.* **13**:422, 1975.

50. B.J. Ainslie et al. Water Impurity in Low-Loss Silica Fibre. *Mater. Res. Bull.* **12**:481, 1977.

51. P. Geittner et al. Low-Loss Optical Fibres Prepared by Plasma-activated Chemical Vapor Deposition (CVD). *Appl. Phys. Lett.* **28**:645, 1976.

52. K. Kuppers et al. Codeposition of Glassy Silica and Germania Inside a Tube by Plasma-activated CVD. *J. Electrochem. Soc.* **123**(7):1079, 1976.

53. D. Hondros & P. Debye. Elektromagnetische Wellen an dielektrischen Drähten. *Ann. Phys.* **32**(4):465, 1910.

54. H.G. Unger. *Planar Optical Waveguides and Fibres.* Oxford, Oxford University Press, 1977.

55. P.J.B. Clarricoats & K.B. Chan. Electromagnetic Wave Propagation Along Radially Inhomogeneous Dielectric Cylinders. *Electron. Lett.* **6**:694, 1970.

56. H. Kirchhoff. Wave Propagation Along Radially Inhomogeneous Glass Fibres. *Arch. Elektr. Übertr.* **27**:13, 1973.

57. C.C. Timmermann. The Influence of Deviations from the Square-Law Refractive Index Profile of Gradient Core Fibres on Mode Dispersion. *Arch. Elektr. Übertr.* **28**:344, 1974.

58. S. Geckeler. Nonlinear Profile Dispersion Aids Optimization of Graded Index Fibres. *Electron. Lett.* **13**:440, 1977.

59. K. Petermann. The Mode Attenuation in General Graded Core Multimode Fibres. *Arch. Elektr. Übertr.* **29**:345, 1975.

60. W.J. Stewart. Fibre Characterisation by Use of the Relation Between the Characteristics of Leaky Modes in Optical Fibres and the Fibre Parameters. First European Conference on Optical Communication, London, 1975.

61. M.J. Adams et al. Correction Factors for the Determination of Optical-Fibre Refractive Index Profiles by the Near-Field Scanning Technique. *Electron. Lett.* **12**:281, 1976.

62. K. Petermann. Uncertainties of the Leaky Mode Correction for Near Square-Law Optical Fibres. *Electron. Lett.* **13**:513, 1977.

63. R. Bergeest & H.G. Unger. Optische Wellen in Hohlleitern. *Arch. Elektr. Übertr.* **23**:529, 1969.

64. L. Jeunhomme et al. Propagation Model for Long Step-Index Fibres. *Appl. Opt.* **15**:3040, 1976.

65. M. Eve. Multipath Time Dispersion Theory of an Optical Network. *Opt. Quantum Electron* **10**:41, 1978.

66. K. Petermann. Theory of Microbending Loss in Monomode Fibres With Arbitrary Refractive Index Profile. *Arch. Elektr. Übertr.* **30**:337, 1976.

67. W.A. Gambling et al. Radiation From Curved Single-Mode Fibres. *Electron. Lett.* **12**:567, 1976.

68. R.A. Sammut. Discrete Radiation from Curved Single-Mode Fibres. *Electron. Lett.* **13**:418, 1977.

69. H.N. Daglish & J.C. North. Watching Light Waves Decline. *New Scientist* **49**:14, 1970.

70. P.J.R. Laybourn et al. A Photometer to Measure Light Scattering in Optical Glass. *Opto-electronics* **2**:36, 1970.

71. J.P. Dakin & W.A. Gambling. Angular Distribution of Light Scattering in Bulk Glass and Fibre Waveguides. *Opt. Commun.* **6**:235, 1972.

72. K.I. White. A Calorimetric Method for the Measurement of Low Optical Absorption Losses in Optical Communication Fibres. *Opt. Quantum Electron.* **8**:73, 1976.

73. B. Hillerich. Pulse-Reflection Method for Transmission-Loss Measurement of Optical Fibres. *Electon. Lett.* **12**(4):92, 1976.

74. J. Guttmann & O. Krumpholz. Location of Imperfections in Optical Glass-Fibre Waveguides. *Electron. Lett.* **11**:216, 1975.

75. D. Williams & K.C. Kao. Pulse Communication Along Glass Fibres. *Proc. IEEE.* **56**:197, 1968.

76. H.J. Heyke. Übertragung Optischer Impulse mit Lichtleitfasern. *Laser & Elektro-Optik* **1**:49, 1969.

77. W.A. Gambling et al. Pulse Propagation Along Glass Fibres. *Electron Lett.* **6**(12):364, 1970.

78. M. Eve et al. Launching Independent Measurements of Multimode Fibres. Second European Conference on Optical Communication, p. 143, Paris, 1976.

79. S. Geckeler & D. Schicketanz. Enhancement of Bandwidth and Additional Loss Caused by Controlled Mode Coupling in Quartz Fibres. *Electron. Lett.* **10**(22): 465, 1974.

80. W.A. Gambling & D.N. Payne. Mode Conversion Coefficients in Optical Fibres. *Appl. Opt.* **14**(7):1538, 1975.

81. W. Eickhoff & E. Weidel. Measuring Method for the Refractive Index Profile of Optical Glass Fibres. *Opt. Quantum Electron.* **7**:109, 1975.

82. B. Costa & B. Sordo. Measurements of the Refractive Index Profile in Optical Fibres: Comparison Between Different Techniques. Second European Conference on Optical Fiber Communication, p. 81, Paris, 1976.

83. G.T. Summer. A New Technique for Refractive Index Profile Measurement in Multimode Optical Fibres. *Opt. Quantum Electron.* **9**:79, 1977.

84. W.J. Stewart. A New Technique for Measuring the Refractive Index Profiles of Graded Optical Fibres. International Conference on Integrated Optics and Optical Fiber Communication, p. 395, Tokyo, 1977.

85. L. Jeunhomme & J.P. Pocholle. Measurement of the Numerical Aperture of a Step-Index Optical Fibre. *Electron. Lett.* **12**(3):63, 1976.

86. F.M.E. Sladen et al. Measurement of Profile Dispersion in Optical Fibres; A Direct Technique. *Electron. Lett.* **13**(7):212, 1977.

87. D.N. Payne & A.H. Hartog. Pulse Delay Measurement of the Zero Wavelength of Material Dispersion in Optical Fibres. *Nachrichtentech. Z.* **31**:130, 1978.

88. J.E. Midwinter. The Prism–Taper Coupler for the Excitation of Single Modes in Optical Transmission Fibres. *Opt. Quantum Electron.* **7**:297, 1975.

89. M. Eve. Novel Technique for the Measurement of Differential Mode Attenuation in Graded Index Fibres. *Electron. Lett.* **13**(24):744, 1977.

90. O. Krumpholz. Mode Propagation in Fibres: Discrepancies Between Theory and Experiment. Conference on Trunk Telecommunications by Guided Waves, p. 56, London, 1970.

91. W.A. Gambling et al. Determination of Core Diameter and Refractive-Index Difference of Single-Mode Fibres by Observation of the Far-Field Pattern. *Microwaves, Opt. Acoust.* **1**(1):13, 1976.

92. M. Börner et al. Lösbare Steckverbindung für Ein-Mode-Glasfaserlichtwellenleiter. *AEU* **26**:288, 1972.

93. V. Vucins. Adjustable Single-Fibre Connector With Monitor Output. Third European Conference on Optical Communication. p. 100, München, 1977.

94. M.A. Bedgood et al. Demountable Connectors for Optical Fiber Systems. *Electr. Commun.* **51**(2):85, 1976.

95. Connector from AEG-Telefunken Aligns Ends of Optical Fibres Automatically. *Electronics* **11**:8E, 1976.

96. P. Hansel. Triple-Ball Connector for Optical Fibres. *Electron. Lett.* **13**(24): 734, 1977.

97. J. Guttmann et al. Multipole Optical-Fibre Connector. *Electron. Lett.*, **11**(24):582, 1975.

98. F. Auracher & K.H. Zeitler. Multiple Fibre Connectors for Multimode Fibres. *Opt. Commun.* **18**(4):556, 1976.

99. D. Kunze et al. Jointing Techniques for Optical Cables. Second European Conference on Optical Communication, p. 257, Paris, 1976.

100. D.G. Dalgoutte. Collapsed Sleeve Splices for Field Jointing of Optical Fibre Cables. Third European Conference on Optical Communication, p. 106, München, 1977.

101. R. Jocteur & A. Tardy. Optical Fibre Splicing with a Plasma Torch and an Oxhydric Microburner. Second European Conference on Optical Communication, p. 261, Paris, 1976.

102. W. Eickhoff et al. Optical Fibre Splicing Using Low Melting-Point Glass Solder Fibre and Integrated Optics, **2**:63, 1979.

103. P. Hensel. A New Fibre Jointing Technique Suitable for Automation. Third European Conference on Optical Communication, p. 103, München, 1977.

104. F. Caspers & E.G. Neumann. Optical-Fibre End Preparation by Spark Erosion. *Electron. Lett.* **12**(17):443, 1976.

105. G.D. Khoe & G. Kuyt. Cutting Optical Fibres with a Hot Wire. *Electron. Lett.* **13**(5):147, 1977.

106. J. Wittmann. Contact-bonded Epoxy-Resin Lenses to Fibre Endfaces. *Electron. Lett.* **11**(20):477, 1975.

107. C.C. Timmermann. Highly Efficient Light Coupling from GaAlAs Lasers Into Optical Fibres. *Appl. Opt.* **15**(10):2432, 1976.

108. G.D. Khoe & G. Kuyt. A Luneburg Lens for the Efficient Coupling of a Laser Diode and a Graded-Index Fibre. Third European Conference on Optical Communication, p. 176, München, 1977.

109. E. Weidel. New Coupling Method for GaAs-Laser-Fibre Coupling. *Electron. Lett.* **11**(18):436, 1975.

110. H.G. Danielmeyer & H.P. Weber. Fluorescence in Neodymium Ultraphosphate. *IEEE J. Quantum Electron.* **QE-8**:805, 1972.

111. G. Winzer et al. Miniature Neodymium Lasers (MNL) as Possible Transmitters for Fiber-Optic Communication Systems. Part I. Stoichiometric Materials. *Siemens Forsch. Entwicklungsber.* **5**:287, 1976.

112. P. Möckel et al. Miniature Neodymium Lasers (MNL) as Possible Transmitters for Fiber-Optic Communication Systems. Part II. YAG:Nd^{3+} Waveguide Lasers. *Siemens Forsch. Entwicklungsber.* **5**:296, 1976.

113. S.D. Hersee & R.C. Goodfellow. The Reliability of High Radiance LED Fibre Optic Sources. Third European Conference on Optical Communication, p. 213, Paris, 1976.

114. R.C. Goodfellow & A.W. Mabbit. Wide-Bandwidth High-Radiance Gallium-Arsenide Light-emitting Diodes for Fibre-Optic Communication. *Electron. Lett.* **12**:50, 1976.

115. J. Heinen et al. Light Emitting Diodes With a Modulation Bandwidth of More than 1 GHz. *Electron. Lett.* **12**:553, 1976.

116. A.W. Mabbit & R.C. Goodfellow. High-Radiance Small Area Gallium-Indium-Arsenide 1.06 μm Light-emitting Diodes. *Electron. Lett.* **13**:291, 1977.

117. W. Harth & M.C. Amman. Modulation Characteristics of Double-Heterostructure Superluminescent Diodes. *Electron. Lett.* **13**:291, 1977.

118. J. Bakker. The Effects of Proton Bombardment on the Electrical and Optical Properties of the Active Layer in GaAs-$Al_x Ga_{1-x}As$ Double Heterostructure (DH) Laser Material. Paper presented at Semiconductor Injection Lasers and their Applications (SILA), Cardiff, March 1977.

119. J.C. Carballes et al. Linear Light-Output Characteristics in DH-GaAlAs Lasers. Conference on International Optics and Optical Communication, Tokyo, July 1977.

120. A.G. Steventon et al. Low Threshold Current Proton-isolated (GaAl)As Double Heterostructure Lasers. *Opt. Quantum Electron.* **9**:519, 1977.

121. P.A. Kirby & G.H.B. Thompson. Channeled Substrate Buried Heterostructure GaAs-(GaAl)As Injection Lasers. *J. Appl. Phys.* **47**:4578, 1976.

122. G.H.B. Thompson et al. Narrow-Beam Five-Layer (GaAl)As/GaAs Heterostructure Lasers with Low Threshold and High Peak Power. *J. Appl. Phys.* 47:1501, 1976.

123. P.J. de Waard. Stripe Geometry DH Lasers With Linear Output/Current Characteristics. *Electron. Lett.* **13**:400, 1977.

124. S. Ritchie et al. The Degradation of (GaAl)As DH Lasers at High Temperatures. Proceedings of the Third European Conference on Optical Communication, p. 117, Munich, September 1977.

125. G. Arnold et al. Investigation of Degradation of Laser Diodes. Proceedings of the Third European Conference on Optical Communication, p. 114, Munich, September 1977.

126. A.R. Goodwin et al. Reliable Semiconductor Lasers for Wide-Band Optical Communication Systems. AGARD Conference Preprint 219, Optical Fibres, Integrated Optics and Their Applications. London, May 1977.

127. M.J. Adams. Rate Equations and Transient Phenomena in Semiconductor Lasers. *Opto-electronics* **5**:201, 1973.

128. W. Harth. Properties of Injection Lasers at Large-Signal Modulation. *Arch. Elektr. Ubertr.* **29**:149, 1975.

129. D. Schicketanz. Large-Signal Modulation of GaAs Laser Diodes. *Siemens Forsch. Entwicklungsber.* **4**:325, 1975.

130. P. Russer & S. Schulz. Direkte Modulation eines Doppelheterostrukturlasers mit einer Bitrate von 2, 3 Gbit/s. *Arch. Elektr. Übertr.* **27**:193, 1973.

131. J. Buus et al. On the Dynamic and Spectral Behaviour of DH GaAlAs Stripe Lasers Modulated With Subnanosecond Pulses. Proceedings of the Second European Conference on Optical Communication, p. 231, Paris, September 1976.

132. P.M. Boers et al. Dynamic Behaviour of Semiconductor Lasers. *Electron. Lett.* **11**:206, 1975.

133. P. Russer. Modulation Behaviour of Injection Lasers With Coherent Irradiation Into Their Oscillating Mode. *Arch. Elektr. Übertr.* **29**:231, 1975.

134. P. Russer et al. High Speed Modulation of DHS-Lasers in the Case of Coherent Light Injection. Proceedings of The Third European Conference on Optical Communication, p. 139, Munich, September 1977.

135. C. Voumard et al. Resonance Amplifier Model Describing Diode Lasers Coupled to Short External Resonators. *Appl. Phys.* **12**:369, 1977.

136. H. Jäckel & G. Guekos. High Frequency Intensity Noise Spectra of Axial Mode Groups in the Radiation From cw GaAlAs Diode Lasers. *Opt. Quantum Electron.* **9**:233, 1977.

137. G. Tenchio. Low-Frequency Intensity Fluctuations of cw DH GaAlAs-Diode Lasers. *Electron. Lett.* **12**:562, 1976.

138. O. Krumpholz & S. Maslowski. Avalanche Mesaphotodioden mit Quereinstrahlung. *Wiss. Ber. AEG-Telefunken* **44**(2):73, 1971.

139. K. Berchtold et al. Avalanche Photodiodes With a Gain-Bandwidth-Product of More Than 200 GHz. *Appl. Phys. Lett.* **26**(10):585, 1975.

140. J. Muller & A. Ataman. Double-Mass Thin-Film Reach-Through Silicon Avalanche Photodiodes With Large Gain-Bandwidth Product. International Electronic Device Meeting, p. 416, Washington, D.C., 1976.

141. J. Muller. Ultrafast Multireflection—and Transparent Thin-film Silicon Photodiodes. International Electronic Device Meeting, p. 420, Washington, D.C., 1976.

142. W. Eickhoff et al. Transparent, Highly Sensitive GaAs/(GaAl) as Photodiode. *Electron. Lett.* **13**(17):493, 1977.

143. K. Mouthaan & R.M. Snoeren. Characteristics of Avalanche Photodiodes Having Nearly-Unilateral-Carrier Multiplication. *Electron. Lett.* **10**(8):118, 1974.

144. G. Ripoche & M. Brilman. Photodiodes a Avalanche au Silicium a Grande Rapidite Pour Systemes de Communications Optiques. AGARD-Conference, Optical Fibres, Integrated Optics and Their Military Application, p. 40–1, London, 1977.

145. T.P. Pearsall et al. A High Performance Avalanche Photodiode at 1.0-1.7 μm, Compatible With $InP/Ga_xIn_{1-x}As_yP_{1-y}$ Integrated Optics Technology. Topical Meeting on Integrated Optics and Guided Wave Optics, Salt Lake City, 1978.

146. H. Haupt & K. Hess. Growth of Large $CuInSe_2$ Single Crystals. International Conference on Ternary Semiconductors, p. A2, Edinburgh, 1977.

147. L.M. Reiber et al. Fabrication of Lithium Niobate Light Guiding Films. Proceedings of the Sixth International Vacuum Congress, p. 697, Kyoto, March 1974.

148. P. Baues. Phase Modulation of Light in Planar Electrooptical Waveguides. *Siemens Forsch. Entwicklungsber.* **5**:153, 1976.

149. M. Papuchon et al. Electrically Switched Optical Directional Coupler: Cobra. *Appl. Phys. Lett.* **27**:289, 1975.

150. C.H. Bulmer & M.G.F. Wilson. Single Mode Grating Coupling Between Thin-film and Fibre Optical Waveguides. *IEEE J. Quantum. Electron.* **QE-14**:741, 1978.

151. M.G.F. Wilson & G.A. Teh. Tapered Optical Directional Coupler *Trans. Microwave Theory Tech.* **MTT-23**:85, 1975.

152. O.D.D. Soares et al. Holographic Elements for Practical Fibre Bundle Couplers. AGARD Conference Preprint 219, Optical Fibres, Integrated Optics and Their Applications. London, May 1977.

153. G. Goldmann & H.H. Witte. Holograms as Optical Branching Elements. *Opt. Quantum Electron.* **9**:75, 1977.

154. M.G.F. Wilson et al. Multimode Optical Systems—Power Coupling Between Waveguides. AGARD Conference Preprint 219, Optical Fibres, Integrated Optics and Their Applications. London, May 1977.

155. F. Auracher. Planar Branching Network for Multimode Glass Fibers. *Opt. Commun.* **17**:129, 1976.

156. R. Ulrich. Image Formation by Phase Coincidences in Optical Waveguides. *Opt. Commun.* **13**:259, 1975.

157. A. Simon & R. Ulrich. Fiber-optical Interferometer. *Appl. Phys. Lett.* **31**:77, 1977.

158. W. Meyer. Star Couplers for Multimode Optical Waveguide Communication Systems. Proceedings of the Third European Conference on Optical Communication, p. 166, Munich, 1977.

159. L. Jeunhomme & J.P. Pocholle. Directional Coupler for Multimode Optical Fibers. *Appl. Phys. Lett.* **29**:485, 1976.

160. C. Stewart & W.J. Stewart. An Adjustable Branching Coupler/Attenuator for Multimode Single Fibre Systems. AGARD Conference Preprint 219, Optical Fibres, Integrated Optics and Their Applications. London, May 1977.

161. H.H. Witte. Optical Tapping Element for Multimode Fibers. *Opt. Commun.* **18**:559, 1976.

162. L. D'Auria & A. Jacques. Bidirectional Central Couplers for Links With Optical Fiber Bundles. AGARD Conference Preprint 219, Optical Fibres, Integrated Optics and Their Applications. London, May 1977.

163. J. Heinen & W. Proebster. A Transparent GaAs Light Emitting Diode for Application in Optical Wavelength-Multiplex Transmission Systems, *Nachrichtentech.* Z. **31**:129, 1978.

164. I.A. Ravencroft. Optical Communication Systems in the UK. Third European Conference on Optical Communication, p. 183, München, 1977.

165. D.R. Hill et al. A 140 MBit/s Field Demonstration System. Third European Conference on Optical Communication, p. 240, München, 1977.

166. P.W. Black et al. The Manufacture, Testing and Installation of Rugged Fiber-Optic Cables. Third European Conference on Optical Communication, p. 50, München, 1977.

167. R.J. Slaughter et al. A Duct Installation of 2-Fibre Optical Cable. First European Conference on Optical Communication, p. 84, London, 1975.

168. M. Eve et al. Transmission Performances of Three Graded-Index Fibre Cables Installed in Operational Ducts. Third European Conference on Optical Communication, p. 53, München, 1977.

169. D. Brace & K. Cameron. BPO 8448 KBit/s Optical Cable Feasibility Trial. Third European Conference on Optical Communication, p. 237, München, 1977.

170. British Post Office Experimental Optical Fiber System. Press Release, June 1977.

171. J. Feldmann & W. Gallenkamp. Prüfung der Einsatzmöglichkeiten Gegenwärtig Verfügbarer Optischer Nachrichtenübertrangungssysteme im Feldeinsatz der DBP. *Nachrichtentech.* Z. **29**:235, 1976.

172. K. Fussganger & H.J. Matt. Optical Fiber Transmissions in the HHI Wide-Band Communications Systems. International Conference on Integrated Optics and Optical Fiber Communication, Tokyo, 1977.

173. R. Bouille. Application of Optical Fibers to Existing Communication Systems. Third European Conference on Optical Communication, p. 231, München, 1977.

174. R. Jocteur. Optical Fiber Cable for Digital Transmission Systems. Second European Conference on Optical Communication, p. 193, Paris, 1976.

175. B. Catania et al. First Italian Experiment With Buried Optical Cable. Second European Conference on Optical Communication, p. 315, Paris, 1976.

176. L. Michetti & F. Tosco. One Year Results on a Buried Experimental Optical Cable. Third European Conference on Optical Communication, p. 56, München, 1977.

Japan

177. T. Nakahara & H. Yanai, Fiber Optics in Japan. *Fiber Integr. Opt.* **1**(1):1977.

178. T. Izawa et al. Continuous Fabrication of High Silica Fiber Preform. *IOOC '77 Technical Digest,* 375, 1977.

179. M. Kawahata. Hi-OVIS Development Project. *IOOC '77 Technical Digest,* 467, 1977.

180. S. Sakurai. Application of Optical Information Systems. *IOOC '77 Technical Digest,* 463, 1977.

181. T. Tokunaga et al. Rod-in-Tube Method for CVD Fiber Manufacture. Japan Electronic and Communication Institute Convention, March 1978.

182. T. Ono et al. Reliability Evaluation of Plastic Clad Fiber. Japan Electronic and Communication Institute Convention, March 1978.

183. T. Nakahara & H. Yanai. Fiber Optics in Japan. *Fiber Integr. Opt.* **1**(1):1977.

184. T. Izawa et al. Continuous Fabrication of High Silica Fiber Preform. *IOOC '77 Technical Digest,* 375, 1977.

185. M. Kawahata. HI-OVIS Development Project, *IOOC '77 Technical Digest,* 467, 1977.

186. S. Sakurai. Application of Optical Information Systems. *IOOC '77 Technical Digest,* 463, 1977.

187. T. Tokunaga et al. Rod-in-tube Method for CVD Fiber Manufacture. Japan Electronic and Communication Institute Convention, March 1978.

188. T. Ono et al. Reliability Evaluation of Plastic-clad Fiber. Japan Electronic and Communication Institute Convention, March 1978.

189. K. Asaya & T. Kimura. Analog Modulation of LED. *J. Appl. Phys. Jpn. 1978.*

190. I. Hayashi. Semiconductor Light Sources. *IOOC '77 Technical Digest,* 81, 1977.

191. Y. Sakakibara & Y. Suematsu. InGaAsP High-Speed Laser Direct Modulation. Japan Electronic and Communication Institute Convention, March 1978.

192. K. Tada & G. Yamaguchi. Optically Activated Thin Film Optical Switch. Conference on Lasers and Electrooptical Systems, San Diego, February 1978.

193. M. Nakamura. Monolithic Integration of Distributed-Feedback Semiconductor Lasers. *IOOC '77 Technical Digest,* 227, 1977.

12 REVIEW AND OUTLOOK

Helmut F. Wolf

Optical communications through waveguides or fibers has advanced from laboratory experiments to a viable technology that is increasingly being employed to serve practical communications needs, particularly in the areas of telecommunications, computers, and military applications. Similar to the basic and applied research and development efforts in optical data transmission, the practical implementations of this most recent transmission technique are dispersed over many geographic areas. The driving force behind this surge in interest is mainly of an economic nature, made possible by significant technological advances that allow unparalleled performance improvements. Few other communications techniques have progressed so rapidly, and few have the potential to carry as much information at as high a data rate in a relatively small volume.

These advances are manifested in the ability to transmit a substantially larger number of channels over the same cables owing to the higher cable bandwidth and in a reduced need for regenerative repeaters because of lower signal attenuation and dispersion. Because of the smaller weight and volume requirement of optical transmission systems, savings in installation and material costs are among the most significant advantages fiber optics communications can offer, in addition to the substantial performance benefits.

Because of the highly promising potential of optical communications, this chapter attempts to place the future of the various systems components into perspective. For reference, it gives some market overview to demonstrate the vitality of this communications concept. The anticipated component and system performance is described on the basis of reasonable extrapolations of present analogs.

The total transfer of information over the various available transmission media, including both optical and nonoptical communications, will show a significant increase in the future, as illustrated in Figure 12.1. For example, the total transfer rate of voice, video, and written data in the United States will increase from about 3×10^{18} bits per year in 1975 to 3×10^{19} bits per year in 1990, assuming 1200 bit/s for voice transmission and an average call dura-

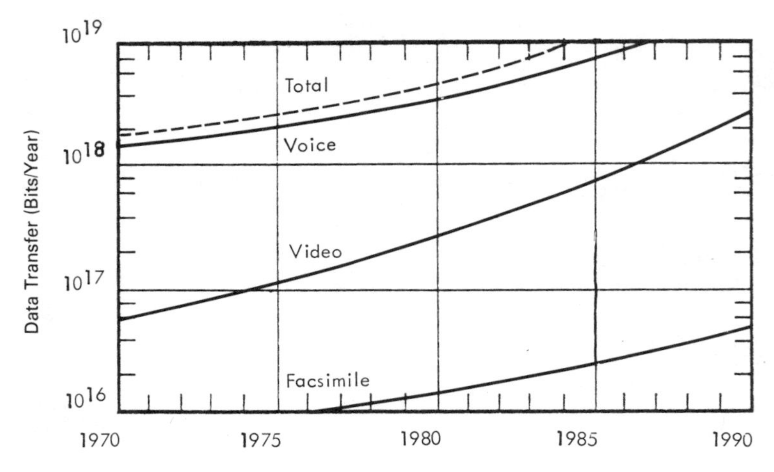

Figure 12.1 Estimated trends in total information transmission needs in the United States.

tion of 1 min. An increasing portion of this information transfer will be accomplished by the use of optical transmission.

As Figure 12.2 indicates, the total worldwide fiber optics systems production is estimated to reach about \$1.5 billion (in current U.S. dollars) by 1990, and it will exceed \$0.8 billion per year in the United States alone. Research and development efforts will level off in the 1980s at almost \$40

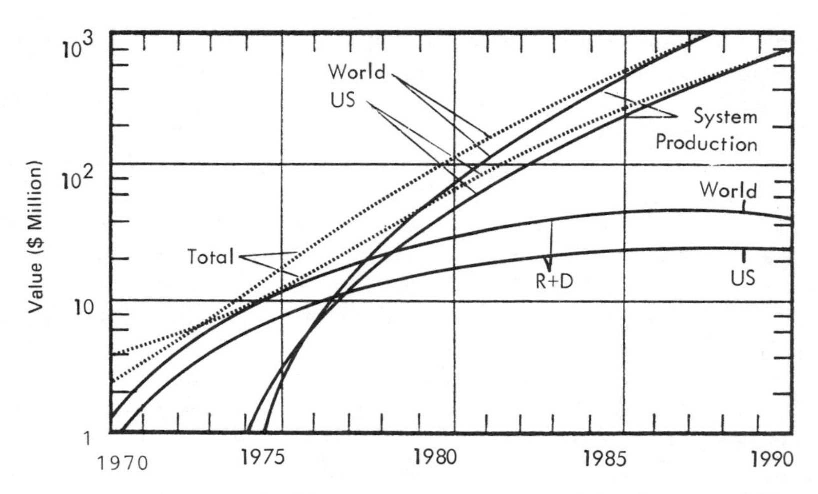

Figure 12.2 Trends in production and in research and development of fiber optics systems. Legend: (——) Totals (World, U.S.); (- - - -) Contributions (R&D, Production).

million worldwide and about $25 million annually in the United States. By that time, the majority of research and development activities will have reached the practical goal of enabling the industry to proceed with the widespread installation of commercial systems. R&D efforts will then be devoted mainly to improved production efficiency and especially to integrated optics. In other words, the development of integrated optical systems out of discrete components will parallel the progression of the semiconductor industry from discrete devices to highly complex and sophisticated integrated circuits.

Digital transmission will become the dominant mode of data transfer, whereas analog information transfer will become almost extinct. This applies particularly to telecommunications, which at present is still largely analog. Data bus transmission, which refers to data transfer over short links, is also a digital form of communication; however, it is used mainly in computer applications. This is illustrated for 1980 and 1990 in Figure 12.3. The shifting trend toward all-digital optical communications is apparent.

Communications applications represent the dominant use of fiber optics systems, as is evident from Figure 12.4. The relative importance of communications use is growing, whereas computer applications are declining in relative but not in absolute terms. In this illustration the category Other Equipment includes business, consumer, industrial, and instrumentation equipment; the category Other Use includes exports, imports, inventory change, and nonproduction use.

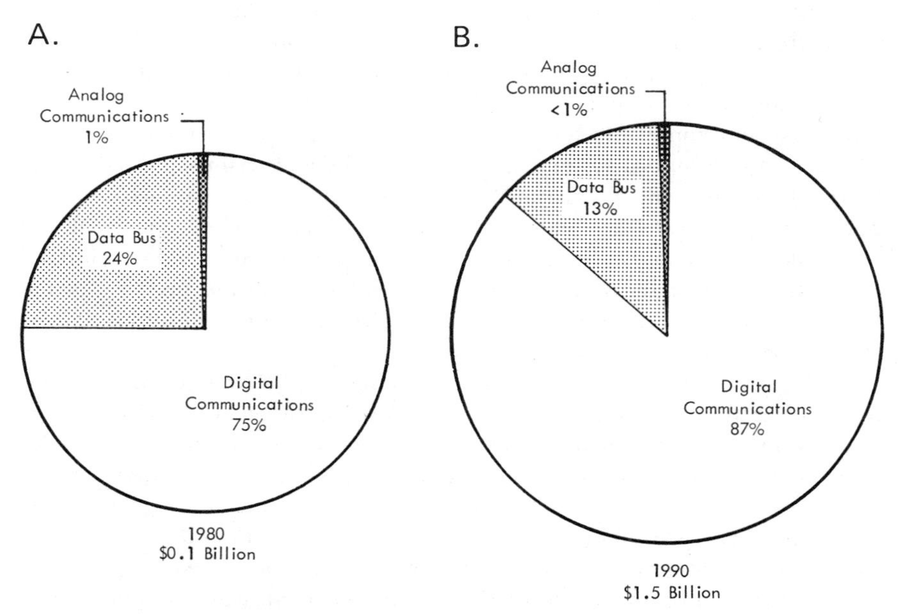

Figure 12.3 Composition of worldwide fiber optics systems market by transmission mode.

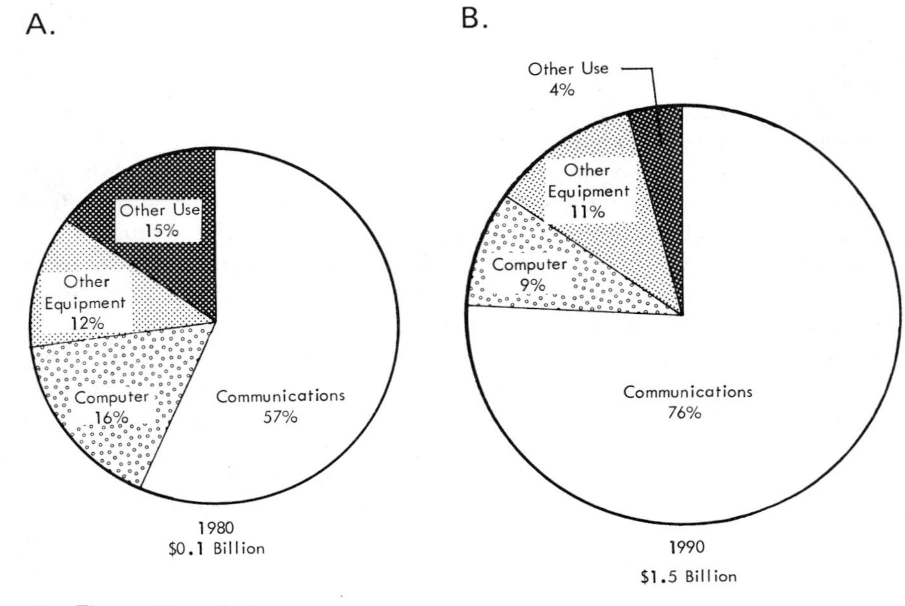

Figure 12.4 Composition of worldwide fiber optics systems market by major applications.

The estimated breakdown of the United States market in 1985 is shown in Figure 12.5. The average distribution of fiber optics systems by components indicates that almost three-quarters of the total value is contributed by fiber cables, whereas the active modules, such as transmitters, receivers, and repeaters, will contribute less than one-quarter. These components are assembled in fiber optics systems in which digital data transmission will be by far the dominant form of data transfer. Optical transmission systems are incorporated further in various types of electronic equipment which consist also of other, nonoptical parts; hence, the total equipment value is substantially larger than that of the purely optical portion. In 1985 the majority of the equipment value will be in computers, followed by communications and by military and other Government equipment.

The largest production value of fiber cables applies to multifiber cables, whereas single-fiber cables constitute less than 10% of the total U.S. fiber production, as is evident from Figure 12.6. Step-index multimode (SI–MM) fibers will continue to be the prominent type, but eventually they will be superseded by graded-index (GI) fibers in terms of total value. Step-index single-mode (SI–SM) fibers will retain their secondary importance, except in special applications which demand extremely high bandwidths.

Below, estimated trends of the various characteristics of fiber optics communications systems are shown in summarized form for the 1970–1990

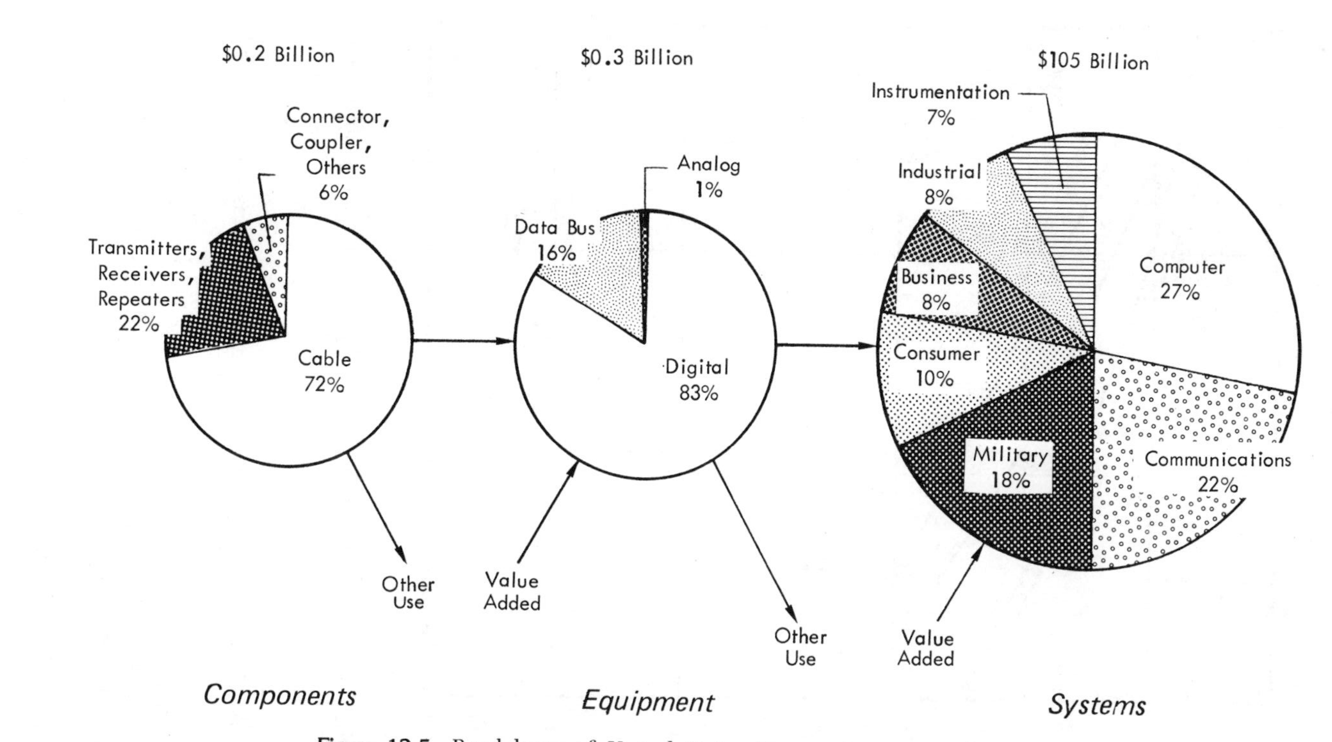

Figure 12.5 Breakdown of United States fiber optics systems by transmission mode and major applications (1985).

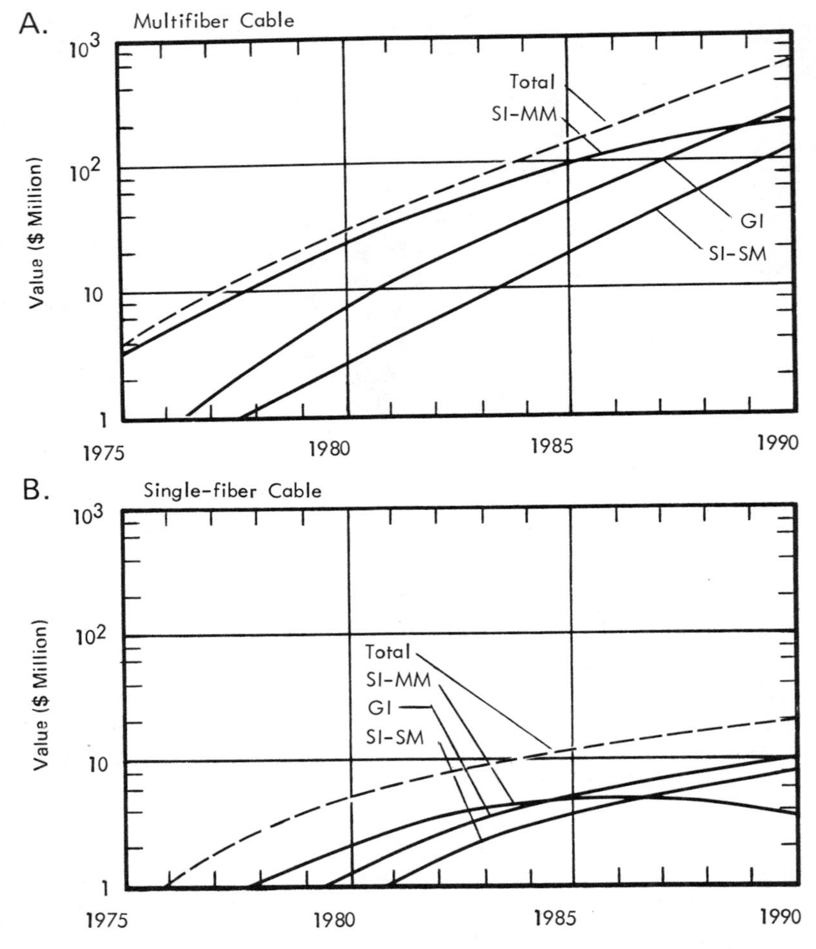

Figure 12.6 Estimated distribution of optical fiber types (United States production).

time period. The information given in these illustrations refers to the following:

Fibers: attenuation, dispersion.

Sources: maximum modulation frequency, mean time to failure, emission angle, optical output power.

Detectors: maximum detection frequency.

Systems: distance between repeaters or length of fiber, maximum data transfer rate, length–bandwidth product.

The trend curves indicate the significant further advances that are expected to be derived from the continued development of optical communications systems. The data and projections given in the following graphs refer chiefly to advanced laboratory systems and not necessarily to production achievements. It is understood that the projections given here are only approximations and do not purport to represent anticipated achievements accurately. Nevertheless, they can serve to familiarize the reader with expected trends in the more general sense.

In addition to the performance constraints that enter the analysis of system cost, the economic constraints imposed by fiber and repeater cost are important design criteria. Although they are usually beyond the control of the system designer, the variation of fiber and repeater cost with data rate has to be considered.

The expected trends in fiber attenuation and dispersion at the two principal wavelengths of interest are shown in Figure 12.7. It is evident that little

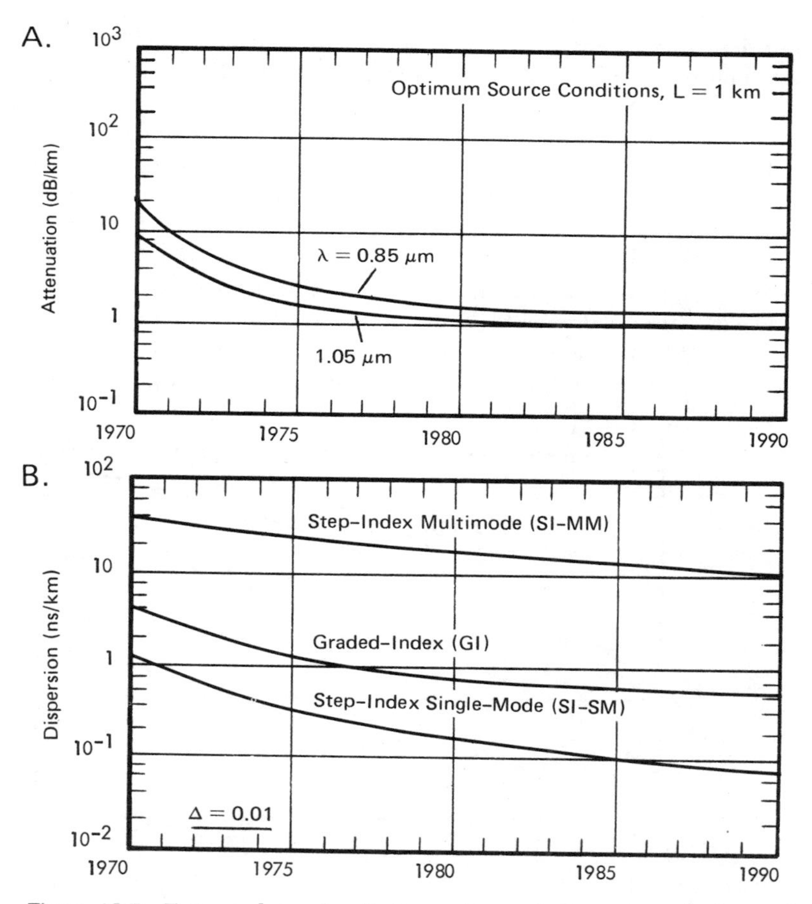

Figure 12.7 Estimated trends of attenuation and dispersion of advanced optical fibers.

improvement in loss can be expected during the next years, partly because the theoretical limit has almost been reached, at least in laboratory fibers. Hence, in the future more attention will be devoted to the achievement of this goal in production fibers and to combination with dispersion and other improvements.

Fiber dispersion of slightly less than about 1 ns/km can be considered to be achievable with graded-index fibers, whereas small-core monomode step-index fibers will allow the attainment of a dispersion that is about one order of magnitude less.

The optical power obtainable from the various optical sources will also display substantial advances in the future, as is evident from Figure 12.8. The most promising sources will continue to be the several versions of semiconductor lasers, whereas nonsemiconductor lasers suitable for fiber optics applications will remain low in power output. Other nonsemiconductor laser types are capable of significantly higher output power, but their dimensions do not permit their use in conjunction with optical fibers. The optical output power of light-emitting diodes falls between that of semiconductor and that of nonsemiconductor lasers.

Another property of optical sources is the frequency at which they can be modulated; this is of major influence on the maximum data transfer rate that can be achieved by a system. Again, laser sources are superior to most other types of optical generators. Light-emitting diodes are largely restricted to frequencies below about 0.1 GHz, whereas other optical sources are expected to allow frequencies in excess of 1 GHz. This is illustrated in Figure 12.9.

Because the performance limit of any system is determined by its weakest link, the frequency response of other components of an optical communications system must also be considered. Although optical detectors are

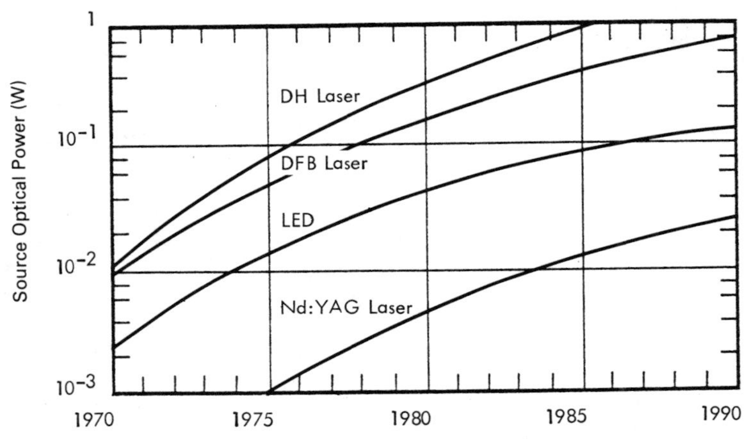

Figure 12.8 Estimated trends of optical power output of various optical sources at optimum wavelength.

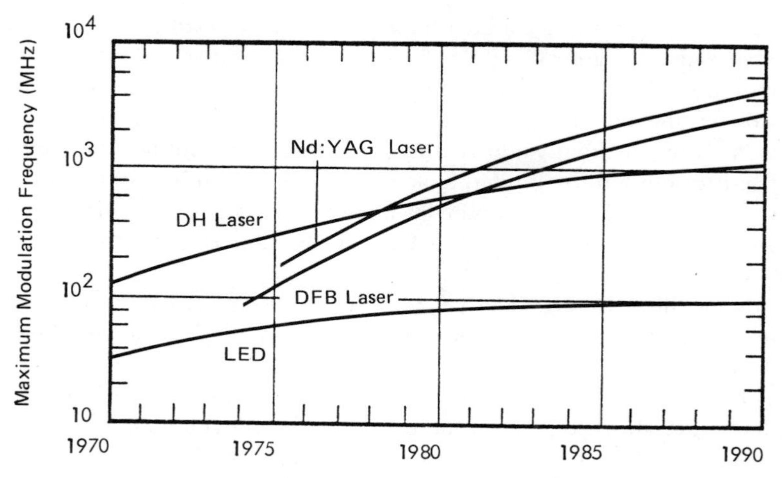

Figure 12.9 Estimated trends of modulation frequency of prominent optical sources.

normally less restricting than optical sources as far as their frequency response is concerned, they represent a limitation, particularly at longer wavelengths, whose sensitivity decrease is inversely proportional to the wavelength. Avalanche photodetectors, although also wavelength dependent, yield substantially better performance than PIN and p-n detectors. This is illustrated in Figure 12.10.

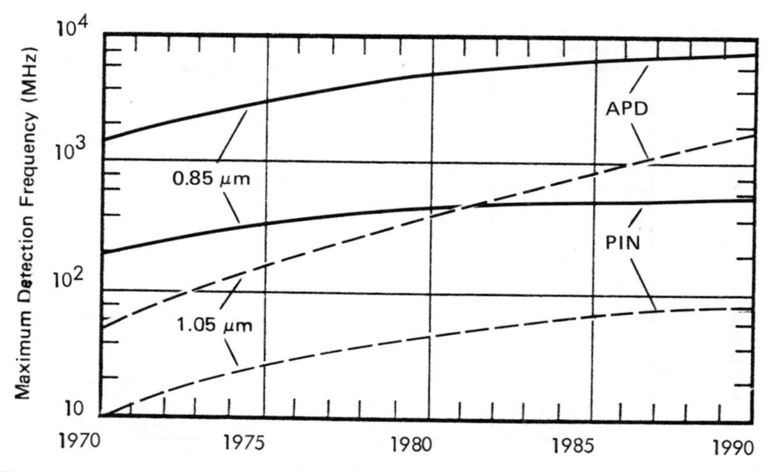

Figure 12.10 Estimated trends in detection frequency of prominent optical detectors.

The projected trends of data transfer rates expected for optical fibers are summarized in Figure 12.11. It is assumed that the most suitable optical sources are being used and that the fibers have an index parameter $\Delta = 0.01$. Obviously, the best performance can be anticipated for a monomode step-index fiber in which only one mode can propagate. In contrast, the multimode step-index fiber will be limited to about 100 Mb/s even by 1990. The graded-index fiber will allow a data transfer rate of about 3 Gb/s and the monomode step-index fiber will be suitable for transmission at a rate of more than 10 Gb/s.

Using these trends, it is possible to arrive at reasonable estimates of future advances in the achievement of increased fiber length-bandwidth products. This is illustrated in Figure 12.12, where a distinction has been made between advanced R&D systems and systems that can be fabricated in a production environment. With increasing system development maturity, the performance differences between R&D and production systems will become smaller. Whereas in 1975 an upper length–bandwidth product of about 3 GHz km was considered to be state-of-the-art, in 1990 this will have increased to about 50 GHz km. It must be appreciated that in early systems, mainly LEDs were used as optical sources, whereas toward the end of the next decade Nd:YAG nonsemiconductor lasers will be employed, with semiconductor lasers being incorporated in practical systems during most of this time.

As a result of the achievable system bandwidth, the distance between repeaters or the total system length of a system that does not contain repeaters can be estimated. This is shown in Figure 12.13 for two specific data transfer rates. Generally, the gap between R&D and production systems tends to decrease with time. Thus, by 1990 it will be possible to achieve a distance

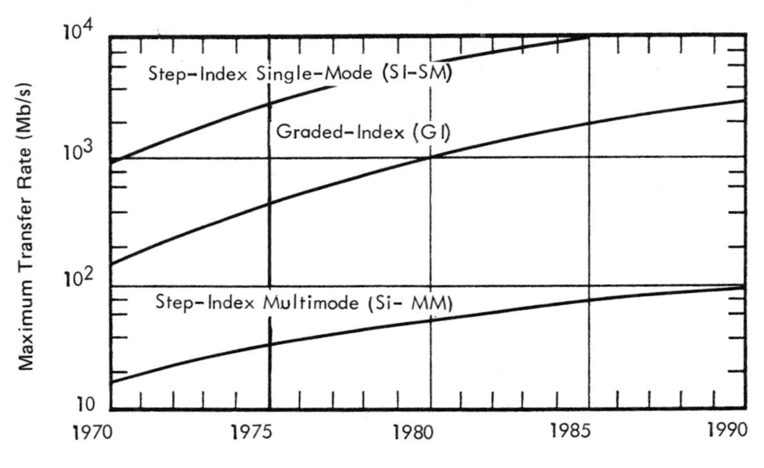

Figure 12.11 Estimated trends in maximum data transfer rate of optical fibers (assuming mode conversion and optimum source conditions; $L = 1$ km).

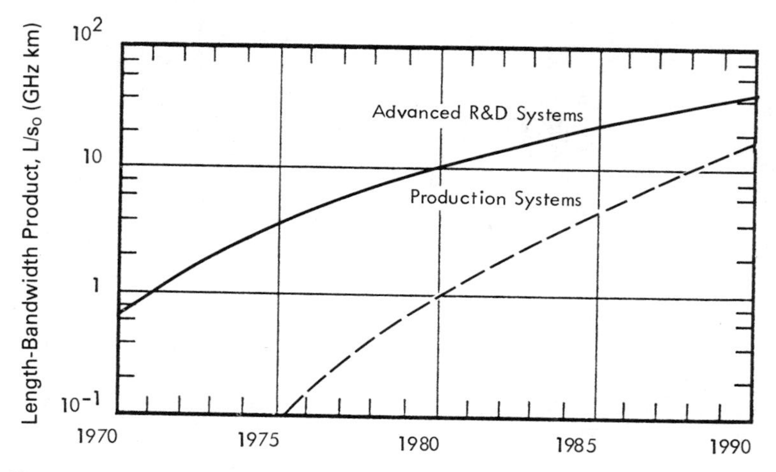

Figure 12.12 Estimated trends in maximum fiber length–bandwidth product.

between repeaters exceeding 100 km for advanced systems and of 50 km for typical systems, assuming a transfer rate of 100 Mb/s.

In a first approximation, the fiber cost (per unit length) can be considered to be independent of data rate. However, with maturing manufacturing process, the fiber cost will be a function of the data rate, since the more expensive monomode fiber is capable of a higher data rate than the less expensive multimode fiber; furthermore, a graded-index fiber of low dispersion is more expensive than a multimode step-index fiber of much higher signal dispersion.

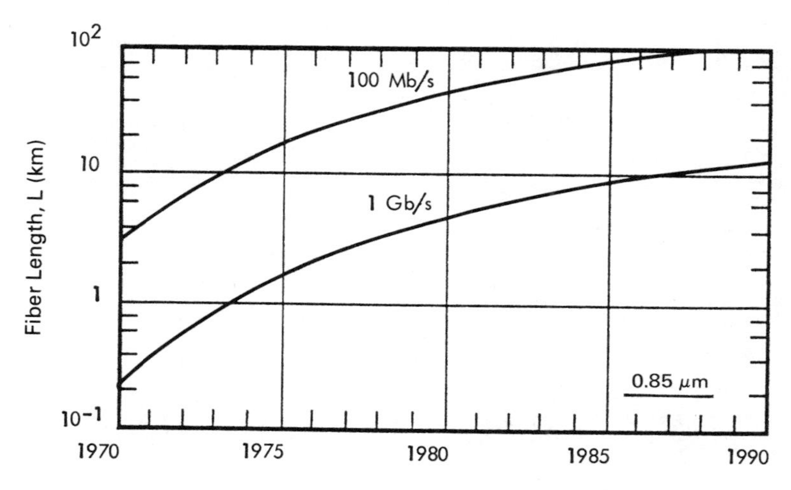

Figure 12.13 Estimated trends in fiber optics system length (distance between repeaters) under optimum conditions.

Appendix

ABBREVIATIONS

ac	Alternating current
A/D	Analog-digital
AGC	Automatic gain control
AM	Amplitude modulation
APD	Avalanche photodetector diode
B	Bandwidth
CATV	Community antenna television
CCD	Charge-coupled device
CER	Cost-estimating relationship
CVD	Chemical vapor deposition
cw	Continuous wave
dc	Direct current
DFB	Distributed feedback (laser)
DH	Double-heterostructure, double-heterojunction (laser)
DLD	Dark-line defect
DPPM	Differential pulse position modulation
EAPD	Electroabsorption avalanche photodiode
EDP	Electronic data processing
EHI	Equalized high input impedance
EMI	Electromagnetic interference
EMP	Electromagnetic pulse
E/O	Electronic-to-optical
EUAC/B	Equivalent uniform annual costs and benefits
EVA	Ethylene vinyl acetate
FDM	Frequency-division multiplexing
FET	Field-effect transistor
FM	Frequency modulation
GI	Graded index
HJ	Heterojunction (laser)
IC	Integrated circuit
ICN	Input capacitance neutralized preamp/detector
IF	Intermediate frequency
ILD	Injection laser diode
IM	Intensity modulation
I/O	Input-output
IR	Infrared
IRR	Internal rate of return
ISF	Improved shunt feedback
LCC	Life-cycle cost
LED	Light-emitting diode
LOC	Large optical cavity (laser)
LPE	Liquid-phase epitaxy
MARR	Minimum attractive rate of return
MCVD	Modified chemical vapor deposition
MM	Multimode
MOS	Metal-oxide-silicon device

MTTF	Mean time to failure
MTTR	Mean time to repair
NA	Numerical aperture
Nd:YAG	Neodynium-doped yttrium-aluminum-garnet (laser)
NEP	Noise equivalent power
NPV	Net present value
NRZ	Non-return to zero
OC	Optical cavity (laser)
OD	Outside diameter
O/E	Optical-to-electronic
O&S	Operating and support
PAM	Pulse amplitude modulation
PCM	Pulse-coded modulation
PCVD	Plasma-activated chemical vapor deposition
PM	Pulse-position modulation
POD	Plasma outside diffusion
PPM	Pulse-position modulation
RC	Resistance-capacitance product
R&D	Research and Development
RDT&E	Research, development, test, and evaluation
rf	Radio frequency
RFI	Radio frequency interference
rms	Root mean square
ROI	Return on investment
RZ	Return to zero
SCH	Separate-confinement heterostructure (laser)
SDM	Space-division multiplexing
SH	Single-heterojunction (laser)
SI	Step index
SIR	Savings-investment ratio
SLD	Superluminescent diode
SLR	Single-lens reflex
SM	Single mode
S/N	Signal-to-noise ratio
SSF	Standard shunt feedback
TDM	Time-division multiplexing
TDR	Time domain reflectometry
TNEP	Total noise equivalent power
TTL	Transistor-transistor logic
VAD	Vapor-phase epitaxial deposition
VCR	Video cassette recorder
VLSI	Very-large-scale integration
VPE	Vapor-phase epitaxy
WDM	Wavelength-division multiplexing

SUBJECT INDEX

aberration, chromatic, 445
absolute cost analysis, 305
absorption coefficient, variation of, 212
absorption loss
  in fibers, 46–47, 76–77
  of glasses, 78–81
absorption spectrum of glass, 45
airborne electronics systems, 375
Airborne Light Optical Fiber Technology, *see* ALOFT cost model
Al content and lifetime, 195–196
ALOFT cost model, 345
amplifier, 19
amplitude-dependent distortion, 406, 408
analog approach to CERs, 323
analog communications in Japan, 492–494
analog information transfer, 515
analog operation vs. digital operation, 380–381
analog receiver, 231
analog systems, 393, 394
analogy judgment method, 326
angular misalignment, coupling loss due to, 246, 261–263
APD, *see* avalanche photodetector
Appendix, 525–526
army field communications, 375–376
arthroscope, 459, 461
  Selfoc, 459
assembly cost, 338–339
attenuation, signal, 5, 58
  estimated trends of, 519–520
  fiber, *see* fiber attenuation
  in glass, 43–47
  mechanisms of, 75–76
  system aspects of, 387
automotive electronics, 372
avalanche photodetector (APD), 32–33
  advantages and disadvantages of, 227
  characteristics of, 391–392
  commercial, 225
  electroabsorption, 221–222
  new materials for, 222–224
  p-n junction, 217–221
  Schottky barrier, 221
  silicon, 229–230
avionic fiber optics systems, 375

bandwidth, system, 385–386
  fiber length and, 423–424
bandwidth capacity, 331
bandwidth increase of digital operation, 381
bandwidth product, signal-to-noise ratio, 398–401
battle damage survival, 331–332
beam merging, efficient, 189
beam splitters, couplers using, 280
beat length, 281
benefit, equivalent uniform annual, 320
benefit-cost ratio, 318–319
benefit identification, cost-, 321–332
benefits of optical communications systems, 330–332
bifurcated fiber bundle, 277–280
biopsy sampling with endoscopy, 454
body, human, portions observed by endoscopes, 452
borosilicate glass, 99
"bottom-up" model, 346–350
Bragg diffraction, anisotropic, 473
Bragg reflectors, 188
breakdown voltage, 220
brillouin scattering, 78
bronchofiberscope, 455–456, 457
bulkhead connectors, 364
bundles, fiber, *see* fiber bundles
buried heterostructure (BHS) device, 389
Burrus LED, 163, 165
business electronics, 368
butt coupling, 243, 271, 486

cable-related losses, 268–271

cable strength, 113–123
cable television, 370
cable types, communications, 4–7
cabling requirements, 108–113
carrier diffusion, 182
channeled substrate planar (CSP)
device, 389
charge-coupled devices (CCDs), 205
chemical vapor deposition fabrication
processes for waveguides, 388, 389
chip placement, improper, 250–251
chromatic aberration, 445
chromatic dispersion, 386
circuit limitation, 417
circuits, hybrid, 359
cladding, 82, 429
coaxial cables, 4–8
channel capacities of, 5
system cost of, 342–346
coding, line, redundancy by, 382
cold light supply in endoscopy, 450–451
colonofiberscope, 435, 454–455, 456
colonoscopic polypectomy, 456–457
commercial applications of optics,
312–314
communications
optical, *see* optical communications
optical systems advantages in, 425
communications cable types, 4–7
communications systems, optical, *see*
optical communications systems
communications transmission media,
characteristics of, 10–11
comparative cost analysis, 305
component costs, trends in, 314–316
components, optical communications,
xi, 357, 362–365
active, 357, 359–362
design criteria of, 17–18
in Europe, 485–487
passive, 33–34
selection criteria for, 20–23
computer systems, 370–372
optical systems advantages in, 425
connection requirements, 123–124
connections, optical, xi
system aspects of, 392
connector-related losses, 253–269
connectors, optical, 241–272, 357,
364–365
future of, 467
consumer areas, optical systems
advantages in, 425
consumer electronics, 372–373
contingency analysis, 309
continuous series present worth,
319–320
core bending, 84–85
core diameter, loss due to mismatch in,
269–270
correlation coefficient, 324
corrugated-waveguide filters, 472
corrugation period, 185
cost
assembly, 338–339
coaxial cable system, 342
equivalent uniform annual, 320
fiber, 419, 523
inspection, 339
maintenance, 341–342
manufacturing, 340
materials, 337
overhead, 339
procurement, 341
production setup, 338
quality control, 339
repeater, 420
special handling and processing, 338
support and test equipment, 340
system aspects of, 419–423
cost analysis, 305–306
cost analysis uncertainty, 333–334
cost-benefit analysis, 310
cost-benefit identification, 321–332
cost-effectiveness models, 310
cost elements selected for sample
analysis, 336
cost-estimating relationships (CERs),
322–325
cost-estimating uncertainty, 333
cost estimation methods for optical
communications system, 316–321
cost factors, life-cycle, 341
cost model, manufacturing technology,
337–340
cost-performance ratio, 380
cost reduction
of digital operation, 381
with production increase, 329
cost sensitivity analysis, 335
cost trends, 380
couplers, optical, xi, 241–242, 272–289,
357, 365
directional multimode, 486

star, *see* star coupler
T, *see* T coupler
using beam splitters, 280
coupling, 131–133
coupling efficiency, 141–143, 243–244
misalignments and, 246–248
coupling length, 67
coupling loss
due to angular misalignment, 246, 261–263
due to lateral misalignment, 247
due to longitudinal misalignment, 247
extrinsic, 126, 128, 129
intrinsic, 125–126, 127
coupling techniques in Europe, 479–481
crack propagation, 114–115
crosstalk, 4
polarization and, 293–294
in switches, 291–292
current densities in optical sources, 195

dark line defects (DLD), 198
with fiber bundle, 443
dark noise, sources of, 397
dark spot with fiber bundle, 443
data bus transmission, 515
data communications systems, business, 369
data processing, electronic (EDP), 370–371
data transfer rate
maximum, 39
estimated trends in, 522
systems cost and, 422
data transmission, guided vs. unguided, 2–3
degradation, signal, 14
degradation tests, 483
delay distortion, 4, 25
Delphi method judgmental technique, 327
demagnification, image, 249–250
depletion-layer photodiodes, 211–217
detection frequency, estimated trends in, 520–521
detector connection, 144
detector noise, 398–400
detector noise limitations, 24
detector nonlinearities, 409
detector quantum efficiency, 385
detectors, optical, xi, 19, 32–33, 203–237
advantages and disadvantages of, 227
basic principle of, 204
characteristics of, 205–206, 226
comparison of, 226–227
criteria for, 204
depletion-layer photodiodes, 211–217
electroabsorption avalanche photodiode, 211–222
in Europe, 484, 485
high-sensitivity, 231
noise characteristics of, 225
noise limitations of, 233–237
nonsemiconductor, 205
p-n junction avalanche photodiode, 217–221
performance of, 225–232
photoconductors, 209–211
PIN, *see* PIN detectors
PN, *see* p-n photodiode
Schottky barrier APD, 221
selection criteria for, 22–23
semiconductor, 205
slow-tail frequency response, 228
speed of, 228
system aspects of, 391–392
types of, 206–225
United States activities in, 471
deterioration, signal, 3
diathermic snares, 457, 458, 459
diffractive switch, 298–299
diffusion, carrier, 182
digital communications in Japan, 490–494
digital long-haul telecommunications link
example of economic analysis, 340–354
digital operation vs. analog operation, 380–381
digital receiver, 231
digital systems, 393–396
digital transmission, 515
diodes
injection laser, *see* injection laser diodes
light-emitting, *see* light-emitting diodes
superluminescent (SLD), 31
direct modulation, 192–193
direct photodetection, 416–417
directivity parameter, 142, 143
discount rate, 320

dispersion
  chromatic, 386
  estimated trends of, 519–520
  fiber, 385–386
  intramodal and intermodal, 86–92, 94
  material, 24, 37, 387, 409
  mode, 59–60, 386–387, 410
  multimode, 37
  in optical fibers, 27
  signal, 44
  system aspects of, 387
  waveguide, 24, 387
  wavelength, 59–60
dispersion limitations, 402–405
  system aspects of, 416
dispersion-limited operation, 418
dissector, image, 432
distance between repeaters, 36–37, 342–343, 403–405
distortion
  amplitude-dependent, 406, 408
  delay, 4, 25
  frequency-dependent, 406–408, 409
  signal, 12, 14
distortion limitations, 405–410
distributed-feedback (DFB) semiconductor lasers, 183–187, 503–504
dome LED, 163–164
double-crucible technique, 102–104, 109
double-heterojunction laser, 173, 177–180

economic analysis
  objective of, 306
  system aspects of, 414–427
economic analysis procedure, 306–309
economic analysis sequence, 309–310
economic life, 308–309; *see also* lifetime
economic techniques, judgmental, 326–328
edge-emitting source, 154
edge LED, 163, 164–165
electrical interference, 372, 373, 375
electroabsorption avalanche photodiode, 221–222
electromagnetic compatibility, 378–379
electron population inversion, 174–175
electronic data processing (EDP), 370–371
electronic switching, 290
electronics
  business, 368
  consumer, 372–373
  industrial, 373–374
  military, 374–376
  office, 368
emission wavelength, variation of peak, 156
end separation loss, 263–265
endface tilt loss, 266–267
endfinish, losses due to imperfect, 268
endoscopes, fiber optics, 445–463
  arthroscope, 459, 461
  bronchofiberscope, 455–456, 457
  colonofiberscope, 435, 454–455, 456
  gastrocamera with fiberscope, 451–453
  gastrointestinal, 453–454, 455
  industrial, 461–463
  laparoscope, 459, 460
  medical flexible, 451
  medical rigid, 458–461
  portions of human body observed by, 452
  structure of flexible, 445, 448–451
endoscopic surgery, 458
endoscopy, xii, 429–463
engineering approach to CERs, 323
environmental resistance, 48
equivalent uniform annual cost or benefit, 320
error rate and signal-to-noise ratio, 233
Ethernet concept, 368
Europe, optical communications activities in, 473–488
  coupling techniques, 479–481
  detectors, 484–485
  fiber-fiber coupling, 479–481
  fiber technology, 475–476
  fiber theory, 476–477
  field trials, 487–488
  injection laser diodes, 482–484
  integrated optics, 485–487
  laser-fiber coupling, 481
  LEDs, 482
  optical components, 485–487
  references, 505–511
  sources, 481–484
evanescent wave coupling, T couplers using, 280–282
excess loss, 276, 287

experience curve, 328–330
experience curve method, 327–328

fabrication processes for waveguides, 388, 389
fabrication techniques for optical fibers, 101–108, 109
Fabry-Perot cavity, 171–172
facet damage, laser, 196–200
  catastrophic, 196–198
  gradual, 198–200
failure, mean time to (MTTF), 199–200
  improved, 351–352
failure mechanisms of optical sources, 194–200
fatigue, static, 114
fiber attenuation, 109–110, 470
  measurements of, 145, 146
  mechanisms of, 46–47
fiber breakage, 108–109
fiber bundles
  bifurcated, 277–280
  classification of, 430–432
  conical coherent, 432
  converter, 432
  dark line, 443
  dark spot, 443
  defects, 441–443
  examples of, 431
  image quality, 439–443
  incoherent, 432
  line resolution, 440–441
  loss at total reflection, 438–439
  nonuniformity, 443
  packing fraction, 435
  requirements, 432–443
  resolution, 439–441
  spectral transmittance, 435, 437, 438
  spot resolution, 440, 441
  transmission through, 430–433
  uniform coherent, 430
  void, 443
fiber cables, categories of, 357, 362–364
fiber cabling, 108–123
fiber characterization in Europe, 477–479
fiber-component connections, 133–144
fiber core, 49–50
fiber cost, 419, 523
fiber-detector connection, 271–272
fiber-detector connection losses, 392
fiber-detector coupler, 16
fiber diameter test arrangement, 148
fiber dispersion, 385–386
fiber dispersion measurements, 145, 147
fiber fabrication, 96–108
  rod-in-tube method for CVD, 496, 497–498
  vapor-phase axial deposition, 495–496
fiber-fiber connection losses, 392
fiber-fiber connections, 123–133, 252–271
  multiway, 480
fiber-fiber coupling, 34
  in Europe, 479–481
fiber-fiber coupling efficiency, 385
fiber geometry, 84–85
fiber geometry measurements, 145, 147, 149
fiber index difference, 93–95
fiber index profiles, loss due to mismatch in, 270
fiber innovations in Japan, 495–498
fiber length, bandwidth and, 423–424
fiber length-bandwidth product, estimated trends in maximum, 522, 523
fiber losses, 36
fiber measurements, 144–150
fiber nonlinearities, 409
fiber optics endoscopes, *see* endoscopes, fiber optics
fiber optics system length, estimated trends in, 522–523
fiber optics systems, *see also* optical communications systems
  advantages of, 8
  costs of, 345
  drawbacks of, 8–9
  potential of, 7–25
  trends in, 514–515
fiber optics technology, trends in, 310–315
fiber optics utilization, trends in, 312–314
fiber propagation, theoretical aspects of, 471
fiber refractive index, 37–39, 50–51
fiber requirements, 47–49
fiber rupture characteristics, 115–118
fiber signal amplitude, modulation frequency and, 410, 411
fiber strength, 113–123

fiber structures, double- and
triple-coated, 490, 491
fiber technology in Europe, 475–476
fiber theory in Europe, 476–477
fiber types, 49–54
fibers, optical, 25–28
definition of, 1
dispersion in, 27
fabrication techniques for, 101–108, 109
graded-index (GI), 27, 363, 516, 618
multimode (GI-MM), 52–53, 402–403
light path in, 54–57
losses of, 25–26
monomode, 417–418
glass, 479
vs. multimode fibers, 61
step-index, 403
multimode, 402, 417–418
graded-index (GI-MM), 52–53, 402–403
individual modes in, 479
vs. monomode fibers, 61
step-index (SI-MM), 51–52, 402–403, 516, 518
parallel, 282
plastic, 363–364
polymer, 74–75, 97
propagation properties of, 61–67
radiation damage to, 332
scattering loss in, 47, 77–78
silica-based, 96, 475–476
silicate, 96–97
step-index (SI), 27, 363, 429
monomode, 403
multimode (SI-MM), 51–52, 402–403, 516, 518
numerical aperture of, 433, 434
properties of, 57
single–mode (SI-MM), 52
types of, estimated distribution of, 517–518
United States activities in, 470–471
W-type, 53–54
fiberscopes, 429–430; *see also* endoscopes, fiber optics
gastrocameras with, 451–453
field trials in Europe, 487–488
France, optical communications activities in, 468
Franz-Keldysh effect, 222
frequency, variation of data transfer function with, 395
frequency-dependent distortion, 406–408, 409
frequency-division multiplexing (FDM), 412–414
frequency modulation, 410
frequency response
LED, 168
slow-tail, 228
Fresnel reflection loss, 433

gastrocamera with fibroscope, 451–453
gastrointestinal fiberscope, 453–454, 455
general systems performance, 377–396
Germany, optical communications activities in, 468
GI fibers, *see* fibers, optical, graded-index
glass, 466
absorption loss of, 78–81
absorption spectrum of, 45
borosilicate, 99
glass fibers, 96–97; *see also* fibers, optical
monomode, 479
glass rod, transmission through, 444–445
glass systems, low-melting-temperature, 475
Government, optical systems advantages in, 425
graded-index fibers, *see* fibers, optical, graded-index
Great Britain, optical communications activities in, 468
ground-level independence, 379
guard ring photodiode, 216

heterojunction devices, 161, 167
hexagonal fiber packing, 435, 436, 437
Hi-OVIS project, 492–494
high-brightness devices, 162
homojunction laser, 172, 177–180
homojunction structure, 171–172
human body, portions observed by endoscopes, 452
hybrid circuits, 359

ILDs, *see* injection laser diodes
image demagnification, 249–250
image dissector, 432

image formation by rod lenses, characteristics of, 446
image quality of fiber bundles, 439–443
implementation difficulties
  digital systems, 393–396
  optical connections, 392
  optical detectors, 391–392
  optical sources, 388–391
  optical systems, 393–396
  optical waveguides, 386–388
  system attenuation, 385
  system bandwidth, 385–386
indirect modulation, 193–194
indirect photodetection, 417
industrial control, optical systems advantages in, 425
industrial electronics, 373–374
industrial endoscope, 461–463
information transmission needs, total, in United States, 513, 514
injection laser diodes (ILDs), 29, 30–31, 171–189
  direct modulation, 192–193
  in Europe, 482–484
  noise limitations, 396–397
  radiance of, 188–189
insertion loss, 265–268
insertion tube, 450
inspection cost, 339
instrument systems applications, 374
instrumentation, optical systems advantages in, 425
integrated optics, 3
  in Europe, 485–487
  in Japan, 503–504
  in United States, 472–473
"intelligent", use of term, ix, 1
intensity modulation (IM), 401
interference, electrical, 372, 373, 375
internal rate of return (IRR), 321
investment
  equations used for recurring and nonrecurring, 348
  return on, 316–319
ionization coefficients of electrons and holes, 217–220
isoelectronic trap, nitrogen, 158

jacketing, 83–84, 110–113
  effectiveness of, 121
Japan, optical communications activities in, 468, 488–504
  analog communications, 492–494
  digital communications, 490–494
  environment effects on optics activities, 488–490
  fiber innovations, 495–498
  integrated optics, 503–504
  LEDs, 498–501
  optical devices, 498–504
  optical fiber communications, 489–494
    features of, 489–490
    major advances in, 490–494
  optical sources, 498–501
  optical switch, 401–403
  predistortion circuits, 499, 500
  references, 512
  rod-in-tube method for CVD fiber fabrication, 496, 497–498
Johnson noise, sources of, 397
judgment methods
  analogy method, 326
  Delphi method, 327
  experience curve method, 327–328
judgmental economic techniques, 326–328
junction depth, 161, 162

kink in output, 173

Lambertian source, 242–246
laparoscope, 459, 460
large-area low-brightness devices, 162
large optical cavity (LOC) laser, 167, 178–180
laser diodes, *see* injection laser diodes
laser facet damage, *see* facet damage
laser-fiber coupling in Europe, 481
laser-fiber joints, 136–137
laser sources, *see* sources, optical
lasers
  distributed-feedback (DFB), 183-187, 503–504
  injection, *see* injection laser diodes
  low-mesa-stripe, 483
  nonsemiconductor, *see* nonsemiconductor lasers
  semiconductor, *see* semiconductor lasers
  solid-state ion, 29, 31–32
  stripe, 173, 181
lateral misalignment losses, 247, 254–260

leakage current noise, sources of, 397
learning curve theory, 327–330
LEDs, *see* light-emitting diodes
life-cycle cost factors, 341
life-cycle costing (LCC) model development, 347
lifetime, Al content and, 195–196
lifetime data of optical sources, 199–200
lifetime tests, 483
light, spurious ambient, 398
light-emitting diode sources, *see* sources, optical
light-emitting diodes (LEDs), 20–21, 29–32, 160–171
  advantages and disadvantages of, 166
  Burrus, 163, 165
  characteristics of, 155, 165
  direct modulation of, 192–193
  dome, 163–164
  edge, 163, 164–165
  in Europe, 482
  -fiber coupling, 137–138
  frequency response of, 168
  in Japan, 498–501
  modulation frequency of, 168
  noise limitations of, 396
  planar, 163
  radiance of, 168
light path in fibers, 54–57
limits, operational, 19
line coding, redundancy by, 382
line resolution of fiber bundle, 440–441
liquid-phase epitaxy (LPE), 159
longitudinal misalignment, coupling loss due to, 247
Luneberg lens, 473

maintainability, optical communications system, 331
maintenance cost, 341–342
manufacturing cost, 340
manufacturing technology cost model, 337–340
material dispersion, 24, 37, 387, 409
material-related losses, 76–86
materials, semiconductor, 154–159
  characteristics of, 157
materials cost, 337
mean time
  to failure (MTTF), 199–200
    improved, 351–352
  to repair (MTTR), 351
mechanical switching, 290
merging, efficient beam, 189
meridional rays, 54–55
mesa diodes, 214
metal ions, 99–100
  absorption loss due to, 81–83
microbending effects, 477
microbending losses, 266
microcomputers, 372–373
microwave waveguides, 5–8
military, optical systems advantages in, 425
military electronics, 374–376
minimum attractive rate of return (MARR), 321
misalignment angle, 261
misalignments, coupling efficiency and, 246–248
mixing rods, 278–280
mode coupling, 67–74
mode dispersion, 59–60, 386–387, 410
mode mismatch loss, 269
modes, number of, 62–67, 68–69
modified chemical vapor disposition for waveguides, 388, 389
modulation
  direct, 192–193
  indirect, 193–194
  intensity (IM), 401
  pulse-position (PM), 401–402
  system aspects of, 410–414
modulation frequency
  estimated trends of, 520, 521
  fiber signal amplitude and, 410, 411
  of LEDs, 168
money, time value of, 308
monofiber-multifiber connections, 33
monomode fibers, *see* fibers, optical, monomode
Monte Carlo simulation, 335–336
MTTF, *see* mean time to failure
multifiber-multifiber connections, 34
multimode couplers, directional, 486
multimode dispersion, 37
multimode fibers, *see* fibers, optical, multimode
multipin connections, 364
multiplexing, system aspects of, 412–414
multiplication factor, 208

NA, *see* numerical apertures

near-field refractive index
measurement, 478
nitrogen isoelectronic trap, 158
noise
detector, 398–400
factors influencing, 236–237
shot, *see* shot noise
thermal, *see* thermal noise
noise equivalent power, total (TNEP),
208
noise immunity of digital operation, 381
noise limitations, 396–402
of optical detectors, 233–237
system aspects of, 416–417
noise-limited operation, 418
nonmodulated transmission, *see*
endoscopy
nonsemiconductor detectors, 205
nonsemiconductor lasers, 153, 189–192
characteristics of, 155
system aspects of, 390–391
numerical apertures (NA), 28, 141–143
optimized, 48
peak, loss due to mismatch in,
269–270
of step-index fiber, 433, 434

office electronics, 368
OH radical impurity, 100, 476
absorption losses due to, 82–83
operational limits, 19
optical cable, 16
optical cavity (OC) laser, 167, 178–180
optical communications
definition of, 1
optimization of, ix
research and development efforts in,
ix-x
optical communications activities, xii,
465–504
in Europe, 473–488
in France, 468
in Germany, 468
in Great Britain, 468
in Japan, 468, 488–504
overview of, 465–468
in United States, 467, 468–473
optical communications components,
see components, optical
communications
optical communications systems, x; *see
also* fiber optics systems
advantages of, in major applications
areas, 425
analytical techniques for, 311
applications of, 425, 515–517
applications tree of, 13
bandwidth capacity of, 331
battle damage survival, 331–332
benefits of, 330–332
changing factors in, 310–311
commercial, 368–370
commercial applications of, 312–314
comparison of, 2–7
component applications of, 357–365
components of, *see* components,
optical communications
considerations for, 34–40
cost analysis of, 305–306, 345
cost and performance trends in, 380
cost estimation methods for, 316–321
cost of, 419–423
cost per kilometer of, 414
digital vs. analog operation of, 14–15,
380–381
disadvantages of, 332
economic viability of, 3–4
electromagnetic compatibility of, 331,
378–379
evaluations of, 469
experimental, 393–396
field trials in Europe with, 487–488
first-generation, 9
future of, 513–523
ground-level independence of, 379
implementation difficulties with,
382–396; *see also* implementation
difficulties
integrated, 35
Japanese environment relations to,
488–489
long-distance, 426, 427
long-distance applications of, 12
loss tolerance limit of, 85–86
maintainability of, 331
performance aspects of, *see*
performance aspects
principal components of, 378
categories of, 379
prototype of, 466
receiver time delay of, 380
reliability of, 331
safety with, 332
schematic illustration of, 358

optical communications systems, (cont.)
second-generation, 9
short-distance, 426, 427
short-distance applications of, 12
size and weight of, 331
size and weight reductions in, 379–380
spectral range of, 24
summary of possible problem solutions in, 382
system applications of, xi-xii, 367–376
system aspects of, 393–396
system economy of, xi, 305–354
third-generation, 12
total loss of, 384
optical connections, *see* connections, optical
optical connectors, *see* connectors, optical
optical couplers, *see* couplers, optical
optical detectors, *see* detectors, optical
optical devices in Japan, 498–504
optical fibers, *see* fibers, optical
optical power, total, 139–141
optical power output, estimated trends of, 520
optical sources, *see* sources, optical
optical switches, *see* switches, optical
optical switching, *see* switching, optical
optical waveguides, *see* waveguides, optical
optics, integrated, *see* integrated optics
optics systems, *see* optical communications systems
oscillations, relaxation, 483
overhead cost, 339

p-n junction depth, 161, 162
p-n photodiode, 214–215
advantages and disadvantages of, 227
avalanche, 217–221
packing fraction, 135–137, 245
fiber bundle, 435
packing fraction loss, 253–254
performance aspects, xii, 377–427
dispersion limitations, 402–405
distortion limitations, 405–410
general systems performance, 377–396
modulation and multiplexing, 410–414
noise limitations, 396–402
system economy, 414–427
performance trends, 380
photocathode, 205
photoconductors, 209–211
photocurrent, 207
photodetectors, *see* detectors, optical
photodiode, *see* p-n photodiode
photon losses, 175–176
pigtailed devices, 251–252, 272
PIN detectors, 32–33, 213–214
advantages and disadvantages of, 227
characteristics of, 391–392
commercial, 224
quantum efficiency of, 228
planar branching network, 486
planar guard ring photodiode, 216
planar LED, 163
plasma-activated chemical vapor deposition for waveguides, 388, 389
plasma outside diffusion for waveguides, 388, 389
plastic fiber, 363–364
plastic optical waveguides, 74–75
PN detector, *see* p-n photodiode
polarization and crosstalk, 293–294
polymer fibers, 74–75, 97
polypectomy, colonoscopic, 456–457
population inversion, electron, 174–175
power, normalized, transverse displacement and, 249
power distribution, modal, 257
power loss in switches, 291
preamplifier, 231–232
predistortion circuits used in Japan, 499, 500
present value method, 319–320
present worth method, 319–320
procurement cost, 341
production, trends in, 514–515
production increase, cost reduction with, 329
production setup cost, 338
propagation angle, 54, 296
propagation properties of fibers, 61–67
pulse broadening, 387
pulse-coded modulation, 410
pulse dispersion, 37
pulse dispersion mechanisms, 86–95
pulse-position modulation (PPM), 401–402
pulse response measurements, 477–478

quality control costs, 339
quantum efficiency, 207–208
  detector, 385
  external, threshold current density and, 184
  of PIN detector, 228
  source, 385
  variation of, 213
quantum noise, 234–235
  sources of, 397

radiance
  injection laser, 188–189
  LED, 168
  source, 187
radiation damage to optical fibers, 332
radiative losses, 77–78
Raman oscillator, tuned-fiber, 473
Raman scattering, 78
Rayleigh scattering, 77–78, 387
Rayleigh scattering loss, 47
receiver noise limitations, 397
receiver time delay, 380
receivers, 357, 361
redundancy by line coding, 382
reflection, loss at total, 438–439
reflective star coupler, 283, 284
refractive index, fiber, 37–39
  effective
    propagation angle and, 296
    and waveguide thickness, 295
refractive index profile, fiber, 50–51, 478
refractive switch, 297–298
regression diagram, linear, 325
relaxation oscillations, 483
reliability, optical communications system, 331
repair, mean time to (MTTR), 351
repeater cost, 420
repeater noise, sources of, 397–398
repeater spacing, 36–37, 342–343, 403–405
repeaters, 16, 357, 361–362
requirements uncertainty, 333
research and development efforts, 467–468
  trends in, 514–515
resolution, fiber bundle, 439–441
response time of photodetector, 209
responsivity of photodetector, 208
return on investment, 316–319
  internal rate of (IRR), 321
  minimum attractive rate of (MARR), 321
risk, 333–336
rod-in-tube technique, 101–102, 109
  in Japan, 496, 497–498
rod lenses
  characteristics of image formation by, 446
  characteristics of typical, 447

safety of fiber optics lines, 332
savings-investment ratio (SIR), 318
scattering loss in fibers, 47, 77–78
Schottky barrier photodiode, 215–217, 221
scrambling rod, 278–280
Selfoc arthroscope, 459
Selfoc fiber, 103–104
semiconductor detectors, 205
semiconductor lasers
  characteristics of, 155
  distributed-feedback (DFB), 183–187, 503–504
  system aspects of, 388–390
semiconductor materials, 154–159
  characteristics of, 157
sensitivity analysis, 309, 334–336
separate-confinement (SCH) laser, 177–180, 187
shot noise, 234, 235–236
  sources of, 397
SI fibers, *see* fibers, optical, step-index
signal attenuation, *see* attenuation, signal
signal degradation, 14
signal deterioration, 3
signal dispersion, 44
signal distortion, 12, 14
signal shaper-decoder, 19
signal shaper-encoder, 14–16
signal-to-noise ratio (S/N), 12
  bandwidth product, 398–401
  of detectors, 233–237
  error rate and, 233
  of photodetector, 208–209
silica
  fused, 98–99, 470
  vitreous, 98
silica fibers, 96, 475–476
silicate fibers, 96–97
silicon, 206

silicon avalanche photodiodes, 299–230
simulation, Monte Carlo, 335–336
single-heterojunction laser, 172–173, 177–180
single-lens reflex (SLR) optics, 452
single-material waveguide, 104–105, 109
skew rays, 54, 244
SLD (superluminescent diodes), 31
snares, diathermic, 457, 458, 459
soda-lime-silicate, 99
solid-state ion lasers, 29, 31–32
source driver, 16
source-fiber connection, 134–138, 242–252
source-fiber connection efficiency, 385
source-fiber connection losses, 392
source-fiber coupler, 16
source modulation of optical sources, 192–194
source nonlinearities, 406–409
source quantum efficiency, 385
source radiance, 187
source radiation pattern, 139–144
source spectral width, 93
sources, optical, xi, 29–32, 153–200
  characteristics of, 155
  current densities of, 195
  development of, 153–160
  in Europe, 481–484
  failure mechanisms of, 194–200
  injection lasers, 171–189; *see also* injection laser diodes
  in Japan, 498–501
  Lambertian, 242–246
  laser, 20–21; *see also* lasers
  lifetime data for, 199–200
  light-emitting diodes, 20–21, 29–32, 160–171
  nonsemiconductor lasers, 189–192
  requirements of, 16
  source modulation of, 192–194
  system aspects of, 388–391
  United States activities in, 470–471
space-division multiplexing (SDM), 412–413
spatial hole burning effect, 191
special handling and processing cost, 338
spectral range, 24
spectral transmittance, fiber bundle, 435, 437, 438
spectral width, 93
splices, 124–131, 357, 364–365, 480–481
spot resolution of fiber bundle, 440, 441
square fiber packing, 435, 436, 437
star coupler, 34, 131–133, 283–286
  compared with T coupler, 286–289
  generalized, 285–286
  transmissive, 283, 284
statistical approach to CERs, 323–325
statistical regression analysis, 324
step-index fibers, *see* fibers, optical, step-index
stress corrosion susceptibility constant M, 114–116
stripe laser, 173, 181
superluminescent diodes (SLD), 31
support and test equipment cost, 340
surface-emitting source, 154
surgery, endoscopic, 458
switches, optical, xi, 289–300
  definition of, 289–290
  diffractive, 298–299
  in Japan, 501–503
  refractive, 297–298
  two-channel, 290–293
switching, optical, 290
  of signal to N locations, 296–297
system-related loss, 255
systems cost
  data transfer rate and, 422
  normalized, 419–423

T coupler, 34, 131–133, 273–283
  compared with star coupler, 286–289
  schematic diagram of generalized, 275
  using evanescent wave coupling, 280–282
TE propagation mode, 293–295
telecommunications link example of economic analysis, 340–354
telecommunications transmission, long-line transcontinental, 370
television, cable, 370
temperature change sensed by optical fibers, 374
termination procedures, 266
thermal noise, 234, 235
  sources of, 297
threshold current density, external quantum efficiency and, 184
throughput loss, 287
tilt-induced loss, 261

time-division multiplexing (TDM), 412–414
time domain reflectometer (TDR), 149
time value of money, 308
TM propagation mode, 293–295
"top-down" model, 346, 350–352
transceivers, 359
transducers, 359
transfer efficiency, total external, 161
transfer function, variation of data, 395
transmission
  through fiber bundles, 430–443
  through glass rods, 444–445
  nonmodulated, *see* endoscopy
transmission medium, dielectric, 1
transmissive star coupler, 283, 284
transmitter noise limitations, 396
transmitters, 357, 360–361
transverse displacement, normalized power and, 249
transverse junction stipe (TJS) device, 389
triangular fiber packing, 435, 436, 437

uncertainty, cost analysis, 333–334
United States, optical communications activities in, 467, 468–473
  detectors, 471
  integrated optics, 472–473
  optical fibers, 470–471
  references, 504–505
  sources, 470–471
  total information transmission needs, 513, 514
user flexibility of digital operation, 381

v groove, 480
v value, 61–66

vapor-phase axial deposition fiber fabrication method, 496–496
vapor-phase epitaxial (VPE) techniques, 105–108, 109, 159
void with fiber bundle, 443
voltage, breakdown, 220

W-type fibers, 53–54
wave coupling, T couplers using evanescent, 280–282
waveguide dispersion, 24, 387
waveguide thickness, effective refractive index and, 295
waveguides
  microwave, 5–8
  optical, x–xi, 43–150
    attenuation mechanisms of, 75–76
    channel capacities of, 6–7
    definition of, 1
    dispersion mechanisms of, 86–95
    fabrication processes for, 388, 389
    fiber cabling of, 108–123
    fiber-component connections of, 133–144
    fiber fabrication of, 96–108
    fiber-fiber connections of, 123–133
    fiber measurements of, 144–150
    material-related losses of, 76–86
    plastic, 74–75
    propagation aspects of, 47–75
    system aspects of, 386–388
  selection criteria for, 22–23
wavelength dispersion, 59–60
wavelength-division multiplexing (WDM), 412–414
wire cables, twisted, 4, 6–8
  channel capacities of, 5

AUTHOR INDEX*

Adams, M. J., *506*,* *509*
Adams, P. L., *302*
Ainslie, B. J., 476, *506*
Alper, T. A., *355*
Amman, M. C., *509*
Anderson, D. B., *201*, 473, *505*
Arnaud, J. A., *300*
Arnold, G., *509*
Asaya, K., *512*
Ataman, A., 484, *510*
August, R. R., *201*, 473, *505*
Auracher, F., 486, *508*, *511*

Bakker, J., 483, *509*
Barnoski, M. K., 247, *300*, *301*
Baues, P., *510*
Beales, K. J., *505*, *506*
Bean, L. H., *355*
Bedgood, M. A., *508*
Berchtold, K., *238*, *510*
Bergeest, R., *506*
Bergh, A., *200*
Bisbee, D. L., *151*
Black, P. W., *511*
Blum, F. A., *201*
Boers, P. M., *509*
Börner, M., 473, *505*, *508*
Bouille, R., *511*
Brace, D., *511*
Bridger, A., *300*
Brilman, M., *510*
British Post Office Experimental Optical Fiber System, *511*
Brown, L. W., *301*
Bube, R. H., *238*
Bulmer, C. H., 486, *510*
Burrus, C. A., *200*, 471, *505*
Buus, J., *509*

Cameron, K., *511*
Carballès, J. C., *200*, *509*
Carter, R. D., *355*
Caspers, F., *508*
Catania, B., *511*
Chan, K. B., 476, *506*
Chu, T. C., *300*
Clarricoats, P. J. B., *150*, 476, *506*
Collier, M. E., *40*, *355*
Comerford, L. D., 473, *505*
Conradi, J., 204, *237*
Cook, J. S., *300*
Costa, B., 478, *507*
Cross, P. S., 472, *505*
Curtiss, L. E., 429, *464*

Daglish, H. N., *507*
Dakin, J. P., *507*
Dakss, M. L., *300*, *303*
Dalgleish, J. F., *301*
Dalgoutte, D. G., *508*
Dalkey, N. C., 327
Danielmeyer, H. G., *508*
Data for Selfoc Single Microlens, *464*
D'Auria, L., *511*
Davies, D. E. N., *427*
Davies, T. W., *151*
Dawson, R. W., 471, *505*
Dean, P. I., *200*
Debye, P., 476, *506*
DeLoach, B. C., Jr., *200*
de Waard, P. J., *509*
DiDomenico, M., Jr., *151*, 471, *505*
Doerbeck, F. H., *200*
Dworkin, L., *355*

Eden, R. C., *238*
Eickhoff, W., 478, 487, *507*, *508*, *510*
Elion, G. R., 204, *237*
Elion, H. A., 204, *237*
Esposito, J. J., xi, 241–303
Ettenberg, M., *200*
Eve, M., 477, 478, 479, *507*, *508*, *511*

*Italicized numbers—Authors cited in a reference section.

Faulstich, M., *506*
Feldmann, J., *511*
Fenton, K. J., *300*
Fleischer, G. A., *355*
Foschini, G. J., *427*
French, W. G., *151*, *504*
Friedrich, H. R., *301*
Fruhmorgen, P., *464*
Fujita, H., *302*
Fulenwider, J. E., *41*, *427*
Fussganger, K., *511*

Gallawa, R. L., *40*, *41*, *150*, *300*, *427*, 471, *505*
Gallenkamp, W., *511*
Gambling, W. A., *302*, 475, 476, 477, 478, *506*, *507*, *508*
Geckeler, S., *151*, *506*, *507*
Geittner, P., *506*
Giallorenzi, T. G., *302*
Glazer, E., *464*
Gloge, D., *150*, *151*, 256, 257, 258, *300*, 471, 481, *504*
Goell, J. E., *41*, 472, *505*
Goldmann, G., *510*
Goodale, R. L., *464*
Goodfellow, R. C., 482, *509*
Goodwin, A. R., *509*
Gordon, T. J., 327
Greenwell, R. A., xi, 305–355
Guekos, G., *510*
Gulati, S. T., *151*
Guttmann, J., *507*, *508*

Hammer, J. M., *201*, 298, *303*, 473, *505*
Hansel, P., 480, *508*
Hansell, C. W., 429, *463*
Harth, W., 483, *509*
Hartog, A. H., 479, *507*
Hasegawa, H., *300*
Haupt, H., *510*
Hawk, R. M., *151*, 269, *300*
Hawkes, T., *301*
Hayashi, I., *200*, *512*
Heinen, J., 487, *511*
Heinen, L., *509*
Helmer, O., 327
Hensel, P., 481, *508*
Hersee, S. D., *509*
Hess, K., *510*
Heyke, H. J., 477, *507*
Hill, D. R., *511*
Hill, K. O., 283, *302*
Hillerich, B., 477, *507*
Hirschman, W. B., *355*
Hirschowitz, B. J., 429, 451, *464*
Hockham, G. A., *150*, 473, 476, *505*
Hondros, D., 476, *506*
Hopkins, H. H., *464*
Hopkins, R. E., *464*
Hotate, K., *152*
Howard, A. Q., Jr., *151*
Hsieh, J. J., 223, *238*
Hsu, H. P., *302*
Hudson, M. C., 283, *302*
Hurwitz, C. E., 223, *238*

Ida, S., *201*
Iga, K., *152*
Ikeda, K., 455, *464*
Inada, K., *151*
Ishikawa, R., *150*
Ito, T., *201*
Izawa, T., *151*, *512*

Jäckel, H., *510*
Jacobsen, A., *427*
Jacques, A., *511*
Jenkins, F. A., *464*
Jeunhomme, L., 477, 478, 486, *507*, *511*
Jocteur, R., *508*, *511*
Johnson, R. J., *355*
Jones, C. R., *355*
Jones, M. W., *151*, 475, *505*

Kaiser, P., *151*, 470, 471, *504*
Kaminow, I. P., *302*, *303*
Kaneda, T., *238*
Kao, K. C., *40*, *150*, *151*, *355*, 473, 475, 476, 477, *505*, *507*
Kapany, N. S., *301*, 429, *464*
Kashiwagi, H., *301*
Kawahata, M., *512*
Kawakami, S., *150*
Kawasaki, B. S., 283, *302*
Kayoma, M., *427*
Keck, D. B., *151*, 471, *505*
Khoe, G. D., *508*
Killinger, G., *427*
Kimura, T., *512*
Kingsley, J. D., *300*
Kingsley, S., *427*
Kirby, P. A., 483, *509*
Kirchhoff, H., 476, *506*

Kita, H., 429, *464*
Kmenta, J., *355*
Knoblich, E. W., *355*
Kobayashi, S., *150*
Kogelnick, H., *300*, 472, *505*
Kokubun, Y., *152*
Kressel, H., *200*, *201*
Krumpholz, O., 238, 473–488, *507*, *508*, *510*
Kruse, P. M., *201*
Kuhn, L., *303*
Kunze, D., *508*
Kuppers, D., *151*
Kuppers, K., 476, *506*
Kuwahara, H., 280, *301*, *302*
Kuyt, G., *508*

Ladany, I., *201*
Lamm, H., 429, *464*
Laybourn, P. J. R., *507*
Lebduska, R. L., *151*, 471, *505*
Lee, A. B., *151*, 279, *301*
Lewin, L., *200–201*
Li, T., *427*
Liu, Y. S., *200*
Logan, R. A., *238*
Loh, K. W., *303*
Love, R. L., *151*, *301*, *302*
Lynch, W. T., 215, *238*

Mabbit, A. W., 482, *509*
McCartney, R. L., *300*, *301*
MacChesney, J. B., 470, *504*
McCormick, A. R., *300*
McGrath, J. M., *355*
McIntyre, P., *238*, *302*
Marcatili, E. A. J., *150*, *300*, *302*, 472, *504*, *505*
Marcuse, D., *150*, 246, 248, 259, 260, 262, *300*, 471, *504*, *505*
Maslowski, S., xii, *150*, *238*, 473–488, *510*
Matsushita, K., *464*
Matt, H. J., *511*
Maurer, R. D., 47, *150*
Mazo, J. E., *427*
Medved, O. B., *427*
Melchior, H., 204, *237*, *238*, 472, *505*
Melngailis, I., *200*
Mettler, S. C., 259, *300*
Metz, S., *238*
Meyer, W., *511*
Michetti, L. *511*
Midwinter, J. E., *507*
Miller, B. I., 471, *505*
Miller, C. M., 255, 256, 257, 259, *300*, *301*
Miller, S. E., *40*, 47, *150*, *302*, 469, *504*
Milton, A. F., *151*, 279, *301*, *302*
Misugi, T., *238*
Mitchell, D. J., *151*
Miyazaki, K., *300*
Möckel, P., *509*
Montgomery, J. D., xi–xii, *40*, *150*, *355*, 357–376
Morokuma, T., xii, 429–464
Mouthaan, K., *510*
Mühlich, A., 475, *506*
Müller, J., 484, 487, *510*
Murata, H., *151*

Nagai, H., *200*
Nakahara, T., xii, *151*, 488–512
Nakamura, M., *200*, *201*, *512*
Nakayama, O., *301*
Nath, G., *464*
Nelson, A. R., *303*
Neumann, E. G., *508*
Newns, G. R., *506*
Noguchi, Y., *200*
North, J. C., *507*
Nuese, C. J., 183, *200*

Oestreich, U. H. P., *151*
Okoshi, T., *152*
Ono, T., *512*
Otsuka, K., *201*
Owaga, K., 283, *302*
Ozeki, T., *201*, *302*

Pace, L. J., *151*
Panish, M. B., 471, *505*
Papuchon, M., 485, *510*
Payne, D. N., 475, 476, 479, *506*, *507*
Pearsall, T. P., 223, *238*, *510*
Personnick, S. D., 238, 471, 472, *505*
Petermann, K., 473–488, *506, 507*
Peters, 429
Philco-Ford/IBM, *355*
Pike, J. N., *464*
Pinnow, D. A., *150*, 470, *504*
Pocholle, J. P., 478, 486, *507*, *511*
Proebster, W., 487, *511*

Racki, J. G., *301*
Ramaswamy, V., *303*
Ramsay, R. M., *302*
Randall, E. N., *151*
Ravencroft, I. A., *511*
Rawson, H., 475, *506*
Reiber, L. M., *510*
Reymond, J. C., *301*
Ripoche, G., *510*
Ritchie, S., *509*
Russer, P., 483, *509*, *510*

Saito, K., *200*
Sakakibara, Y., *512*
Sakurai, S., *512*
Salz, J., *427*
Sammut, R. A., 477, *507*
Samson, J., *427*
Sandbank, C. P., *40*
Saxena, A. N., xi, 203–240
Schicketanz, D., *301*, 478, *507*, *509*
Schroeder, J., *151*, 470, *504*
Schulz, S., 483, *509*
Schumacher, W. L., *301*
Scifres, D. R., *200*
Scott, B., 475, *506*
Shannon, J. M., *238*
Shinya, H., 456, *464*
Simon, A., *511*
Sladen, F. M. E., 478, *507*
Slaughter, R. J., *511*
Smith, D. A., *200*
Smith, R. G., 471, *505*
Snoeren, R. M., *510*
Snyder, A. W., *151*, *302*
Soares, O. D. D., 486, *510*
Someda, C. G., *151*
Sordo, B., 478, *507*
Soref, R. A., 298, *303*
Special Issue on Optical
  Communication, *41*
Standley, R. D., *303*
Steinberg, R. A., *302*
Steventon, A. G., 483, *509*
Stewart, C., 486, *511*
Stewart, W. J., *152*, 486, *506*, *507*, *511*
Stillman, G. E., 204, 222, *237*
Stolen, R. H., 473, *505*
Stoll, H. M., 187, *201*
Strack, H., *200*
Stulz, L. W., *303*
Suematsu, Y., *201*, *512*
Summer, G. T., 478, *507*
Suzuki, Y., *301*
Sze, S. M., 204, *238*

Tada, K., *512*
Takanashi, H., *238*
Takasaki, Y., *427*
Takusagawa, M., *201*
Tanaka, M., *427*
Tardy, A., *508*
Tasker, G. W., *504*
Taylor, H., *302*
Teh, G. A., *510*
Tenchio, G., *510*
Thiel, F. L., *151*, 269, 283, *300*, *301*, *302*
Thompson, G. H. B., 483, *509*
Timmermann, C. C., *300*, 476, *506*, *508*
Tokunaga, T., *512*
Tosco, F., *511*
Tsai, C. S., 473, *505*
Tsuchiya, H., 261, 265, *300*
Tynes, A. R., 470, *504*

Uchida, T., *150*, 429, *464*
Ulrich, R., 486, *511*
Unger, P. H., 476, *506*

van Ass, H. M. J. M., *506*
van Heel, A. C. S., 429, *464*
Voumard, C., *510*
Vucins, V., *508*

Watanabe, M., 459, *464*
Webb, P. P., *238*
Weber, H. P., *508*
Weidel, E., *200*, 478, 481, *507*, *508*
White, H. E., *464*
White, K. I., *507*
Williams, D., 477, *507*
Wilson, M. G. F., 486, *510*
Winzer, G., *508*
Witte, H. H., 486, *510*, *511*
Wittmann, J., *508*
Wolf, H. F., ix-xii, 1–240, *355*, 377–*427*,
  465–473
Wolfe, C. M., 204, 222, *237*
Wolff, W. I., *464*

Yamada, T., *151*, *201*
Yamaguchi, G., *512*
Yamamoto, T., *200*
Yamane, T., *355*

Yamanishi, M., *200*
Yanai, H., *512*
Yang, K. H., *300*
Yariv, A., *302*
Zeitler, K. H., *508*
Zemon, S., *301*